*This book
is dedicated to
my wife and family*

Preface to the Fourth Edition

The primary purpose of this book continues to be the guiding of the dental student in the study of those parts of human anatomy which are most relevant to the practice of dentistry. It is also intended that the text will be a useful resource for dentists in clinical speciality training and for practitioners involved in continuing education.

Extensive additions and revisions have been made to achieve improved sequencing of material, to give greater emphasis in clinical and functional anatomy, and to update certain sections in light of recent developments in the field. A number of new illustrations have been added and others deleted. Similar changes have been made in the selected reading list.

I continue to be grateful to Professor G. A. G. Mitchell and Dr E. L. Patterson for permitting reproduction of illustrations from their *Basic Anatomy* and to Professor N. B. B. Symons for permission to use several illustrations from his *Introduction to Dental Anatomy*. It is a pleasure to acknowledge the invaluable assistance of Ms Adelaide Achtmeyer with preparation of the manuscript; Ms Irene Petravicius for the new drawings, and Ms Michele Kirsch for her typing skills. I am grateful to Churchill Livingstone who took their customary care with the publication of this new edition.

Los Angeles, 1978   A.D.D

Preface to the First Edition

This book has been written with the aim of guiding the dental student in the study of anatomy. It is presumed that the chief method of learning anatomy will continue to be the dissection of the relevant parts of the human body, with the assistance of a reliable dissection manual, and that the larger text-books will continue to be used as books of reference. These are, however, written primarily for medical students and, while a certain proportion of the contents of this volume will be found in them, we have included a good deal of material of special dental interest from other sources.

The present volume has also been designed as a companion to *Introduction to Dental Anatomy* by Scott and Symons, but it can of course be used independently. For this reason a certain amount of the description of such matters as the development and growth of the face is common to both books as they form the subject matter of both general and dental anatomy.

In the Appendix certain of the illustrations are reproduced without captions and are to be used, if desired, for revision purposes.

The majority of the illustrations have been prepared from material in the Anatomy Departments in Manchester and Belfast and we would record our indebtedness for the invaluable artistic help given by Miss D. Davison and Miss M. Jenner. The photographic and radiographic illustrations are the result of the technical skill of Mr P. Howarth and Mr D. W. McKears respectively. In addition, we have been fortunate to be able to make use of a number of illustrations from other sources and acknowledge our cordial thanks to those who have so generously accorded this privilege. We are especially grateful to Professor G. A. G. Mitchell and Dr E. L. Patterson for permitting us to reproduce illustrations from their *Basic Anatomy*; to Miss M. Gillison from her *A Histology of the Body Tissues*; and to Professors Hamilton, Boyd and Mossman for the use of blocks from their *Human Embryology*. All the figures obtained in this way are individually acknowledged in the text.

We also wish to acknowledge our thanks to the journals in which some of our own diagrams and photographs appeared for permission to reproduce these in our text.

Finally we would like to offer our sincere thanks to Dr A. Young of Glasgow University for his constructive criticism, to Miss M. E. Gaffikin for preparing the typescript and Mr Charles Macmillan and Mr James Parker of Messrs E. & S. Livingstone Ltd. for the great care which they have taken in the publication of this book.

1959
JAMES H. SCOTT, Belfast
ANDREW D. DIXON, Manchester

Contents

1. Introduction

The human body is a multicellular organism which in its development has an axial notochord (therefore a member of the Phylum Chordata), which is later replaced by a vertebral column (therefore a member of the Subphylum Vertebrata). Human beings are warmblooded, give birth to living young which are fed at the mammary glands of the female (therefore are members of the Class Mammalia), and within the Mammalia they occupy with lemurs, monkeys and anthropoid apes the Order Primates. Anatomically, man differs from other living primates chiefly in the large size and complexity of his brain and in having attained the fully upright posture.

Anatomy is the exploration of the structure of the human body. The historical technique of anatomy is dissection but in its contemporary form the study of anatomy involves many methods which take our anatomical knowledge far beyond the limitations of the naked eye. For example, the use of the electron microscope has revealed fine details of tissue structure which were unknown in the comparatively recent past. Members of the health professions must understand the structural organization of the human body as a foundation for a proper knowledge of normal and abnormal function, in health and disease.

Gross or macroscopic anatomy is the study of the structure, form, and organization of the human body as revealed by the method of dissection. From the structure and organization of the parts of the body, information can be derived in regard to their function. This knowledge is completed by the study of Human Physiology. The microscopic examination of human cells and tissues is the subject matter of Histology, while the developmental history of the human organism from fertilization of the ovum to birth is the concern of Embryology. Studies of the similarity and differences between parents and offspring is the subject of Genetics. Growth of the human body begins at fertilization and ends with the attainment of adult life. Its study covers the foetal period as well as childhood and adolescence. Anthropology involves the detailed comparison of the structure and function of the human body with the bodies of other primates, the investigation of the characteristic structural features of human races, both living and extinct, and speculation on the evolution of the human species.

In so far as the teeth and jaws have forms characteristic for mankind, a developmental history in each individual, a complex microscopic structure, play a part in the digestive process, and have an evolutionary history, their study involves all the phases of Biology mentioned above. They cannot be fully studied in isolation and the dentist requires a general knowledge of the human body as a functional, biological, evolving unit as well as an accurate and detailed knowledge of the anatomy of the orofacial region.

THE MAMMALIAN SKELETON

In a typical mammal the axial skeleton, or vertebral column, is horizontal in position with the head carrying the special senses at the fore, or cranial, end (Fig. 1.1). The body is lifted off the ground by the fore and hind limbs. These are also responsible for the locomotion of the animal. The vertebral column commences with the cervical region immediately behind the skull. The cervical vertebrae make up the skeleton of the neck, which permits a wide range of movement of the head.

Behind (caudal to) the cervical region comes the thoracic region of the vertebral column. The thoracic vertebrae give attachment to the ribs, which, with the breast-bone (sternum) on the ventral surface, form an expansile cage regulating the process of respiration, whereby air is drawn into and driven out of the lungs. The thoracic cage also contains and protects the heart and great vessels.

Caudal to the thoracic region of the vertebral column is the lumbar region. Here the vertebral column and posterior ribs support the abdomen. Among mammals the abdominal cavity is completely separated from the thoracic cavity by a muscular partition, the diaphragm, which, with the muscles between the ribs, plays an important part in the function of respiration. The abdominal cavity contains the stomach, the large and small intestine, two large glands of the digestive system (the liver and pancreas), the kidneys, spleen and adrenal glands.

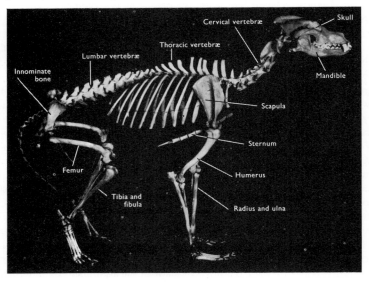

Fig. 1.1 Lateral view of the skeleton of a large quadruped animal—the wolf.

Behind the lumbar segment is the sacrum consisting of a number of vertebrae which have united in evolution to form a rigid segment of the vertebral column. The sacrum, with the innominate bones on each side, makes up the pelvic skeleton. Each innominate bone is formed by the union of pubic, ischial and ilial elements. Within the pelvic cavity, which is continuous with the abdominal cavity, are the terminal parts of the alimentary and urinary apparatus and the sex organs. The hind limbs are attached to the pelvis at the hip joints.

The terminal part of the vertebral column consists of a variable number of coccygeal vertebrae, which form the skeleton of the tail.

The skeleton of the head consists of the skull, which protects the brain and the organs of special sense, the eyes, ears, nose and tongue. The ears are situated in the floor of the cranial part of the skull; the other sense organs in the facial skeleton. The nasal cavity is the commencement of the respiratory system, and the mouth the beginning of the digestive system. Behind the nasal and oral cavities is a common chamber, the pharynx, where the food and the air passages cross one another. From the back of the pharynx the food passage continues as the oesophagus, and the air passage as the larynx and trachea (windpipe) (Fig. 3.32).

The neck contains the cervical vertebrae; groups of muscles which move the neck and regulate the processes of mastication, swallowing and voice production; cranial and spinal nerves; the great vessels carrying blood to and from the head and brain; the pharynx, oesophagus, trachea and larynx (Fig. 4.8).

The oesophagus continues through the thorax, pierces the diaphragm, and ends in the stomach, which is situated in the fore part of the abdominal cavity. The trachea divides within the thorax in two tubes, one passing to each lung (the right and left bronchus). Great vessels entering and leaving the heart (veins and arteries) enter or leave the neck and upper limb at the cranial end of the thoracic cavity (the thoracic inlet). They communicate with the abdominal and pelvic cavities and the lower limbs by passing through the diaphragm.

The fore limbs are not directly attached to the vertebral column as are the hind limbs at the pelvis, but to the two movable plates of bone (the scapulae or shoulder blades), which are attached to the vertebral column and thoracic cage by powerful muscles. While the hind limbs in quadruped animals thrust the body forwards in locomotion, the fore limbs draw the body forwards. Each limb consists of four segments:

1. *A proximal segment* containing a single strong bone (the humerus in the fore limb and femur in the hind limb).

2. *A proximal intermediate segment* containing two bony elements placed side by side. These are the radius and ulna in the fore limb, and the tibia and fibula in the hind limb. In primitive mammals the radius or ulna had the capability of rotating around the other bony element. In some specialized mammals this movement is lost, and in others, part of, or all of the ulna and fibula are missing. In man the primitive range of movement is maintained in the fore limbs.

3. *A distal intermediate segment* made up of a number of small bones: the carpal bones in the fore limb, the tarsal bones in the hind limb. These form either the skeleton of the wrist or instep in man, but in many animals they lie above the distal segment which alone makes contact with the ground.

4. *The distal segment* in each limb in the primitive (and human) skeleton consists of five digits (fingers or toes), which form the distal or terminal part of the hands and feet. In many animals they are reduced in number and carry claws or hoofs.

Each limb segment is separated from the next by one or more joints at which various movements can take place. In the fore limb the chief joints are the shoulder joint between the scapula and humerus, the elbow joint between the humerus and the radius and ulna; the wrist joint between the radius and ulna and the proximal row of carpal bones, the joint between the proximal and distal carpal bones (mid-carpal joint), carpo-metacarpal joints, and finally the joints between metacarpals and phalanges (the digital joints).

In the hind limb the chief joints are the hip joint between the side of the pelvis and the femur, the knee joint between the femur and the tibia, the ankle joint between the tibia and the fibula and the talus (one of the tarsal bones), the mid-tarsal joint between the proximal and distal tarsal bones, tarso-metatarsal joints; and the digital joints between the metatarsals and phalanges.

Movement of the limb segments is produced at the joints by the contraction and relaxation of powerful muscles. The greater part of the substance of each limb consists of bones and muscles. Between the muscles run the blood vessels (arteries and veins), from which they obtain nutrients and oxygen and to which they give up their waste products; and the nerves carrying the impulses, which determine the timing, extent and force of their contraction and relaxation.

HUMAN ADAPTATIONS FOR UPRIGHT POSTURE

Man differs noticeably from other primates in the curvatures formed by the vertebral column in the sagittal plane, particularly the anterior convexities of the cervical and lumbar regions.

These curvatures contribute to erect posture and the musculature of the trunk is adapted for balancing the body on the lower limbs as well as for movements of the head and trunk. The thorax is flatter from before backwards than in other animal forms and this shape moves the centre of gravity backwards in line with the feet. The lower limbs are specialized for bipedal locomotion and the forearm is long and capable of a marked degree of rotation. The human pelvis is greatly expanded in its upper part compared with other primates to provide adequate support for the abdominal organs and to give attachment for the gluteal muscle mass. The gluteal (hip) muscles, the muscles on the front of the thigh, the quadriceps femoris, and the calf muscles, the gastrocnemius and soleus, are especially well developed in man as a result of acquired upright posture.

The human skull is distinctive in the large capacity of the cranial cavity, the lack of projection of the face and the relatively small size of the teeth, jaws and jaw musculature. Despite the flattened appearance of the face, the human lower jaw is distinctive in possessing a prominent chin.

With the evolution of upright posture, which has reached its fullest development in man (Fig. 1.2), important anatomical changes took place. The forelimbs were freed from the requirements of body support and locomotion. Much of the work carried out by the teeth in the obtaining and preliminary preparation of food was transferred to the hands. However, the human fore limb is not a specialized structure, for in the rotary movements between the radius and ulna and the number of the digits, it has retained primitive features which developed among the amphibians and have been lost in many mammalian species. The human hand is uniquely versatile because of the high degree of specialization of the brain and the refined movements of the hands which this makes possible.

The brain in primitive mammals was made up largely of areas concerned in establishing body relexes in relation to the skin surface, muscles and the special sense organs, especially the sense of smell. Later in evolution sight became predominant and the sensory inflow from the whole of the body surface increased. The characteristic feature of the human brain is the great development of association centres in the cerebral cortex. These regulate body movements, both in relation to immediate sensory impressions and past experiences. This ability is the basis of the process of learning, planning and decision making.

In man the head is more fully balanced than in the anthropoid apes. The better balancing is produced in part by placing the foramen magnum, through which the spinal cord leaves the skull, further forward and in part by reducing the size and degree of forward projection of the facial part of the skull. The human face lies below the cranium rather than in front of it.

In four-footed animals the muscles which pass between the thoracic cage and the fore limb skeleton are used in respiration. They move the thoracic cage in relation to the forelimbs. In man, with the change in function on the part of the forelimbs, the muscles can only assist respiration if the arms are fixed by holding on to some support as an asthmatic person holds on to the back of a chair.

In quadrupeds the anterior abdominal muscles, especially the strap-like rectus abdominis muscles situated on either side of the middle line, play an important part in the support of the body viscera. In the upright position this function is transferred to the lower part of the anterior abdominal wall, a region which is weakened by the passage of various vessels through the inguinal and femoral canals and is a common site of hernia (Fig. 3.29). Likewise, in the human pelvis, muscles which are concerned with the movements of the tail in lower animals form a pelvic diaphragm to support the pelvic contents in man. The remains of the tail, the coccyx, become tucked in between the buttocks and help to give firm attachment to muscles of the pelvic diaphragm.

Some of the most extensive changes in body structure which follow the attainment of upright posture are related to the lower extremities which carry the whole of the body weight and have become the chief agents in locomotion. The lower part of the upright vertebral column has the greatest load to bear and slipped disks and lower back pain are common clinical indications of a breakdown in its efficiency. Body weight is transferred from the vertebral column through the pelvic skeleton to the femora. The massive gluteal muscles of the hips and the associated expansion of their sites of origin, especially from the iliac bones, are characteristic features of human anatomy. The knee joint shows adaptation to the upright position in the flattening of the upper surface of the tibia on which the lower end of the femur is supported by the powerful quadriceps femoris and hamstring muscles.

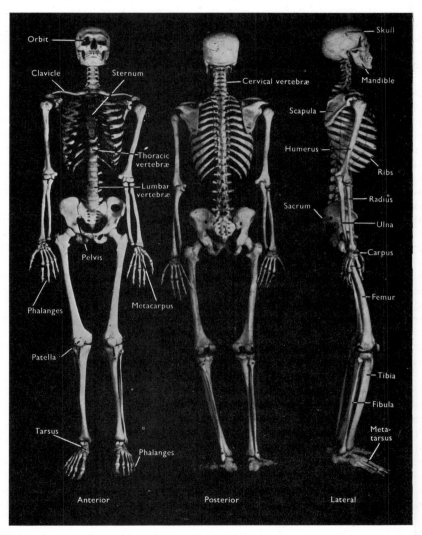

Fig. 1.2 Anterior, posterior and lateral views of an adult human skeleton.

GENERAL FEATURES OF BODY SYSTEMS

A characteristic feature of mammalian anatomy is the establishment of the *cardio pulmonary system* (Fig. 1.3). In fishes the heart is a muscular tube-like structure which propels the blood through a linear series of chambers. After passing through the heart the blood is oxygenated in the gills. Beginning in the amphibians and continuing through the reptiles to reach comple-

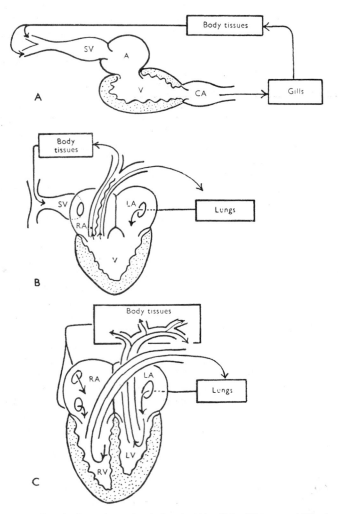

Fig. 1.3 Diagram comparing the heart-lung circulation in (A) a fish, (B) an amphibian, and (C) a mammal. A, atrium; V, ventricle; SV, sinus venosus; CA, conus arteriosus.

tion in mammals, a phylogenetic change takes place in the relationship between the heart, as the organ of circulatory propulsion, and the mechanism whereby the blood is oxygenated and purified. The pulmonary circulation becomes interpolated in the cardiac circulation. The right atrium receives the venous blood from all parts of the body and passes it to the right ventricle, which by its contraction drives the blood through the pulmonary circulation. The essential feature of the pulmonary circulation is the relationship between the pulmonary capillaries and the terminal air sacs (alveoli). Between the blood in the capillaries and the air in the alveoli there is a minimum of tissue consisting of little more than the endothelial lining of the blood vessels. An endothelium consists of a thin layer of cells in the form of a pavement epithelium supported by delicate connective tissue. The oxygenated blood is returned from the lungs to the left atrium of the heart and the left ventricle provides the driving force for the general (systemic) circulation.

By interposing the pulmonary circulation between the right and left sides of the heart, a complex and extensive capillary system can develop in the lungs and the resulting loss in driving force behind the circulation is made good for the general circulation by returning the blood to the cardiac pump.

A further characteristic of the mammalian circulatory system, which also reaches its full development in man, is the heart–brain circulation (Figs 4.28, 5.10). From the aorta a direct vascular channel (the common and internal carotid arteries) runs to the brain. At the base of the brain a circle of arteries (the circle of Willis) is established which is fed by the two internal carotid vessels and also by the vertebral arteries which form an alternative route from the subclavian arteries. From this arterial circle, the brain obtains a constant and adequate supply of oxygenated blood. A series of blood circulation regulating mechanisms, sensitive both to alterations in blood pressure and to the carbon dioxide content of the blood, situated on the aortic arch and carotid vessels and working through a series of nerve reflexes, regulates the heart and respiratory rates to meet the metabolic requirements of the brain cells. Fainting is the result of a temporary breakdown of this mechanism.

The *nervous system* is composed of nerve cells and their protoplasmic processes in which two properties common to all living protoplasm, namely excitability and conductivity, are highly developed. The nerve fibres can be short or extend for considerable distances, reaching a length of three feet or more in man, and they have the primary responsibility of conducting nervous impulses. By this means a stimulus which affects one part of the body may lead to a response of some motor mechanism in a distant part. Because the nerve impulses travel at high speed, the evoked response may occur with great rapidity.

The nervous system is divided into peripheral and central parts. The *peripheral nervous system* consists of bundles of nerve fibres which are distributed throughout the body as 'nerves' and their constituent fibres are either afferent (sensory) in function, or efferent (motor). The sensory fibres are concerned with impulses initiated at the periphery and, in that sense, they are the information mechanism of the nervous system whereby changes in the external or internal environment are recorded and generate appropriate responses. Not all the impulses from sensory receptors give rise to conscious sensation and receptors take many forms and have a variety of functions, including those for the special senses of sight and hearing. Most receptors are specific for particular stimuli so that the nervous system is able to respond most appropriately. The motor fibres of peripheral nerves constitute the effector mechanism and nerve impulses passing along a motor nerve lead to the contraction of muscles or to gland secretion.

The coordination of sensory and motor impulses and their most effective transmission is the responsibility of the *central nervous system*, comprising the brain and the spinal cord. It is a mass of nerve cells and nerve fibres imbedded in a special connective tissue called neuroglia. The nerve fibres transmit incoming impulses to various parts of the central nervous system and pass them through a series of nerve cell relays so that they are guided to the correct efferent fibres of peripheral nerves. These communications within the central nervous system are extremely complicated and involve many millions of nerve cells.

The brain and spinal cord are protected by the bones of the cranial part of the skull and the vertebral column. They are also safeguarded by being suspended in a fluid shock absorbing medium, the cerebrospinal fluid. The fluid circulates in a space between two of the membranes which cover the central nervous system, the pia mater and the arachnoid. This subarachnoid space is more accurately described as a delicate sponge-like tissue through which the fluid slowly circulates (Fig. 8.22).

The *locomotor system* consists of the bones, joints and muscles of the limbs, the vertebral column, shoulder girdle and bony pelvis (Figs. 1.2, 3.58), with their associated nerves and blood vessels. In man, the fore limbs have taken on functions other than locomotion and the lower limbs become specialized for weight bearing and to carry the body upright, as well as for movement.

Locomotion involves not only movement of the body but balance of the body, that is, of the vertebral column on the lower limbs during the alternating phases of limb movement during walking, as well as in standing or any other postural position. In standing, the muscles acting on the various joints must maintain them in a fixed position against the forces of gravity with the least possible muscle fatigue. In the upper limb the muscles of the whole limb skeleton are involved in bringing and maintaining the hands in the positions necessary for the skilled movements of the thumb and fingers. Again the vertebral column may be involved, both to alter the position of the body and to maintain it in a position of balance.

The bony elements making up the limb skeleton were referred to on page 3 and they give attachment to the muscles which produce and control limb movements, and ligaments which provide joint stability.

The shoulder or pectoral girdle, unlike the pelvic girdle, is freely movable. The scapula can glide forward and backward as well as upward and downward at the side of the thoracic cage round the strut of the clavicle. This adds greatly to the mobility of the upper limb.

Both the shoulder joint, between the head of the humerus and the glenoid cavity of the scapula, and the hip joint, between the head of the femur and the acetabulum of the hip bone (Fig. 3.55), are ball and socket joints allowing a wide range of movements. At the elbow and knee joints movements are limited to flexion and extension with a limited amount of rotation at the latter.

In the forearm a considerable degree of movement of the radius around the ulna is possible during pronation and supination of the hand. There is no corresponding movement between the tibia and fibula. At both the wrist and ankle the main movements are flexion and extension with some degree of abduction and adduction. These movements are associated with corresponding movements among the carpal and tarsal elements of the skeleton of the hand and foot.

Except for the joint at the base of the thumb, there is only a limited amount of movement between the carpal and metacarpal and the tarsal and metatarsal elements. The carpometacarpal joint of the thumb permits extensive movement including the ability to place the thumb across the palm to the base of any of the fingers, *opposition*, which is a distinctive human characteristic.

Within the *digestive system* the chief functions are the movement of food along the alimentary canal, its disruption by mechanical means and by the activity of the chemical contents (enzymes) of various secretions, and the absorption of the products of digestion by the blood and lymph vessels of the gut.

The initial process of bringing food to the mouth, masticating it and swallowing it, which involves the digestive system from the lips to the middle of the oesophagus, are habitual but voluntary acts. After the food passes into the oesophagus and for the greater part of the remainder of its journey, its movement and digestion are under the involuntary control of muscles and glands whose activity does not normally enter consciousness. Emotional disturbances can alter the action of these muscles and glands to a considerable extent, and also the activity of the heart, vascular system and the respiratory system.

Digestion, the circulation, and to a more limited extent respiration, are largely under the control of the involuntary or autonomic nervous system. Like the somatic part of the nervous system, which regulates the movements of the voluntary muscles, the autonomic system consists of a series of complicated reflex arcs built up of sensory, intercalated, and motor neurons, but their higher level reflexes do not directly involve the cerebral cortex. They work through a number of centres in the basal parts of the brain, including the hypothalamus and medullary portion of the brain stem. These centres regulate the heart rate, blood pressure, respiratory rate, the peristaltic or wave-like movements of the gut, the secretion of urine and the secretory activity of glands.

The *glandular system* has two important components. The *endocrine* system is composed mainly of glands that have lost their connection with the epithelium from which they originated, for example, the thyroid. Endocrine glands, also called ductless glands, produce secretions or hormones which pass directly into the blood stream or lymph. *Exocrine* glands, for example the sweat glands and salivary glands, possess ducts along which the glandular secretions pass to an epithelial covered surface (e.g., skin, mucous membrane of the mouth). Hormones subserve many functions including the control of skeletal growth, the onset of sexual maturity, the mobilization of the body's reserves to meet emergencies, the fluid balance in the tissues and the rate of body metabolism. The pituitary gland, as well as manufacturing a number of hormones each with its own definite functions in the body economy regulates the interaction of the endocrine system as a whole and has been termed 'the conductor of the endocrine orchestra.' It in turn is regulated by the hypothalamic centres of the brain.

Other important gland-like structures are the spleen and the kidneys. Strictly speaking the spleen is not a true gland. Its function is the manufacture of new blood cells and the destruction of old cells. It is also a storehouse for blood cells which are passed into the circulation as they are required. The kidneys eliminate fluid waste products, just as the lungs are concerned with the elimination of gaseous metabolites, and the lower end of the alimentary canal with the elimination of the solid faeces which remain when all that is useful is abstracted from the food. The kidneys work on the principle of first removing from the blood that passes through them all the substances in solution, and then returning to the body the necessary elements. What remains is excreted as urine. As well as regulating the removal of waste products, the kidneys play an important part in the control of the fluid balance of the tissues of the body. erythropoietin renin Vit D

THE TEETH AND BODY SYSTEMS

In considering the full application of anatomical and physiological knowledge to the understanding of the human dentition, we must not limit ourselves to a knowledge of the structure of tooth, bone and gum or to their developmental history, or to the growth changes involved in facial growth, the change from the deciduous to the permanent dentition and the maintenance of normal occlusion. The pulp of each tooth and its supporting tissues contain parts of the circulatory system, parts of the nervous system, are influenced by the secretions of endocrine glands, depend on the proper function of the heart and lungs and digestive system for their health, and are involved in disease processes which may have their primary sites in remote parts of the body. A minute tumour composed of cells of the pituitary gland may produce drastic effects on the facial skeleton; a disturbance in the complex mechanism of calcium metabolism involving absorption from the gut, solution in the blood and deposition in certain tissues, may have a profound effect on the structure of both teeth and bones. On the other hand, bacterial infections developing in relation to decayed or carious teeth and diseased gums may have a harmful effect on the total health of the body, and gross irregularities of the teeth may produce deep psychological trauma.

COMMON DESCRIPTIVE TERMS USED IN ANATOMY

For descriptive purposes the human body is regarded as standing erect with the arms by the sides and the palms of the hands facing forwards. The following terms are used to express the relationship of any structure to another within the body:

Anterior, or ventral, or in front = nearer to the front surface of the body.

Posterior, or dorsal, or behind = nearer to the back surface of the body.

Superior, or cranial, or above = nearer the crown of the head.

Ascending, or upwards, denotes passing towards the head.

Inferior, or caudal, or below = nearer the soles of the feet.

Descending, or downwards, denotes passing towards the feet.

Outer surface and *inner surface* are applied to bony cavities such as the cranium, thorax and pelvis, and to the elements composing them, and also to other body cavities and organs such as the nose, mouth, pharynx, abdominal cavity, alimentary canal, heart, and urinary bladder.

Superficial and deep denote nearness to and remoteness from the skin surface.

Internal and external describe relative distances from the centre of an organ or body cavity.

The three major planes used in describing sections of the body organs are:

Median sagittal (dividing into left and right) cutting the surface of the body at the anterior and posterior median lines. Passes through the sagittal suture of the skull. Any section parallel to this is in a *parasagittal* plane.

Coronal or frontal (dividing into front and back), and passing through the coronal suture of the skull.

Both the sagittal and coronal planes are vertical but are at right angles one to another.

Transverse or horizontal planes or cross section (dividing into above and below). They are at right angles to the sagittal and coronal planes.

Medial = nearer the median or mid-sagittal plane of the body which divides it into right and left halves.

Lateral = further from the median or mid-sagittal plane.

Proximal and distal are used in describing the parts of the limbs and of the limb bones and refer to their nearness or to their remoteness from the trunk. The hand is distal to the elbow, and the shoulder is proximal to both. In the dentition, the distal surface of a tooth is that which faces away from the mid-sagittal interval between the central incisors, and the *mesial* surface is that which faces towards this central point.

Flexion denotes bending of a limb at a movable joint. *Extension* is the movement which brings the limb into or towards the straight condition. Equates also with forward and backward movements of the head on the neck.

Abduction and *adduction* imply movement of a limb away from and towards the midline of the body, respectively.

Pronation is the act of turning the palm of the hand downwards, while *supination* is the reverse movement.

Rotation is the action of turning around a centre or axis, such as rotation of the head on the neck.

2. Some Elements of Body Structure

CELLS AND EPITHELIA

Reference must be made to some of the features of typical cells about which our knowledge has greatly increased over the past twenty years with the use of the electron microscope and the details of cellular fine structure which this technique has made possible. Cells vary widely in their shape, size, structure and chemical composition and they have many different functions as a result. Figure 2.1 is a schematic representation of some of the features of the internal structure of a cell as would be seen in electron micrographs. All cells have a limiting membrane

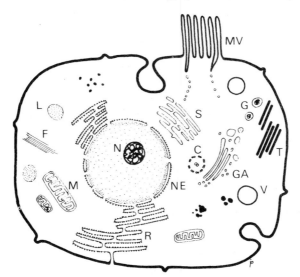

Fig. 2.1 Diagrammatic representation of a cell to show some important organelles and inclusions. C, centriole; F, filaments; G, granules; GA, Golgi apparatus; L, lysosome; M, mitochondrion; MV, microvilli; N, nucleolus; NE, nuclear envelope; R, rough endoplasmic reticulum; S, smooth endoplasmic reticulum; T, tubules; V, vesicle.

or *plasma membrane* and almost all of them have a *nucleus* which lies in the *cytoplasm*. The nucleus is enclosed in its own nuclear membrane or envelope and is essential for the life and reproduction of the cell. Within the nucleus is found an amorphous material (nucleoplasm) which contains a dense region, the *nucleolus*, consisting of a mass of one form of nucleic acid (ribonucleic acid or RNA). Filaments of chromatin material in the nucleus (chiefly deoxyribonucleic acid or DNA) are the genetic material of the cell and become the chromosomes during cell division (Fig. 2.2). The structures within the cell may be classified as organelles or inclusions. *Organelles* include the endoplasmic reticulum (rough and smooth), ribosomes, the Golgi apparatus, mitochondria, lysosomes, fibrils, microtubules and centrioles. Cell *inclusions* are nonliving components, for example, secretion granules in glandular cells, storage granules such as glycogen and fat, or pigments such as melanin which is found in skin and mucous membranes.

One granule of chromatin—the *sex chromatin*—is particularly conspicuous in the cells of females, close to the nuclear membrane. It is probably formed in cells, which are not undergoing division, from the union of the two large sex chromosomes (the X chromosomes) which are two of the full complement of 46 chromosomes (Fig. 2.3). It is interesting to note that while man has 46 chromosomes, many of the apes have 48. The chromosome complements of their cells are very similar, the main difference being in the relative position of a small constriction on each chromosome called the *centromere*. Various workers have speculated on

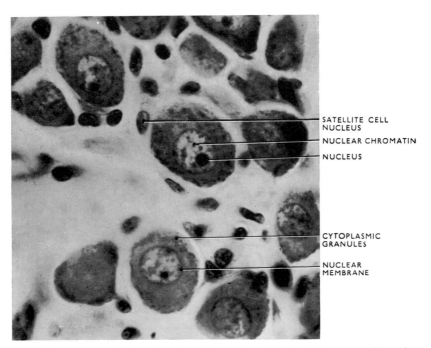

Fig. 2.2 A group of large nerve cells from the trigeminal ganglion stained with haematoxylin and eosin to show some of the features illustrated by the diagram in Fig. 2.1 × 650. (By courtesy of *Journal of Dental Research*.)

the mechanisms involved in evolution of the human and other primate karyotypes. It seems probable that at some time in a prehuman ancestral population the reduction in the chromosome number from 48 to 46 must have occurred. In male cells, the sex chromosomes are known as the X and Y chromosomes. As the Y chromosome is of small size, the sex chromatin mass is not apparent in the intermitotic (resting) phase.

Human somatic cells have 22 pairs of autosomes and 2 sex chromosomes. Each human chromosome has been designated a number, largely based on the size of the chromosome pair, and they have been arranged into groups by geneticists and cytologists. Although there are differences of opinion among scientists about the identification, they have agreed on a simple method for designating the chromosomes so that much confusion is avoided (Fig. 2.3). In some diseases in man one or even two chromosomes are characteristically missing.

In recent years the presence or absence of sex chromatin has become a reliable method of determining the sex of tissues, and helps in our understanding of abnormal sexual develop-

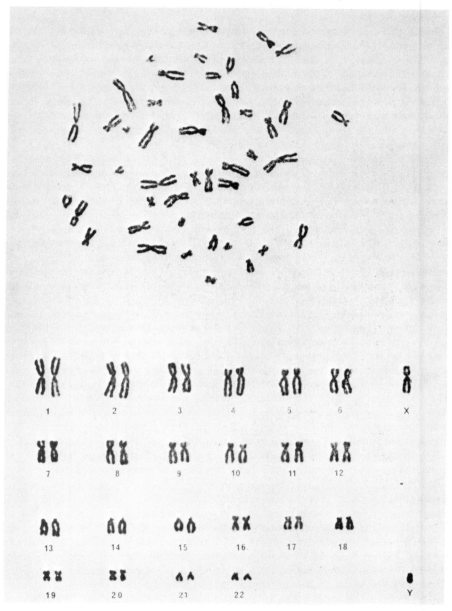

Fig. 2.3 Chromosomes of a male somatic cell in a culture of peripheral blood. Top—The chromosomes, each comprised of a pair of elongated chromatids arranged side by side, are seen at metaphase. Bottom—Karyotype; the 22 pairs of somatic chromosomes have been arranged in numerical order. The X and Y sex chromosomes are shown at the right × 3,400. (By courtesy of Professor B. Cruickshank, Mr T. C. Dodds and Dr D. L. Gardner.)

ment. From the microscopic examination of cells taken from the lining of the cheek, the proportion of cells which contain the sex chromatin body can be estimated and the genetic sex determined.

Mature human sex cells contain half the full number of chromosomes (haploid number) including an X or Y chromosome in the male and an X chromosome in the female. At fertilization the full chromosome number (diploid number) is restored but the sex chromosomes may group as either X + Y or X + X, depending on the chromosome content of the male germ cell (the sperm). When the gonad (ovary or testis) is abnormally differentiated during its development, this grouping of the sex chromosomes may be altered. For example, in the condition known as Klinefelter's syndrome, in which there is a failure of development of the tubules which produce the spermatozoa, the sex chromosome pattern is XXY or even a mixture of XXY and XY (the normal male grouping).

Each chromosome is composed of two genetically identical *chromatids* which can be seen clearly during cell division, at the stage of metaphase. Photographs of chromosomes can be arranged in a sequence for study called a *karyotype* (Fig. 2.3) in which the 44 somatic chromosomes (or autosomes) can be numbered and the sex chromosomes identified. Because of the considerable doubt in identifying chromosomes accurately, for example, chromosomes 11 and 12 appear so similar, workers in this field of investigation have adopted a letter grouping scheme, as follows: Group A₁ 1 to 3; B₁ 4, 5; C₁ 6 to 12 and the X chromosome; D₁ 13 to 15; E₁ 16 to 18; F₁ 19, 20 and G₁ 21, 22 and the Y chromosome.

Each chromosome is made up of a number of *genes*, the individual active units of heredity. These are strung together in a linear fashion, each gene having its particular position or locus within the chromosome.

As component parts of the chromosomes, genes consist of a segment of a deoxyribonucleic acid (DNA) molecule. Each DNA molecule of a mammalian chromosome has been shown to consist of two long thin strands wound together in a double helix formation. Each strand consists of a linear arrangement of smaller nucleotides united chemically with a corresponding adjacent although dissimilar series on the other strand. The genetic information contained by any DNA molecule depends on the number and sequence of amino acids along its strands. The DNA molecules act as patterns for the molecules of messenger RNA, which form beside them, which then move to the cytoplasm with the necessary information to synthesize appropriate protein macromolecules for the given cell type. Thus, genes appear to act by determining the nature and number of proteins formed in cells. During cell division or mitosis the chromatids separate to reconstruct their associated strand in each of the resulting cell nuclei. In this manner every cell in the body comes to contain the same genetic (hereditary) constitution.

The chromosomes contain the physical and chemical characteristics of specific cells, coded into the DNA fraction. The information must be conveyed from the nucleus to the cytoplasm where metabolic function takes place. It is believed that the nucleus manufactures RNA which is then transported to the cytoplasm through pores in the nuclear membrane, conveying the necessary information of the nucleus. This form of nucleic acid is called *messenger* RNA and becomes attached to certain structures within the cytoplasm where it functions as a template for the manufacture of cellular products, which often are extremely specific for the cell type. There are a number of theories concerning the interaction of these chemical constituents of the cell and absolute proof of the correctness of any of these theories has yet to be demonstrated.

Using the light microscope it is possible to observe the following structures in the cytoplasm of the cell: the *centrosome* or centrosphere which is concerned with cell division, *vacuoles* containing lipids or fat, *granules* or inclusions of other food materials such as glycogen, and, using special staining techniques, rod-like bodies, the *mitochondria*, concerned with cell metabolism.

The development of the electron microscope has enabled us to visualize these features of the cell in much greater detail. This is possible, not only because the electron microscope provides a much greater magnification than the light microscope, but chiefly because the resolving power of the electron microscope is in the realm of molecular size. Very special techniques are used to prepare biological materials for electron microscopy, but essentially the procedures are similar to those used for material which is to be examined in the light microscope. The resolving power, that is the ability to distinguish two points in a tissue which are extremely close together, is for the light microscope approximately 0·2 of a micron. In the electron microscope the resolving power is approximately 3 to 5 Ångström units, 1 Ångström unit being 10^{-7} mm (10^{-1} nm).

Some of the features of the cell which can be seen readily using the electron microscope include the endoplasmic reticulum, the Golgi complex, mitochondria and lysosomes.

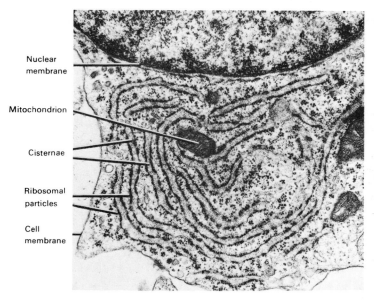

Nuclear
membrane

Mitochondrion

Cisternae

Ribosomal
particles

Cell
membrane

Fig. 2.4 Electron micrograph of a fibroblast in the developing palate, showing the form of the endoplasmic reticulum, consisting of cisternae and ribonucleoprotein particles or ribosomes × 26,000.

The *endoplasmic reticulum* (Fig. 2.4) appears as a network of small canals, flattened sacs, vacuoles or cisternae which have a limiting membrane. On the surface of the living membranes are frequently found dense aggregations of ribonucleoprotein (*ribosomes* or RNP granules). In three dimensions the cisternae of the endoplasmic reticulum communicate with one another and form parallel layered sacs which have a characteristic form, depending on the function of the cell. It has been shown that many substances are synthesized by the endoplasmic reticulum, for example the secretory products of the exocrine cells of the salivary glands. It is thought also that many of the products of the endoplasmic reticulum are transported to the Golgi complex where they are concentrated into granules or droplets. This type of secretory process has been suggested also for salivary tissue.

The *Golgi complex* or apparatus (Fig. 2.4) is a specialized form of endoplasmic reticulum in which the membranes are not studded with protein granules. The flattened sacs of the Golgi complex are small in size compared with those of the endoplasmic reticulum and are associated with small vesicles and vacuoles of varying size. The presence of the Golgi complex has been known for many years but its precise part in secretory activities of cells has been shown more recently. The Golgi complex is also well developed in nerve cells and the reason for this is not yet fully understood.

The third cytoplasmic organelle which will be described briefly is the mitochondrion (Fig. 2.5). Mitochondria are sites of intense biochemical activity which function to convert chemical

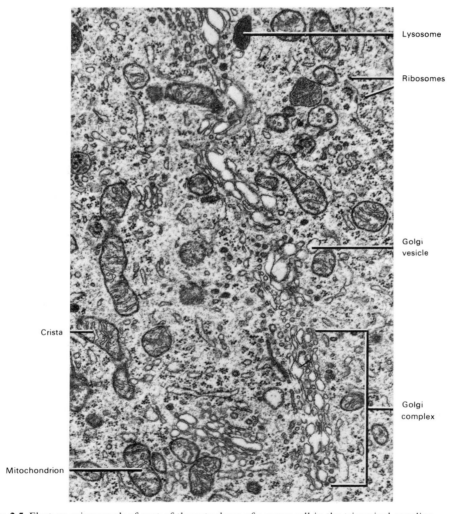

Fig. 2.5 Electron micrograph of part of the cytoplasm of a nerve cell in the trigeminal ganglion showing typical organelles × 37,000.

energy into forms which the cell can utilize for its metabolic activities, such as the manufacture of enzymes and the chemical changes associated with muscle contraction. Much was known concerning the structure and function of mitochondria before the advent of the electron microscope. However, as with other organelles and cell inclusions our knowledge has been considerably advanced with the use of the techniques of electron microscopy. Mitochondria vary in shape from circular or ovoid structures to elongated cylinders. They are limited externally by a double membrane, separated by a space of about 100 Ångströms (10 nm) in thickness.

From the inner membrane, partitions, or *cristae*, pass into the depth of the mitochondrion to subdivide its cavity into a series of smaller compartments. The mitochondria vary in number according to the activity of the cell and are a sensitive indicator of either cell well-being or disease.

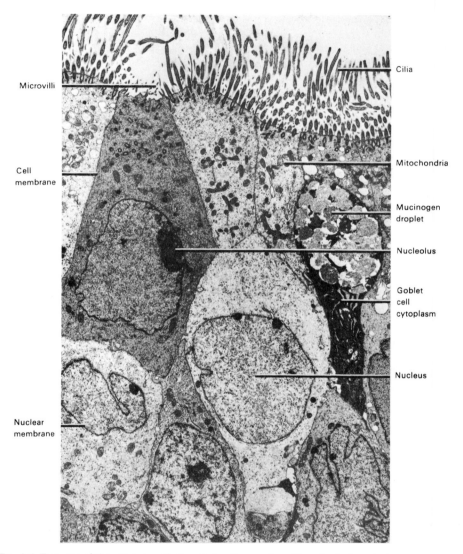

Fig. 2.6 Features of the ciliated epithelium lining the nasal cavities of a foetus, electron micrograph × 5,000.

Lysosomes contain hydrolytic enzymes. They are opaque to the electron beam and are surrounded by a limiting membrane that serves to isolate their enzymes from the remainder of the cytoplasm, preventing autodigestion of the cell structure. It is thought that lysosomes contain the enzymes for digestion of some of the substances which are taken up into the cell, by modifications of the cell membrane which form vacuoles. In the case of fluids these vacuoles are called *pinocytotic vesicles* and in the case of solid materials the process of uptake is called *phagocytosis*. Many different enzymes have been identified in the lysosome, for example, ribonuclease.

Many other features of cells can be seen in electron micrographs but the details of these are outside the scope of this text and for information on them the student should refer to standard texts on histology.

The external or internal surfaces of the body are covered by layers of cells forming *epithelia* resting on a basal lamina. The epithelia are of different types:

1. *Squamous epithelium* consisting of a single layer of flattened attenuated cells, e.g. the endothelium of vessels (Fig. 2.35).

2. *Cuboidal epithelium* consisting of a single layer of cubical shaped cells, e.g. the lining of the ducts of salivary glands and thyroid gland vesicles (Fig. 5.38).

3. *Columnar epithelium* consisting of a single layer of pillar-like cells, e.g. the lining of parts of the alimentary tract. This type may be *ciliated* (possessing hair-like processes which project from the surface) as is the lining of much of the respiratory system (Fig. 2.6).

4. *Stratified epithelium* consisting of several layers of cells. Usually the deeper layers are cuboidal and the surface layers flattened or squamous, e.g. the mucous membrane of the mouth (Fig. 6.1).

5. *Transitional epithelium* consisting of a small number of layers of which the surface layers are not flattened but ovoid thus permitting a considerable degree of stretching, e.g. the lining of the urinary bladder.

Mucous membranes

The inner surfaces of both the alimentary canal and respiratory system, like other body surfaces which communicate with the exterior, are lined by a mucous membrane consisting of a superficial layer, or layers, of epithelium and a deeper layer of connective tissue, the lamina propria (Fig. 2.16). Within the lamina propria and supported by its connective tissue fibres are small blood vessels, nerves, nerve endings, lymphatics and various cellular elements including fibroblasts, histiocytes and cells derived from the blood stream. The lamina propria also contains mucous glands, the ducts of which pierce the overlying epithelium. The epithelium may consist of a single layer of high columnar cells or, as in the oral cavity, of a number of cell layers which become more flattened towards the surface to form a stratified squamous epithelium (Fig. 6.1). The cells adjacent to the basal lamina, which separates the epithelium from the lamina propria, form the stratum Malpighii or stratum germinativum, undergo continual division to replace the cells lost from the surface. A certain amount of pigment may be deposited in the deeper layers of the epithelium. The epithelium depends for its nutrition and sensitivity on the blood vessels and nerves of the underlying lamina propria. In the mouth cavity the lamina propria in some places may be directly continuous with the underlying bone. In other regions it is continuous with a submucosal layer containing larger blood vessels and nerve trunks. The development of the teeth depends on the downgrowth of the epithelium of the oral cavity into the adjacent lamina propria (see page 289).

BODY TISSUES

The details of tissue structure are the subject of Histology, but because it is difficult to appreciate the gross structure of the human body without some understanding of the microscopic anatomy of its constituent parts, some essential features are described in the following pages.

The body is made up of a great number of cells derived from the single fertilized ovum. During development the cells of the body become grouped into tissues. In each tissue the characteristic cells which compose it have acquired a particular structure and form, and a specialized function which is often a specialization of one of the general cellular functions such as secretion, contraction or conduction.

The tissues of the body can be divided into:

1. The more *specialized tissues*, which may be gathered together to form organs such as glands and muscles, or may be more diffuse such as the blood cells; and

2. The *connective tissues*. The latter consist of collagenous and elastic fibres and cells such as fibroblasts, histiocytes, fat-storing cells and pigment cells. Connective tissue, the unit of which is a *collagen fibre*, is found in nearly all parts of the body as a general matrix of body structure, forming capsules for glands and sheaths for muscles and nerves, and within the various organs supporting their specialized cellular elements. This type of diffuse connective tissue is sometimes called 'connective tissue proper.' Collagenous fibres also form the matrix of bone and cartilage and of the dentine and cement of teeth. These are sometimes described as special forms of connective tissue in virtue of the specialized functions of their formative cells: the osteoblasts, chondroblasts, odontoblasts, and cementoblasts, respectively.

Collagen is a fibrous protein which is an essential component of connective tissues and has a unique chemical composition. It is synthesized by the fibroblast in the form of intracellular macromolecules called procollagen. These consist of helical polypeptide chains of amino-acids, including proline and lysine. When secreted by the fibroblast the procollagen macromolecules become aggregated to form collagen fibrils, which in turn undergo further precise aggregation and cross-linking to form the mature collagen fibre. The process of collagen synthesis is complex and the subject of intensive research. The collagen of connective tissue undergoes turnover and remodelling which are sensitive to change. Vitamin C deficiency interferes with the intracellular phase of the process and the halting of procollagen synthesis can result in eventual loss of collagen, for example, from the periodontal ligaments of the teeth.

Connective tissue is of *mesodermal origin*. It does not contribute to the formation of enamel or the central nervous system, which are of ectodermal origin. The connective tissues have the ability of repairing themselves after injury by the activity of fibroblasts, an ability which is less marked or even absent in the case of the more specialized tissues. Among connective tissues the connective tissue proper, bone and cement have the power of repair better developed than cartilage or dentine. The epithelial cells of skin and the mucous membranes, which are of *ectodermal* origin, have also a high degree of ability of self-replacement and repair, for the surface cells are constantly shed during life and must be recruited from deeper layers, to avoid rapid loss of epithelium. In the central nervous system, however, and in enamel, which are also of ectodermal origin, the capability is extremely limited or even absent.

Cartilage

Cartilage is a specialized form of connective tissue consisting of cells, the chondrocytes, and extracellular fibres embedded in an amorphous, gel-like matrix. Many parts of the skeleton, including the base of the skull, are first formed as cartilage models which are later replaced by bone. Cartilage assumes different structural features and properties in different parts of the body. The characteristics of the chief kinds of cartilage—hyaline, fibrous and elastic—are determined by the nature of the fibres found in the intercellular substance. The hyaline variety is found extensively in the embryo and foetus, in the larger laryngeal cartilages and in the cartilage of the air passages. Elastic cartilage develops in the external ear, the pharyngo-tympanic tube, the epiglottis and the smaller laryngeal cartilages. Fibrous cartilage develops at the attachment of ligaments and tendons to bone and in some joint disks, including the disk found in the joint between the skull and the mandible, the temporomandibular joint.

The cartilage which is found in the embryo and foetus:

1. Develops into definitive structures, such as the cartilages of the nose, external ear, larynx and trachea.

2. Forms ligaments, such as the sphenomandibular ligament between the mandible and the skull.

3. Provides a temporary 'skeletal' scaffolding for bone development, e.g. Meckel's cartilage which precedes the development of the mandible.

Much of the cartilage found during early development of the individual is replaced by bone prior to birth, by intracartilaginous or endochondral ossification. Some cartilage persists, for instance between the shaft and the ends of growing long bones, or as articular cartilage covering the extreme end of a bone where a joint cavity exists. The cartilaginous base of the developing skull is eventually replaced by bone but some cartilage persists for a number of years in certain locations and contributes to the further growth of the skull by a process of continuous cartilage matrix formation and its replacement by bone.

Bundles of collagenous fibres form a thick continuous sheet, the perichondrium, on the surface of cartilage, except where the cartilage forms a joint surface. Damage to or tears in adult cartilage undergo connective tissue repair and there is evidence that the cells in the deeper layers of the perichondrium can manufacture new cartilage matrix. Hyaline cartilage is usually avascular and nutritive materials reach the chondrocytes by simple diffusion through the cartilage matrix. Blood vessels are sometimes present in cartilage which will later be replaced by bone. Calcification of adult cartilage is common in elderly persons.

Bone

This specialized connective tissue consists of collagen fibres embedded in a calcified ground substance which contains the bone cells or osteocytes in small spaces called lacunae. The inorganic material makes up approximately 70 per cent of dried bone and consists of apatites of calcium and phosphate. The calcified ground substance combined with a helical arrangement of successive layers of collagen fibres gives bone its great strength. Despite its outward appearance, bone is a dynamic substance continually undergoing renewal and remodelling throughout the life span. While bone in all parts of the body has the same chemical composition, individual bones show variations in their form and internal architecture to meet the demands of function. Internal stress stimulates bone formation. The thickness of bones can increase if they are called upon to bear excessive weight and muscular paralysis produces losses in the internal architecture of a bone without any obvious changes in its external form. Bones which become bent as a result of pathological change thicken in the concavity of the bend to compensate for the altered direction of stress. The ability of bones to adapt to new conditions is best seen following fracture, especially if the bony segments are improperly set. The internal structure of the new bone which unites the fragments closely resembles the original pattern and the external surfaces become smooth in an attempt to restore the bone to its original form.

Bones are not absolutely rigid but have some degree of elasticity which allows them to undergo transient deformation with recovery to their original form when the stress is removed. The elastic property of bone decreases with age so that stresses which will not produce fractures in a young person will do so in an elderly individual. The breaking stress of bone is comparable to that of cast iron and has a lower value in old persons due to the lowered content of calcium salts and thinning of the individual struts or trabeculae of the bone. The bones of children and adolescents contain proportionally more collagen than those of adults and are less resistant to tension forces. Excessive bending of the long bones, such as the femur or radius, frequently produces an incomplete fracture of the bone limited to the side to which tension was applied. This is called a green-stick fracture.

Bones vary considerably in shape and can be classed as long, short, flat or irregular. Long bones are typical of the appendicular skeleton and the humerus and femur are the best examples. Each consists of a shaft, or diaphysis, which is a thick walled cylinder of compact bone with a central bone marrow cavity. At the ends of the shaft, the epiphyses consist mainly

Epiphysis - end
Diaphysis = shaft
meataphysis = part of the diaphysis adj to the epiphyseal plate containing the
actively growing bone
SOME ELEMENTS OF BODY STRUCTURE 21

of spongy bone covered by a thin layer of compact bone. During the growing cycle, the diaphysis and the epiphyses are separated by cartilaginous epiphyseal plates which are important growth sites for increase in the total length of the bone. Short bones are generally cuboidal in shape as in the wrist and foot. Flat bones have a large area relative to their thickness; the scapula and bones of the vault of the cranium are good examples. Many bones are irregular in shape, such as the mandible and maxilla. Some of the skull bones can be classified further as pneumatic bones because they contain air-filled spaces lined by mucous membrane, for example the air sinus within the maxilla and the mastoid air cells in the temporal bone. Accessory or supernumerary bones occur occasionally and create difficulty in the interpretation of radiographs. Such bones occurring in the sutures of the skull are termed Wormian bones.

Bones make up the skeleton of the body and serve many functions:

1. They provide attachment for muscles and ligaments.

2. They act as levers which, by their movement at joints under the control of muscles, permit locomotion and other body movements.

3. They protect organs such as the brain, eyeballs, the inner ear mechanism and the contents of the thoracic and pelvic cavities.

4. They give support to the body and enable it to maintain its characteristic form.

5. They play an important part in the regulation of growth.

6. They act as storehouses for calcium and other minerals used in body metabolism.

7. They give attachment to the teeth through the periodontal ligaments.

8. The ear ossicles play a part in sound conduction.

Adult bone has a lamellar structure and two types can be distinguished, *compact* and *cancellous*. The cortex or outer layer of a bone consists of compact bone, made up of a series of *osteones* or *Haversian systems*. Each osteone is a three-dimensional structure formed of a central Haversian canal containing blood vessels, surrounded by a series of concentrically arranged *lamellae*, between which are small spaces for the *osteocytes* (Figs. 2.7, 2.8.). The collagen fibres in the individual lamellae can be seen with the polarizing microscope, when they are found to be arranged in alternating spirals with collagenous fibres passing also in a radial direction from one lamella to the next. Such an arrangement gives the bone considerable strength. Because the blood vessels of compact bone are numerous and branch extensively, so the Haversian systems branch and merge with one another to form a continuum. Many blood vessels communicate with those found in the periosteum through channels in the compact bone called *Volkmann's canals*. The largest of these blood vessels can be seen readily by the naked eye and are the *nutrient arteries*, typically found near the central part of the shaft of a bone. The Haversian canals are usually oriented in the long axis of a bone and, in addition to small arteries and veins, they contain lymphatic vessels and nerve filaments, which may be confined to the outermost Haversian systems. The osteocytes, or resting bone cells, have numerous processes which radiate into the adjacent lamellae in tiny channels called *canaliculi*. Thus circulation of tissue fluid is permitted between the bone cells and the blood vessels and the nutrition of the bone is maintained. On cross section the Haversian systems are circular in outline (Fig. 2.8) and the spaces between adjacent Haversian systems are filled by interstitial lamellae which are remnants of complete Haversian systems which existed at an earlier stage of bone development. The outer surface of a bone is made up of circumferential lamellae which do not show the tubular Haversian pattern (Fig. 2.7).

Cancellous or spongy bone consists of bony plates or *trabeculae* in the form of interlocking sheets which make up a loose network of bone tissue surrounding spaces which contain the bone marrow (Fig. 2.9). Each trabecula has a lamellar structure similar to that of compact bone but they tend to be flat and thus do not form the concentric tubular arrangement typical of Haversian systems. The arrangement of the trabeculae of spongy bone is not as random as it might first appear, for the direction, thickness and arrangement of the bony plates is related to the mechanical needs of the bone to resist stress and strain. The trabeculae form a series of internal struts whose architecture is ideally suited to the functions of the particular bone.

Fibrous or *embryonic bone* is found during the development of many bones. It is characterized by bundles of osteogenic fibres which are continuous with the collagenous or reticular fibres of the surrounding mesenchyme at the earliest stages of bone development and later with the fibrous tissue of the periosteum. With further development and growth fibrous bone becomes reorganized to form spongy bone and subsequently differentiates into either the compact or spongy bone characteristic of the adult.

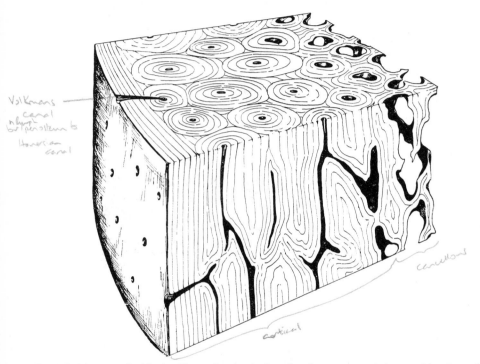

Fig. 2.7 Diagram of a block of bone showing its lamellated structure and the transition from the bone surface (on the left) through the cortex of compact bone to the cancellous tissue (on the right). Haversian systems are shown in both transverse and longitudinal section. (By courtesy of Miss Margaret Gillison.)

Where bones form contributing elements for a joint, they are covered at their articular ends by hyaline or articular cartilage (Fig. 2.15). Articular cartilage is characterized by the cartilage cells, or chondrocytes, being arranged in rows or columns at right angles to the principal joint surfaces and, in the case of long bones, roughly parallel to the long axis of the bone. Except for the joint surfaces, bones are surrounded by a sheath of fibrous tissue which is quite vascular, forming a periosteum consisting of a deeper cellular layer immediately adjacent to the bone surface and a more superficial fibrous layer (Fig. 2.10). Collagen fibres derived from the outer fibrous or capsular layer of the periosteum pass obliquely in to the subjacent bone (Sharpey's fibres of bone) attaching the periosteum to the bone and thus providing attachment for muscle tendons and ligaments to the bone matrix (Fig. 2.11).

While this is an important feature of adult bones, Sharpey's fibres are not present at every site of muscle attachment in children. When growth is taking place the attachment of muscles

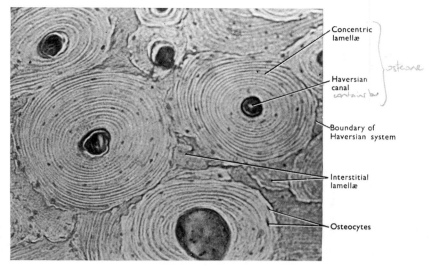

Concentric
lamellæ

osteone

Haversian
canal
contains bv

Boundary of
Haversian system

Interstitial
lamellæ

Osteocytes

Fig. 2.8 Transverse section of compact bone showing Haversian systems × 78.

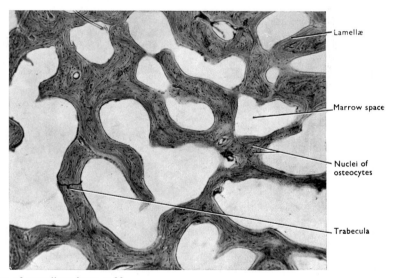

Lamellæ

Marrow space

Nuclei of
osteocytes

Trabecula

Fig. 2.9 Section of cancellous bone × 30.

is primarily to the periosteum and the absence of Sharpey's fibres permits the shifting or sliding of the periosteum on the underlying bone, which must occur during growth as the size of the bone increases or its bony processes change their relationship to the rest of the bone. The special processes (tubercles, trochanters, plates and ridges) which develop for the attachment of muscles form the surface features which aid the identification of individual bones. The coronoid and angular processes of the mandible are good examples of bony processes which develop specifically for the attachment of the muscles of mastication and undergo constant remodelling during growth of the mandible.

The cellular layer of periosteum (Fig. 2.10) contains fibroblasts and osteoblasts, bone-form-

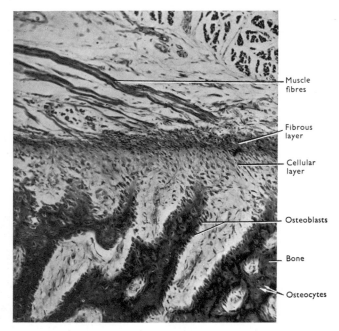

Muscle fibres

Fibrous layer

Cellular layer

Osteoblasts

Bone

Osteocytes

Fig. 2.10 Section from the developing mandible of a 135 mm C-R length human foetus showing the layers of the periosteum and related tissues × 100.

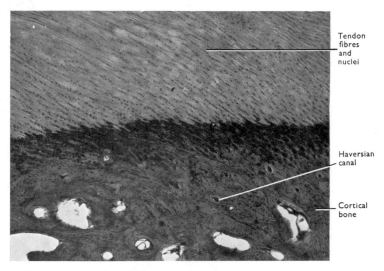

Tendon fibres and nuclei

Haversian canal

Cortical bone

Fig. 2.11 Section of compact bone through the area of attachment of a muscle tendon. Many tendon fibres pass obliquely into the substance of the bone as Sharpey's fibres. × 30.

ing cells which are in direct contact with the underlying bone matrix. The latter cells are particularly noticeable during stages of bone development or bone repair. The two layers of the periosteum are of a special importance in consideration of the structure of fixed joints in the skull (*sutures*) and they are described on page 355.

Bone is the product of cellular activity. Bone-forming cells, *osteoblasts* (Fig. 2.10), and bone-removing cells, *osteoclasts*, in the balance of their activity determine the growth of bone, the thickness of its cortical layer and the structural arrangement of its lamellae. Bone is continually altering its internal structure to meet the requirements of function and these alterations are brought about by the relative activity of osteoblasts and osteoclasts. Bone in its developmental history is usually classified into two types:

1. *Membrane* or *intramembranous bone* which develops directly in a mesodermal condensation of connective tissue, e.g. the mandible and the bones of the vault of the skull.

2. *Cartilage* or *endochondral bone* which replaces cartilage, e.g. during formation of the base

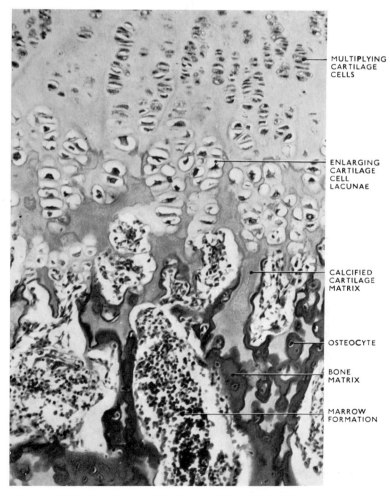

MULTIPLYING CARTILAGE CELLS

ENLARGING CARTILAGE CELL LACUNAE

CALCIFIED CARTILAGE MATRIX

OSTEOCYTE

BONE MATRIX

MARROW FORMATION

Fig. 2.12 Photomicrograph showing the zone of replacement of cartilage by bone (endochondral ossification) in the finger of a human foetus × 150.

of the skull. The same cells produce both types of bone and in their final structure there is no difference between them.

Intramembranous ossification begins within the embryonic connective tissue with the differentiation of some mesodermal cells into *osteoblasts*. These cells deposit collagenous reticular fibres and the cells become enclosed by primitive bone matrix, or *osteoid*, and develop lacunae. The osteoid becomes calcified and the outermost differentiated cells take on the characteristics of a cellular layer of the periosteum. These several changes constitute a *centre of ossification* which appears for individual bones with remarkable consistency in location and time.

Endochondral ossification begins in the perichondrium which covers the cartilaginous

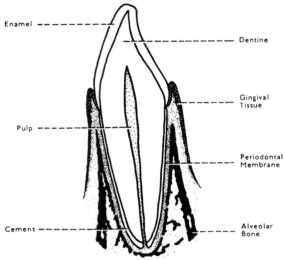

Enamel
Dentine
Gingival Tissue
Pulp
Periodontal Membrane
Cement
Alveolar Bone

Fig. 2.13 Diagram of a longitudinal section through a premolar and its socket showing the arrangement of the collagen fibres which form the fibrous joint (gomphosis) between the surface of the root of the tooth and the alveolar bone.

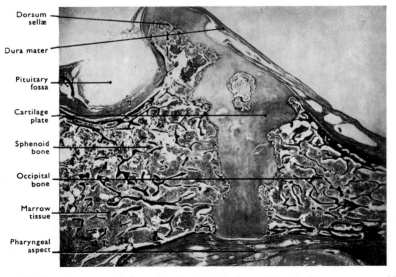

Dorsum sellæ
Dura mater
Pituitary fossa
Cartilage plate
Sphenoid bone
Occipital bone
Marrow tissue
Pharyngeal aspect

Fig. 2.14 The spheno-occipital synchondrosis in the base of the skull in a one-year-old child.

model of a bone. About the centre of the model some of the cells of the perichondrium differentiate into osteoblasts, at which stage this part of the perichondrium becomes a periosteum,
and a *subperiosteal collar* of bone matrix is formed. Zones of ossification rapidly appear within
the cartilage, probably by a vascular invasion from the periosteal region, which carries in
osteoblasts to form a *primary centre* of ossification. The process of ossification extends towards
either end of the cartilage model, that is, towards the eventual diaphyses until only the terminal
parts of the cartilaginous model remain. *Secondary centres* of ossification appear at specific
times within the ends of the model, the epiphyseal centres. The spread of endochondal ossification is characterized by distinctive changes in the cartilage cells and their matrix, followed
by the laying down of bone matrix on the surface of calcified cartilage (Fig. 2.12). The body
and greater part of the mandible develops by intramembranous ossification but the condyle
development and growth is the result of a process closely resembling endochondral ossification
(see page 360).

The growth of bone depends on genetic and environmental factors, including the effects
of hormones, diet and mechanical forces. A bone increases in length by continued proliferation
of the epiphyseal cartilages which are constantly being invaded and replaced by bone matrix
of the metaphysis (Fig. 2.12). The rate of growth is not everywhere the same; for example,
it is more rapid at the proximal than at the distal end of the humerus. Just as the internal
architecture of spongy bone depends on the direction of stresses which the bone has to bear,
the direction in which new bone formation takes place in the region of an epiphyseal plate
is determined by the direction and distribution of stress lines. Increase in thickness or width
of a bone is the result of the deposition of new bone in the form of circumferential lamellae
beneath the periosteum. Later, as the growth of bone continues, these lamellae become sub-
merged beneath a new bone surface and are replaced by Haversian canal systems (Fig. 2.7).

Bone is a highly vascular tissue, especially during its growth when it depends upon an
adequate blood supply for its raw materials and for the hormones which regulate its growth.

In old age the bones of the body become less dense, the cortical layers thinner, the lamellae
less numerous, the Haversian canals wider. Such bones are more liable to fracture.

Joints

The movements which take place at a joint are determined by the relationship of the bony
parts, the strength and attachment of the ligaments, and the arrangement of the muscles. The
relationships of these structures at a joint are often of such a nature as to prevent unwanted
movements and therefore to make more efficient the pattern of movements which are produced. Muscles produce movements at joints which are very similar in their structure and
form in man and the other primates, but it is the pattern of movements under the control
of the central nervous system which is important, not the activity of individual muscles at
particular joints.

The mechanical requirements of *movable (diarthrodial) joints* (Fig. 2.15), as well as the limitation of unwanted movements, are a minimum of friction between the moving parts, a minimum
of wear and tear, and in many cases a position of maximum stability, especially in weight-
supporting joints. The articular (hyaline) cartilage, which covers the bony elements, presents
a smooth surface, is capable of repair by the proliferation of its deeper cells and can withstand
pressure in a manner which would be impossible if joint surfaces consisted of naked bone.
The synovial fluid, secreted by the synovial lining of the joint capsule, lubricates, provides
nutrition and removes any products of tissue disintegration from the joint cavity. In all diarthrodial joints, but especially in the large weight-supporting joints such as the hip joint and
knee joint, there is a position in which there is a maximum area of contact between the joint
surfaces and there are special muscular locking mechanisms for holding the parts in this position. The shoulder joint, however, is an example of a joint where there is little need for such
a mechanism and here the emphasis is on the range of movement rather than on its limitation.

Unlike the hip joint, the head of the humerus is not embraced by a socket-like structure, but is capable of a very wide range of movement on the almost flat surface of the articular fossa of the scapula. However, this means that the joint is much more liable to dislocation. This danger is overcome not, as might be expected, by the presence of strong ligaments. These would in any case limit movement, and like all ligaments would be liable to gradual stretching unless supported by muscles. Instead at the shoulder joint a special group of muscles have taken over the function of regulating the mobility of the joint, acting as a dynamic capsule. By their relaxation in co-ordination with the prime movers, they permit the necessary movements of the joint; by their regulated contraction they prevent the displacement of the humerus from the glenoid fossa.

Joints are defined as sites of union between two or more elements of the skeleton. They can be divided into three main categories depending on the way the bones are held together:

Fibrous joints. The skeletal elements (bone or cartilage) are firmly united by fibrous tissue. There is no joint cavity and movement is very limited, except between the bones of the vault of the skull in the infant, which can be moulded to overlap one another during childbirth. Examples of fibrous joints are the periodontal ligaments of the teeth and the sutures between the bony elements of the skull (Figs 2.13, 7.30).

Cartilaginous joints

1. Primary cartilaginous joints in which the skeletal elements are united by a continuous plate of cartilage (synchondrosis). Examples are the epiphyseal plates between the diaphyses and epiphyses of long bones (Fig. 2.15) and the important synchondrosis between the occipital and sphenoid bone at the base of the skull (Fig. 2.14). At such joints there is no movement but they play an indispensable part in the growth of the skeleton.

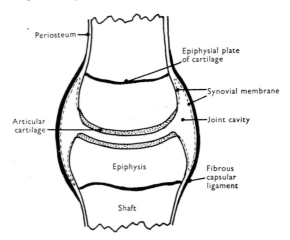

Periosteum

Epiphysial plate of cartilage

Synovial membrane

Articular cartilage

Joint cavity

Epiphysis

Fibrous capsular ligament

Shaft

Fig. 2.15 Diagram showing the parts of a diarthrodial (movable) joint.

2. Secondary cartilaginous joints are those in which the skeletal elements are covered by cartilage and interconnected by fibrocartilage. Examples are the joints between the bodies of the vertebrae, the symphysis pubis, and the joint between the two halves of the mandible which disappears completely during the first year after birth. A certain amount of movement is permitted at these joints. Flexion and extension of the vertebral column is most marked in the cervical and lumbar regions where the fibrocartilage disks are thick. Much of the movement between the bodies of the vertebrae depends on the capability of the intervertebral disks to respond to compression and tension forces and is restrained by strong ligaments which

run the length of the vertebral column. Rotation to the right or left is most pronounced between the atlas and axis vertebrae. Twisting movement is limited in the rest of the vertebral column, except in the thoracic part. Routinely no movement occurs at the symphysis pubis, but changes in the fibrocartilage during the later stages of pregnancy allow for some separation of the pubic bones at the time of child birth. The mandibular symphysis (Fig. 7.49) is analogous to the joint which persists throughout life in other animal forms. Movement at this joint in the human infant has no great functional significance but needs to be distinguished from fracture lines in radiographs.

Synovial (diarthrodial) joints (Fig. 2.15). The several characteristic features of these joints are:

1. The bones are held together by a fibrous tissue cuff or *joint capsule*. The capsule consists of interwoven bundles of collagen and it is extremely strong. The degree of tightness of the capsule dictates the extent of movements which may occur at the joint, for instance, it is slack in front of the temporomandibular joint but quite taut on the lateral side of the joint, where it is thickened to form the temporomandibular ligament. Thus, the head of the mandible can move forwards freely within the joint capsule during mouth opening but any side to side movement at the joint is limited. This is due in part to the configuration of the bony elements.

2. The opposing skeletal elements are covered with a layer of *articular hyaline cartilage*, or fibrocartilage in the case of the temporomandibular joint, to reduce friction to a minimum and withstand the wear and tear which the moving parts exert on one another.

3. A potential space or *joint cavity* exists between the skeletal parts and is lined by an endothelial synovial membrane and contains a small amount of synovial fluid secreted by the lining cells. The fluid acts as a lubricant and is an additional source of nutrition for the superficial layers of the articular cartilage. The synovial membrane ends at the margins of the articular cartilage.

4. The *ligaments* of a joint are usually situated outside the joint capsule (extracapsular ligaments) but may lie within it. If the ligaments are intracapsular they are covered by synovial membrane. In some joints the cavity is partly or completely divided into two compartments by an articular disk, which is complete in the temporomandibular joint (Fig. 4.22) so that there are two joint cavities, each associated with its own range of movement.

The synovial membrane has a rich capillary blood supply, ligaments and articular disks have a limited blood supply, while articular cartilage is usually avascular. The joint capsule, its ligaments, the synovial membrane and articular disks have a well-developed sensory nerve supply, related to the reflexes that regulate the coordinated activity of the muscles which produce movements at the joint. This is muscle-joint sense or *proprioception*.

The articular cartilage covering the opposing bones is quite smooth although this quality begins to decrease after the third decade. Changes in the surface contours of the articular cartilage, the blood supply to the joint and the amount of synovial fluid which is secreted contribute to the slow onset of chronic joint inflammation or arthritis. In its more severe forms these pathological changes are visible to the naked eye.

In addition to the structural classification of joints, synovial joints can be classed according to the shape of the joint surfaces or the movements which can take place at them. The most common types are:

a. *ball and socket*, for example the shoulder joint at which movements in all planes are possible

b. *condyloid*, with two pairs of articular surfaces whose long axes are parallel to one another, for example, the temporomandibular and knee joints at which movements take place predominantly in one direction or plane

c. *hinge*, for example the ankle joint at which movement is restricted to one plane

d. *pivot*, for example, the joint between the odontoid process of the axis vertebra and the anterior arch of the atlas

e. *plane*, for example the joints between the articular processes of the vertebra where the joint surfaces are flat and movements are sliding, tilting or rotational.

Movements at joints are limited by shape of the joint surfaces, the arrangement of the capsule and joint ligaments, as well as the body surfaces coming into contact. Nerve reflexes cause contraction of muscle groups which prevent overstretching the joint capsule or its ligaments and the part of the capsule involved and the contracting group of muscles have the same nerve supply (Hilton's Law).

Joint development

The contiguous ends of developing bones at the region of a future articulation are connected initially by undifferentiated mesenchyme. If this primitive connective tissue becomes chondrified, a synchondrosis is formed. The joint cavity of diarthrodial joints does not appear until after the cartilaginous models of the constituent bones have differentiated.

When the cartilaginous models have been laid down the mesenchymal tissue between the adjacent surfaces becomes arranged to form an *intermediate zone* which is continuous at its periphery with the perichondrium of the cartilage models. The cells in the central part of the intermediate zone become compressed and one or more primitive joint cavities appear. The central cells then disappear so that the cartilaginous ends of the developing bones are in contact and there is a recognizable joint cavity. The cells derived from the intermediate zone which line the joint cavity give rise to a thin synovial membrane. The connective tissue around the developing joint, which was formed from mesoderm, becomes the fibrous joint capsule and is continuous with the perichondrium of the cartilage models. Some diarthrodial joints, and the temporomandibular joint is a good example, develop intra-articular disks which are composed of dense fibrous tissue or fibrocartilage. In the temporomandibular joint two primitive joint cavities appear and the intervening tissue becomes the intra-articular disk. In other joints, such as the knee joint, the intra-articular fibrocartilages develop as ingrowths from the joint capsule but do not completely divide the joint cavity into two parts. The development of the temporomandibular joint is described in more detail on page 366.

Muscles

There are three types of muscle in the human body (Figs 2.17, 2.20):

Striated, voluntary (skeletal) muscle in which the muscle fibres show transverse striations and is under the control of the will, through the somatic nervous system.

Smooth, involuntary (visceral) muscle in which only longitudinal striations are present and which is not under the control of the will but is regulated by autonomic nervous control.

Striated involuntary (cardiac) muscle is found in the myocardium of the heart and is regulated by autonomic nervous control.

All muscle tissue is built up of muscle cells (fibres) containing myofibrils which impose contractability on the cells. Each fibre has its own sheath, the sarcolemma. In the case of skeletal and visceral muscle the individual fibres are united by a delicate connective tissue matrix, the endomysium. Cardiac muscle fibres appear to be directly united to one another by continuity of muscle tissue forming a continuous protoplasmic network or syncytium (Fig. 2.20). However, studies using the electron microscope show that cardiac muscles fibres are independent units which adjoin one another in a complex manner at the intercalated disks.

Striated, voluntary, muscle depends upon its nerve connections for its ability to contract and to maintain itself in various degrees of contraction (tone). Smooth muscle and cardiac muscle, however, have the power of contractility inherent in their fibres and depend upon the nervous system only for the regulation and co-ordination of their activity.

Muscle tone can be defined as that slight degree of muscle contraction present even at rest, but which does not result in movement. Muscle tone is under nerve reflex control, maintains it in a state of preparedness and allows movement to take place more rapidly and smoothly.

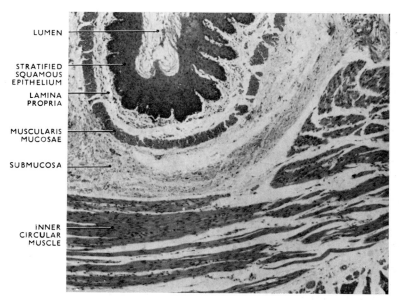

LUMEN

STRATIFIED
SQUAMOUS
EPITHELIUM

LAMINA
PROPRIA

MUSCULARIS
MUCOSAE

SUBMUCOSA

INNER
CIRCULAR
MUSCLE

Fig. 2.16 Photomicrograph showing the layers in the wall of the oesophagus, in particular the structure of the mucous membrane × 54.

In voluntary muscle each nerve fibre passing to the muscle supplies a number of muscle fibres making up a physiological muscle unit. The power of a contraction of a muscle depends on the number of muscle units in synchronized action at the time of contraction. The tone of a resting voluntary muscle depends on the fact that a certain percentage of the muscle units are in contraction even when the muscle is at rest. Muscle units succeed one another in shifts; each has a period of activity followed by a period of rest. The continual activity of voluntary muscle even during sleep is associated with a continual activity of the nerve pathways and reflexes which maintain muscle contraction.

Voluntary muscle is made up of muscle fibres arranged in bundles (Fig. 2.17). Each fibre is enclosed within a cell membrane, the *sarcolemma*. Outside the sacrolemma is a fine connective tissue sheath, the *endomysium*, and adjacent fibre bundles are held together by a somewhat denser layer, the *perimysium*. The whole muscle is covered and contained by the *epimysium*, the fascial sheath of the muscle which is continuous with the looser intermuscular connective tissue.

The principal blood vessels and the nerves to a muscle usually pierce the connective tissue sheath and enter the muscle at a definite region called the *neurovascular hilum*. Within the muscle the vessels anastomose and terminate in a fine network of capillaries around the individual muscle fibres. Muscle tissue depends on its rich blood supply for food and oxygen which it converts into energy, and for the removal of the waste products of its activity.

Striated voluntary muscle fibres consist of numerous myofibrils which have a banded or striated appearance when seen in the light microscope. These are called the A bands (anisotropic) and the I bands (isotropic) which appear dark and light alternately when seen with transmitted light in the microscope (Fig. 2.18). In the middle of the I band there is a dark zone called the Z band. The unit of the myofibril between two adjacent Z bands is known as the *sarcomere*. The myofibril is composed of even smaller units, the *myofilaments*. Thick and thin myofilaments can be seen in electron micrographs and studies of the chemical composition of myofilaments indicate that the contractile protein actin is located in the thin filament, while myosin is associated with the thick filament (Fig. 2.19). Between the myofibrils

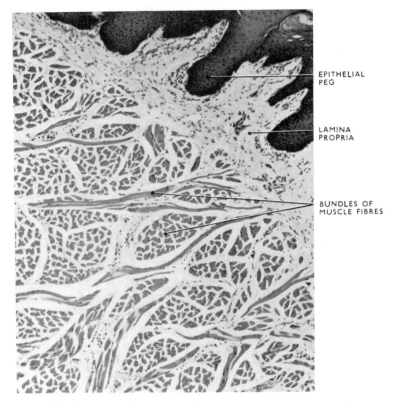

EPITHELIAL
PEG

LAMINA
PROPRIA

BUNDLES OF
MUSCLE FIBRES

Fig. 2.17 Photomicrograph showing bundles of striated, voluntary (skeletal) muscle in the tongue × 60.

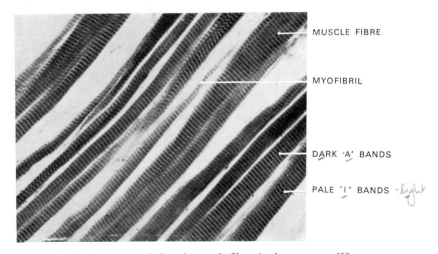

MUSCLE FIBRE

MYOFIBRIL

DARK 'A' BANDS

PALE 'I' BANDS *-light*

Fig. 2.18 Photomicrograph showing cross striations in muscle fibres in the tongue × 400.

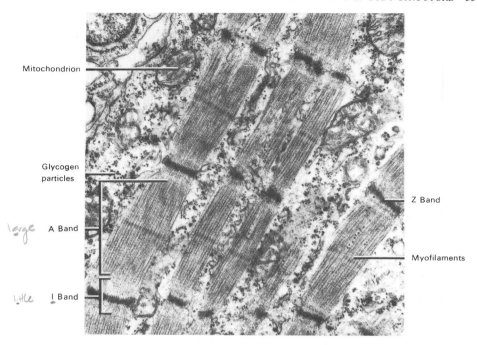

Mitochondrion

Glycogen
particles

large   A Band

Z Band

Myofilaments

little   I Band

Fig. 2.19 The fine structure of striated muscle fibres in the developing tongue, electron micrograph × 21,600.

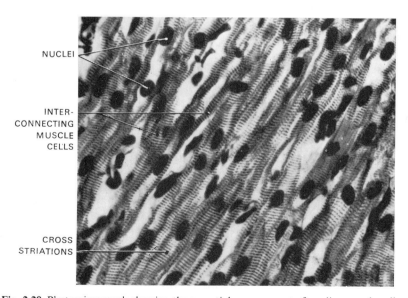

NUCLEI

INTER-
CONNECTING
MUSCLE
CELLS

CROSS
STRIATIONS

Fig. 2.20 Photomicrograph showing the syncytial arrangement of cardiac muscle cells × 540.

there is an extensive system of smooth ~~membrane~~ endoplasmic reticulum and many mitochondria. It is believed that this endoplasmic reticulum functions in the transmission of the
nerve impulse from the nerve fibre to the units of muscle contraction. The mitochondria, which
in striated muscle are also called *sarcosomes*, are carriers of the enzymes which are essential
in the chemical changes which result in muscle contraction. The shortening of the myofibrils
is thought to be due to a sliding mechanism between the thick and thin myofilaments.

Voluntary muscles are usually, but not always, attached to parts of the bony or cartilaginous
skeleton. The attachment may be in the form of the direct union of the perimysium of the
muscle with the fibrous layer of the periosteum (e.g. the insertion of the masseter) or the muscle
may be attached through a cord-like tendon or a membrane-like aponeurosis. These are usually

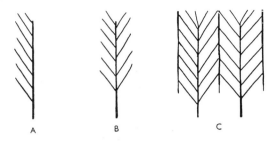

Fig. 2.21 Diagrammatic representation of voluntary muscles showing the arrangement of muscle fibre
bundles within them. A. Unipennate. B. Bipennate. C. Multipennate.

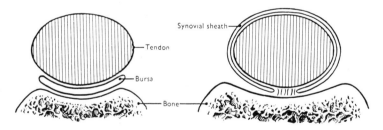

Fig. 2.22 Diagrams of a muscle tendon in transverse section showing anti-friction devices: a bursa
and a synovial sheath. (Modified from Mitchell and Patterson.)

more firmly attached to the cortical layer of the bone. Tendons consist of numerous bundles
of collagen fibres tightly packed together with fibroblast cells lying between the fibres and
bound together in a connective sheath, the interfascicular tissue, which contains the small
blood vessels within and on the surface of the tendon. An aponeurosis may extend on one
or more surfaces of the muscle or take the form of leaf-like extensions into its substance from
one or both its attachments (*unipennate, bipennate* or *multipennate muscles*). In the latter, the
muscle fibre bundles run between part of the aponeurosis of origin and part of the aponeurosis
of insertion, giving the muscle a complex structure and enabling a greater number of fibres
to occupy a limited space, therefore producing a more powerful muscle (e.g. masseter, temporal, deltoid—the muscle over the shoulder which helps to raise the arm) (Figs 2.21, 2.23).

The *origin* of a muscle is usually taken as the less movable of its attachments and the *insertion*
as the more movable. However, it is very common for muscles to act from their insertions
when these are fixed by the contraction of other muscles (e.g. the action of the geniohyoid
in moving the mandible when the hyoid bone is fixed).

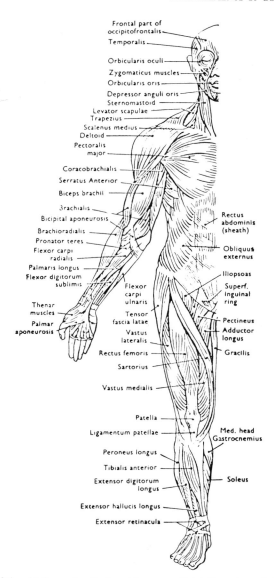

Frontal part of
occipitofrontalis
Temporalis
Orbicularis oculi
Zygomaticus muscles
Orbicularis oris
Depressor anguli oris
Sternomastoid
Levator scapulae
Trapezius
Scalenus medius
Deltoid
Pectoralis
major
Coracobrachialis
Serratus Anterior
Biceps brachii
Brachialis
Bicipital aponeurosis
Brachioradialis
Pronator teres
Flexor carpi
radialis
Palmaris longus
Flexor digitorum
sublimis
Thenar
muscles
Palmar
aponeurosis

Rectus
abdominis
(sheath)
Obliquus
externus
Iliopsoas
Superf.
inguinal
ring
Flexor
carpi
ulnaris
Tensor
fascia latae
Vastus
lateralis
Rectus femoris
Sartorius
Pectineus
Adductor
longus
Gracilis
Vastus medialis
Patella
Ligamentum patellae
Med. head
Gastrocnemius
Peroneus longus
Tibialis anterior
Extensor digitorum
longus
Extensor hallucis longus
Extensor retinacula
Soleus

Fig. 2.23 Drawing of the chief muscles displayed in a superficial dissection of the body. (By courtesy of Professor G. A. G. Mitchell and Dr E. L. Patterson.)

Muscles which during a movement are by their contraction responsible for that particular movement are known as *prime movers*. Movement at a joint also usually involves a relaxation of one or more muscles. These are the *antagonists* for that particular movement. When the movement is reversed the antagonists become the prime movers. If movement is to to take place in one joint and other joints are to be maintained in a fixed position, this fixation is produced by the activity of the *fixation muscles*. *Synergist muscles* are those which by their activity prevent or minimize those movements which can be produced by the prime movers but which are not required in a particular movement. Some muscles act on more than one joint and the same muscle is often capable of producing more than one movement owing to the disposition of its fibres in relation to the joint. The deltoid (Fig. 2.23), whose fibres are related to the anterior, lateral and posterior aspects of the shoulder joint, takes part in flexion, extension, abduction and rotary movements of the arm. The arm can be abducted above the level of the shoulder joint due to the accompanying rotation of the scapula, which allows the glenoid cavity to face progressively further upwards and outwards as the arm swings above the head. The temporalis muscle is a good example of one which has two distinct actions on its associated joint. The anterior fibres have the function of elevating the mandible while the posterior fibres pull the lower jaw backwards when they contract.

Muscles often lie in very close relationship to one another. In life, however, the intervening connective tissue (deep fascia) is a highly mobile, fluid-containing layer which permits easy movement of one muscle in relation to another. In some regions special fluid-containing sacs (*bursae*) are situated between adjacent muscles and between muscles and adjacent bones, to act as anti-friction devices. Muscle tendons are often covered by *synovial-lined sheaths* which enable them to move readily and without friction in constricted areas (Fig. 2.22). For the development of muscles see page 331.

Muscles vary greatly in shape, ranging from straight, strap-like muscles, to sheet-like forms. The sternomastoid and platysma are examples of these extremes in the neck region. Some muscles have more than one head of origin, e.g. the biceps brachii, or they may consist of two bellies united by a tendon, e.g. the digastric muscle in the upper part of the neck. Many muscles are named from their position, e.g. temporalis, occipitalis, orbicularis oris, intercostalis, gluteus and tibialis. Still others have names determined by their functions, e.g. depressor anguli oris, tensor palati, pronator teres, adductor longus. Many of these muscles are shown in Figure 2.23 and fuller descriptions of them will be found elsewhere in the text.

Nerves

The nervous system has for its chief function the correlation of body functions, their initiation and regulation. It consists of many millions of nerve cells (neurons), possessing protoplasmic processes in the form of nerve fibres which vary considerably in length (Fig. 2.24). Nerve fibres make up nerve bundles, and by virtue of their functional continuity with one another, nerve tracts or nerve pathways and nerve reflexes.

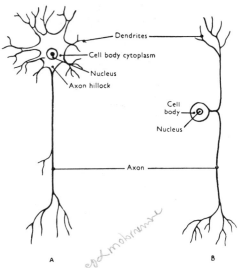

Fig. 2.24 The parts of nerve cells: A. A multipolar neuron. B. Unipolar neuron. These typify motor and sensory neurons respectively.

Within the central nervous system the nerve cells are grouped to form masses of cellular tissue (grey matter or nuclei), while their nerve fibres form ascending, descending and connecting tracts (white matter). Outside the central nervous system the nerve fibres form bundles of nerve tissue surrounded by and supported by connective tissue (*epinerium, perineurium* and *endoneurium*) (Figs 2.26, 2.27). The central nervous system as a whole (brain and spinal cord) is protected by a series of three membranes, the dura mater, the arachnoid and the pia mater. The cerebrospinal fluid circulates between the arachnoid and the pia mater (Fig. 8.22). The cells of the central nervous system are supported and maintained by a specialized group of

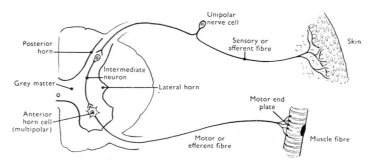

Fig. 2.25 Nerve fibre arrangement in a simple spinal reflex arc.

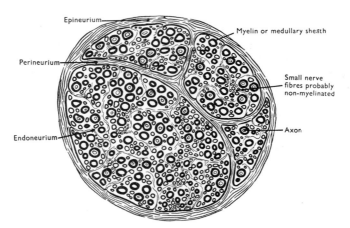

Fig. 2.26 Transverse section of a peripheral nerve and the connective tissue layers which surround groups of myelinated and non-myelinated nerve fibres.

cells, collectively known as the *neuroglia*. There are three distinct kinds of neuroglial cell; microglia, astrocytes and oligodendroglia:

1. *Microglial cells* resemble phagocytic, or scavenging cells and are of mesodermal origin.

2. *Astrocytes* are closely related to blood capillaries which supply the brain or spinal cord tissue, and appear to be concerned in the nutrition of the nervous tissues. They are of ectodermal origin.

3. *Oligodendroglial cells* are also concerned with the nutrition of the nerve cells and their processes and are thought to play a part in the maintenance of the myelin sheath. They are of ectodermal origin.

MUSCLE
FIBRES

EPINEURIUM

MYELIN
SHEATHS

PERINEURIUM

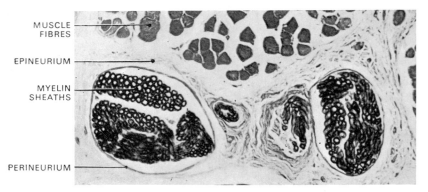

Fig. 2.27 Photomicrograph showing a small bundle of nerve fibres in the tongue. The myelin sheaths have been stained by osmium tetroxide × 150.

Nerve cells or neurons (Figs 2.24, 2.28)

These are the units of the nervous system and consist of a *body or perikaryon* together with a variable number of processes. Of these processes one conducts impulses away from the cell body and is called the *axon*. It is usually the longest process and the branches by which it establishes functional connections with other neurons are limited to its terminal portion. The shorter processes; the *dendrites*, conduct impulses towards the cell body and branch freely throughout their length. The cell body contains:

1. fine protoplasmic filaments or *neurofibrillae* which may have to do with the conduction of the nerve impulse through the cytoplasm from the dendrites to the axon and

2. granules of *Nissl substance*, or chromophil material, which is readily stained with basic dyes. Nissl granules consist of aggregates of ribonucleoprotein (RNP granules, or ribosomes) arrayed on and between the cisternae of the endoplasmic reticulum, which are somewhat irregularly arranged in sensory neurons (Fig. 2.29). Mitochondria, elements of the Golgi complex and other customary cytoplasmic features can be demonstrated also. When the axon of the nerve cell is damaged the Nissl substance accumulates at the periphery of the cell and the large nucleus migrates from the centre of the cell to become eccentrically placed (Fig. 2.28).

Normal
cell

Satellite
cell nucleus

Normal cell

Nissl
substance

Nucleolus
of
chromatolytic
cell

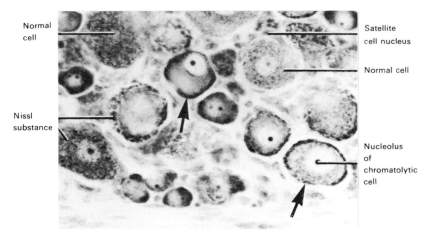

Fig. 2.28 Photomicrograph of nerve cells from the trigeminal ganglion showing the Nissl substance and chromatolytic cells (arrows) × 500.

These changes which reach a maximum about two weeks after axon damage constitute *chromatolysis* and have been extensively used experimentally to study the connections of the various parts of the nervous system.

The nerve impulses are conducted within the nervous system along a series of neurons and are passed from the axon of one cell to the dendrites of another, where these processes are in close contact. There is no actual structural continuity at these points of *synapse* and the

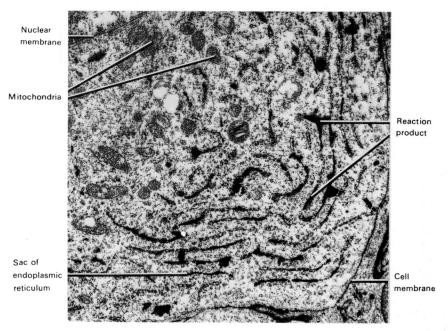

Fig. 2.29 The fine structure of the Nissl substance in a nerve cell from the trigeminal ganglion. A histochemical method has been used to demonstrate the enzyme acetylcholinesterase in the cisternae of the endoplasmic reticulum as a black reaction product, electron micrograph × 12,000.

view that every neuron is structurally independent of every other neuron, although in virtual functional continuity, forms the basis of the *neuron theory*.

Synapses take many morphologic forms but essentially consists of *pre- and postsynaptic membranes*, belonging to the terminal part of the axon and the surface of the cell to which the impulse is being transmitted, respectively. These membranes are separated by a small gap— the *synaptic interval* or *cleft*. Mitochondria and microvesicles are numerous in the terminal part of the presynaptic element. The microvesicles contain acetyl choline, which is essential for the transmission of the nerve impulse across the synaptic cleft. Propagation or conduction of the nerve impulse along intact nerve fibres seems to depend on a combination of chemical and electrical potential changes.

Types of neurons
Neurons are of several types depending on the number and arrangement of their processes (Fig. 2.30).

1. *Multipolar neurons* have angulated cell bodies with one axon and several dendrites. The axon is of variable length but may extend from the spinal cord to the tips of the finger or toes. This type of neuron is typified by the efferent or motor neuron in the reflex arc (Fig. 2.25) and they make up the greatest number of nerve cells within the central nervous system.

2. *Unipolar (pseudo-unipolar)* neurons have apparently a single process which almost immediately bifurcates into a process conducting the nerve impulse towards the cell body (dendrite) and a process conducting away from the cell body (axon). These neurons have an afferent or sensory function and their cell bodies accumulate in the posterior root ganglia of spinal nerves (Fig. 2.30). The cells of the trigeminal ganglion fall into this category.

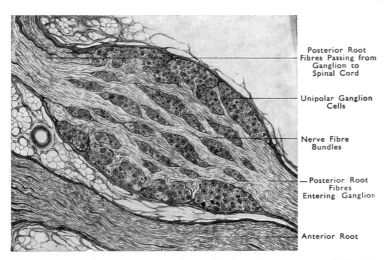

Posterior Root
Fibres Passing from
Ganglion to
Spinal Cord

Unipolar Ganglion
Cells

Nerve Fibre
Bundles

—Posterior Root
Fibres
Entering Ganglion

Anterior Root

Fig. 2.30 Section of ganglion on posterior root of a spinal nerve. (By courtesy of Miss Margaret Gillison.)

Other types of neurons are *pyramidal, bipolar* and *Purkinje* neurons. These are found respectively in the cortex (outer layer) of the cerebral hemispheres, in special sense organs such as in the retina of the eye and olfactory epithelium of the nasal cavity, and in the cerebellum which fills the posterior cranial fossa.

The *nerve fibres* make up the white matter of the brain and spinal cord and form the chief part of the bulk of the peripheral nerves. Each fibre contains an axon which may or may not be ensheathed by a layer of fatty material—the *myelin sheath*. Fibres which have a myelin sheath are said to be *myelinated* or medullated and the layer of lipoid substance is contained within a thin membrane, the *neurilemma*. The neurilemma cells (or *Schwann cells*) possess flattened nuclei which are a feature of longitudinal sections of peripheral nerves (Fig. 2.31). These cells are important in nerve fibre regeneration and are absent as such in the central parts of the nervous system.

Fibres which do not have a myelin sheath are said to be *non-myelinated* or unmedullated. By various means, for instance by the use of polarized light or electron microscopy, it can be argued that these fibres do have a very thin coating of myelin but it is convenient to retain the time honoured terminology. In any event the non-myelinated nerve fibres still possess the equivalent of a neurilemmal sheath.

The myelin sheath, when present in demonstrable amount, as in the nerve fibres belonging to the peripheral exteroceptive conducting mechanism, does not form a continuous layer but at regular intervals is absent or greatly reduced so that the neurilemma dips inwards to the axon. These constrictions, called the *nodes of Ranvier* (Fig. 2.31), are important in the conduction of the nerve impulse along the axon and are also found in the central nervous system.

By means of electron microscopy it has been shown that the myelin sheath of myelinated nerve fibres is formed as a spiral wrapping of the surface membrane of the Schwann cell around the central axon, in a Swiss-roll type of arrangement (Fig. 2.32). The layers of myelin, called

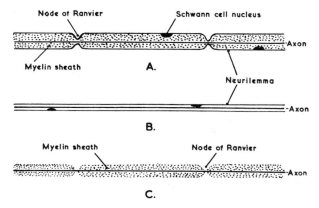

Fig. 2.31 Representation of the structure of nerve fibres: A, in a peripheral spinal nerve (voluntary); B, in a postganglionic fibre of the autonomic nervous system (involuntary); C, in the central nervous system.

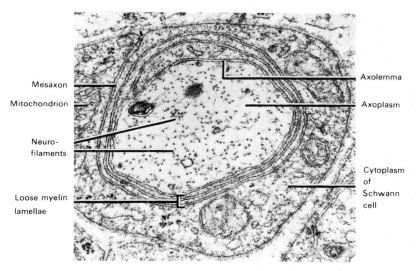

Fig. 2.32 Electron micrograph of a developing nerve fibre in the oral mucosa showing an early stage in the formation of the myelin sheath × 34,000.

lamellae, have a lipoprotein structure and are evenly spaced, being approximately 100 Ångstrom units (10 nm) apart (Fig. 2.33). The outer surface of the myelin has a thin covering of Schwann cell cytoplasm between it and the cell membrane of the Schwann cell and it is this combination of cytoplasm and cell membrane which is equivalent to the neurilemmal sheath of light microscopy. In the central nervous system the arrangement is a little different and the cytoplasm of the Schwann cell forms a small tongue-like process at one aspect of the myelin sheath, when it is seen in cross section.

Myelination of nerve fibres commences about the fourth month of foetal life and usually appears first in those tracts which function earliest or that are phylogenetically the oldest, e.g. association and intersegmental tracts (page 394). Projection tracts, e.g. the pyramidal system, myelinate at a later time, the first myelin sheaths appearing about full term. Some tracts in the nervous system myelinate much earlier, e.g. the dorsal and ventral roots of the spinal

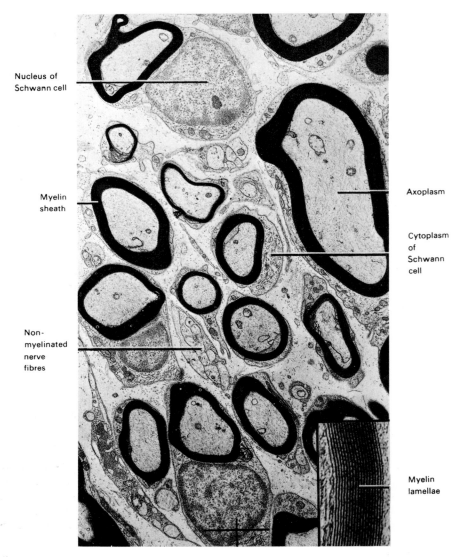

Nucleus of
Schwann cell

Myelin
sheath

Non-
myelinated
nerve
fibres

Axoplasm

Cytoplasm
of
Schwann
cell

Myelin
lamellae

Fig. 2.33 Electron micrograph of myelinated and nonmyelinated nerve fibres in the trigeminal ganglion × 5,000. The insert shows a small segment of a myelin sheath at higher magnification × 57,500.

cord show myelin sheath formation between the sixteenth and twentieth week of foetal development. Our knowledge about the origin and course of nerve fibre tracts in the nervous system has been advanced by studies of the times of myelination.

Non-myelinated nerve fibres seen in electron micrographs have a distinct appearance (Fig. 2.34). Customarily, they are in groups and each axon is enclosed in simple fashion by lip-like extensions from the Schwann cell, which in this case occupies a central position in the group. Thus, electron microscopy has helped to differentiate myelinated and non-myelinated fibres in a positive way, an identification which is always difficult in routine histologic preparations.

When a nerve fibre is accidentally or purposely cut the axon degenerates on the distal side of division and the myelin sheath fragments to form lipoid droplets (*Wallerian degeneration*). The neurilemma does not degenerate and, provided the cut ends of the nerve are accurately approximated, the axon will sprout from the proximal side of division and go out to the periphery within the neurilemmal sheaths. This process, *regeneration*, may take several months to reach completion.

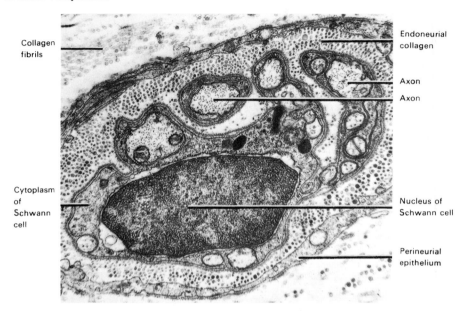

Fig. 2.34 Electron micrograph of a group of nonmyelinated nerve fibres which form a nerve bundle in the mucosa of the hard palate. The nerve fibre bundle is enclosed by the perineurial epithelial sheath × 21,000.

Sensory nerve fibres have been classified according to their size. Large myelinated fibres up to 20 μm in diameter conduct at speeds of up to 100 metres a second. Moderate sized fibres are less than 3 μm in diameter and conduct impulses at a rate of 3 to 14 metres per second. The smallest fibres, less than 1 μm in diameter, are unmyelinated or very poorly myelinated and conduct at less than 2 metres per second. On this basis fast and slowly conducted pain has been distinguished. The former arises chiefly from skin and mucosal surfaces; the latter from deeper tissues, blood vessels and viscera (non-myelinated sympathetic fibres). For the pathways taken by pain and other sensations in the spinal cord and brain stem see page 413.

Somatic nervous system

The system is composed of sensory or afferent fibres arising from nerve terminations in the skin, the deeper structures, and special end organs (eye, ear, nose, tongue); tracts and centres in the central nervous system; the cerebral cortex; and motor or efferent fibres to the general body musculature. The somatic nervous system is under voluntary control, although many of its activities are not always at the conscious level but have become the basis of habits.

It is important to remember in relation to the somatic nervous system that, as well as the peripheral and special sensory organs and nerve terminations, there are deeper end organs (receptor organs) situated in the muscles, joints and ligaments, which are stimulated by the contraction of muscles and the movements of joints. These make up the proprioceptive subsystem and, through central nervous reflexes the nerve impulses initiated by these activities play an important part in the control and regulation of body movements and posture. They coordinate the activity of prime movers, antagonists and synergists, the degree of muscle contraction required for the particular function and the coordination of adjacent or distant muscle groups.

The functional unit of the nervous system is the reflex circuit or arc (Fig. 2.25), and the nervous system functions by the interaction and integration of a multitude of nerve reflexes. Each reflex pathway has at least three essential components. These are a sensory end organ or *receptor* which receives and responds to a stimulus; and a sensory or *afferent nerve fibre* which enters the spinal cord through the dorsal root of a spinal nerve. The afferent fibre makes synaptic contact with a motor or *efferent* fibre whose axon passes to an *effector organ*, such as striated muscle fibres. This is the simplest form of reflex arc and is exemplified by the well-known knee-jerk reflex. A series of one or many intermediate or *intercalated neurons* are interposed between the afferent and efferent components in the spinal cord. These may conduct the nerve impulses to some other part of the central nervous system thus dispersing the impulses to a variety of locations; or they may have the effect of concentrating a number of afferent impulses on single motor neurons leading to an intensified response. Intercalated neurons are responsible also for carrying reflex arc information to higher levels of the brain and bringing these activities into consciousness. The multiple routes which nervous impulses may take within the nervous system through intercalated neuron pathways means that the motor neurons will be affected at different times owing to the delay in transmission which occurs at synapses. Another effect is that a stimulus which gives rise to one kind of motor response may on other occasions evoke a different response or none at all. Again, neurons can be sensitized by a stimulus from one source so that they will react to subliminal stimuli from another, a phenomenon called facilitation. In the long run, the complexity of the communications between neurons determines that the behaviour of a person in response to his or her environment is the result of all of the afferent stimuli, past or present, to which he or she is exposed. The response is to patterns of excitation rather than individual stimuli.

The functional units of the nervous system are collected together to form the central or peripheral nervous systems. The *peripheral nervous system* consists of those parts which lie outside the brain and spinal cord and includes the cranial and spinal nerves with their associated ganglia, as well as the peripheral parts of the autonomic nervous system (page 46). A typical *spinal nerve* contains two major groups of nerve fibres, somatic and autonomic (or visceral), each with its own afferent and efferent subgroups. Thus a functional analysis of a spinal nerve discloses that it contains somatic efferent, somatic afferent, general visceral efferent, and general visceral afferent fibres. The somatic efferent fibres arise from cell bodies in the anterior part of the spinal cord, enter the spinal nerve through the ventral root and terminate on striated muscle fibres. The cell bodies for somatic afferent fibres are found in a ganglion on the dorsal root of a spinal nerve (Figs 2.25, 2.30). The processes of these cells constitute the dorsal root and are concerned with general sensation from the skin and underlying tissues. General visceral afferent fibres arise also from cell bodies in the dorsal root ganglia and their peripheral processes are distributed through the sympathetic part of the autonomic nervous system. The general visceral efferent fibres arise from cell bodies in the lateral part of the spinal cord and they pass to sympathetic ganglia where they synapse with postganglionic neurons.

There are thirty-one pairs of spinal nerves and each is attached to a segment of the spinal cord by the ventral, or anterior root and a dorsal, or posterior root. The roots leave the confines of the vertebral canal by passing through the intervertebral foramina and unite to form a mixed spinal nerve. Successive roots of spinal nerves pass outwards and downwards with increasing obliquity so that the last ten spinal nerves descend almost vertically in the vertebral canal. During growth of the individual the vertebral column lengthens much more than the spinal cord and the termination of the spinal cord comes to be at the level of the lower border of the first lumbar vertebra in the adult. There are eight pairs of cervical, twelve thoracic, five lumbar and five pairs of sacral nerves, all of which leave the vertebral canal immediately below the corresponding vertebra. As soon as the spinal nerve has pierced the dura mater and emerges from the vertebral canal, it divides into dorsal and ventral *primary rami*. The dorsal ramus passes backwards to supply a segment of skin and the post-vertebral musculature. The ventral primary ramus passes forwards and supplies corresponding tissues on the lateral and anterior aspect of the body. A thoracic spinal nerve is a typical example and it supplies the muscles between two adjacent ribs and an overlying segmental strip of skin, called a *dermatome*. The patterns of dermatomes over the body is of considerable clinical importance. Adjacent dermatomes overlap extensively so that damage to the afferent fibres of a spinal nerve may produce little loss of sensation. In the disease shingles, caused by a herpes virus infection of a dorsal root ganglion, the typical pain which results is confined to the corresponding dermatome. The banded pattern of dermatomes over the body is distorted in the limbs which have rotated during development and over which skin has migrated as the limb buds lengthened. Thus, for example, the skin covering the shoulder is innervated by the fourth cervical segment and the skin over the backs of the index and middle fingers is supplied by the seventh cervical segment.

The ventral primary rami are much larger and clinically more important than the dorsal rami and they form the cervical, brachial, lumbar and sacral plexuses. These are regroupings of nerve fibre bundles to provide the most effective innervation patterns for muscle groups in the limbs. The *cervical plexus* is formed by the ventral rami of the first four cervical nerves and, in addition to supplying muscles in the neck, the most important muscular branch is the phrenic nerve which is the motor nerve to the diaphragm. The *brachial plexus* supplies the major nerves to the upper limb and is formed from the ventral rami of the lower four cervical and the first thoracic nerve to form a complex arrangement of roots, trunks, divisions, and muscular branches. The latter include the median, ulnar and radial nerves to the forearm and hand. Injuries to the brachial plexus are important, particularly those which involve the nerve supply to the hand. Injury to the radial nerve above the wrist produces paralysis of the extensor muscles of the forearm and the hand drops at the wrist. Division of the ulnar nerve leads to paralysis of some of the small muscles in the hand and there is a characteristic spreading of the thumb and little finger. Injury to the median nerve at the wrist results in the loss of ability to oppose the thumb to the other fingers and the accompanying loss of skin sensation on the palmar surface of the thumb and index finger makes a grasping movement of the hand even more difficult. Obviously, nerve injuries to these spinal nerves has much significance for the practising dentist.

The *lumbar* and *sacral* plexuses are involved in the nerve supply to the lower limb. The lumbar plexus receives contributions from the first to fourth lumbar nerves. Its chief branches are the femoral and obturator nerves which are limited largely in their distribution to the front of the limb above the knee joint. An exception is the long saphenous nerve which is sensory for the skin on the inner side of the leg as far as the ankle. The sacral plexus receives contributions from the fourth and fifth lumbar and the upper three sacral nerves. Its main branches are the sciatic nerve, which splits into the peroneal and tibial nerves, the posterior femoral cutaneous and the gluteal nerves. Their cutaneous distribution is mainly to the back of the thigh and the leg below the knee joint, except for the area of skin supplied by the saphenous nerve, a branch of the femoral nerve (page 434). The sciatic nerve supplies the muscles of the back of the thigh, all the muscles of the leg and foot, and contributes filaments to the joints of the lower extremity.

The autonomic nervous system

Many of the responses produced by activity of the autonomic nervous system are of an involuntary nature, controlled by reflex centres in the brain and spinal cord and coordinated by the hypothalamus, a region on the inferior aspect of the brain close to the mid-line above the optic tracts (Fig. 8.8). The system has its own sensory and motor pathways and regulates the visceral functions of the body such as digestion, circulation and glandular activity. The efferent side of the system innervates smooth muscle, including that of the vascular, respiratory and alimentary systems; cardiac muscle and glands. The arrangement of the motor neurons of the autonomic nervous system differs from that found in the somatic motor pathway in that their cell bodies are located outside the central nervous system in *autonomic ganglia*. These are the postganglionic neuron cell stations and their activity is controlled by preganglionic neurons whose cell bodies are found in the brain stem and spinal cord. Collectively the neurons of the efferent or motor pathway are referred to as the *visceral* efferent neurons. As in the somatic nervous system many of the structures which are supplied by the autonomic nervous system have a sensory of afferent nerve supply through visceral afferent neurons.

It is usual to subdivide the autonomic nervous system into sympathetic and parasympathetic components, each with its own nerve pathways and each exerting opposing effects on the viscera; for example, contraction versus relaxation, mobility versus rest. Normally we are un-aware of the continual functioning of the autonomic nervous system, but its abnormal activity in disease is appreciated usually as some form of visceral pain or discomfort. The autonomic nervous system is very readily affected by emotional states which often have quite distinctive somatic and autonomic aspects.

The cell bodies of the preganglionic neurons of the *sympathetic subdivision* lie in the lateral grey column of the spinal cord between the first thoracic and the second lumbar segments. The axons of these neurons emerge through the ventral roots of the corresponding spinal nerves as the *thoracolumbar outflow*. They travel in the primary rami of spinal nerves and split off as white *rami communicantes* to enter the sympathetic trunk on the same side of the body. The sympathetic trunk shows a series of enlargements which are the ganglia containing the cell bodies of postganglionic sympathetic motor neurons. Their axons form the *grey rami* com-municantes of the spinal nerves through which they are distributed to peripheral blood vessels, sweat glands, the smooth muscle of the skin and the gastro-intestinal tract.

The *parasympathetic component* has a similar arrangement of its effector pathway but the visceral efferent cells originate in either the brain or the sacral part of the spinal cord, collec-tively called the *craniosacral outflow*. In the brain stem visceral efferent fibres contribute to the third, seventh, ninth, and tenth cranial nerves. In the spinal cord the cell bodies of parasym-pathetic effector neurons are found in the second, third and fourth sacral segments, where they form a column of cells similar in position to those for the sympathetic thoracolumbar outflow. Parasympathetic ganglia in the head region do not form a chain comparable to the sympathetic trunk but are more widely dispersed and lie close to the innervated viscera, so that the postganglionic neurons are extremely short.

Despite these distinct differences in the layout of the sympathetic and parasympathetic divisions, the same chemical transmitter substance, *acetylcholine*, is involved in synaptic trans-mission between the pre- and postganglionic neurons. However, the transmitter substances at the neuroeffector junctions, that is between the postganglionic neuron and the affected organ, are different. The sympathetic transmitter is *noradrenaline*, the parasympathetic trans-mitter is acetylcholine. Additionally, the effects of stimulation of the sympathetic component are of a much more generalized nature than results from parasympathetic stimulation. This is because the preganglionic sympathetic neurons terminate on a very much larger number of postganglionic neurons than do their parasympathetic counterparts. Also, sympathetic activity is enhanced by the release of adrenaline and noradrenaline by the adrenal medulla, which has a direct preganglionic sympathetic innervation. While the sympathetic system is distributed widely throughout the body, especially to the smooth muscle of blood vessels, the parasympathetic system is more restricted in its distribution and does not supply any structures in the limbs or body wall. In some viscera stimulation of the sympathetic and parasympathetic components have opposite effects, for example stimulation of parasympathetic fibres to salivary glands causes them to produce a copious flow of saliva, while stimulation of sympathetic fibres inhibits secretion by a vasoconstrictor effect on their blood vessels. Similarly, sympathetic stimulation accelerates the heart rate and parasympathetic action causes slowing of the heart.

Blood vessels

The walls of arteries and veins consist of three coats (Fig. 2.35). These are from within outwards:

1. The *tunica intima*, which is lined on the interior by a layer of flattened endothelial cells united at their edges by a cementing substance. Outside the endothelial layer is a thin layer of loose connective tissue and an internal elastic lamina which, in the larger vessels, forms a fenestrated tube.

2. The *tunica media*, which in small and moderate sized arteries and in veins is made up

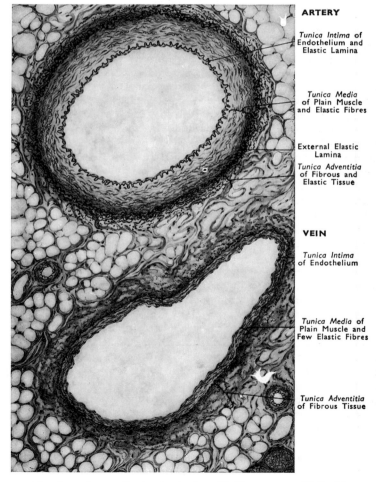

ARTERY

Tunica Intima of
Endothelium and
Elastic Lamina

Tunica Media
of Plain Muscle
and Elastic Fibres

External Elastic
Lamina

Tunica Adventitia
of Fibrous and
Elastic Tissue

VEIN

Tunica Intima
of Endothelium

Tunica Media of
Plain Muscle and
Few Elastic Fibres

Tunica Adventitia
of Fibrous Tissue

Fig. 2.35 Transverse section through a small artery and vein × 90. (By courtesy of Miss Margaret Gillison.)

chiefly of smooth muscle fibres arranged in a circular or spiral manner. With increase in the size of the vessels and especially in the arteries, the amount of elastic tissue predominates. Thus arteries can be classified as *muscular*, or distributing arteries, e.g. the facial artery, and *elastic*, or conducting arteries, e.g. the aorta. In veins the amount of muscle in the tunica media is less and therefore their walls are thinner than are those of arteries.

3. The *tunica adventitia*, which consists for the most part of collagenous fibres forming an external sheath for the vessels. It also contains elastic fibres, which in medium sized vessels may form an external elastic lamina on the outer side of the tunica media.

The muscle tissue of arteries and veins is supplied by networks or plexuses of nonmyelinated nerves of the autonomic nervous system. Most of these nerves belong to the sympathetic system and when stimulated produce contraction of the muscle tissue or vasoconstriction, especially in the smaller arteries (arterioles). There are also vasodilator nerves belonging to the parasympathetic system. These vasomotor nerves play an important part in the regulation of blood pressure and the control of peripheral circulation of the blood. As well as these motor nerves some medullated sensory nerve fibres are found which may commence in either the tunica intima or tunica adventitia. The larger vessels have their own blood supply, which reaches them from branches of neighbouring small vessels, and in the adventitia tissue these are called the *vasa vasorum*.

Many of the larger veins show longitudinal muscle bundles in the tunica adventitia. The veins of the limbs have numerous valves in their interior which, especially in the lower limb, aid in maintaining the circulation against the force of gravity. Valves are not found in the great veins close to the heart. The valves are semilunar in form and consist of a connective tissue core, continuous with the tunica intima and covered by endothelial cells.

Capillaries, which complete the circulation between the smaller arteries (arterioles) and veins, consist of a single layer of endothelial cells cemented together at their edges. They have no tunica media or adventitia but are usually supported by a delicate connective tissue network or reticulum. Through the endothelial walls of the capillaries (diameter 8–10 microns), tissue fluids leave and enter the circulation, gaseous exchange takes place and waste products are removed. These processes occur through the formation of pinocytotic vesicles, via small pores or openings in individual endothelial cells, or through gaps between adjacent endothelial cells. The latter route is taken by blood cells when they migrate out of the blood stream, as when an infection of surrounding tissue has taken place.

Rich capillary plexuses are found in the skin and mucous membranes, in muscle tissue and bone, on the surface of the brain and in the pulps of teeth. In the alimentary canal they play an important part in the absorption of food, in the lungs they give up carbon dioxide and receive oxygen from the air; in the kidneys they regulate the water balance of the body and lose various non-gaseous waste products; in ductless glands they take up hormones produced by the glandular tissue.

In certain organs such as the spleen, the bone marrow, the liver and the adrenal glands, the capillaries open into blood spaces or sinusoids in which the endothelial lining is incomplete and the slowly circulating blood comes into direct contact with the cells of the organ.

In some parts of the body such as the skin, the tongue, and nasal mucous membrane, alternative pathways are provided for the blood between the arterioles and venules. These are the *arteriovenous anastomoses* whose walls contain a large amount of smooth muscle and receive a rich nerve supply. They seem to act as nerve-regulated short-circuiting mechanisms working in close relation to the capillary system.

The main artery to the upper limb is the axillary artery, a continuation of the subclavian artery (Fig. 4.28). Beyond the axillary region it continues as the brachial artery to the elbow joint where it divides into the radial and ulnar arteries. In the hand these form a superficial and deep palmar arch in front of the carpal bones and a dorsal carpal arch on the back of the hand, from which branches pass to the fingers. The venous drainage is by the superficial cephalic and basilic veins and their tributaries and deeper veins which run with the arteries (venae comitantes). All these veins join the axillary vein which continues with the subclavian vein. Superficial veins on the front of the elbow joint, namely the median cubital, cephalic and basilic, are used frequently to withdraw blood from the circulation or to administer fluids directly into the bloodstream.

The chief artery of the lower limb is the femoral, a continuation from the external iliac below the inguinal ligament (Fig. 3.29). Behind the knee it becomes the popliteal artery which in turn continues in the leg as the anterior and posterior tibial and the peroneal arteries. The tibial arteries enter the foot to form plantar and dorsal arches from which branches pass to the toes.

The superficial venous drainage of the lower limb is through the great and small saphenous veins and deeper veins which run with the arteries. The great saphenous vein joins the femoral vein just below the inguinal ligament, through the saphenous opening in the fascia lata of the thigh (page 88), which is the site of femoral hernia, while the small saphenous joins the popliteal vein at the back of the knee joint. The great saphenous vein is subject to the development of varicosities if its valves become incompetent.

Lymphatics

The lymphatic system has two main components, an elaborate system of lymphatic vessels containing lymph and masses of *lymphoid tissue* which are most often found in close relation to the lymphatic vessels. Lymph is an almost colourless fluid which contains a small number of red and white blood cells. Lymph from the lining of the intestine, especially after a meal which has a high fatty content, contains fat particles and is called *chyle*. This gives the lymphatic capillaries draining the intestinal villi a whitish or milky appearance and, for this reason, they are called *lacteals*.

The lymphatic vessels begin as a closed network of capillaries in close relation to the blood capillaries. The smaller lymph vessels (capillaries) consist of a single layer of endothelial cells united by a cementing substance. Although they resemble closely the structure of blood capillaries, they exhibit a much greater degree of permeability which permits the passage of water and protein. Lymphatic vessels are difficult to demonstrate except by injecting a dye substance into the tissues to make use of the permeability of the vessels and hence show the paths taken by the lymphatics as a fine tracery of dye particles. Gaps between the endothelial cells of lymphatic capillaries appear readily under altered tissue conditions and play an important part in controlling the amount of water and protein in the general connective tissue spaces of the body. In the larger vessels the endothelial lining is surrounded by a coat of connective tissue and some smooth muscle. These vessels contain numerous valves which are arranged to direct the flow of lymph in a central direction, for example, from the head region towards the neck and thorax. The direction of flow is thus similar to that in the veins and lymphatic vessels may be regarded as accessory to them.

Lymphatic vessels develop as outgrowths of the venous system with which they have only limited connections in man. The importance of the lymphatic system in maintaining a correct fluid content in the tissues is demonstrated dramatically after surgical removal of components of the lymphatic system to arrest the spread of cancer. If the lymphatics and lymph nodes in the axilla (arm pit) are dissected out the venous system in the arm has difficulty compensating for the loss of the pathway for the return of lymph to the circulation. The tissues of the forearm and hand become water logged, swollen or oedematous. The condition is relieved by elevating the arm so that gravity assists the flow of tissue fluid towards the trunk. While lymphatic capillaries have a wide distribution throughout the body and are found in most tissues which have a blood supply, there are important exceptions. The most notable are the brain and spinal cord. The tissue fluids of the central nervous system drain into the cerebrospinal fluid along perivascular spaces and lymphatic vessels are not required for these spaces communicate directly with the subarachnoid space (see page 410).

Lymph nodes, incorrectly called lymph glands for they have no true secretory function, consist of a delicate connective tissue meshwork or reticulum within a fibrous capsule which supports masses of lymphoid tissue interspersed between lymph sinuses (Fig. 2.36). The lymph percolates slowly through the reticular compartments in close contact with lymphocytes which

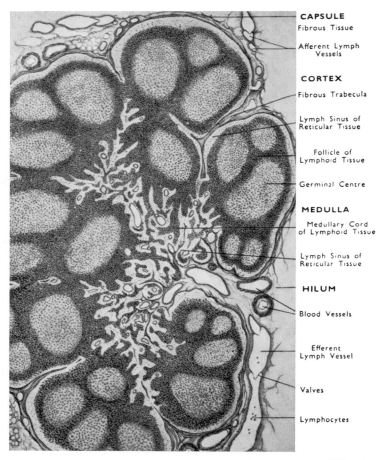

CAPSULE
Fibrous Tissue

Afferent Lymph Vessels

CORTEX
Fibrous Trabecula

Lymph Sinus of Reticular Tissue

Follicle of Lymphoid Tissue

Germinal Centre

MEDULLA
Medullary Cord of Lymphoid Tissue

Lymph Sinus of Reticular Tissue

HILUM
Blood Vessels

Efferent Lymph Vessel

Valves

Lymphocytes

Fig. 2.36 Longitudinal section through a lymph node. (By courtesy of Miss Margaret Gillison.)

are abundant in the lymphoid tissue, especially in the outer cortex of the node. Lymph nodes readily become involved in infective conditions (lymphadenitis) and in the spread of certain forms of cancer. Lymphoid nodules develop after birth in response to immunological stimulation and are therefore not confined to the nodes but exist where antigenic stimulation has occurred. Therefore, they are common in the lymphoepithelial tissues of the intestine, occurring as single nodules or in aggregated nodules (*Peyer's Patches*) immediately under the mucous membrane. Lymphoid nodules are also characteristic of the internal structure of the spleen.

Lymph nodes filter tissue fluid passing along the lymphatic vessels, remove particulate matter, and are important sites for the production of lymphocytes essential for the control and prevention of the spread of infections. Lymph nodes are important in the immune response to bacteria and foreign proteins from the external environment. One immune reaction in the nodes, the humoral response, involves the synthesis of antibodies by plasma cells and an increase in the number and size of the germinal centres producing new lymphocytes. The importance of lymphocytes in a cell-mediated immune response is known from research on tissue and organ graft rejection. Mechanisms are being evolved to selectively destroy the host lymphocytes, thus preventing graft rejection for which they are largely responsible.

Substances in the lymph are carried back to the venous system by a series of lymphatic channels. These include the lymphatic capillaries, the lymphatic vessels with their lymph nodes, and two lymphatic ducts, the thoracic duct and the right lymphatic duct (Fig. 3.26).

Lymph nodes are often arranged in groups, either superficial or deep, lying along the course of the lymph vessels which open into them. Main groups of lymph nodes are found at the junction of the head and neck in a series of small clusters from the chin to the back of the head (page 151); at the roots of the limbs (axillary and femoral lymph nodes); along the great vessels of the neck, thorax, abdomen and pelvis; and in relation to blood vessels supplying internal organs, such as the alimentary canal and lungs.

The lymphatic system is important for the production of lymphocytes by lymph nodes, and in infective or invasive processes which threaten any part of the body, as well as sites of chronic inflammation. Lymphoid tissues are generally most important in the production of antibodies which neutralize the toxic effects of bacterial infections. Lymph nodes act as filters for particulate matter and prevent dust, carbon particles and bacteria from passing into the venous system. Malignant growths of epithelia spread by lymphatic vessels and the lymph nodes have a protective role, by filtering out the malignant cells and becoming sites of secondary growths. Surgical procedures for eradicating these forms of cancer involve removing related groups of lymph nodes, as well as the primary tumour, in attempts to prevent widespread involvement by the tumour cells.

Glands

The glands of the body can be divided into three categories according to their method of secretion:

1. *Exocrine glands* which discharge their secretions onto the body surface, e.g. sweat glands, or into the alimentary, respiratory or genito-urinary passages, e.g. salivary glands.

2. *Endocrine* or *ductless glands* which discharge their products directly into the blood stream, e.g. thyroid gland.

3. *Mixed glands* which are composed of both exocrine and endocrine elements, e.g. the pancreas.

Exocrine glands may be *unicellular* as are the mucus-secreting goblet cells of the alimentary canal, or *multicellular*, which vary in complexity from the simple mucus-secreting glands of the lips and cheeks (Fig. 4.19) to the much larger complex glands such as the salivary glands, the pancreas and liver, which have compound tubular duct systems. Some of the larger glands such as the parotid and submandibular glands secrete via a single large duct; others, e.g. the sublingual and lacrimal glands, discharge along a series of separate smaller ducts. Exocrine glands have a good blood supply to provide the raw material from which they manufacture their secretions and many are innervated by both sympathetic and parasympathetic nerves which regulate the rate and quality of secretion. In certain cases, as in the common bile duct and the pancreatic duct, the terminal openings (or orifices) of the ducts are guarded by a smooth muscle sphincter mechanism which regulates the flow of the fluid secretions into the intestine.

Endocrine glands are derived from epithelial covered surfaces with which continuity is lost subsequently so that there are no ducts or channels to carry the products of secretion to a surface. The secretions are chemically and biologically specific and these *hormones*, which depend on the blood stream for their distribution throughout the body, play an important role in body metabolism, growth and regulation of the reproductive system. The endocrine tissue of the sex glands, the adrenal glands and the pancreas are situated in relation to the abdominal cavity.

The *adrenal glands* lie on the upper poles of the kidneys and are bound to them by the renal fascia. The cortex of the human adrenal gland synthesizes a number of steroids including cortisol which has a marked effect on carbohydrate metabolism. Aldosterone controls the salt-water balance by acting on kidney tubules. The adrenal medulla produces and stores adrenaline and noradrenaline which influence blood capillary circulation. Removal of both adrenal glands results in death and destructive lesions of the cortex lead to the clinical condition known as Addison's disease, in which there is disturbance of the inorganic salt content of tissues and a characteristic bronzing of the skin. The responses of the adrenal medulla are associated with the activity of the autonomic nervous system and it reacts rapidly to stressful situations.

The endocrine cells of the pancreas produce hormones which control the level of blood sugar. If these scattered aggregations of cells are removed or damaged the lack of *insulin* synthesis results in the condition called diabetes mellitus. This can be counteracted clinically by injections of insulin. If the production of insulin is excessive, as in tumours of pancreatic endocrine cells, hypoglycaemia follows with important neurological signs and symptoms. More detail about these glands will be found on page 111.

An important group of endocrine glands develops in the neck region from the pharyngeal pouches (page 340). These include the thyroid and the parathyroid glands. They develop as outpouchings or diverticula of the pharyngeal wall and later become cut off from the pharynx. Epithelial remnants of their connecting stalks may produce cysts in the neck, especially in the case of the thyroid. The *thyroid gland* is a bilobed structure which lies in the lower part of the neck in front of the larynx and the upper part of the trachea. It produces thyroxine which controls the rate of metabolic activity. Deficiency of thyroid hormone in the infant results in cretinism, which is characterized by retarded mental development and a failure of growth. Timely administration of thyroid hormone corrects the condition. Another group of thyroid cells, the C-cells, secrete calcitonin which has a marked effect on the levels of calcium and phosphate in the blood. The *parathyroid glands* are small structures which are closely related to the thyroid gland. They regulate the exchange of calcium and phosphorus between the bones and the blood stream. The *pituitary gland* (hypophysis cerebri) is a small oval-shaped body which lies in the sagittal plane in the pituitary fossa (sella turcica) of the sphenoid bone. It is a complex gland which regulates skeletal growth and water metabolism, as well as coordinating the activities of the other endocrine glands.

It develops in two parts: an anterior portion (pars anterior) as an upgrowth from the roof of the primitive mouth cavity (page 347) and a posterior part (pars nervosa) which is a derivative of the fore-brain. The pineal gland, situated at the dorsal surface of the adult mid-brain, also develops as a diverticulum of the fore-brain. Although its endocrine function is not certain it may influence the processes of growth and ageing.

Endocrine glands are highly vascular, possessing a complex system of vascular channels in close proximity to the special cells which manufacture the hormones and discharge them into the blood stream, by which they reach the target cells or tissues that depend on them for specific activity. The bone-forming cells and those which produce the dental tissues, that is, enamel, cement and dentine, are regulated by the pituitary growth hormone and by the parathyroid hormone which controls calcium metabolism. In some animals such as the wild boar and the anthropoid apes the sex hormones determine the size of the teeth, especially the canines, which are much larger in the male animal.

3. Anatomy of the Thorax and Abdomen

The dental student has to be especially concerned with the structure and function of those parts of the anatomy adjacent to the teeth and mouth. However, as the whole body makes up a functional unit, the dentist cannot restrict himself to the care of the mouth region and neglect the remainder of the body. This is particularly true of the various body systems associated with the functions of respiration, blood circulation, digestion, the endocrine glands and the nervous system. Therefore, it is necessary that students of dentistry have a good knowledge of the major organs and systems in the thoracic and abdominal cavities and their relationships to one another. The purpose is to attain a complete understanding of the structure and functional details of the oral cavity and related parts, including the ability to be concerned with the care of the total individual.

THE WALL OF THE THORAX

The thoracic cavity is protected by the ribs which form the greater part of its skeleton. Each *rib* is attached behind to the *vertebral column* and in front either directly, or indirectly through the lower costal margin, with the *sternum* or breast-bone (Fig. 3.1), except for the lower two ribs.

The *vertebral column* (Fig. 1.2) is composed of 33 vertebrae—7 cervical, 12 thoracic, 5 lumbar, 5 sacral and 4 coccygeal—firmly connected to one another by intervertebral disks of fibrocartilage, ligaments and muscles. The sacral and coccygeal vertebrae are fused to one another to form the *sacrum* and *coccyx*, whilst the remaining vertebrae are separated and individually movable. The vertebral column functions as a support; it resists the stresses and strains caused by compression, tension and torsion which are the results of body movement and muscular activity; it acts as a buffer or shock absorbing mechanism; it provides protection for the contained spinal cord; it affords attachment to many muscles, and transmits weight forces to the pelvis and lower limbs.

The vertebral column, which is about 70 cm long in the adult male, is curved in the sagittal plane. At birth the spine is generally convexly curved in a dorsal direction (primary curvature) but with postnatal development, when the infant learns to hold up its head and to sit up, secondary curvatures (ventral convexities) develop in the cervical and lumbar regions. The dorsal convexities which remain in the thoracic and sacral regions serve a useful purpose in that they increase the capacity of the thorax and pelvis respectively.

The vertebrae

These show similar common features throughout the vertebral column in possessing a *body* which is anteriorly placed; a *neural* or *vertebral arch* posteriorly placed enclosing and protecting the spinal cord in the *vertebral foramen*; *transverse processes* which project from the vertebral arches and provide muscular and ligamentous attachments; *articular processes* by which one vertebra articulates with the next; *pedicles* uniting the body to the neural arch; *spinous processes* which give attachment to the muscles and ligaments; *laminae* which attach the

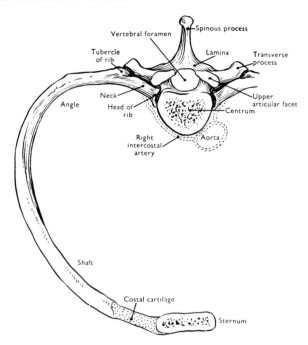

Fig. 3.1 The parts of a typical thoracic vertebra and rib.

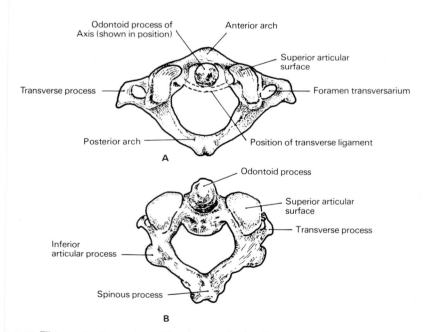

Fig. 3.2 The atlas and axis vertebrae. The pivot mechanism between the odontoid process and the ring formed by the anterior arch of the atlas and the transverse ligament are shown in A. (By courtesy of Professor G. A. G. Mitchell and Dr. E. L. Patterson.)

spinous processes to the transverse processes and which, with the pedicles, complete the neural arch.

Typical vertebrae show important regional differences:

1. The *atlas* (*first cervical*) lacks a body and articulates superiorly by cup-shaped depressions with the occipital bone (condyloid type of joint).

2. The *axis* (*second cervical*) possesses an odontoid process (or peg) projecting upwards from the body. This peg is derived from the body of the atlas. It forms a pivot joint with the anterior arch of the atlas.

3. The *remaining cervical vertebrae* have small rectangular bodies, stunted transverse processes, each perforated by the *foramen transversarium* for the vertebral artery and vein and a short *spine* which is forked or bifid, except in the case of the seventh which has a single prominent process. This is the *vertebra prominens* which gives attachment to the lower end of the ligamentum nuchae and may be felt readily at the base of the neck. However, in many instances, the first thoracic vertebral spine is even more prominent than that of the seventh cervical vertebra (Fig. 3.2).

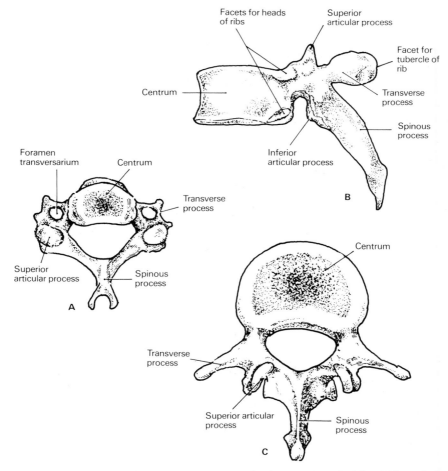

Fig. 3.3 Drawings of typical vertebrae. The articular facets are outlined heavily. A, A cervical vertebra viewed from above; B, A thoracic vertebra viewed from the left side; C, A lumbar vertebra viewed from above. (Courtesy of Professor G. A. G. Mitchell and Dr. E. L. Patterson.)

4. The *thoracic vertebrae* have heart-shaped bodies, costal facets for articulation with the heads of the ribs, costal facets on the well marked transverse processes for articulation with the tubercles of the ribs (Fig. 3.3) except T. 11 and T. 12, and long downwardly directed spines.

5. The *lumbar vertebrae* have large kidney-shaped bodies, strong transverse processes and thick, blunt spinous processes.

Accessory vertebrae or parts of vertebrae may exist in any region of the vertebral column. A typical case is where a vertebra is represented by only one half of its body. The result is a bending of the spinal column to one side; the condition is called scoliosis and it is difficult to correct. Sometimes the vertebral arches are incomplete, especially in the caudal part of the column, and the spinal cord is not protected by bone in the affected region (spina bifida). The fifth lumbar vertebra is often joined partly or entirely to the body of the sacrum. This is called sacralization and can be present without producing any symptoms.

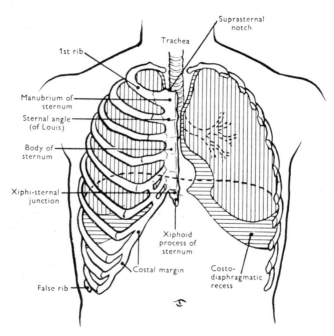

Fig. 3.4 The bones of the thoracic cage, the outline of the lungs (vertical shading) and pleura (horizontal shading).

The ribs (Fig 1.2, 3.4.)

The ribs number twelve on each side. The upper seven are *true ribs* for they are joined to the sternum by the *costal cartilages*. The lower five ribs are *false ribs* for they end without direct attachment to the sternum. Indeed the lowest two ribs usually end freely in the body wall and are known as *floating ribs*.

Each rib is a flat bone having a curved *shaft*, sternal and vertebral extremities, and an *angle*, an abrupt curvature near the posterior end of the rib. The upper surface of the ribs is smooth and blunt; the lower border is sharp and contains the shallow *costal groove* on its inner aspect in which lie the intercostal vessels and nerve (Figs 3.5, 3.8).

At the vertebral extremity is the *head*, for articulation with the vertebral body; the *neck*, attached by ligaments to the transverse process of the corresponding vertebra; and the *tubercle*

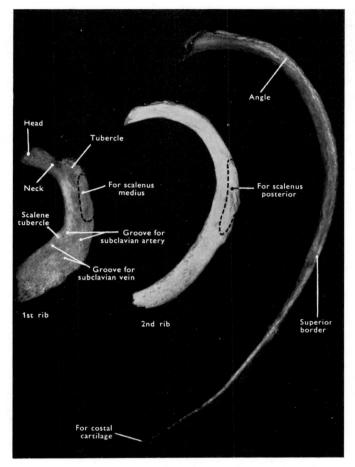

Fig. 3.5 The first and second ribs and a typical midthoracic rib viewed from above.

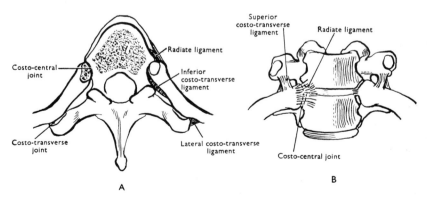

Fig. 3.6 Costovertebral joints viewed from above (A) and from the front (B). (By courtesy of Professor G. A. G. Mitchell and Dr E. L. Patterson.)

which articulates with the tip of the transverse process of the vertebra (except T. 11 and T. 12) at the costotransverse joint. The costovertebral joints are illustrated in Figure 3.6).

In a small number of people the last cervical vertebra has a small rib associated with it, the *cervical rib*. The most common abnormality takes the form of an extra large costal process of the seventh cervical vertebra, fused with the transverse process and prolonged downwards as a fibrous band attached to the first rib behind the scalene tubercle. When it it is present the subclavian artery and brachial plexus cross it on their way to the axilla and are lifted by it. The lowest part of the brachial plexus is abnormally stretched and there may be interference with nerve conduction. There is numbness of the ulnar side of the forearm and atrophy of the small muscles of the hand. Vascular symptoms may also occur due to pinching or compression of the subclavian artery.

The sternum or breast-bone (Figs 1.2, 3.4).
This consists of three parts, the manubrium, the body and the xiphoid process. The upper border of the *manubrium* is concave (the suprasternal notch) and its lower border joins the body or *mesosternum* at the *sternal angle* (of Louis). The second rib is attached at this level, where there is limited hinge movement at the manubriosternal joint. The joint frequently loses its mobility from the twentieth year onwards.

The *xiphoid process* is cartilaginous but may show an increasing content of bone with age. It is easily felt and indicates the level of the lower border of the heart and the dome of the diaphragm. The seventh costal cartilage is attached to the side of the xiphisternal junction.

Intercostal musculature
The intercostal muscles between the ribs, and uniting one rib to another, are arranged in three layers: the *external*, *internal* and the *transversus thoracis* group of muscles (Fig. 3.7). The fibres of the external intercostal muscle run obliquely downwards and forwards; those of the other muscle layers downwards and slightly backwards. The external layer is completed anteriorly by membranous tissue which extends from the costochondral junction to the side of the sternum as the anterior intercostal membrane. The internal layer becomes membranous behind at the angle of the ribs, while the transversus thoracis muscle layer is divided into three parts: the subcostales, the intercostales intimi and the sternocostalis.

The subcostales are in the form of slips that pass between the inner surfaces of the lower ribs close to the angles. The intercostales intimi are better developed in the lower intercostal spaces and are attached to the inner surface of the ribs, usually passing over more than one space. The sternocostalis part of the transversus thoracis muscle lies behind the sternum and adjacent costal cartilages. It arises from the xiphoid process and the lower part of the body of the sternum and radiates upwards to insertion into the second to the sixth costal cartilages.

The three sheets are remnants of a continuous layer of muscle found in lower vertebrates. All are depressors of the ribs and the sternocostalis is the most powerful of the group. The intercostal muscles, together with the diaphragm, play an active role in the movements of respiration (see page 67).

The intercostal nerves and blood vessels run forwards between the internal and innermost (transversus thoracis) layers in the costal groove (Fig. 3.8).

The costal groove is placed on the lower part of the inner surface of the rib and is best seen in the posterior part of the rib near the angle. The intercostal vein, artery and nerve lie in the groove in that order from above downwards, so that the nerve is more exposed than the blood vessels in the intercostal space. In the lateral wall of the thorax the nerve and vessels give off collateral branches which run forwards between the muscles along the upper border of the rib which forms the lower boundary of the intercostal space (Fig. 3.8).

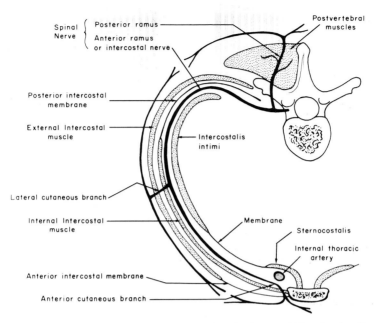

Fig. 3.7 The intercostal muscles and the distribution of a spinal nerve. (After Pernkopf.)

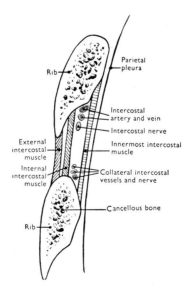

Fig. 3.8 Vertical section through an intercostal space showing the intercostal muscles and vessels.

Clinical and applied anatomy

Fractures of the ribs may result from either direct or indirect trauma. They cause more pain than fractures in other bones due to respiratory movements of the chest and the sensory nerve supply to the underlying pleura. The pain is most severe on deep breathing, coughing or sneezing and it is abolished by the injection of local anaesthetic agents around the intercostal nerve in the posterior part of its course. The fragments of the ribs may pierce the pleura and injure the lung, or they may damage the diaphragm and injure the liver or the spleen. This can result in haemorrhage (haemothorax) or leakage of air (pneumothorax) into the pleural cavity or lung tissue. When the thorax is compressed in an anterior posterior direction the ribs may break in an indirect way, just in front of the rib angle. The bone fragments are forced outwards so that there is less danger of injury to the deep structures of the chest. The most dangerous type of rib fracture is when a series of ribs are broken near to the sternum and the chest is 'stove-in', as may be caused by crushing or automobile accidents. The ability of the lungs to ventilate adequately is seriously impaired and surgical stabilization of the sternum is essential.

When fluid is being drawn off, or aspirated from the pleural cavity, the arrangement of the intercostal nerves and vessels determines the direction in which the needle is passed through the intercostal space. When the chest is aspirated from behind, the needle is kept close to the upper border of the rib below the space, to avoid injury to the intercostal nerve. Further forwards, because the collateral vessels and nerve run in close relation to the upper border of the rib that forms the lower boundary of the intercostal space, the needle is inserted through the middle of the intercostal space (Fig. 3.8).

Major surgical procedures involving the contents of the thorax usually call for wide access, major *thoracotomy*, either from the side through an intercostal space, through the bed of a rib, or through the sternum. In lateral thoracotomy the surgeon prefers to enter through the bed of a rib because the tough periosteum of the bone makes closure of the wound easier than when the incisions pass through the intercostal space. Major cardiac surgery involves dividing the sternum along its length and spreading the two halves by retractors or, in some cases, transverse division of the sternum is employed.

THE CONTENTS OF THE THORACIC CAVITY

The chief contents of the thoracic cavity are the lungs and heart. Leading to the lungs from the larynx (which lies in the throat and is concerned with voice production) is the trachea or windpipe. Within the thoracic cavity it divides into a left and a right branch or bronchus to enter the corresponding lung (Fig. 3.4). Each *lung* (page 62) occupies and almost entirely fills, a large compartment of the thoracic cavity known as the *pleural cavity*. Each pleural cavity is lined by a smooth endothelial membrane which at one place (the lung root or hilum) is reflected onto the surface of the lung. The membrane lining the pleural cavity (parietal pleura) is separated from the membrane covering the lung (visceral pleura) by a potential space containing a small quantity of serous watery fluid (Fig. 3.9). As the pleural cavities expand with expansion of the thoracic cavity and the descent of the diaphragm, the lungs expand with them and air is drawn into the lungs along the respiratory passages (nose, pharynx, larynx, trachea, bronchi, and intrapulmonary tubes). With recoil of the thoracic cage and ascent (relaxation) of the diaphragm the lungs contract by virtue of their elasticity and air is expired.

Between the pleural cavities of each side of the body lies the heart (page 73) in a fibrous tissue sac, the *pericardium*. This middle region of the thorax is called the *mediastinum*. Through it there pass a number of major structures, entering or leaving the thorax (Figs 3.9, 3.10). These include the oesophagus (food passage), trachea, the great vessels leading to and coming from the heart, the vagus (tenth cranial) nerves, the nerves to the diaphragm (phrenic nerves), the thoracic duct, and the thoracic segments of the sympathetic nerve chain. At the lower

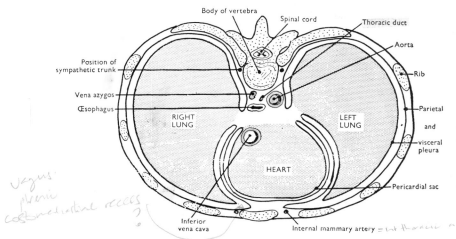

Fig. 3.9 The relationships of some intrathoracic structures as seen on transverse section at a midthoracic level.

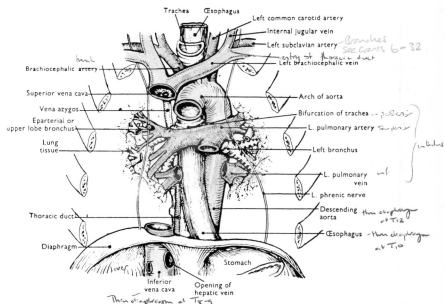

Fig. 3.10 Contents of the mediastinum viewed from the front, after removal of the heart.

end of the thorax certain of these structures pierce the diaphragm to enter or leave the abdominal cavity. These include the aorta, the oesophagus, the vagus nerves, the sympathetic nerve chain, its branches the splanchnic nerves, the inferior vena cava and the thoracic lymph duct (see page 82).

THE RESPIRATORY SYSTEM

Is concerned with conveying air rich in oxygen to the functional area of the lungs (the alveolar sacs and alveoli) where oxygen is taken up by the blood and carbon dioxide is removed (Fig. 3.11). The respiratory system commences with the nose and nasal cavities from which the inspired air passes through the pharynx to the beginning of the trachea, or wind pipe. The detailed anatomy of the nasal cavities and pharynx is dealt with elsewhere (pages 179 and 191).

The trachea and extrapulmonary bronchi

The trachea commences in the neck where it is continuous with the larynx. It descends in front of the oesophagus into the thoracic cavity where, at the level of the 5th or 6th thoracic vertebral body, it divides into a right and a left main bronchus. The trachea is about 15 cm long. It is highly elastic so that it can adjust its length to the movements of the neck and lungs. Its structure is described on page 68. In the neck it is closely related to the oesophagus, the thyroid gland, the infrahyoid muscles and the contents of the carotid sheaths (pages 136, 138). At the thoracic inlet it is separated from the first rib by the apices of the lungs and the great vessels entering and leaving the thoracic cavity (Fig. 3.10). Within the thoracic cavity the trachea is related to the oesophagus, behind; the lungs, especially on the right side; the arch of the aorta, which lies anteriorly at first and then on the left side of the trachea; the thoracic duct; the left brachiocephalic vein; superior vena cava and the vena azygos; and the right vagus nerve (Fig. 3.10).

Of the main extrapulmonary bronchi, the right is wider, shorter and more nearly vertical than the left. Because of this feature foreign bodies, such as fragments of extracted teeth, which may enter the trachea accidentally, are more likely to lodge in the right bronchus or its branches. In front of both bronchi lie the pulmonary arteries, below them the pulmonary veins. The left bronchus crosses the oesophagus on its way to the hilum of the left lung. Before entering the lung the right main bronchus gives off a large segmental branch to the upper lobe of the right lung (eparterial bronchus).

The lungs

Each lung shows the following external features:

1. An *apex*, which rises some distance above the first rib as seen from in front, and occupies the cervical portion of the pleural cavity in the root of the neck where it is closely related to the subclavian artery.

2. A *base*, or *diaphragmatic surface*, related to the upper surface of the diaphragm.

3. A *costal surface* related to the ribs and the sides of the bodies of the thoracic vertebrae.

4. The *mediastinal*, or *visceral*, *surface* related to the heart, the great vessels, the oesophagus and other mediastinal structures. These structures form markings in the form of grooves and depressions on the lung surface (Fig. 3.13).

On the mediastinal surface of the *right lung* the chief markings are produced by: the right atrium of the heart (cardiac impression); the superior vena cava and the azygos vein; the inferior vena cava; the right subclavian artery; the trachea; and the oesophagus. On the mediastinal surface of the *left lung* the chief markings are produced by: the ventricle (cardiac

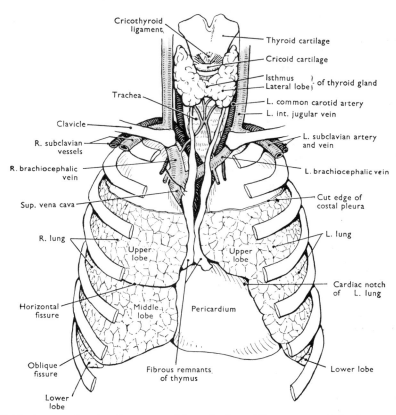

Fig. 3.11 The lungs and pericardium exposed from the front. The remnants of the thymus gland and some of the great vessels in the superior mediastinum are visible also. (By courtesy of Professor G. A. G. Mitchell and Dr E. L. Patterson.)

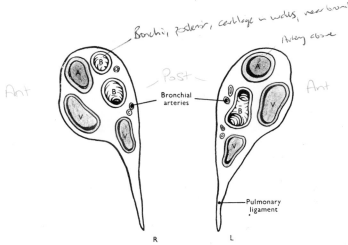

Fig. 3.12 Structures at the root of the right (R) and left (L) lungs.

impression); the aortic arch and descending aorta; the left subclavian artery; the trachea; and the oesophagus.

Sometimes a part of the upper medial surface of the right lung is separated off from the rest of the upper lobe by a fold of the pleura which contains the azygos vein in its free margin, in other words the azygos vein is suspended by a mesentery from the roof of the pleural cavity. This forms an accessory or *azygos lobe* of the right lung which is important because its presence may cause unusual appearances in radiographs of the chest. It is often the site of lung disease and it is important for the surgeon or the pathologist to recognize its significance.

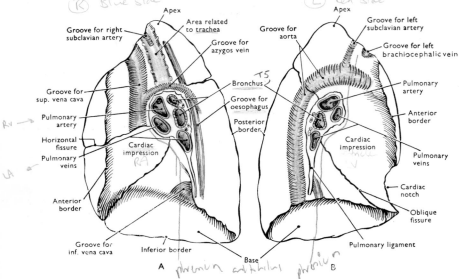

Fig. 3.13 (A) the right lung, and (B) the left lung rotated to display their mediastinal (medial) surfaces: some of the chief relationships of these surfaces are indicated. (By courtesy of Professor G. A. G. Mitchell and Dr E. L. Patterson.)

The right lung is somewhat heavier than the left and is wider, because the heart bulges to the left side. It is shorter than the left lung because the right dome of the diaphragm is higher than the left. It usually consists of three lobes: upper, middle and lower, separated by deep fissures (interlobular fissures). The left lung has two lobes, upper and lower in position. The lungs are everywhere in contact with the pleura lining the pleural cavities except at the costodiaphragmatic and costomediastinal recesses. In these regions the layers of the parietal pleura are sharply reflected from the ribs to the diaphragm and from the costal cartilages to the anterior mediastinal wall so as to lie in contact with one another for a short distance, forming potential spaces which are not fully opened up and occupied by lung tissue even during the deepest inspiration (Fig. 3.4).

Each lung is attached to the mediastinum at its *hilum* where a number of structures enter or leave the lung (Fig. 3.17). These include:

The *bronchus*. On the left side this enters as a single tube; on the right side as two tubes. Within the lung the bronchus becomes subdivided so that finally a complex radiating system of tubes (bronchioles) carries air to and from the smallest units (alveolar sacs) of the lung substance (Fig. 3.12). This system is often called the bronchial tree. The larger bronchioles have cartilage in their walls, as have the trachea and bronchi. These cartilaginous rings and plates keep the air passages open. The smaller bronchioles lose their cartilage but have, however, a relatively greater amount of smooth muscle in their walls. The contraction and dilatation of this muscle, which is not under the control of the will, determines the flow of air into the terminal air passages. In the alveolar sacs rapid exchange takes place between the oxygen in the inspired air and the carbon dioxide in the blood.

The *pulmonary artery*. This large blood vessel carries venous blood (poor in oxygen, rich in carbon dioxide) from the heart to the lungs. It is called an artery because the blood is passing away from the heart. Within the lung it divides into smaller branches and finally into a greater number of capillaries which lie close to and between the walls of the air sacs. Through the thin basement membrane dividing the air sacs from the capillaries the gaseous interchanges of respiration take place. Oxygenated blood from the capillaries is collected by veins which form larger vessels, the pulmonary veins, and return to the heart.

At the hilum of each lung one to three *pulmonary veins* leave the lung and pass to the left atrium of the heart.

At each hilum there are also:

1. *Branches of the pulmonary nerve plexus*. This is composed of branches of the sympathetic and parasympathetic (vagus) divisions of the autonomic nervous system. They supply the smooth muscle in the walls of the smaller bronchioles and arteries.

2. *One or two bronchial arteries*. These small vessels are usually branches of the thoracic aorta and supply the lung tissue. The right bronchial artery frequently arises from the third right intercostal artery.

3. *Lymphatics* from the surface and interior of the lung leave the lung at the hilum and enter the group of lymph nodes around the bronchus and the bifurcation of the trachea. These nodes are involved in such conditions as pulmonary tuberculosis and lung cancer.

Within each lung are a number of *bronchopulmonary segments* each of which is fed by a third order intrapulmonary bronchus. The first branch is a main bronchus, the second order branches supply individual lobes of the lung. The bronchopulmonary segments are (Fig. 3.14):

RIGHT LUNG

Upper lobe   .   .   Apical segment
Posterior segment
Anterior segment

Middle lobe .   .   Lateral segment
Medial segment

Lower lobe .   .   Superior segment
Medial basal segment
Anterior basal segment
Lateral basal segment
Posterior basal segment

LEFT LUNG

Upper lobe
Upper division .   Apical-posterior segment
Anterior segment

Lower division .   Superior segment
Inferior segment

Lower lobe .   .   Superior segment
Anterior basal segment
Lateral basal segment
Posterior basal segment

The functional subdivision of the lung into a series of bronchopulmonary segments each with its own bronchus, its own blood supply from the pulmonary artery and with its lung tissue distinct from that of adjacent segments, has revolutionized clinical treatment of lung

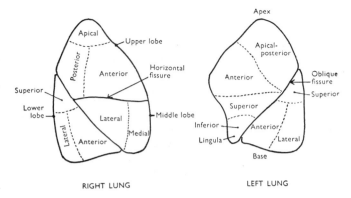

RIGHT LUNG   LEFT LUNG

Fig. 3.14 The bronchopulmonary segments and fissures on the lateral (costal) surface of the lungs. The medial and posterior segments of the lower lobe in the right lung and the medial segment of the lower lobe in the left lung are not seen from this aspect.

conditions. Modern lung surgery, postural drainage and the diagnosis of chest conditions from X-rays are based on the detailed anatomy of bronchopulmonary segments. The surgeon is assisted by the potential planes of separation which exist between adjacent segments during resection of parts of a lung for chronic infective conditions such as tuberculosis, and in cancer. The segmental anatomy of the lungs is also extremely valuable during direct examination of the bronchial tree (bronchoscopy) for the site of a disease or a foreign body can be determined precisely. A cause of lung abscess is the introduction of septic material or even teeth into the bronchi during operations in the mouth cavity under general anaesthesia. The posterior bronchopulmonary segment of the upper lobe of the right lung is considered to be the most important part of this lobe and probably of both lungs from a clinical viewpoint; it is often a primary site of tuberculosis or lung abscess.

The primary unit of the lung is the *lobule*. These are roughly pyramidal in form. The bases of many lobules lie immediately beneath the pleura forming a pattern of polygonal areas on the surface. Others, however, are placed in the interior of the lung substance. Each lobule receives at its apex a small terminal bronchiole, which gives rise to a number of branches (respiratory bronchioles) within the lobule. These open in turn into alveolar ducts and alveoli or sacs, the terminal segments of the bronchial tree (Fig. 3.17), where the process of gaseous interchange takes place.

The pleura

Each lung is covered by a smooth serous membrane, the *visceral pleura*, except at the hilum where the pleura covering the lung is reflected to become continuous with the *parietal pleura* clothing the inner surface of the thoracic cavity and the upper surface of the diaphragm. On the lungs the pleural membrane is reflected into the fissures between the lobes to line the adjacent surfaces. The parietal pleura is highly sensitive with a rich nerve supply derived from the intercostal and phrenic nerves.

The cervical dome of the pleura and the apex of the lung which lies immediately below it, project above the level of the first rib through the upper aperture of the thorax. They are protected by a fascial diaphragm, called Sibson's fascia or the *suprapleural membrane*, which extends like a fan from the transverse process of the seventh cervical vertebra to the inner border of the first rib. *is muscular = Scalenus minimus ?*

The parietal pleura which lines the thoracic cage (the costal pleura) is reflected sharply on to the mediastinum (Fig. 3.9) where it is continuous with the mediastinal pleura. Between the layers of this reflection is the *costomediastinal recess*. Similarly, the costal pleura is continuous beneath the lung with the diaphragmatic pleura and its reflection creates the *costodia-phragmatic* recess (Fig. 3.4), a likely place for the accumulation of excessive amounts of pleural fluid. Because lung tissue does not fill this recess completely, especially during expiration, it is a suitable site for the withdrawal of fluids from the pleural cavity with little risk of perforation of lung tissue by the hypodermic needle.

The visceral and parietal pleurae form a closed sac, separated only by a thin film of fluid. Because the parietal pleura is closely adherent to the inner aspect of the ribs and intercostal musculature and the visceral pleura adheres closely to the surface of the lung, there is a negative pressure within the pleural sac. This prevents the collapse of the lung, which takes place immediately if air or other substances are allowed to enter the pleural cavity. An accumulation of pus in the pleural cavity (purulent effusion) may be local or widespread. The most common location for the pus is between the costal aspect of the lung and the chest wall. The pleural cavity may be opened inadvertently in surgical approaches to the kidney or the adrenal gland through the bed of the twelfth rib.

RESPIRATORY MOVEMENTS

The upper seven ribs gain direct attachment to the sternum through their costal cartilages, the following three or four ribs are attached to the costal margin, while the last one or two ribs have no attachment (floating ribs). The ribs together with the sternum make up a cage-like structure capable of producing an expansion or reduction in the size of the thoracic cavity. Expansion of the thoracic cavity is brought about by an upward and outward movement of each rib at its vertebral attachment like a bucket handle being raised (Fig. 3.15), a rotation

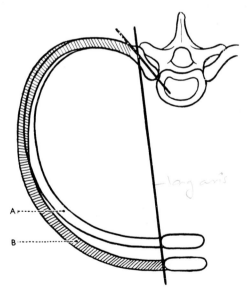

Fig. 3.15 Diagram to show the change in the position of the ribs when the lungs are expanded. The unshaded rib outline (A) represents the position of the rib at the end of expiratory movements. The shaded outline (B) indicates the rib position when the lung is filled with air as a result of inspiration. The rib moves upwards and outwards along the two axes shown by the solid straight lines. (By courtesy of Professor G. A. G. Mitchell and Dr E. L. Patterson.)

of each rib in its long axis, and a hinge-like elevation of the lower part (body) of the sternum on the upper part (manubrium). These movements are produced by the muscles of respiration which are made up of: the intercostal muscles, which form three layers of muscle tissue in each intercostal space (Fig. 3.8) and are innervated by the thoracic spinal nerves (page 390); muscles which pass to the thoracic cage from the vertebral column (scaleni and posterior serrati muscles); and those from the skeleton of the upper limb, innervated by the cervical nerves (Fig. 4.35). These latter muscles are collectively referred to as the accessory muscles of respiration. They are used in deep breathing and in respiratory emergencies such as asthma. Under normal conditions they move the neck and upper limbs in relation to the thoracic cage. When acting as respiratory muscles, they reverse their direction of action.

The principal muscle of respiration in man, however, is the *diaphragm*, which forms a muscular partition between the thoracic and abdominal cavities (Figs 3.4, 3.10, 3.18). It is dome-shaped, rising high into the thoracic cage to the level of the tenth or eleventh thoracic vertebra from its peripheral attachment to the upper lumbar region of the vertebral column, costal margin, and the lower end of the sternum. Because of the presence of the liver the right side of the diaphragm is higher than the left by about 1 cm, a detectable amount on X-ray examination. The lumbar, costal and sternal parts are inserted into the central tendon, a thin non-muscular trefoil-shaped structure which lies immediately below the heart. The tendon transmits the inferior vena cava, the right phrenic nerve and lymphatic vessels from the liver. Between the lumbar parts of the diaphragm, which form two muscular crura (pillars), is the aortic opening, through which pass the aorta and, usually, the thoracic duct (page 83), and the greater splanchnic nerves (page 86). The oesophagus and vagus nerves pass through the right crus of the diaphragm, attached to the upper three or four lumbar vertebrae.

Congenital or acquired diaphragmatic hernias, in which an abdominal organ or structure protrudes through a defect in the diaphragm into the thoracic cavity, occur most often via the oesophageal opening. Because of the anteroposterior curvature of the diaphragm the aortic opening is at the lowest level (T.12), the oesophageal opening is found at the level of the tenth thoracic vertebra and the opening for the vena cava between the eighth and ninth thoracic vertebrae (Fig. 3.10). When the costal margin is fixed by the abdominal muscles, the contracting diaphragm descends like a piston into the abdominal cavity and in this manner increases the capacity of the thoracic cavity. Pressure on the abdominal contents unless required for purposes of defaecation or childbirth, is relieved by a co-ordinated relaxation of the anterior abdominal muscles. The diaphragm receives its motor nerve supply from the neck region, where it develops, through the long phrenic nerves which descend through the thoracic cavity to reach their termination.

During inspiration the root of the lung moves downwards and forwards and allows the apex and posteromedial part of the lung to be expanded. Any pathological fixation of the root of the lung interferes with the normal degree of lung expansion. Although the lung appears to have a uniform structure throughout its substance, the zone of lung tissue adjacent to the costosternal and diaphragmatic surfaces is the only part actively involved in normal inspiration. Even when respiratory movements are forced, as in strenuous exercise, only the peripheral lung substance is functionally significant.

STRUCTURE OF PARTS OF THE RESPIRATORY SYSTEM

Trachea and extrapulmonary bronchi (Fig. 3.16)

The trachea contains sixteen to twenty U-shaped rings of hyaline cartilage. Their open ends are directed backwards and the gap between the two limbs is filled in with connective tissue and smooth muscle. Adjacent rings are united by connective tissue containing both collagenous and elastic fibres. The lining mucous membrane is a pseudo-stratified ciliated columnar epithelium containing numerous mucus-secreting goblet cells. The lamina propria contains a large amount of elastic tissue in the form of a dense membrane or lamina, and both serous and mucus-secreting glands (the tracheal glands).

Posteriorly these airways have relatively poor support. Outpouchings, or diverticula, may occur in the gaps between the smooth muscle bundles as they pass between their cartilage attachments. When a series of such diverticula occurs along each side of the posterior tracheal wall it gives a striking picture in radiographs. Collapse of large airways is seen in obstructive lung disease, such as chronic bronchitis and emphysema, with resulting respiratory deficiency.

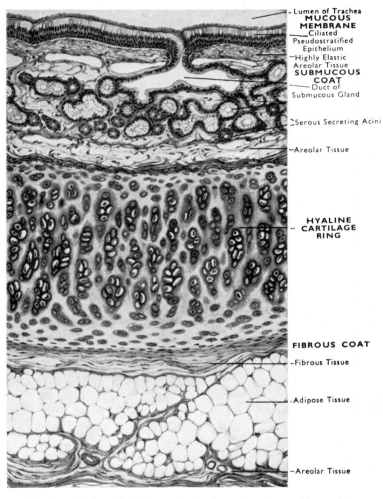

Labels on image:
Lumen of Trachea
MUCOUS MEMBRANE
Ciliated Pseudostratified Epithelium
Highly Elastic Areolar Tissue
SUBMUCOUS COAT
Duct of Submucous Gland
Serous Secreting Acini
Areolar Tissue
HYALINE CARTILAGE RING
FIBROUS COAT
Fibrous Tissue
Adipose Tissue
Areolar Tissue

Fig. 3.16 Transverse section through the wall of the trachea in the region of one of its cartilage rings × 115. (By courtesy of Miss Margaret Gillison.)

Intrapulmonary bronchi
In these the cartilage is very irregular in its disposition, consisting of plates of cartilage of irregular size and shape and also some forming complete rings. The muscle tissue forms a layer deep to the cartilaginous plates which are united to·one another by deep connective tissue. The muscle layer is not complete but is arranged in two (right and left) spirals. The elastic tissue of the lamina propria does not form a definite lamina as in the trachea and extra-pulmonary bronchi, but consists of a diffuse network. The epithelial lining is the same as in the trachea. The larger bronchioles have a similar structure (Fig. 3.17).

Small bronchioles (Fig. 3.17)

These are usually less than 1 mm in diameter. They run through the lung tissue without being accompanied by connective tissue septa. They contain no cartilage in their walls; the muscle coat is more continuous and relatively better developed. There are no mucous glands. The lining epithelium becomes nonciliated and cuboidal in type.

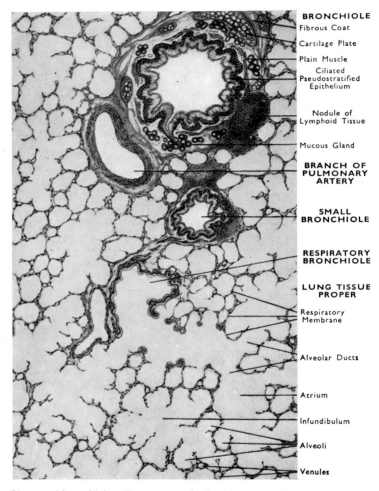

BRONCHIOLE
Fibrous Coat

Cartilage Plate

Plain Muscle

Ciliated
Pseudostratified
Epithelium

Nodule of
Lymphoid Tissue

Mucous Gland

**BRANCH OF
PULMONARY
ARTERY**

**SMALL
BRONCHIOLE**

**RESPIRATORY
BRONCHIOLE**

**LUNG TISSUE
PROPER**

Respiratory
Membrane

Alveolar Ducts

Atrium

Infundibulum

Alveoli

Venules

Fig. 3.17 Section of lung and bronchioles. (By courtesy of Miss Margaret Gillison.)

Alveolar sacs or alveoli (Fig. 3.17)

For many years there was some doubt as to whether or not the alveolar sacs were lined by continuous epithelium. Electron microscopy has shown conclusively that several types of cells form the thin interalveolar septum which separates adjacent alveoli:

    a. The respiratory or *alveolar epithelium*, separated by a basal lamina from the

    b. *Capillary endothelium*,

    c. Mesodermal cells (fibroblasts), often called *septal cells.*

The alveolar epithelium forms a continuous layer throughout the alveoli and the capillary endothelium is never directly exposed to the alveolar air. It seems, therefore, that gaseous exchange takes place through three tissue layers: alveolar epithelium, basal lamina and capillary endothelium. It is probable that adjacent air sacs communicate through pores in their walls. For development of the respiratory system see page 343.

Radiographic appearance of the lung
The air which the lung contains allows X-rays to pass readily through it compared to the more solid structures of the mediastinum and thoracic wall (Fig. 3.18). The more the lung is expanded, the greater the translucency so that it is customary to require a person to inspire deeply and hold the breath during the taking of X-rays for the diagnosis of lung disease. The pulmonary blood vessels, because of their contained blood and relative resistance to the passage of X-rays, stand out against the surrounding lung fields as a series of linear shadows radiat-

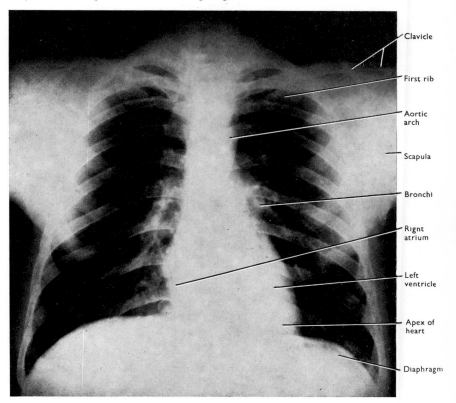

Clavicle
First rib
Aortic arch
Scapula
Bronchi
Rignt atrium
Left ventricle
Apex of heart
Diaphragm

Fig. 3.18 Postero-anterior radiograph of the thorax.

ing from the hilus. It is usual to take radiographs of the chest with the subject upright, so the shadows of the vessels are more noticeable in the lower half of the lung field. Only the larger vessels are seen, unless special radiographic techniques are used, so the peripheral parts of the lung field appear to be devoid of blood vessels. The bronchi are not seen unless they are filled with a radio-opaque substance. Lymph nodes at the root of the lung may be visible in radiographs as shadows. Foreign objects, tumours, or sites of tubercular infection have similar appearances. Fractures of the ribs will show as displacement of the fragments or as a dark line running through the rib shadow. Radiographs have to be taken in more than one plane to determine the precise position of foreign bodies or tumours, for in a posterior–anterior X-ray they could lie anywhere between the sternal and vertebral limits of the thorax.

THE CIRCULATORY SYSTEM

This system consists of the heart, vessels taking blood away from the heart (arteries), the vessels which bring blood to the heart (veins), and the capillary vessels where exchange of nutrients, oxygen, carbon dioxide, waste products, hormones and cellular elements takes place between the blood stream and the tissues. The movement of the blood is maintained by the contractions of the heart. An important adjunct of the closed vascular circulatory system is the lymphatic system which, having received tissue fluid from such areas as the connective tissue spaces, the alimentary canal, the pleural and peritoneal cavities, deposits its lymph into the vascular system after it has been filtered through the lymph nodes.

THE FUNCTIONS OF THE CIRCULATORY SYSTEM

The function of the circulatory system is to carry the blood to those tissues and organs which require it. Some organs such as the heart, the brain and central nervous system, the retina of the eye, the endocrine glands and the kidneys require a fairly constant and rich blood supply. Other organs such as voluntary muscle during periods of active contraction, the alimentary

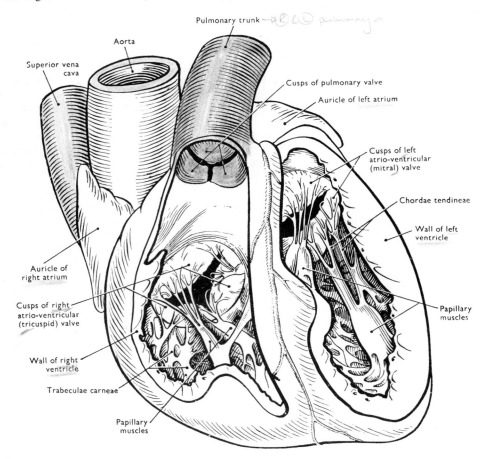

Fig. 3.19 The ventricles of the heart opened from the front to show the papillary muscles, chordae tendineae and the cusps of the atrio-ventricular and pulmonary valves. (By courtesy of Professor G. A. G. Mitchell and Dr E. L. Patterson.)

canal during digestion and the skin when it is losing body heat require a greatly fluctuating blood supply. This is produced by the vasomotor autonomic nerve control of the arterioles and the dilatation of the local capillary network. If all the capillary networks in the body were open at the same time the individual would bleed to death within the vastness of his own circulatory system. Red blood corpuscles are concerned with the carriage of oxygen, and carbon dioxide; blood plasma with food, waste products of tissue breakdown, hormones and mineral salts especially those required to build and maintain the skeleton. White blood corpuscles are concerned with the reaction of the body to bacterial infection, which presents as acute and chronic inflammation, and the complex processes of tissue repair. After a severe haemorrhage blood corpuscles are replaced from the blood-forming tissue of the spleen and bone marrow. In severe infections bacteria are carried to various parts of the body from the primary focus of infection such as an abscess developing in relation to a tooth. These circulating bacteria may destroy the heart valves or form secondary abscesses in the brain. Certain forms of cancer spread via the blood stream, while others spread along lymphatic vessels. Blood clots (thromboses) develop over areas where the lining of blood vessels has been damaged. Sometimes clots come adrift in the blood stream and finally block a smaller blood vessel. In certain organs such as the brain, the heart and the kidneys the result is death of the area normally supplied by that particular vessel (infarction).

THE HEART

The heart is somewhat asymmetrically placed in the middle of the thorax, for it lies more to the left of the middle line than to the right. It is contained within a fibrous tissue sac, the pericardial sac. This is covered on its outer surface on each side by part of the pleural lining of the pleural cavities (Figs 3.9, 3.20). The phrenic nerve, one on each side, descends through the thorax in front of the hilum of the lung along the surface of the pericardial sac to reach the diaphragm (Fig. 3.10). The pericardial sac is united to the surface of the great afferent and efferent vessels of the heart a little distance from where they enter or leave the heart. Below, the pericardial sac is firmly adherent to the upper surface of the diaphragm in its central region.

The right border of the heart lies close to the right margin of the sternum, extending from the third to the sixth costal cartilage. The left border commences at the second intercostal space close to the left margin of the sternum and extends downwards, and to the left, as far as the fifth left intercostal space about 3 inches ($7\frac{1}{2}$ cm) from the sternal margin ($3\frac{1}{2}$ inches from the middle line). Here, over the apex of the heart, the heart sounds can best be heard with a stethoscope.

The heart consists of four chambers: right and left atria, which receive veins entering the heart; and the right and left ventricles, which give origin to the pulmonary artery and the

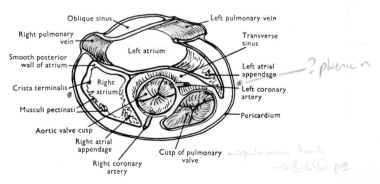

Fig. 3.20 Transverse section through the atria of the heart. (After Jamieson.)

aorta respectively (Fig. 3.19). The muscular walls of the atria are thin as they have only to drive the blood into the corresponding ventricles through the atrio-ventricular orifices. Each atrio-ventricular orifice is guarded by an atrio-ventricular valve to prevent the blood from being driven back into the atrial cavities during contraction of the ventricles. The valve of the right atrio-ventricular orifice has three main cusps and is sometimes called the *tricuspid valve*; that of the left atrio-ventricular orifice has two main cusps and is called the *mitral valve*.

The walls of the ventricles are considerably thicker than those of the atria. The right ventricle has to drive the blood stream through the lungs; the left ventricle has to drive the blood to the remainder of the body and has the greatest amount of muscle in its walls. The orifices leading out of the ventricles into the pulmonary artery and the aorta are guarded by the pulmonary and aortic valves. Each has three 'pocket-like' cusps which prevent the blood from flowing back into the ventricles at the end of their contraction phase (Fig. 3.20).

The atria

The *right atrium* lies on the right side of the heart and is related to the sternum, the third to the sixth costal cartilages of the right side, and right lung. It receives blood from the superior vena cava which enters it from above, from the inferior vena cava which enters it from below, and from the coronary sinus into which open many of the veins draining the heart wall, from behind. The coronary orifice is usually guarded by a valve; the valve of the inferior vena cava is absent or vestigal in the adult. On the back of the septal wall of the right atrium is a depression (*fossa ovalis*) which may show a small slit-like opening. This is the remains of the foramen ovale which in foetal life allows most of the blood entering the right atrium to pass directly to the left atrium, short-circuiting the lungs which are not used as organs of respiration until after birth. The right atrial appendix (the true atrial chamber) is a small cul-de-sac of the atrial cavity lying in front of the origin of the aorta. The interior of the atrium shows numerous muscular ridges (musculi pectinati) which end behind in a crest—the crista terminalis (Fig. 3.20).

In the early development of the heart, the veins entered a sinus venosus, which in turn communicated with the atrial cavity proper. Later in development the sinus venosus became incorporated in the right atrium (see page 333). That part of the atrial cavity derived from the sinus venosus lies behind the crista terminalis, is smooth and shows no muscular ridges on its interior.

The *left atrium* lies at the back of the heart behind the origins of the pulmonary artery and aorta and in front of the oesophagus (Fig. 3.20). It receives blood from the lungs through the pulmonary veins and passes it to the left ventricle through the mitral orifice. Although there are usually two pulmonary veins from each lung, the number of pulmonary veins can vary between two and four, because the posterior wall of the left atrium was included during development in the same way as the sinus venosus was included into the right atrium, to form part of the definitive atrial wall. The left atrial appendix (auricle) is closely related to the pulmonary trunk (Fig. 3.19). The atria are separated from the aorta and pulmonary trunk by the transverse sinus of the pericardium (Fig. 3.20).

The ventricles

The ventricles are separated from each other by a septum which runs obliquely backwards and to the right from the anterior to the posterior wall of the heart (Fig. 3.21). As a result the right ventricle lies somewhat in front of the left ventricle. Within each ventricle are a number of muscular ridges and processes. From the latter (papillary muscles) there pass a number of slender tendon-like cords (the chordae tendineae) which gain attachment to the ventricular surfaces of the cusps of the atrio-ventricular valves. These cords prevent the valve cusps from being 'turned inside out' during ventricular contraction (Fig. 3.19). The part of

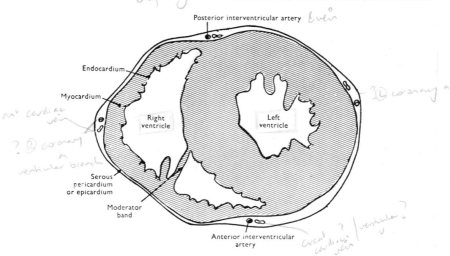

Fig. 3.21 Transverse section through the ventricles of the heart.

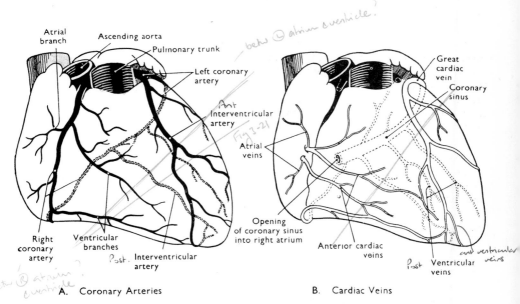

A. Coronary Arteries   **B. Cardiac Veins**

Fig. 3.22 Scheme of the blood supply (A) and venous drainage (B) of the heart wall; in each case the positions of some of the vessels are indicated as if seen through a transparent heart. (By courtesy of Professor G. A. G. Mitchell and Dr E. L. Patterson.)

each ventricle adjacent to the pulmonary or aortic orifice is smooth and shows no muscular processes.

Each valve cusp consists of a central core or connective tissue covered on each surface by endothelium, which lines the heart cavities and is continuous with the endothelial lining of the afferent and efferent blood vessels and the entire vascular system. The right or anterior cusp of the mitral valve, situated between the atrio-ventricular and aortic orifice, is particularly vulnerable to injury. In the condition called *mitral stenosis*, often a sequel to rheumatic heart disease, the aperture of the mitral valve is narrowed and the cusps are fused. The reduction in the flow of blood from the left atrium into the left ventricle may cause sufficient interference with the circulation to require surgical intervention. The cusps of the valve are separated by a combination of instruments and the finger inserted through incisions in the heart wall (valvotomy).

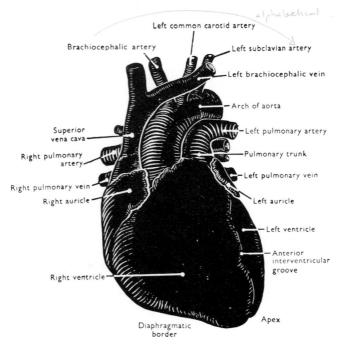

Fig. 3.23 Anterior view of the heart and great vessels

The form of the heart

The general form of the human heart is that of a blunt cone (Fig. 3.23):

1. The *base* formed by the right and left atrial chambers faces backwards and slightly upwards and lies opposite the fifth to the eighth thoracic vertebrae in front of the oesophagus and descending aorta and separated from them by the pericardial sac. This part of the pericardial cavity is known as the oblique sinus.

2. The *blunt apex* points forwards, downwards and to the left, lying deep to the fifth left intercostal space about 3½ inches (9 cm) from the middle line.

3. The *sternocostal* or *anterior surface* looks forwards, upwards and to the left. It lies behind the sternum and the third to the sixth costal cartilages of each side, but chiefly on the left. It is formed by part of the right atrium, the right ventricle and a small part of the left ventricle. Between the right atrium and right ventricle is part of the coronary (atrio-ventricular) sulcus containing the main stem of the right coronary artery. The sternocostal surface is covered in large part by the pleura and lungs, with the pericardial sac intervening.

4. The *diaphragmatic* or *inferior surface* is almost horizontal in position and rests above the central tendon of the diaphragm. It is made up of the walls of the left and right ventricles separated by the posterior interventricular sulcus.

5. The *right surface* (or right margin) of the heart is in contact with the mediastinal surface of the right lung with the pericardial sac intervening. It consists of the wall of the right atrium.

6. The *left surface* of the heart (sometimes called the margo obtusus) is in contact with

the left lung and pleura, the pericardial sac intervening. It consists of the wall of the left ventricle.

The relationship of the cardiac chambers to the surfaces are as follows:

1. The right atrium forms the right surface, part of the sternocostal surface, and a small part of the base of the heart.

2. The left atrium forms the greater part of the base of the heart.

3. The right ventricle forms the greater part of the sternocostal surface, and part of the diaphragmatic surface.

4. The left ventricle forms a small part of the sternocostal surface, the whole of the left surface and part of the diaphragmatic surface.

The structure of the heart

The heart is made up of three layers of tissue (Fig. 3.21):

The *epicardium*, an external serous membrane reflected at the place of entrance of the great vessels from the inner surface of the pericardial sac. The amount and distribution of the subepicardial fat varies considerably from one individual to another. It increases with age.

The *myocardium*. This makes up the greater part of the cardiac wall, especially in the ventricles. It consists of cardiac muscle. In the atria the muscle is arranged in a superficial and deep layer. The superficial layer is common to both atria; the deeper layer is separate for each atrial wall. In the ventricles the superficial and deep layers form a complex system of spirals. The greater part of the interventricular septum is muscular but the upper part consists of connective tissue (pars membranacea septi). The cardiac muscle is attached to the 'cardiac skeleton'. This takes the form of four connective tissue rings situated around the atrioventricular and the arterial orifices and a fibrous tissue septum forming the membranous part of the interventricular septum. The cardiac muscle, unlike the endothelium, is not continuous with

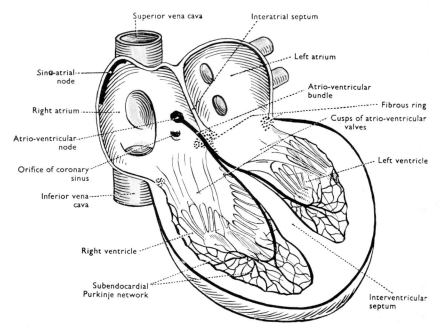

Fig. 3.24 Diagram of the conducting system of the heart. The antero-superior wall of the heart has been removed. The position of the fibrous 'cardiac skeleton' between the atria and ventricles is indicated by stippling. (By courtesy of Professor G. A. G. Mitchell and Dr E. L. Patterson.)

that of the great vessels. It is different in its histological structure from the smooth muscle of the blood vessels, forming a branching syncytium of striated muscular tissue. Although the muscles of the atrial and ventricular cavities are also separated from one another at the atrioventricular fibrous rings, they are maintained in functional continuity by the atrioventricular *bundle of His* composed of modified cardiac muscle tissue (Purkinje fibres) and which conducts nervous impulses co-ordinating the contraction of the heart chambers (Fig. 3.24). These impulses are initiated in the atrioventricular node situated in the lower part of the inter-atrial septum, immediately above the opening for the coronary sinus. The atrial contractions are initiated by the sino-atrial node situated in the upper part of the right atrium near the opening of the superior vena cava.

The *endothelial layer* is made up of smooth endothelial cells in the form of a squamous epithelium lining the atrial and ventricular cavities, continuous with the endothelium of the great vessels leaving and entering the heart and supported by a thin layer of delicate connective tissue.

Blood vessels of the heart (Fig. 3.22A)

The right coronary artery arises from the root of the aorta above the anterior aortic cusp and passes between the pulmonary artery and the right auricular appendix. It runs downwards and to the right in the coronary sulcus (atrioventricular groove) around the right margin of the heart to its posterior surface (base), where it terminates by running in the posterior inter-ventricular sulcus. The chief branches are:

1. One or two branches running on the costosternal surface of the right ventricle, and a larger branch running along the lower margin between the costosternal and diaphragmatic surfaces.

2. Small branches to the walls of the right atrium.

3. Small branches to the roots of the pulmonary artery and aorta.

The left coronary artery, usually somewhat larger than the right, arises from the root of the aorta above the left posterior cusp. It appears on the anterior surface between the pulmonary artery and left auricular appendix and sends a large branch into the anterior interventricular sulcus. It continues to the back of the heart in the coronary sulcus. The chief branches are:

1. A number of branches running on the surface of the left and right ventricles derived from the main stem, and the large anterior interventricular branch.

2. Small branches to the walls of the left atrium.

In some cases the left coronary artery reaches the posterior interventricular sulcus and provides the posterior interventricular branch in place of the right artery.

DISTRIBUTION OF CORONARY ARTERIES

The right coronary artery supplies the whole of the right ventricle except for the left third of the anterior wall. It also usually supplies the back part of the interventricular septum and part of the posterior wall of the left ventricle. The left coronary artery supplies the greater part of the left ventricle, the anterior half of the septum and part of the anterior wall of the right ventricle. There is considerable individual variation in the extent of anastomoses between branches of the coronary arteries.

The vascular requirements of the heart muscle are great and if a major branch of a coronary artery is suddenly occluded this is followed by necrosis, or death, of the part of the heart wall which it supplies. The anterior interventricular artery (Fig. 3.22), which is a branch of the left coronary artery, is more frequently blocked than the others. Anastomoses between the coronary arteries allows for the development of a satisfactory collateral circulation if the occlusion of a main coronary vessel is gradual. A *coronary occlusion* may be rapidly fatal, but not always so; the heart may continue to function quite well in spite of the fact that part of its wall has been replaced by fibrous scar tissue, representing the dead area of cardiac muscle. Nowadays the signs and symptoms of coronary artery occlusion, such as breathlessness and chest pain, lead surgeons to undertake corrective procedures to offset a heart attack. This involves the anastomosis of vessels around the heart to shunt the blood in the coronary system around the region of the obstruction, the so-called 'coronary bypass procedure'. This operation has become quite routine and has been remarkably successful in alleviating the symptoms of heart disease.

CARDIAC VEINS (Fig. 3.22B)

The coronary vein commences where the ascending interventricular vein enters the coronary sulcus from the anterior interventricular sulcus. It passes as the coronary sinus to the left and then between the base and diaphragmatic surfaces of the heart to end close to the right margin of the heart by entering the right atrium. It receives tributaries from the walls of both the atrial and ventricular cavities. The longest branches are from the left surface of the heart (left marginal vein) and from the anterior and posterior interventricular sulcus.

As well as the tributaries which join the coronary sinus, a number of moderate-sized veins (venae cordis parvae) from the anterior surface of the right atrium, and the right marginal vein draining the right atrium, open directly into the right atrium. A number of smaller veins (venae cordis minimae) also open directly into the right and left atrial cavities.

The pericardium

The pericardium is a fibrous sac which encloses the heart and the roots of the great vessels and has an outline which corresponds to that of the heart.

The fibrous layer of the pericardium rests on the central tendon of the diaphragm and fuses with it; above and behind it is connected with the adventitial layers of the great vessels which enter and leave the heart. The fibrous pericardium is lined by the serous pericardium which has two layers. The outer or parietal layer adheres to the fibrous pericardium; the inner or visceral layer ensheaths the heart and forms its covering, the epicardium. The serous pericardium is a closed sac and the pericardial cavity which lies between its layers is a capillary space containing a small amount of serous fluid which serves as a lubricant between the opposing surfaces of the sac during movements of the heart. The pericardial cavity may be obliterated by connective tissue adhesions as a sequela of tuberculosis and, if the tissue contracts like other scar tissue, the heart is enclosed in a progressively rigid sheath so that there is progressive obstruction of the veins returning to the heart, diminished cardiac output and eventually cardiac failure. Inflammation of the serous layer of the pericardium is painful and often accompanied by accumulation of fluid which interferes with the normal efficiency of heart action.

Nerves of the heart

In common with other viscera and vascular organs the heart and its great vessels are supplied by the autonomic nervous system, by both motor and sensory fibres. The sympathetic supply is via cardiac branches of the cervical and upper thoracic parts of the sympathetic trunks (Fig. 3.27). The parasympathetic supply reaches the heart through cardiac branches of the vagus nerves which descend mainly from the neck, for developmentally the heart forms in the neck region (Fig. 4.36).

All these nerves converge to an area below the arch of the aorta where they form the *cardiac plexus*. Branches from the plexus are distributed to all parts of the heart along the coronary arteries. Nerve cells are found in the heart wall, chiefly in the atria, and they are thought to be of parasympathetic origin.

THE FUNCTION OF THE HEART

The heart is the power-house of the circulatory system (Fig. 3.25). The incoming venous blood is received by the right atrium. When the atrium contracts it drives the blood through the right atrio-ventricular orifice into the right ventricle as the cusps of the tricuspid valve slowly separate. When the right ventricle commences to contract the atrio-ventricular orifice is first closed and for a while the blood is contained in the contracting ventricle. The pulmonary valves then open and the blood is driven towards the lungs along the pulmonary arteries.

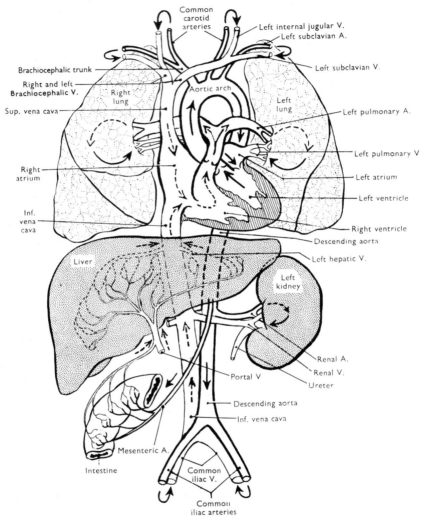

Fig. 3.25 Diagram of the heart and great vessels showing the pulmonary, systemic and portal circulations. (By courtesy of Professor G. A. G. Mitchell and Dr E. L. Patterson.)

At the same time as the right atrium contracts the left atrium also contracts to drive the oxygenated blood from the lungs into the left ventricle. The left ventricle contracts at the same time as the right ventricle and the aortic valve opens as the blood from the left ventricle is flung with great force into the aorta. The aorta responds by expanding and this wave of expansion travels along the arteries as the pulse wave. The co-ordination of the heart chambers is regulated by the conducting mechanism of the atrio-ventricular node and the bundle of His (page 78). The rate of the heart beat is controlled by the autonomic nervous system.

Causes of Failure of cardiac muscle (heart failure) — *Blood vol ↓ — ext haemorrhage / vascular shock (blood → dilated caps) / Voluntary control / Blood supply ↓ — CT / Acute pericarditis / Toxaemia / Hypertension / Poison / Aneurysm — rupture of weakened wall / Valve disease*

Sympathetic stimulation will quicken the heart, while stimulation of vagal branches will slow the rate of contraction. The power of the heart depends on the degree of hypertrophy of the cardiac muscle. There is only one fundamental heart disease: failure of the cardiac muscle.

It may fail against the back pressure of peripheral vascular resistance (high blood pressure), or of diseased valves. It may fail from a breakdown in its blood supply. It may fail under the embarrassment of acute pericarditis. It may fail when there is insufficient blood in the circulatory system after severe external haemorrhage or the condition of vascular shock when the blood is lost in the dilated capillaries. It may fail from toxaemia or poisoning, but it is improbable that it often fails as a result of emotional sorrow or distress. It can rupture if part of its wall is weakened by disease, and it is possible that certain individuals can stop the action of their heart at will through the central nervous system.

Cardiac massage
When the heart stops beating during a heart attack or under general anaesthesia it may be stimulated mechanically by cardiac massage. This can be carried out indirectly by the technique of cardiopulmonary resuscitation, in which the heart is compressed rhythmically between the sternum and the vertebral column. With the person lying prone a single blow of the clenched fist from about 4 inches above the lower part of the sternum provides the initial stimulus of the heart and rhythmic compressions of the sternum against the posterior thoracic wall, together with mouth-to-mouth breathing at appropriate intervals is often sufficient to restore heart action. When applying pressure to the sternum with the heel of the hand the pressure is also borne by the ribs and one has to be careful not to apply excessive pressure which may fracture the ribs, especially in older persons. *you are providing the pressure to squeal the heart & using its valves to prevent backflow*

THE GREAT VESSELS OF THE THORAX

(Figs 3.10, 3.19, 3.23)

The aorta
This commences at the aortic orifice behind the sternum at the level of the second intercostal space. It ascends for about 4 cm (ascending part) and then arches backwards and to the left (aortic arch) to reach the left side of the vertebral column immediately to the left of the oesophagus. From this position it descends through the thoracic cavity and leaves it by passing through the aortic opening (hiatus) in the diaphragm. As the aorta descends it inclines towards the middle line, to lie behind the terminal part of the oesophagus in the lower part of its thoracic course (Fig. 3.10).

Branches of the thoracic aorta. The following branches are given off the thoracic aorta:

1. The two coronary arteries. These are branches of the ascending aorta (see page 78).

2. The brachiocephalic (innominate) artery, the left common carotid, and the left subclavian arteries (Fig. 3.23). These are branches of the arch of the aorta. The brachiocephalic artery after a short course divides into the right subclavian and right common carotid arteries (Fig. 3.10).

3. The branches of the descending aorta are:
   a. Nine pairs of intercostal arteries to the intercostal spaces below the second, and a pair of subcostal arteries running beneath the last ribs. The first two spaces are supplied by a branch of the subclavian artery (page 171).
   b. Two or more bronchial arteries to the lungs.
   c. Four or five oesophageal branches to the oesophagus.
   d. Small branches to the tissues and lymph nodes of the posterior part of the mediastinum.
   e. Small branches to the pericardial sac.
   f. Branches to the upper surface of the diaphragm (superior phrenic arteries).

The intercostal arteries run between the ribs and supply the intercostal muscles and the parietal pleural lining the pleural cavities (Figs 3.1, 3.8). In front they join a vertical artery which runs at the side of the sternum behind the costal cartilages. This is the internal thoracic or mammary artery, a branch of the subclavian artery and so named because branches from it pass through the upper intercostal spaces to supply the mammary gland (Fig. 4.34).

The aortic intercostal arteries are accompanied in their course by intercostal veins and nerves. The veins drain into the superior vena cava via the azygos veins. As the thoracic aorta lies somewhat on the left of the middle line for the greater part of its course, most of the right intercostal arteries cross the vertebral column (Fig. 3.1) and are thus longer than the left intercostal arteries.

The pulmonary trunk

This continues from the infundibulum at a slightly higher level and is at first a little in front of the aorta (as it comes from the more anteriorly placed right ventricle). It soon passes to the left side of the ascending aorta below the aortic arch where it divides into right and left terminal branches, the pulmonary arteries.

The *right pulmonary artery* passes behind the ascending aorta and the superior vena cava to reach the hilum of the right lung. The *left pulmonary artery* passes in front of the descending thoracic aorta to reach the hilum of the left lung. It is connected to the aortic arch by the ligamentum arteriosum (see page 337).

Sudden occlusion of the pulmonary trunk by a blood clot dislodged into the circulation (embolus) may result from clotting in the deep veins of the leg following surgery or immobilization. If the blockage is complete, death is rapid; but sometimes the condition is more gradual and surgical removal of the clot is attempted through a vertical incision in the wall of the occluded vessel.

The superior vena cava

This drains blood from the upper half of the body and begins behind the costal cartilage of the first rib on the right side close to the sternum, by the union of the right and left brachiocephalic (innominate) veins. The left brachiocephalic vein is larger than the right vein and passes behind the manubrium of the sternum and crosses in front of the summit of the aortic arch and its three great branches (the brachiocephalic, the left common carotid, and left subclavian arteries). The superior vena cava descends vertically for about 5 cm behind the right sternal margin and in front of the hilum of the right lung. The lower half of the vessel is inside the pericardium and it ends in the upper part of the right atrium opposite the upper border of the third right costal cartilage (Figs 3.23, 3.25). The right phrenic nerve runs along its right side as it descends through the upper part of the thorax. On its left side are the commencement of the brachiocephalic artery and the ascending aorta, the latter overlapping it. At the upper end of the hilum of the lung the vena cava receives the azygos vein, which arches across the top of the hilum as the vein passes from the back of the mediastinum to enter the vena cava (Fig. 3.13A).

The inferior vena cava

This pierces the diaphragm (at the level of the eighth thoracic vertebra) slightly to the right of the middle line and almost immediately enters the pericardial sac and the base of the right atrium. Further details are given on page 109.

OTHER IMPORTANT STRUCTURES IN THE THORAX

The *thoracic duct* commences in a sac-like receptacle (receptaculum or cisterna chyli) in the upper part of the abdominal cavity (Fig. 3.26). It enters the thorax through the aortic orifice of the diaphragm and ascends through the posterior part of the mediastinum lying to the

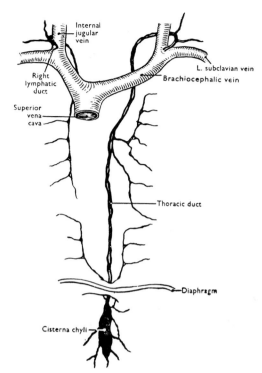

Fig. 3.26 The thoracic duct and its branches.

right of the mid-line in front of the vertebral column and the right intercostal arteries and behind the oesophagus (Fig. 3.10). In the upper part of the thoracic cavity it lies on the left side of the oesophagus, having crossed the mid-line at the level of the fourth thoracic vertebra. In the neck it joins the venous system at the junction of the left subclavian and left internal jugular veins, as these unite to form the left brachiocephalic vein, by arching laterally behind the carotid sheath contents. It drains all the lymph nodes of the body except those of the right side of the head and neck, the right upper limb and the upper part of the right side of the mediastinum, which drain into a shorter *right lymphatic duct*. This vessel joins the origin of the right brachiocephalic vein. The thoracic duct also drains the greater part of the alimentary canal, where it plays an important part in the absorption of fat.

The *oesophagus* (Fig. 3.10) is a muscular tube which descends through the thorax in front of the vertebral column, the thoracic duct and right intercostal arteries. It lies behind the trachea in the upper part of the thorax, behind the pericardium and the heart (left atrium) in the middle of the thorax, and behind the ascending posterior wall of the dome-like diaphragm in its lower part (Figs 3.9, 3.10). It is slightly constricted where it is crossed by the left bronchus opposite the fifth thoracic vertebra. Where it enters and leaves the thorax the oesophagus lies somewhat to the left of the mid-line and is here closely related to the left lung. In the middle of its thoracic course it lies more to the right and comes into relation with the right lung. In the lower part of the thorax it is covered by a plexus derived from the vagus nerves, which pass into the abdominal cavity with the oesophagus through the oesophageal opening in the diaphragm (Fig. 3.31), to become the right and left gastric nerves (Fig. 3.27).

The wall of the oesophagus is composed of two layers of muscle, a lining mucosa which is quite tough and an external fascial layer which merges with the fascia of the posterior medias-

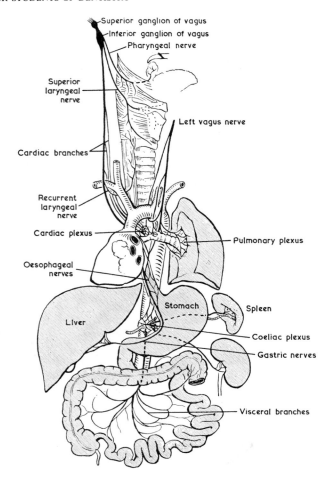

Fig. 3.27 The distribution of the vagus nerve.

tinum. The muscle coats are arranged in outer longitudinal and inner circular directions. In the upper third of the oesophagus both layers are formed of striated muscle; in the middle third the striated muscle is mixed with smooth muscle; and in the lower third of the oesophagus both layers are formed of smooth muscle. The circular muscle coat is usually thicker and at the level of the diaphragm remains contracted except during swallowing, to form a functional sphincter.

The mucous membrane of the oesophagus is quite strong and is formed of stratified squamous epithelium which changes abruptly to columnar epithelium at the oesophagogastric junction. Sometimes the terminal part of the oesophagus has a lining similar to the gastric mucosa and then is subject to stomach lesions, such as peptic ulcer and adenocarcinoma. Due to the type of muscle in the oesophageal wall, items of food or foreign bodies can be regurgitated voluntarily from the upper part of the oesophagus, but not from the lower part where the smooth muscle is controlled by the autonomic nervous system and voluntary muscle contractions are not possible.

The *right vagus nerve* enters the thoracic cavity lying to the right side of the trachea. As it crosses the right subclavian artery it gives off the right recurrent laryngeal nerve which turns below the artery and passes up behind it into the neck region to supply the muscles of the larynx. The right vagus nerve is crossed by the arch of the azygos vein and passes behind the hilum of the right lung to reach the oesophagus. The *left vagus nerve*, with the left phrenic nerve, crosses the arch of the aorta and immediately gives off the left recurrent laryngeal nerve which hooks round the arch of the aorta and returns to the neck in the groove between the trachea and oesophagus. The vagus then passes downwards behind the hilum of the left lung to reach the oesophagus. Both vagus nerves in the thoracic cavity contribute to the pulmonary and cardiac nerve plexuses, which lie below the arch of the aorta and above the bifurcating pulmonary artery (Fig. 4.36). The vagus nerves carry parasympathetic fibres to the thoracic and upper abdominal organs (Fig. 3.27).

The oesophageal plexus is formed by branches from the pulmonary plexus and contains fibres from both vagus nerves. Near the lower end of the plexus an anterior and posterior vagal nerve trunk is formed, the anterior trunk containing fibres predominately from the left vagus. Surgical division of the vagal nerve trunks is used in the treatment of duodenal ulcer to lessen the quantity and the acid content of the gastric secretions. The procedure is called vagotomy.

The *phrenic nerve* arises in the neck and enters the thorax by passing downwards across the dome of the pleura behind the subclavian vein. It runs in front of the root of the lung in contact with the mediastinal pleura. The right phrenic nerve lies lateral to venous structures throughout its thoracic course; the right brachiocephalic vein, the superior vena cava, the right atrium and the inferior vena cava. It passes through the caval opening in the diaphragm, to supply it from its abdominal surface. The left phrenic nerve lies lateral to the major branches of the aortic arch, passes laterally across the pericardium towards the apex of the heart, and pierces the diaphragm to supply the muscle on its abdominal surface (see also page 139). Sometimes, avulsion of the phrenic nerve is carried out to paralyse the diaphragm permanently on one side. The procedure raises the diaphragm on the operated side and considerably diminishes the volume of the lower lobe of the lung. This is one method used to immobilize the lung tissue in the treatment of tuberculosis or other chronic lung diseases, for example, bronchiectasis in which there is destruction of the elastic and muscular elements of the medium sized bronchi.

The *sympathetic trunk* lies on either side of the vertebral column beneath the parietal pleura and can be seen as a vertical chain of nervous tissue with ten or eleven enlargements (ganglia) along its course (Fig. 4.36). The sympathetic nerve trunk reaches upwards into the neck (cervical part) as far as the base of the skull and downwards into the abdominal cavity. The *thoracic part of the sympathetic trunk* lies on the necks of the ribs in the upper part of the thorax, over the costovertebral joints at the mid-thoracic level and on the sides of the vertebrae in the lower part of the thorax. The alteration in position relative to the ribs is the result of the progressive widening of the thoracic vertebrae from above downwards. The beginning of the thoracic part of the sympathetic trunk is marked by the *stellate ganglion*, formed by the fusion of the inferior cervical ganglion (see page 176) with the uppermost thoracic ganglion. Due to the different planes in the neck and thorax occupied by the cervical and thoracic parts of the sympathetic trunk there is a sudden change in direction of the trunk at the neck of the first rib and the stellate ganglion (or inferior cervical ganglion if the stellate ganglion is not present as such) lies with its long axis not vertically but anteroposteriorly. Preganglionic myelinated fibres (white rami communicantes) reach the trunk on each side along the intercostal nerves. These fibres arise in the *thoracic segments* of the spinal cord. Within the sympathetic chain they ascend or descend to end in one of the ganglia. From the ganglia postganglionic nonmyelinated fibres (grey rami communicantes) arise and pass in their distribution to all parts of the body. Some ganglia of the sympathetic system lie in the abdominal cavity in closer relation to the organs which receive fibres from them, e.g. the coeliac ganglia. The preganglionic fibres to these ganglia descend as separate nerve fibres through the lower part of the thoracic cavity and behind the diaphragm (greater and lesser splanchnic nerves). Further details of the sympathetic system are given on pages 45 and 416.

THE ABDOMINAL CAVITY

The abdominal cavity can be divided into nine regions by two vertical and two horizontal planes (Fig. 3.28). These are used clinically to describe the location of pain and tenderness, associated with disease of abdominal viscera. For example, the point of intersection between the transpyloric plane and the right lateral plane is a site of maximum tenderness when the gall bladder is inflamed. The *transpyloric plane* passes horizontally through the body midway between the xiphisternum and the umbilicus, frequently coinciding with the level of the pylorus

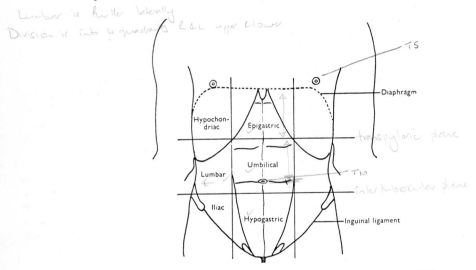

Fig. 3.28 Abdominal regions. The upper horizontal plane is the transpyloric plane which passes through the tip of the ninth costal cartilage; the lower horizontal plane is the intertubercular plane; the vertical or lateral planes pass through the tips of the ninth costal cartilages.

of the stomach. Various systems of region classification are employed and that given here is one which is used frequently. The most important regions of the abdomen are:

1. An *epigastric region*, which lies above the level of the transpyloric plane and to a large extent within the dome of the diaphragm.
2. An *umbilical region*, bounded at the sides by the vertical planes, above by the transpyloric plane and below by the intertubercular plane.
3. A *hypogastric region*, bounded at the sides by the innominate bones of the pelvis and iliac regions, between the intertubercular plane above and the brim of the true pelvis below.
4. The *pelvic cavity* is bounded at the sides, in front and behind by bone and below by the muscular pelvic diaphragm.

The lumbar vertebrae lie behind the three upper regions (abdomen proper), while the sacrum and coccygeal part of the vertebral column bound the true pelvis behind. The anterior abdominal wall from the sternum to the symphysis pubis of the pelvis contains the two vertically placed rectus abdominis muscles within their fibrous tissue sheaths. The umbilicus (navel) marks the place of attachment of the umbilical cord during foetal life.

The muscles bounding the abdominal and pelvic cavities are:

1. The *diaphragm*, forming the roof of the abdominal cavity.

2. The *bilateral rectus abdominis muscles* in front, extending from the anterior surface of the sternum and lower costal cartilages to the symphysis pubis.

3. The *external oblique, internal oblique and transversus abdominis* (corresponding to the three layers of intercostal muscles in the thorax) forming the remainder of the anterior and lateral walls of the abdomen.

4. The *pelvic diaphragm* (levator ani, coccygeus and piriformis muscles) forming the floor of the pelvic cavity.

The anterior abdominal wall

The muscular wall of the abdominal cavity between the lower thoracic margin and the bony pelvis consists of three layers of muscle laterally and a pair of vertical muscles on each side of the middle line.

The lateral muscles are, from without inwards, the external oblique, the internal oblique and the transversus abdominis, and directly correspond to the arrangement of the intercostal muscle layers (page 58).

Each anterior vertical *rectus abdominis muscle* extends from the front of the symphysis pubis and upper border of the pubic bone to the seventh, sixth, and fifth costal cartilages and is contained within a fibrous tissue envelope or *rectus sheath* formed from the aponeurotic portions of the lateral muscles. Below the umbilicus, the posterior wall of the sheath is incomplete and all the aponeuroses of the abdominal muscles pass in front of the rectus muscles. The muscles here lie directly on the *fascia transversalis* which everywhere lines the abdominal cavity external to the parietal peritoneum.

The *external oblique muscle* arises from the outer surfaces of the lower eight ribs. Its posterior surface is free. It has a wide insertion into the anterior half of the iliac crest of the innominate bone and to the pubic crest above the pubic symphysis. Between these bony insertions the free lower edge of its aponeurosis forms the *inguinal ligament* reaching from the anterior superior iliac spine to the pubic tubercle. The aponeurotic fibres of the external oblique muscle contribute to the strength of the anterior wall of the rectus sheath by interdigitating with those from the other side, along the mid-line *linea alba*.

The *internal oblique muscle* arises from the outer two-thirds of the inguinal ligament, the anterior two-thirds of the iliac crest and the lumbar fascia. It is inserted above into the lower three ribs and the costal margin. In front it divides to embrace the rectus abdominis to reach the linea alba except in its lower part, where it passes in front of the rectus muscle. Its lower part arches across the inguinal canal to be attached with the lower part of transversus abdominis to the crest of the pubic bone, as the *conjoint tendon*.

The *transversus abdominis* muscle arises from the inner surfaces of the lower six ribs, the lumbar fascia, the iliac crest and outer half of the inguinal ligament deep to the external oblique muscle. It is inserted into the linea alba, contributing to the posterior wall of the rectus sheath except in its lower part, where it passes in front of the muscle. It also contributes to the conjoint tendon in the inguinal region.

The terminal parts of the lower intercostal nerves and iliohypogastric and ilio-inguinal branches of the lumbar plexus run between the lateral muscle layers before entering the rectus sheath and the rectus abdominis muscle.

Superficial to the abdominal muscle lies a layer of deep fascia, superficial fascia and skin. The deeper layer of the superficial fascia is continuous below with the superficial perineal fascia (page 119). The fascia lata of the thigh is attached to the full length of the inguinal ligament. These fascial attachments and relationships determine the extent of abnormal accumulation of fluids in the lower abdominal wall, such as follows rupture of the urethra in the perineal region.

The *fascia lata*. The major muscle groups of the thigh are enclosed by this dense stocking-like arrangement of deep fascia. The upper end is attached to the inguinal ligament and anterior superior iliac spine in front, the external lip of the iliac crest laterally, the ilium, sacrum and ischial tuberosity on the posterior aspect (Fig. 3.55). The attachments are completed on the medial side to the body of the pubic bone.

The upper part of the fascia lata on the front of the thigh, immediately below the inguinal ligament, is 'split' and the margins overlapped, so that an oblique opening, the *saphenous opening*, exists which permits the great saphenous vein, a major leg vein, to pass from the superficial fascia to the deep aspect of the fascia lata, where it joins the femoral vein. A loop of intestine which has become herniated through the femoral canal may present as a swelling in the upper part of the thigh by projecting through the saphenous opening.

In the lower part of the abdominal wall the intermuscular *inguinal canal* passes obliquely through the layers of the lateral musculature anterior to the conjoint tendon (Fig. 3.29). Along

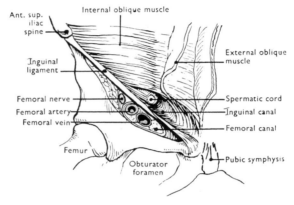

Fig. 3.29 Diagram showing important structures in the male inguinal region. The spermatic cord has been cut to show the inguinal canal which passes downwards and medially towards the pubic symphysis. The structures which emerge from the abdominal cavity below the level of the inguinal ligament are shown in cross section. In the living subject they pass vertically downwards into the upper part of the thigh. The femoral and inguinal canals are common sites of hernia.

it passes the spermatic cord of the male and the round ligament of the uterus in the female. Its anterior wall is chiefly formed by the external oblique muscle; its floor by the inturned lower edge of the inguinal ligament; its roof by the lower edges of the internal oblique and transversus abdominis muscles which arch over the spermatic cord or round ligament; its posterior wall by the conjoint tendon medially and transversalis fascia laterally.

The lower parts of the anterior abdominal wall are less strong than elsewhere because of the passage through it of certain structures into or out of the abdominal cavity, particularly in three regions:

The *inguinal region* (lateral to the lower attachment of the rectus muscles). Here the *vas deferens* and accompanying vessels in the male (making up the spermatic cord passing from the testis) or the less well developed *round ligament* of the uterus in the female, pass through the wall of the abdominal cavity along the inguinal canal (Fig. 3.29). This is sometimes the site of an *inguinal hernia* (more common in the male) when a portion of the gut or greater omentum may extend through the body wall pushing the coverings of the spermatic cord or round ligament in front of it.

The *groin* where the femoral vessels and nerve pass from the abdominal cavity to enter the lower limb. *Femoral hernia* through the femoral canal is more common in females, because the pelvis is wider than in the male (Fig. 3.58). ♂ 18 inguinal hernia : 1 femoral

The *umbilicus* is the third relatively weak region, between the rectus muscles. It may be the site of an *umbilical hernia*, which is especially frequent in newborn infants. At a certain stage of foetal life it is normal for a loop of the developing gut to protrude into the umbilical stalk at this position. This loop returns to the body cavity during the tenth week of development, but may not do so, and then persists as an umbilical hernia after birth.

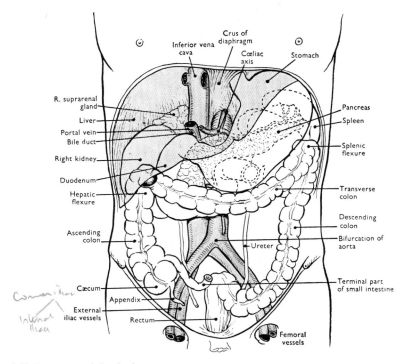

Fig. 3.30 Important abdominal contents.

The posterior abdominal wall (Fig. 3.31)
This lies between the diaphragm above and the pelvis below and consists of the bodies of the lumbar vertebrae in the middle line with the *psoas major* muscles on either side. These muscles and the iliacus muscles of the pelvis pass beneath the inguinal ligament to gain attachment to the femur in the thigh (page 87). Lateral to the psoas of each side lies the *quadratus lumborum*. This muscle arises from the transverse process of the fifth lumbar vertebrae, the ilio lumbar ligament and the posterior part of the iliac crest, is attached above to the last rib and gives slips of attachment to the transverse processes of the upper four lumbar vertebrae. It fixes the rib in contraction of the diaphragm. On the lateral side of each quadratus lumborum muscle lies the posterior part of transversus abdominis. All these muscles form posterior relations for retroperitoneal abdominal structures such as the kidneys, ureters, ascending and descending parts of the colon (Fig. 3.59).

Contents of the abdomen
The inner surface of the abdominal and pelvic cavities is lined by a smooth endothelial membrane, the *peritoneum*, forming the *greater sac* of the peritoneum. Certain parts of the alimentary canal and abdominal organs project into the abdominal cavity and carry the peritoneal covering with them so that they are connected to the anterior or posterior abdominal walls

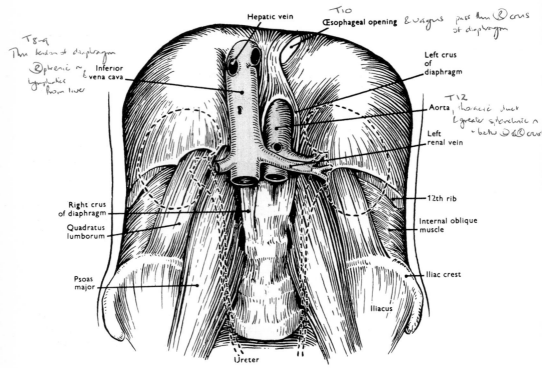

Fig. 3.31 The posterior abdominal wall. The position of the kidneys and ureters is outlined.

by two-layered folds of peritoneum called *mesenteries*. These permit a wide range of movement to certain parts of the alimentary canal and help to maintain certain organs in their proper positions in the greater sac of the peritoneal cavity.

The contents of the abdominal and pelvic cavities are:

1. The alimentary canal (gut). It is divided into the following parts (Fig. 3.30):
   a. The oesophagus—which, after it pierces the diaphragm, enters
   b. The stomach, at the gastro-oesophageal junction
   c. The duodenum, which is the first part of the small intestine
   d. The greater part of the small intestine (jejunum and ileum)
   e. The large intestine made up of
      (i) Appendix
      (ii) Caecum
      (iii) Ascending colon
      (iv) Transverse colon
      (v) Descending colon
      (vi) Pelvic colon
      (vii) Rectum
   f. The anal canal.

2. The *liver and pancreas* which are two large glands closely related in their development and function with the alimentary canal. Closely related to the liver is the gall bladder, a storage chamber for bile, produced by the liver cells and which is poured into the second part of the duodenum.

3. The *spleen*, a large vascular and lymphoid structure concerned with the formation and destruction of worn-out red blood cells.

4. The *kidneys*, their draining tubes the *ureters*, and the *urinary bladder*.

5. The *suprarenal ductless glands* and the endocrine gland tissue of the *pancreas* (islets of Langerhans).

6. The *sex organs*.

    a. In the male: the testes (which lie outside the abdominal cavity in the scrotal sacs), their ducts, the vas deferens; the seminal vesicles, the prostate gland and part of the urethra.

    b. In the female: the ovaries (inside the peritoneal cavity), their ducts, the oviducts; the uterus and the vagina.

7. *Vessels, nerves and lymphatics.* The abdominal aorta and its branches, the inferior vena cava and its branches, the portal venous system, the lumbar and sacral nerve plexuses, parts of the sympathetic nerve trunk and parasympathetic nerves the (hypogastric and vagus nerves) lymph nodes, lymphatics and the commencement of the thoracic duct.

In transverse section the abdominal cavity proper is bean-shaped or kidney-shaped owing to the mid-line projection of the vertebral column. In front of the vertebral column the antero-posterior diameter of the cavity is limited, but is much greater on either side of the column (Figs 3.36, 3.60).

THE DIGESTIVE SYSTEM

Commences at the oral cavity which is closed in front by the lips and behind by the palato-lingual seal between the soft palate and the posterior surface of the tongue (Fig. 3.32). After leaving the mouth the food passes through the middle part of the pharynx (oropharynx) to

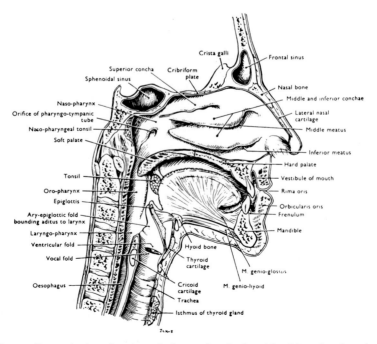

Fig. 3.32 Paramedian sagittal section through the nasal and ral cavities. Note that the cricoid cartilage lies one vertebra above its normal level. (By courtesy of Professor G. A. G. Mitchell and Dr E. L. Patterson.)

the oesophagus, or gullet, which leads through the thoracic cavity into the abdominal cavity, where it enters the stomach (Fig. 3.34).

The oral cavity and pharynx are described in detail later (pages 191, 272).

The alimentary canal

The alimentary canal from the oesophagus to the rectum consists of the following layers of tissue as seen in cross section, from the outside inwards (Figs 2.16, 3.33):

1. Adventitia
2. External muscle layer (inner circular, outer longitudinal fibres)
3. Submucosa
4. Mucous membrane:
   a. Muscularis mucosae
   b. Lamina propria
   c. Lining epithelium

Glands derived from the lining epithelium may be situated:

1. In the lining epithelium
2. In the lamina propria
3. In the submucosa
4. Outside the digestive tube, i.e. pancreas and liver

There is some variation of the disposition and degree of development of the various layers in different parts of the gut.

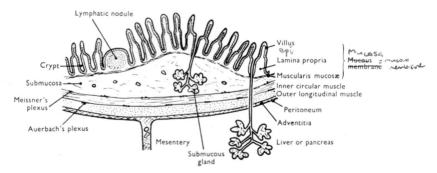

Fig. 3.33 Diagram illustrating the structure of the wall of the alimentary canal.

Adventitia
The outer covering of the alimentary canal consists of loose connective tissue in which ramify the larger vessels and nerves on the surface of the muscle layer. In certain regions the adventitia is covered in part, or almost completely, by the mesothelium of the pleural or peritoneal membrane. The peritoneal covering is reflected from the gut at the sides or, if the covering is complete, along the line of reflection of the mesentery.

External muscle layer
This consists of an outer layer of longitudinally arranged fibres and an inner layer of circularly disposed fibres. The fibres of the inner layer pursue a somewhat spiral course. In the stomach the circular layer of muscle is divided into an outer part where the fibres are circular in direc-

tion, and an inner layer of oblique fibres. The longitudinal muscle fibres are best developed along the lesser and greater curvature; the fundus has no circular fibres but these increase towards the pyloric end where they contribute largely to the formation of the pyloric sphincter.

Between the two muscle layers lies a nerve plexus (the myenteric or Auerbach's plexus) containing numerous small ganglia. The plexus is made up of fibres derived from the sympathetic and parasympathetic divisions of the autonomic nervous system (Fig. 8.4). The sympathetic fibres are postganglionic and derived from the paravertebral ganglia of the sympathetic chain (page 174). Preganglionic fibres of the parasympathetic system (vagus and sacral nerves) enter the plexus and their neurones terminate in the ganglia of the plexus. From these various ganglia postganglionic fibres contribute to the formation of the plexus and from the plexus autonomic fibres pass to terminate in the circular and longitudinal muscle. This nerve plexus regulates the activity of the smooth muscle.

When the gut is quiescent the muscle is not completely relaxed but is maintained in a state of sustained contraction (muscle tone). This is a property inherent in smooth muscle, but is regulated by nerve action. Movements of the gut, consisting of spontaneous phases of contraction and relaxation, are also characteristic of smooth muscle. These movements keep the contents in a state of constant local motion. There also occur waves of contraction followed by, and perhaps preceded by, waves of muscle relaxation which involve segments of various size along the gut. These *peristaltic movements* are responsible for moving the contents along the alimentary canal. The process is regulated by nerve impulses through the intramuscular plexus. Normally peristaltic movements pass towards the hind end of the gut, but in certain disease conditions they may occur in the opposite direction (reverse peristalsis).

Impulses reaching the plexus along the sympathetic fibres inhibit muscle tone and peristaltic movements; those reaching the plexus along the parasympathetic fibres increase tone and muscle activity. A certain amount of gastro-intestinal disturbance is the result of emotional disharmony affecting the movements of the bowel through the autonomic nervous system.

Submucosa

This layer, situated between the circular muscle and the muscularis mucosa of the mucous membrane, consists of areolar tissue containing some elastic fibres; moderate sized blood vessels forming a vascular plexus from which branches pierce the muscularis mucosa to reach the mucous membrane; and a second autonomic nerve plexus (the submucous or Meissner's plexus), which is similar in structure to Auerbach's plexus. It regulates the activity of the muscularis mucosa. In the duodenum the submucosa contains numerous glands (Brunner's glands) which manufacture an alkaline secretion and help to neutralize the acid contents.

Mucous membrane ᵎ Mucosa ₌ ₑₚ + basement membrane + Lamina propria + muscularis mucosae

1. *Muscularis mucosa*. This usually consists of an inner and outer layer of fibres. In the outer layer the wider distribution of the muscle spirals gives them a longitudinal appearance on cross section; in the inner layer the closer spiral arrangement gives the appearance of a circular layer. The smooth muscle fibres permit localized movements of the mucous membrane.

2. *Lamina propria*. This consists of areolar tissue, some elastic tissue, and, in certain sites (small intestine—Peyer's patches; and the appendix), large amounts of lymphoid tissue. It contains a vascular plexus and a lymphatic plexus, from which supplies the mucous membrane and its glands and from which branches enter the villi to take part in the absorption of food material from the lumen of the gut.

3. *Epithelial lining*. This is stratified squamous epithelium in the oesophagus, where its function is protective; and columnar epithelium elsewhere, where its function is secretory and absorptive. Among the columnar cells, especially in the large intestine, are large numbers of mucus-secreting goblet cells. Other surface cells form part of the lining of tube-shaped glands

(stomach and small intestine) and manufacture various enzymes used in the process of food digestion.

The functional units of food absorption are in the *villi*. These are found in the duodenum and the small intestine (jejunum and ileum). The surface of each villus is covered by a single layer of columnar cells resting on a basement membrane. A branch of the arterial plexus in the lamina propria enters the base of each villus and breaks up into a delicate capillary network beneath the epithelial lining. This plexus is drained by veins which join the venous plexus in the lamina propria. Within each villus are one or more *lacteals* draining into a plexus of lymph vessels in the mucous membrane. The villi also contain some smooth fibres and branches of the submucosal nerve plexus.

The epithelium lining the alimentary canal undergoes continual replacement. It is the first tissue in the stomach and duodenum to be involved in ulcer formation and, especially in the oesophagus, stomach and rectum, the lining cells may become the site of cancer involving the deeper layers and adjacent organs by spreading along the lymphatics. The glands of the stomach open in groups of two or more at the bottom of shallow gastric pits. They secrete

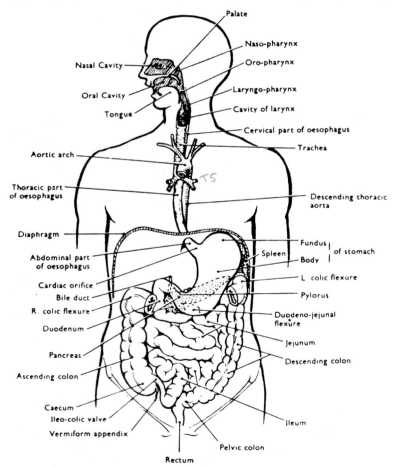

Fig. 3.34 Diagram showing the disposition of the alimentary canal and its relationship to the air passages. Most of the transverse colon has been removed to display the duodenum and the liver is not shown. (By courtesy of Professor G. A. G. Mitchell and Dr E. L. Patterson.)

mucus, hydrochloric acid, and pepsinogen, the precursor of pepsin. The mucus-secreting cells are found chiefly at the cardiac (oesophageal) and pyloric regions, while the acid- and enzyme-secreting cells are found in the body of the stomach.

The oesophagus (Figs 3.10, 3.22, 3.34)

The oesophagus descends through the neck and thoracic cavity, lying in front of the vertebral column. It pierces the diaphragm in front of, and at a somewhat higher level than the aorta (at the level of the tenth thoracic vertebra) and slightly to the left of the middle line. On its surface are fibres of the oesophageal plexus of the vagus nerves. After a short abdominal course it enters the stomach on the left side of the vertebral column at the cardiac orifice. Its total length is about 25 cm.

The stomach

The stomach is the most dilated part of the alimentary canal. It is described as consisting of three parts, a. the fundus which lies above the level of the oesophageal opening, b. the

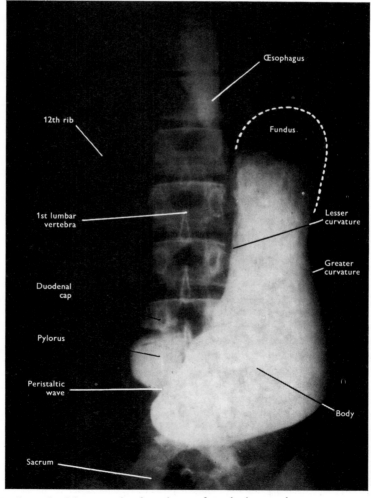

Fig. 3.35 Radiograph of the stomach a few minutes after a barium meal.

body, and c. the pyloric region (Fig. 3.35). The pyloric region is sometimes divided into a pyloric antrum and a pyloric canal. The latter ends at the pyloric sphincter, a thickening of the circular smooth muscle fibres, which marks the junction between the stomach and the first part of the duodenum. In long narrow bodies the stomach is J-shaped with the lower border lying low in the abdominal cavity. In broad wide bodies the stomach passes more transversely across the abdomen. The most fixed part of the stomach is the oesophageal entrance (the cardiac orifice). The fundus lies beneath the left side of the dome of the diaphragm, which separates it from the base of the left lung and the heart. The fundus contains a large gas-bubble which can be seen on X-ray examination (Fig. 3.35). The right side of the stomach is called the lesser curvature, and here the layers of peritoneum covering the anterior and posterior surfaces of the stomach come together to form the *lesser omentum* (stomach mesentery) which passes to the under surface of the liver. The left side of the stomach forms a greater curvature and along this the peritoneal layers of the *greater omentum* (meaning 'apron') are attached. When traced to the left from the fundus and body of the stomach, the greater omentum passes first to the spleen (gastrosplenic ligament), which it surrounds, and then to the posterior abdominal wall in the region of the left kidney as the lienorenal ligament (Fig. 3.36). When traced downwards from the pyloric region of the stomach the great omentum forms a loop descending as an apron in front of the abdominal contents before turning upwards to support the tranverse colon and reach the posterior abdominal wall in front of the pancreas (Fig. 3.37).

In front of the stomach lies the left lobe of the liver and, at a lower level, the anterior abdominal wall. Behind the stomach are the diaphragm and part of the left kidney, the spleen (in the greater omentum), the pancreas, and if the stomach is dilated, the transverse colon and some coils of the small intestine (Fig. 3.38). The relationship of these organs to the stomach is important in the spread of cancer for it is a common site of cancer, especially in males. The posterior wall of the stomach does not lie in direct contact with these organs, but is separated from them by a diverticulum of the peritoneal cavity called the *lesser sac*. The opening of the lesser sac lies above the first part of the duodenum and below the hilum of the

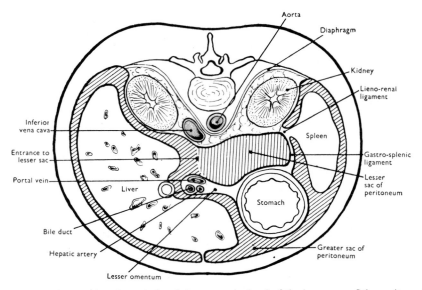

Fig. 3.36 Transverse section through the abdomen at the level of the lesser sac of the peritoneal cavity.

liver. In front of the opening is the free right edge of the lesser omentum containing the portal vein, bile duct and hepatic artery. The bile duct lies at the right of the free edge, the hepatic artery towards the left and the portal vein behind. Posterior to the opening lies the inferior vena cava covered by the peritoneum (Fig. 3.36). These relations are extremely important in operations on the bile duct, because they are all vital structures and trauma to them results

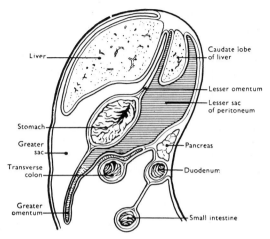

Fig. 3.37 Sagittal section through the abdomen.

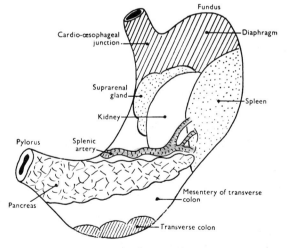

Fig. 3.38 The stomach bed: structures related to the posterior aspect of the stomach. —as seen through transparent stomach

in massive haemorrhage. The lesser sac acts as a bursa permitting free movement of the stomach. Because of the nature of the attachment of its mesenteries (lesser and greater omenta), the stomach is freely movable and capable of a considerable degree of dilatation except at the oesophageal end.

The stomach receives the masticated and swallowed food. The food is acted upon by the gastric secretions which, in association with a churning action on the part of the stomach musculature, are responsible for the further breakdown of the stomach contents. Food is not absorbed by the gastric mucous membrane. The stomach contents are acid owing to the secretion of hydrochloric acid, and absence of the protective coating of mucus produced by the mucous membrane results in erosion of the stomach lining (gastric ulcer). Psychological factors, such as mental stress, also appear to play a part in the onset of ulcer formation, by causing excessive acid secretion and autodigestion of the gastric epithelium. The stomach has a rich blood supply derived from the coeliac axis of the abdominal aorta. Most of its veins drain into the portal circulation (which passes through the liver). These matters are discussed more fully elsewhere (page 110 and Figs. 3.46, 3.50).

The duodenum

25cm — *Same as esophagus*

This is a C-shaped section of the alimentary canal and is about ~~10 inches~~ long. It is divided into four parts for descriptive purposes (Fig. 3.39).

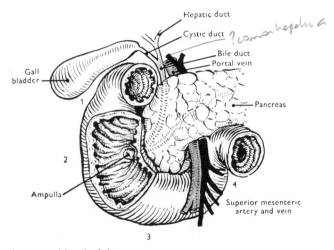

Hepatic duct
Cystic duct ? *common hepatic*
Bile duct
Portal vein
Gall bladder
Pancreas
Ampulla
Superior mesenteric artery and vein

Fig. 3.39 The duodenum and head of the pancreas.

1. The *first part* passes backwards and to the right from the pyloric sphincter, and lies below the right lobe of the liver and the neck of the gall bladder. It is the most movable part of the duodenum. Behind it pass the common bile duct, the portal vein and the gastro-duodenal artery, a branch of the hepatic artery.

2. The *second part* is vertical in position, lying to the right side of the vertebral column in front of the right kidney and its vessels and on the right psoas muscle. At about the middle of its course the common bile duct (from the liver and gall bladder) and the main pancreatic duct, open into it by a common opening, the duodenal papilla, or ampulla. In front of this part of the duodenum there are, from above downward, the right lobe of the liver, the beginning of the transverse colon, and coils of the small intestine. On its left side is the head of the pancreas (see Fig. 3.39).

3. The *third part* of the duodenum cross the vertebral column from right to left. The duodenum lies on the inferior vena cava and abdominal aorta as it crosses the vertebral column below the head of the pancreas. In front of it the superior mesenteric vessels pass to the wall of the small intestine in the root of the mesentery.

4. The *fourth part* ascends on the left psoas muscle and left renal vessels to terminate at the duodeno-jejunal junction.

The first part of the duodenum is attached to the liver by a continuation of the lesser omentum, and behind and below to the posterior abdominal wall by a short mesentery (Fig. 3.41).

The remainder of the duodenum lies behind the parietal layer of the peritoneum covering the posterior abdominal wall and therefore has no mesentery.

The part of the duodenum above the level of the opening of the bile and pancreatic ducts (first part and the upper half of the second part) along with the oesophagus and stomach belongs to the foregut in its developmental history. Its blood supply is therefore derived from branches of the coeliac axis (foregut) artery. The duodenum below the opening of the ducts belongs to the midgut along with the small intestine, ascending colon and right half of the transverse colon, and its blood supply is from the superior mesenteric (midgut) artery. The mucous membrane lining the duodenum is thrown into numerous overlapping circular folds which greatly increase its surface area.

The duodenal secretions and the bile are alkaline and neutralize the acid stomach contents which are discharged periodically into the first part of the duodenum, with the periodic opening of the pyloric sphincter. These small quantities of stomach contents form the 'duodenal cap' seen in X-rays (Fig. 3.35). The foregut region of the duodenum is sometimes the site of duodenal ulcers. The midgut region of the duodenum is the beginning of the section of the alimentary canal from which food is absorbed into the portal system of veins (all of which drain through the liver) and the lymphatics situated in its mucous membrane. Food absorption takes place chiefly through numerous small finger-like projections, the villi, which give a velvety texture to the mucous membrane. Each *villus* contains a capillary plexus drained by a small vein, and a central lymphatic capillary or lacteal.

The small intestine

This is sometimes divided into an upper part, the jejunum, and a lower part, the ileum, but the division is rather arbitrary. It varies in length from 10–25 feet (3–7½ metres) and lies in coils within the middle part of the abdominal cavity. Some coils may descend into the pelvis and come in contact with the pelvic organs.

For the whole of its length the small intestine has a posterior mesentery attached to the posterior abdominal wall (Figs 3.37, 3.60). The line of attachment of the root of the mesentery runs from the left of the vertebral column at its upper end, at the duodenal-jejunal junction, to the right of the vertebral column at its lower end where the ileum joins the caecum of the large intestine. Close to its line of attachment, the mesentery contains the superior mesenteric artery and vein, and branches pass from them into the mesentery to reach the wall of the intestine, forming a complex series of anastomosing loops. The mesentery also contains fat, lymph vessels draining the gut, and nerve fibres derived from the sympathetic and vagus nerves which control the activity of the smooth muscle of the gut wall.

The interior of the lower end of the small intestine contains fewer folds, fewer villi, and more patches of lymphoid tissue, the greatest amount of lymphoid tissue being found in the appendix.

The large intestine

The first part of the large intestine is the sac-like diverticulum, the *caecum* (Fig. 3.40). This is joined on its left side by the terminal part of the small intestine. The opening of the ileum into the caecum is guarded by a sphincteric valve, the *ileocaecal valve*, which regulates the entrance of fluid contents into large intestine. Below the ileal opening is that of the *appendix*, a blind finger-like diverticulum of the caecum which varies greatly in length and in its position. It usually has a mesentery and, in about 65 per cent of individuals, lies in direct contact with the posterior abdominal wall behind the caecum and ascending colon. Acute appendicitis is the condition in which the organ is severely inflamed. Later an abscess may form which, if it should rupture into the greater sac of the peritoneal cavity, gives rise to the more serious condition of *acute peritonitis*.

The caecum leads directly upwards to the ascending colon, which ascends in the right colic gutter on the right side of the vertebral column (Fig. 3.60). This part of the colon has no mesentery and lies directly in front of the muscles of the posterior abdominal wall (transversus abdominis and quadratus lumborum) and on the anterior surface of the lower part of the right kidney. It ends at the right-angled hepatic flexure beneath the right lobe of the liver and in front of the right kidney (Fig. 3.30).

From this position the large intestine passes towards the left side of the body as the *transverse colon*. The first half inch of the transverse colon is in direct contact with the kidney, the second part of the duodenum and the head of the pancreas. Beyond this the transverse colon comes away from the posterior abdominal wall with the greater omentum (mesocolon). The transverse colon ends on the left side of the abdominal cavity at the splenic flexure in contact with the visceral surface of the spleen, at a somewhat higher level than the hepatic flexure (Fig. 3.30).

The *descending colon* begins at the splenic flexure and descends in the left colic gutter, lying on the anterior surface of the left kidney and the muscles of the posterior abdominal wall.

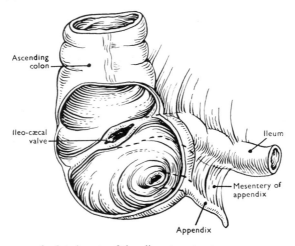

Ascending colon

Ileo-cæcal valve

Ileum

Mesentery of appendix

Appendix

Fig. 3.40 The caecum and related parts of the alimentary tract.

At the pelvic brim it becomes the *pelvic colon* (̲Sigmoid), a section of varying length attached by a Λ-shaped mesentery to the posterior abdominal wall. The pelvic colon enters the rectum or pelvic part of the large intestine, which in turn pierces the pelvic diaphragm to become the *anal canal*.

The whole of the large intestine is 5–6 feet (1½–2 metres) in length. The vermiform appendix, transverse colon, and pelvic colon have mesenteries; the caecum is usually free to some extent, but the ascending colon, descending colon and rectum have no mesenteries. As far as the middle of the transverse colon the large intestine (with the small intestine and part of the duodenum) is derived from the midgut of the foetus and therefore receives its blood supply from the superior mesenteric (midgut) artery, and its parasympathetic nerve supply from the vagus nerve. The remainder of the large intestine is derived from the hindgut; its blood supply is from the inferior mesenteric branch of the aorta and its parasympathetic nerve supply from the sacral nerves.

The chief function of the large intestine is to absorb water from its contents, which it receives from the small intestine in a highly fluid state. Most of this fluid is derived from the various secretions of the stomach, liver, pancreas and small intestine, and if it were lost the body would suffer severely from dehydration. There is also a limited amount of absorption of the products of cellulose digestion from the large intestine. It contains, especially in its distal end, a large number of mucus-secreting glands which, by their lubricating action, assist in the passing of the faeces.

THE FUNCTION OF THE ALIMENTARY CANAL

Physiologically and developmentally speaking it should be remembered the alimentary canal is an interpolated segment of the body surface from which food is absorbed. In it the food is also prepared for the process of absorption by the secretion of various enzymes. Enzyme secretion is greatest in the stomach and duodenum; food absorption is greater in the small intestine. The main function of the large intestine is to reabsorb fluids from the intestinal contents into the circulation. Glands for the secretion of the surface protecting mucus are most frequent in the oesophagus (receiving end) and the large intestine (rejecting end). Lymphoid tissue, which plays a part in the regulation of the bacterial flora of the alimentary canal, is most fully developed around the junction of the mouth and pharynx, in the lower end of the small intestine, and in the appendix. In an acute bacterial crisis the tonsils and appendix are most likely to become infected and may require to be removed surgically.

The liver

This large gland develops as an outgrowth from the terminal part of the foregut. In the adult it is somewhat wedge-shaped with the base facing to the right and the apex to the left (Fig. 3.30). The wedge form, however, is considerably modified by the conformation of the liver to adjacent structures, such as the forward projection of the vertebral column, the under surface of the dome of the diaphragm, and various abdominal organs (Fig. 3.36). It can also be described as having two surfaces: a parietal, related to the diaphragm and anterior abdominal wall, and a visceral surface related to the following abdominal organs:

1. The fundus, part of the body of the stomach and the terminal part of the oesophagus.
2. The first and part of the second portions of the duodenum.
3. The right colic (hepatic) flexure.
4. The upper pole of the right kidney and the right suprarenal gland.

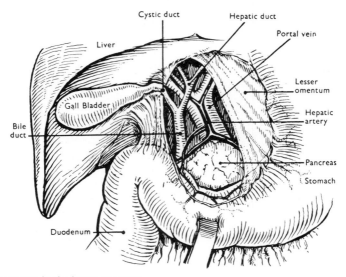

Fig. 3.41 Structures in the lesser omentum.

Closely related to the visceral surface of the liver is the gall-bladder which with the bile and cystic ducts acts as a storage mechanism for the bile secreted by the liver cells.

The following structures enter or leave the liver at the hilum (Fig. 3.41).

The *portal vein*. This carries venous blood loaded with food material from the greater part of the alimentary canal to the liver. It also drains the spleen (Fig. 3.50).

The *hepatic artery*. This artery carries blood to supply the liver tissue. It is a branch of the coeliac axis of the aorta (Fig. 3.46).

The *left and right hepatic ducts*. These unite at the hilum to form the common hepatic duct which is joined by the cystic duct from the gall bladder (Fig. 3.41), to form the common bile duct. The common bile duct descends alongside the portal vein and hepatic artery in the lesser omentum and passes behind the first part of the duodenum and the upper part of the head of the pancreas to enter the second part of the duodenum with the pancreatic duct. It is about 3 inches (7·5 cm) in length.

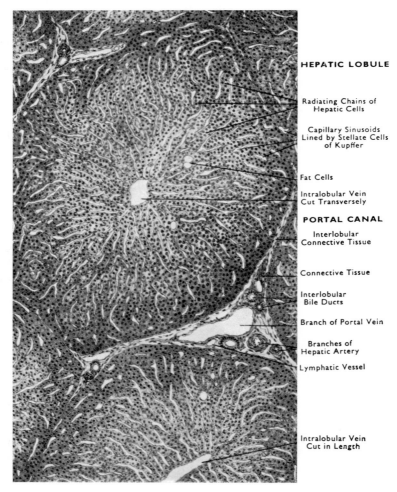

HEPATIC LOBULE

Radiating Chains of
Hepatic Cells

Capillary Sinusoids
Lined by Stellate Cells
of Kupffer

Fat Cells

Intralobular Vein
Cut Transversely

PORTAL CANAL

Interlobular
Connective Tissue

Connective Tissue

Interlobular
Bile Ducts

Branch of Portal Vein

Branches of
Hepatic Artery

Lymphatic Vessel

Intralobular Vein
Cut in Length

Fig. 3.42 Section of liver (× 75) showing one complete lobule and parts of several others. The central vein of each lobule is also called an intralobular vein. (By courtesy of Miss Margaret Gillison.)

The liver is made up of numerous *lobules*. Each lobule consists of a vast number of paren-chymal or *hepatic cells* and opens into bile canaliculi, which radiate towards the periphery of the lobule where they empty their bile into bile capillaries. At the periphery are also branches of the portal vein and hepatic artery. Groups of these three vessels are called the *portal triads*. From branches of the portal vein sinusoidal channels pass between the gland units towards the centre of the lobule where they are collected in a central (intralobular) vein which drains into two or more hepatic veins. These in turn join the inferior vena cava (Fig. 3.42). The termi-nal part of the inferior vena cava lies in a deep groove at the back of the liver.

The greater part of the liver is covered by peritoneum reflected above and in front to the under surface of the diaphragm and anterior abdominal wall. The reflection to the diaphragm from the upper surface of the right lobe exposes a fairly large area of liver substance in direct contact with the diaphragm (the bare area). On its under surface the peritoneum is reflected from the visceral surface of the liver to the lesser curvature of the stomach and the first part of the duodenum as the lesser omentum.

The pancreas

This is the second of the large glands derived from the foregut portion of the alimentary canal. The pancreas contains endocrine cells in addition to its more obvious exocrine elements (page 113). It is described as having a head embraced by the duodenum, a neck above the duodenal-jejunal flexure, a body lying in front of the left kidney and a tail which enters the lienorenal ligament to reach the spleen (Fig. 3.43). It lies in front of the following structures from right to left:

a. The inferior vena cava
b. The portal vein and common bile duct
c. The aorta and crura of the diaphragm
d. The left psoas muscle
e. The left renal vessels
f. The left kidney
g. The spleen.

The *crura of the diaphragm* are made up of the muscle fibres attached to the upper two or three lumbar vertebrae. They embrace the aorta as it enters the abdominal cavity, at the level of the twelfth thoracic vertebra (Fig. 3.31).

In sagittal section the pancreas is wedge-shaped with a posterior surface related to the structures enumerated above; an antero-superior surface situated above the line of attachment of the greater omentum to the posterior abdominal wall (mesocolon), and forming one of the structures of the stomach bed; an antero-inferior surface below the line of attachment of the omentum, which is related to coils of the small intestine (Fig. 3.37).

The main duct passes through the substance of the pancreas from the tail to the head and enters the duodenum with the bile duct (Fig. 3.43B). There is usually a secondary accessory duct draining the upper part of the head. It enters the duodenum at a higher level than the main duct.

Obstruction to the pancreatic duct, which may be due to a stone or an infection, leads to a serious inflammation called pancreatitis, which is well known for its tendency to recur. Chronic pancreatitis is a cause of great discomfort and ill health. Stones or calculi are some-times visible in radiographs and they are removed surgically by an anterior approach to the gland.

The spleen

This large vascular organ belongs to a special category. It is concerned not with the manu-facture of any secretions or hormones but with the regulation of blood cell formation and

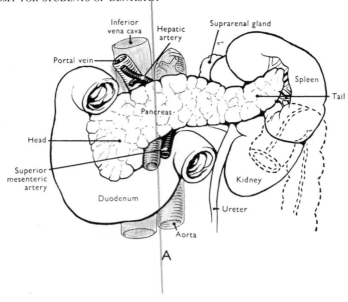

midline

Inferior
vena cava

Hepatic
artery

Suprarenal gland

Portal vein

Spleen

Tail

Pancreas

Head

Superior
mesenteric
artery

Kidney

Duodenum

Ureter

Aorta

A

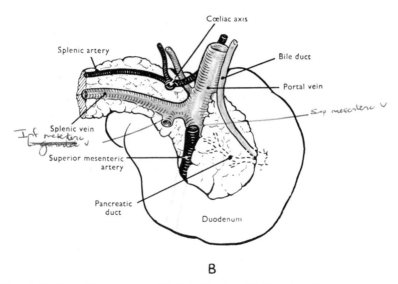

Cœliac axis

Splenic artery

Bile duct

Portal vein

Sup mesenteric V

Splenic vein

Inf mesenteric,
gonadal V

Superior mesenteric
artery

Pancreatic
duct

Duodenum

B

Fig. 3.43 The relationship of the pancreas. (A) Anterior view. (B) Posterior view.

destruction. The same function is also performed by the foetal liver and the red bone marrow
and in a specialized way (in the manufacture of lymphocytes) by the lymph nodes. All these
organs and tissues collectively make up the *reticulo-endothelial system*, which plays an important role in the protective response of the body to vascular and inflammatory emergencies.

The spleen has a connective tissue capsule containing elastic fibres and some smooth muscle
beneath its peritoneal covering. From the capsule fibrous trabeculae pass into the splenic pulp.
The pulp contains arteries, capillaries, sinusoids and veins, lymphatic nodules and masses of

blood cells in varying stages of development (Fig. 3.44). The lymphoid tissue is called the white pulp of the spleen; the remainder is the red pulp.

The *splenic artery* passes across the posterior abdominal wall and enters the hilum on the visceral surface of the spleen as a number of large branches. From the spleen emerge veins forming the splenic vein, which in turn drains into the portal vein. The surface of the spleen is covered by peritoneum except at the hilum. Its visceral surface lies in contact with the left kidney, the left (splenic) flexure of the colon, and the stomach (Fig. 3.43). The tail of the pancreas reaches the hilum region between the layers of the lienorenal ligament (Fig. 3.36). The parietal surface of the spleen lies in contact with the diaphragm separating the spleen

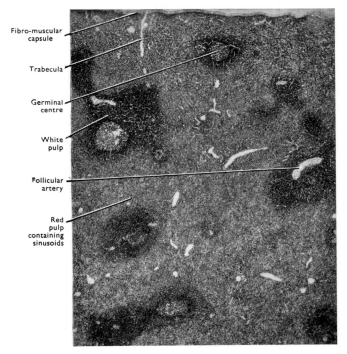

Fibro-muscular capsule

Trabecula

Germinal centre

White pulp

Follicular artery

Red pulp containing sinusoids

Fig. 3.44 Histological features of the spleen × 18.

from the lower part of the left pleural cavity. The spleen occupies an area the size of the fist between the ninth to eleventh ribs in the mid-axillary line. The spleen becomes enlarged in certain diseases of the blood forming (reticulo-endothelial) system and may project below the left costal margin. Enlarged spleens are sometimes ruptured as the result of minor accidents and are removed surgically. The spleen is not essential to life.

ABDOMINAL BLOOD VESSELS

The abdominal aorta
This gives off four groups of branches (Fig. 3.45):

1. *Mid-line, single, branches.* These are the coeliac trunk or axis (foregut artery), the superior mesenteric (midgut) and the inferior mesenteric (hindgut) arteries. They supply the alimentary canal.

2. *Bilateral visceral branches*, which supply on each side of the body the adrenal glands, the kidneys and the sex glands.

3. *Parietal branches* to the muscles of the posterior abdominal wall. These are the bilateral inferior phrenic arteries, the bilateral lumbar arteries (four on each side), and the median sacral artery which descends in the mid-line into the pelvic cavity in front of the sacrum.

4. *Large terminal branches* which are the right and left common iliac arteries. At the pelvic brim each common iliac artery divides into an external iliac branch which leaves the abdominal cavity to enter the lower limb (and become the femoral artery below the inguinal ligament), and the internal iliac branch which, after a short course in the pelvis, gives off a number of branches to the pelvic organs, muscles, and to the gluteal region (buttocks).

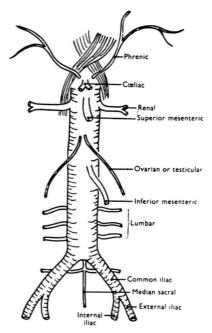

Fig. 3.45 Branches of the abdominal aorta.

The coeliac axis

This is a short trunk given off the aorta immediately after it enters the abdominal cavity (Fig. 3.45). It is surrounded by a dense gangliform plexus of nerve fibres, the coeliac plexus, consisting of sympathetic fibres (from the splanchnic nerves—page 85), and parasympathetic fibres from the vagus nerves. Branches from the plexus and its ganglia are distributed to the stomach, duodenum, liver and gall bladder. The coeliac axis divides into three large branches (Fig. 3.46):

The *splenic artery* runs to the left on the posterior abdominal wall above the pancreas. At the hilum of the spleen it gives off two or three branches which enter the spleen, and the left gastro-epiploic branch, which continues in the greater omentum to reach the greater curvature of the stomach; and the short gastric branches to the fundus of the stomach.

The *left gastric artery* passes to the lesser curvature of the stomach a short distance below the oesophageal opening.

The *hepatic artery* enters the lesser omentum behind and above the first part of the duodenum and runs close to its free edge to the hilum of the liver. It gives off a right gastric branch to the lesser curvature of the stomach and there completes an arterial loop with the left gastric branch of the coeliac axis. The hepatic artery also gives a gastroduodenal branch

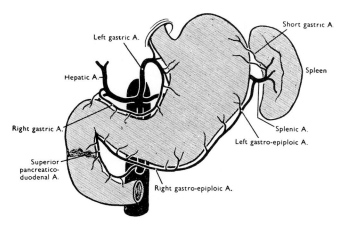

Fig. 3.46 Blood supply of the stomach.

which passes behind the first part of the duodenum. One branch of this (the right gastro-epiploic artery) passes to the greater curvature of the stomach and forms an arterial loop with the left gastro-epiploic branch of the splenic artery. Another branch runs in the groove between the second part of the duodenum and the head of the pancreas to supply these structures. At the hilum of the liver the hepatic artery gives a small branch to the gall bladder (cystic artery) and usually divides into two branches which enter the liver.

The superior mesenteric artery
This large vessel is given off the front of the abdominal aorta a short distance below the coeliac axis (Fig. 3.47). At its origin it lies behind the neck of the pancreas and the portal vein and in front of the left renal vein as this crosses the aorta. The superior mesenteric artery then passes in front of the third part of the duodenum with the superior mesenteric vein on its right side. Here it gives off a pancreaticoduodenal branch to supply the duodenum below (distal to) the level of the entrance of the common bile and pancreatic ducts (midgut portion). The superior mesenteric artery for the remainder of its course runs downwards and to the right in the root of the mesentery of the small intestine. It gives off the following branches:

A number of *mesenteric branches* (16–18) which run forwards between the layers of the mesentery of the small intestine to reach the gut. Within the mesentery these branches form a series of communicating loops and from these the terminal branches arise and penetrate the gut wall.

Ileocolic and right colic branches which run across the posterior abdominal wall behind the parietal peritoneum. Some branches of the ileocolic artery reach the terminal part of the ileum via the lower segment of the mesentery. One of these branches reaches the appendix and runs in its mesentery. Caecal and colic branches of the ileocolic artery and of the right colic artery supply the large intestine from the caecum to the hepatic flexure.

The *middle colic artery* reaches the transverse colon in the mesocolon (that part of the mesentery between the transverse colon and the posterior abdominal wall).

The inferior mesenteric artery
This branch is given off the front of the abdominal aorta a short distance above its bifurcation (Fig. 3.48). It runs downwards on the left psoas muscle crossing the left ureter. It gives off:

A *superior colic branch* which supplies the descending colon and sends a communicating branch into the mesocolon to anastomose with the middle colic artery.

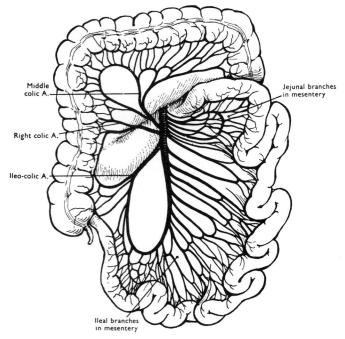

Fig. 3.47 The distribution of the superior mesenteric artery.

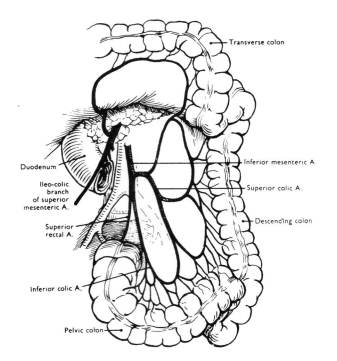

Fig. 3.48 The distribution of the inferior mesenteric artery.

Inferior colic branches enter the mesentery of the pelvic colon and supply this part of the gut.

The *superior rectal artery* descends into the pelvic cavity to supply the rectum. The rectum also receives a blood supply from branches of the internal iliac artery.

The iliac arteries

The *common iliac* arteries, terminal branches of the aorta, after a short course across the lower lumbar vertebrae, bifurcate at the pelvic brim opposite the sacroiliac joint into *external* and *internal iliac vessels*. The former continues forward to leave the lower part of abdominal cavity below the inguinal ligament where it becomes the femoral artery of the thigh (page 49). As each external iliac artery leaves the abdomen it gives branches to the abdominal muscles, including the *inferior epigastric* artery to the muscles within the rectus sheath (page 87).

The internal iliac artery descends into the pelvis and gives off several parietal branches to the pelvic walls and visceral branches to pelvic organs. The parietal branches include the ilio-lumbar, the lateral sacral, the internal pudendal (to the perineal region and external genitalia), the obturator (to the medial side of the thigh), the superior and inferior gluteal branches (to the hip region).

The visceral branches common to both sexes are the superior vesical (to the fundus of the bladder) and the middle rectal (chiefly prostate gland). In the male the inferior vesical artery supplies the seminal vesicles and prostate; in the female the uterine and vaginal vessels supply the pelvic sex organs. Iliac veins accompany the iliac arteries and the common iliac veins, uniting in front of the fifth lumbar vertebra to form the inferior vena cava.

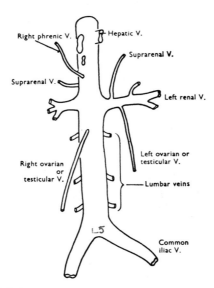

Fig. 3.49 Branches of the inferior vena cava.

The abdominal veins

The *inferior vena cava* receives the following tributaries (Fig. 3.49):

1. The left and right common iliac veins.
2. The lumbar veins.
3. The right ovarian or testicular vein.
4. The renal veins.

5. The right suprarenal vein.
6. The right, and sometimes the left, phrenic vein.
7. The hepatic veins.

The left ovarian or testicular veins and the left suprarenal vein usually enter the left renal vein and not the vena cava directly. The left phrenic vein more often ends in either the left suprarenal or the left renal vein.

THE PORTAL VENOUS SYSTEM

The blood from the greater part of the alimentary canal, the pancreas and the spleen, does not drain directly into the inferior vena cava but into the portal venous system. The chief tributaries of the portal vein are (Fig. 3.50):

The *splenic vein* draining the spleen, pancreas and part of the greater curvature of the stomach.

The *superior mesenteric vein* draining the alimentary canal from the duodenum (lower half) to the transverse colon and the pancreas.

The *inferior mesenteric vein* draining the alimentary canal from the transverse colon to the rectum.

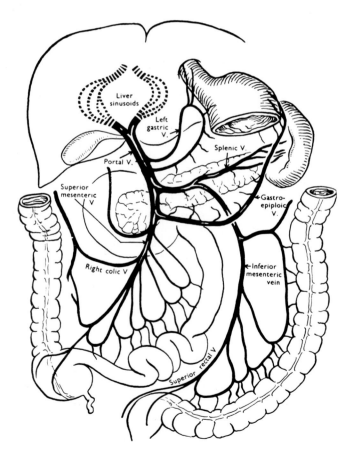

Fig. 3.50 The portal system of veins. (Note: the pancreas lies anterior to the veins in relation to it. See Figure 3.43.)

The portal vein is formed by the junction of the superior mesenteric and splenic veins behind the pancreas. The inferior mesenteric joins the latter. The portal vein lies behind the neck of the pancreas and the first part of the duodenum, reaching the liver in the free margin of the lesser omentum. It receives tributaries from the stomach, duodenum (upper half), the pancreas and the gall bladder.

Within the liver the portal vein breaks up into capillaries and sinusoids (page 103). The venous blood is then collected by the hepatic veins and drains into the inferior vena cava before it pierces the diaphragm to enter the right atrium of the heart.

Around the lower end of the oesophagus veins draining downwards into the portal system communicate with veins draining upwards into the thoracic azygos system. The rectum is drained upwards into the portal system (inferior mesenteric vein) and downwards into the internal iliac veins. These regions (and others) where the portal venous system communicates with the systemic venous system (inferior and superior vena cava) are important in those diseases of the liver in which the portal circulation is obstructed. They provide alternative routes for the blood to reach the heart and are called the *portalsystemic anastomoses*.

ABDOMINAL AND PELVIC NERVES

Within the abdominal cavity the *vagus nerves* distribute branches along blood vessels to the stomach, liver, kidneys and alimentary canal as far as the splenic plexus. The constituent fibres are parasympathetic (cranial outflow).

The abdominal part of the *sympathetic nervous system* consists of bilateral lumbar ganglionated trunks continuous with the thoracic sympathetic trunks and a midline plexus around the aorta. This extends from the coeliac plexus, which surrounds the origin of the coeliac artery, to the bifurcation of the aorta, as the *abdominal aortic plexus*, giving branches along the renal, superior mesenteric and inferior mesenteric arteries to abdominal region. The aortic plexus continues into the pelvis as right and left *hypogastric nerve trunks* which, with contributions from the sacral parasympathetic outflow emerging through S2, 3 and 4 (the pelvic splanchnic nerves or nervi erigentes), form the *pelvic plexuses* embedded in the pelvic fascia and supplying pelvic organs and vessels (page 117).

Some fibres relay in small ganglia within these plexuses and others pass on through them to terminate by forming synapses with nerve cells situated in or near the viscus or vessels which they innervate. These nerve cells form ganglia of miscroscopic size. The arrangement conforms to the cranial parasympathetic outflow (page 417), with long preganglionic fibres extending out to the vicinity of the structure supplied and correspondingly short postganglionic fibres which have a well-defined distribution in a small area. This is one explanation why the physiological effects of this system are quite localized as compared with the effects of stimulation of the sympathetic component of the autonomic nervous system.

ENDOCRINE GLANDS OF THE THORAX AND ABDOMEN

The endocrine or ductless glands produce internal secretions or hormones which diffuse into the adjacent capillaries and are carried into the blood stream to all parts of the body. These hormones exercise a specific and well controlled influence over the activities of certain tissues and organs and, therefore, have considerable physiological importance. For example, those which are found in the thorax and abdominal cavities influence cell metabolism, the level of sugar in the blood stream, and certain reproductive system activities. (For information about the endocrine glands of head and neck see page 265.)

The adrenal (suprarenal) glands
These ductless glands lie in close relation with the upper pole of each kidney and in contact with the diaphragm (Fig. 3.59). In front of the right gland is the bare area of the liver and

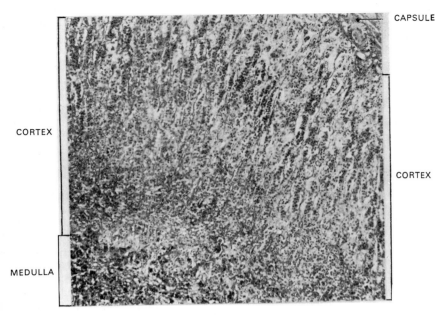

CAPSULE

CORTEX

CORTEX

MEDULLA

Fig. 3.51 Photomicrograph showing the structure of the adrenal gland × 54.

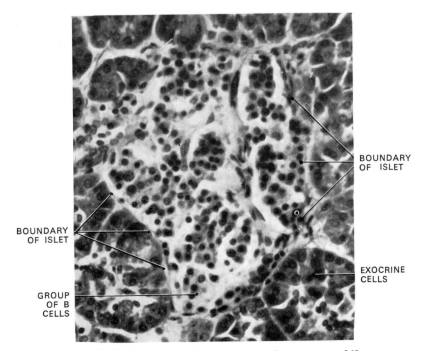

BOUNDARY
OF ISLET

BOUNDARY
OF ISLET

EXOCRINE
CELLS

GROUP
OF B
CELLS

Fig. 3.52 Photomicrograph showing an Islet of Langerhans in the pancreas × 360.

medially the inferior vena cava; in front of the left gland is the stomach, separated from the gland by the lesser sac of the peritoneal cavity. Its lower part is crossed by the pancreas (Fig. 3.43).

Each gland receives a rich blood supply from the aorta and from the renal artery. The right suprarenal vein drains into the inferior vena cava, the left vein into the left renal vein. The suprarenal gland consists of a cortex and a central medulla which manufactures the hormones adrenaline and noradrenaline. These are closely related in their actions to the sympathetic part of the autonomic nervous system. The medulla contains some ganglion cells and the abundant nerve supply from the sympathetic fibres of the splanchnic nerves is unusual, in that the nerve fibres pass mainly to the medulla as preganglionic fibres.

The cortex of the gland can be divided into three distinct zones (Fig. 3.51) based on the arrangement of the cells which compromise its thickness: an outer glomerular zone (ovoid groups of parenchymal or cortical cells); a fascicular zone (parallel cords of cells); and, nearest to the medulla, a reticular zone (a network of thin cords of cells).

Function. The adrenal cortex manufactures a number of steroid compounds, including cortisone, and some forty-one steroids have been isolated from cortical tissue. The cortex is essential for life and, of the various functions attributed to it, the most noteworthy are the production of hormones concerned with water balance, carbohydrate metabolism, and certain male and female sex hormones. At present it is not understood completely if specific steroids are produced in certain zones of the cortex, or if the presence of three zones indicates a cell replacement mechanism, whereby new cells are produced in the outer zone to pass inwards during their secretory life and degenerate in the inner regions of the cortex.

The islets of Langerhans

Groups of endocrine cells within the pancreas do not pass their secretions into the alimentary system, as do the pancreatic exocrine cells (Fig. 3.52). They pour out their secretions into the blood stream and play an important role in the control of glucose level in the circulating blood. Using special staining techniques three types of cells can be recognized in the islets:

1. A or alpha cells, thought to produce a hormone *glucagon*, which raises the level of blood sugar (hyperglycaemia).

2. B or beta cells, which are the most numerous type and are known to produce *insulin*, an excess of which will reduce the level of blood sugar (hypoglycaemia). Without this hormone the body cells are unable to utilize available glucose and the condition of diabetes results. This condition can be controlled by the administration of insulin.

3. D or delta cells, which are few in number and whose significance is unknown.

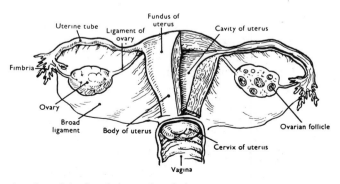

Fig. 3.53 Posterior view of the female internal genital organs.

The thymus gland

In infants and children the thymus is a relatively large bilobed structure, of varying size and shape, which lies below the level of the thyroid gland. It is situated partly in the lower part of the neck, anterior to the trachea, above the suprasternal notch. Although it develops from the pharyngeal region (page 340) it moves rapidly in development to its final position in the anterior part of the mediastinum of the thorax, where the greater part of the gland is found after birth. It continues to grow until puberty, after which time it appears to atrophy (i.e. wastes away) and is represented in the adult by bands of connective tissue behind the upper part of the sternum.

Each lobe of the thymus consists of lobules interconnected by fibrous tissue. Histologically, it possesses a cortex and a medulla and resembles closely the structure of lymph nodes. Within the medulla are found circular masses of epitheliod cells, *Hassall's corpuscles*, apparently derived from the pharyngeal endoderm (Fig. 3.54).

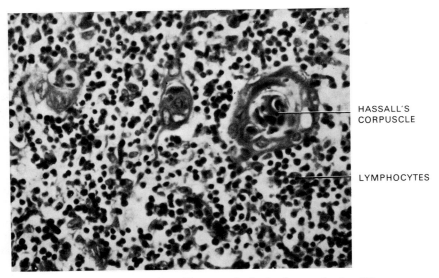

HASSALL'S
CORPUSCLE

LYMPHOCYTES

Fig. 3.54 Photomicrograph showing Hassall's corpuscles in the thymus gland × 400.

The thymus produces lymphocytes and plasma cells, particularly during foetal life. Its functions are obscure but it seems to play a vital role in the formation of antibodies, which are synthesized in response to the presence in the blood stream of foreign substances called antigens. The thymus has also been associated with the mechanism of rejection of skin grafts and a clinical condition involving muscular weakness, called myasthenia gravis. Although it does not produce a specific hormonal substance, the thymus is interrelated physiologically with hormone producing glands, such as the thyroid, adrenal and gonads. For this reason, the thymus gland is excluded sometimes from the group of endocrine glands, to be included rather with the lymphatic system.

The endocrine cells of the testis and ovary

The connective tissue elements which support the coiled seminiferous tubules of the testis contain characteristic cells known as the interstitial cells (of Leydig). There is conclusive evidence that these cells produce and secrete the male sex hormone, *testosterone*, which is responsible for the onset of secondary sex characters at puberty, such as growth of facial hair, deepening of the voice and marked muscular development.

The ovary, under the control of the anterior lobe of the pituitary gland, or hypophysis cerebri (see page 265), produces two hormones which are also steroid. *Oestrogen* is probably formed by the interstitial cells of the ovary, which correspond to those of the testis. They lie between the ovarian follicles, the sites of oogenesis (formation of the ova) (Fig. 3.53). *Progesterone* is produced by the *corpus luteum*, a temporary glandular structure which is the remnant of a ruptured ovarian follicle after the ovum has been released into the ovarian tube by the process of ovulation during the menstrual cycle. If the ovum is fertilized the corpus luteum persists and continues its production of progesterone until the later months of pregnancy. Otherwise the corpus begins to degenerate one week after ovulation and ceases to manufacture the hormone.

As in the male, the ovarian hormones are responsible for the changes associated with puberty in the female, including the commencement of menstruation and the enlargement of the mammary glands.

THE PELVIC CAVITY

The pelvic cavity is bounded by the iliac, pubic and ischial bones, which make up the right and left innominate or hip bone, and behind by the united sacral and coccygeal vertebrae (Figs 3.55, 3.56). Covering the inner surfaces of these bones are the iliacus and psoas muscles which pass beneath the inguinal ligament to gain attachment to the femur; the internal obturator muscle, which passes through the lesser sciatic notch; and the piriformis, which passes

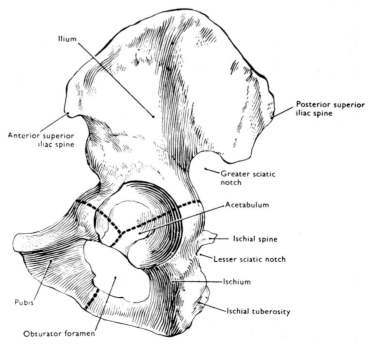

Fig. 3.55 The left hip bone viewed from the lateral side. The demarcation of the constituent bones is indicated by dotted lines. (By courtesy of Professor G. A. G. Mitchell and Dr E. L. Patterson.)

through the greater sciatic notch; both these latter muscles are also attached to the femur; the levator ani and coccygeus muscles, making up the pelvic diaphragm; and the pelvic fascia. Within the pelvis are the following organs (Figs 3.56, 3.57):

1. In front, the *urinary bladder*. Into its base open the ureters which come from the kidneys, and at its base commences the *urethra*, which pierces the front of the pelvic diaphragm, passing between the levator ani muscle of each side. The male urethra is made up of

a. A prostatic segment receiving the opening of the ejaculatory ducts and the ducts of the prostate gland (the part of the urethra that lies within the pelvic cavity);

b. A short membranous part passing through the perineum below the pelvic diaphragm; and

c. A larger spongy part which passes to the exterior through the body and glans of the penis.

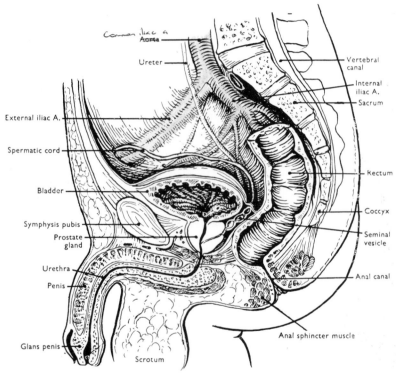

Fig. 3.56 Median sagittal section through the male pelvis.

In the female the urethra is much shorter; after passing through the pelvic diaphragm and perineum it opens in front of the vagina (Fig. 3.57).

The bladder is covered on its upper surface by the peritoneum and when full rises into the abdominal cavity. In this condition it can be approached in surgical operations through the lower part of the anterior abdominal wall above the symphysis pubis without opening into the peritoneal cavity.

The ureters and the bladder are lined by a transitional epithelium. Emptying of the bladder is brought about by contraction of its smooth muscle and relaxation of the sphincter situated at the beginning of the urethra. The nerves regulating this process are derived from the hypo-

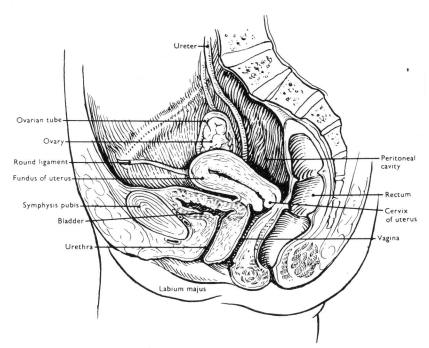

Fig. 3.57 Median sagittal section through the female pelvis.

gastric and pelvic plexuses (of the autonomic nervous system) made up of sympathetic nerve fibres and parasympathetic (sacral) nerves.

2. At the back of the pelvis lies the rectum and the terminal part of the large intestine. It lies in front of the sacrum and passes through the pelvic diaphragm to end in the anal canal. It receives its blood supply from the inferior mesenteric and internal iliac arteries (from the middle rectal arteries) and the middle sacral artery (the terminal branch of the aorta). The veins drawn upwards into the inferior mesenteric vein (portal system) and downwards into the inferior rectal veins which drain into the internal iliac veins (vena caval circulation).

3. Between the rectum and anal canal behind and the bladder and urethra in front are the pelvic sex organs and their ducts. These are in the male: the bilateral vas deferens (ductus deferens) coming from the testes, the seminal vesicles, the prostatic and ejaculatory ducts opening into the urethra. The female sex organs are entirely within the pelvis and consist of the ovaries, the ovarian (Fallopian) tubes, the uterus and, piercing the pelvic diaphragm, the vagina, which opens between the anal canal and urethra.

During pregnancy the enlarging uterus (womb) rises into the abdominal cavity. In order to permit of the passage of the child during labour, the female pelvis is of a somewhat different shape from that of the male and study of the pelvis is the most reliable method of determining the sex of a skeleton (Figs 1.2 and 3.58).

All the pelvic organs (except the ovary) and muscles are supplied by branches of the internal iliac artery which lies on either side towards the back of the cavity. From it branches pass forwards through the pelvic fascia beneath the peritoneum to the bladder (vesical branches), the uterus (uterine artery), the rectum (middle rectal artery). Another branch passes beneath the pelvic diaphragm (internal pudendal artery) to supply the anal canal, the vagina in the female and, in the male, the external genitalia.

On the posterior wall of the lower abdomen and pelvic cavity are the lumbar and the sacral plexuses of nerves derived from the lumbar, sacral and coccygeal branches of the spinal cord. These plexuses supply the abdominal and pelvic muscles and enter the lower limb either anteriorly beneath the inguinal ligament, or posteriorly through the greater and lesser sciatic notches between the innominate bone and the sacrum.

The smooth muscle of the rectum, uterus, and bladder receives its nerve supply (autonomic nervous system) from the hypogastric plexus. The lymphatic vessels of the pelvis drain into lymph nodes along the iliac and middle sacral vessels and from these upwards to the *receptaculum chyli* which lies in front of the first and second lumbar vertebrae (Fig. 3.26). At its upper

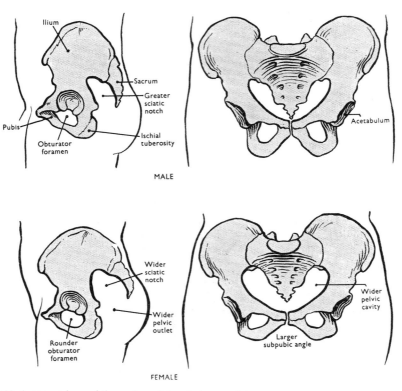

Fig. 3.58 A comparison of the male and female bony pelvis.

end commences the thoracic duct which passes into the thoracic cavity through the aortic orifice. The rectum and uterus are common sites of cancer in the pelvis.

The *pelvic diaphragm*, consisting of the *levator ani* and *coccygeus* muscles, forms a muscular, basin-like floor for the pelvis and convex roof for the perineal region. The *coccygeus* muscle fans out from the spine of the ischium to the side of the sacrum and coccyx covering the *sacrospinous ligament*. This ligament is derived from the muscle which wags the tail in lower animals. The *levator ani* arises from the inner surface of the body of the pubis, the spine of the ischium, and between these bony attachments from a linear thickening of the fascia covering the *obturator internus*. The muscles of each side meet at a midline tendinous raphe reaching back to the coccyx. The raphe is interrupted by the passage from pelvis to perineum of the anorectal canal, the urethra and the vagina, in the female. In front there is a space between the edges of two muscles where the covering fascia forms the superior fascia of the *urogenital diaphragm*.

The *perineum* consists of a posterior region related to the anus (*anal triangle*), and an anterior region related to urethra and genital organs (*urogenital triangle*). The latter contains two membrane bound compartments. The upper (deep) compartment is bounded above by areolar tissue on the under surface of the levator ani muscle and below by the *perineal membrane* (inferior fascia of the diaphragm), forming a closed space containing the *sphincter urethrae* muscle. The lower compartment, the *superficial perineal pouch*, is bounded above by the perineal membrane (attached at the sides to the ischiopubic rami) and below by the membranous layer of the superficial fascia (of Scarpa). The superficial pouch contains the *bulb* of the penile urethra, *corpus spongiosum*, and *corpora cavernosa* of the penis in the male; the body of the *clitoris* and associated corpora cavernosa of the female.

THE UROGENITAL SYSTEM

This system consists of the kidneys, the ureters and the urinary bladder, the sex glands and the sex organs. Although their functions are quite distinct the components of this system share a common developmental history and remain closely related to one another throughout life (Fig. 3.59).

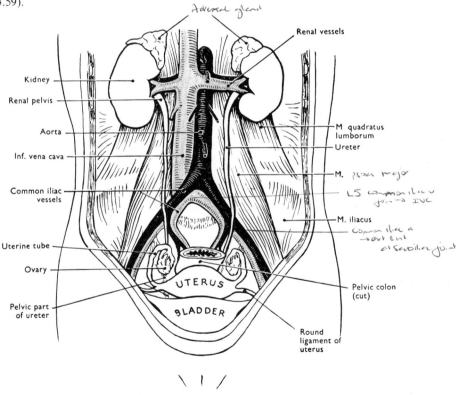

Fig. 3.59 The female urogenital system. (By courtesy of Professor G. A. G. Mitchell and Dr E. L. Patterson.)

The kidneys

The primary function of the kidneys is to excrete various waste products of metabolism in the urine and maintain the correct water balance in the body tissues. Each kidney lies at the

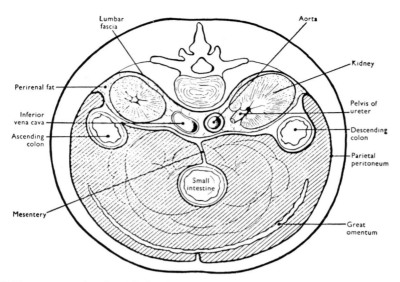

Fig. 3.60 Transverse section through the abdominal cavity at the level of the kidneys.

back of the abdominal cavity on either side of the vertebral column, resting behind on the lower part of the diaphragm, and on the psoas, quadratus lumborum and transversus abdominis muscles (Figs 3.31, 3.59). It is separated from these structures by a mass of fat and a connective tissue capsule, the renal fascia, which helps to hold the kidney in position. The kidney also has its capsule proper consisting of connective tissue and a few smooth muscle fibres. The hilum of each kidney faces antero-medially and here the renal arteries (branches from the aorta) enter the organ and the renal veins leave it to join the inferior vena cava. As the vena cava lies to the right of the middle line, the left renal vein is a long vessel crossing in front of the aorta just below the origin of the superior mesenteric artery. At the hilum lies the renal *pelvis* into which the urine is collected from the kidney substance before its passage along the ureter to the urinary bladder (Fig. 3.62).

The kidney is made up of a great number of *renal corpuscles*, each consisting of the blind end of a kidney tubule invaginated by a clump of blood capillaries (the *glomerulus*) (Fig. 3.61). The glomerular capillaries filter a large volume of protein free fluid across their endothelium, then through a basement membrane and the epithelium of the glomerular capsule (*Bowman's capsule*). After leaving the corpuscle each tubule runs through a second capillary plexus where much of the secreted fluid (the glomerular filtrate) and certain of the contained constituents, such as sodium and chloride ions, are reabsorbed into blood stream. That which remains in the tubules is urine and is passed through a system of collecting tubules which open at the summits of the *renal papillae*. These project into diverticulae of the *renal pelvis* called the minor *calyces* (Fig. 3.61A).

In front of the right kidney lie the ascending colon, the hepatic flexure, and the second part of the duodenum. In front of the left kidney lie the descending colon, the pancreas, part of the stomach, and coils of the small intestine. The spleen is in contact with the upper part of the lateral border of the left kidney and each kidney is capped by an adrenal gland (Fig. 3.62).

The urinary bladder

The urinary bladder is situated in the pelvic cavity above the muscles of the pelvic diaphragm, in front of the rectum and behind the symphysis pubis. In the female the body of the uterus

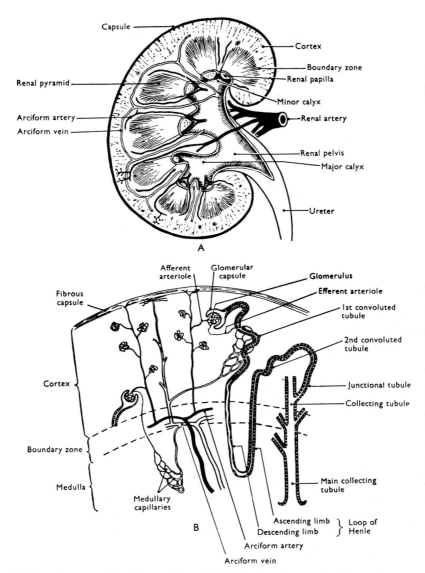

Fig. 3.61 The structure of the kidney: A. Longitudinal section through the kidney calyces and renal pelvis. B. Diagram of a nephron (kidney tubules) and its blood supply. (By courtesy of Professor G. A. G. Mitchell and Dr E. L. Patterson.)

lies between the bladder and the rectum and is supported by ligaments which pass between these three structures (see also page 123).

The sex glands

The *testes* lie in the scrotal sacs. From the testis a duct, the vas deferens, passes through the inguinal canal (page 88) to enter the pelvic cavity. Behind the bladder each vas receives the short duct draining the seminal vesicle and opens as the ejaculatory ducts into the prostatic part of the urethra (Fig. 3.63). The testis as well as being responsible for the manufacture

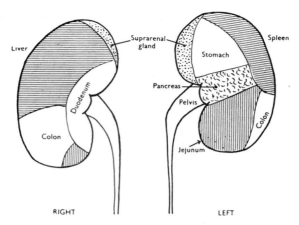

Fig. 3.62 Anterior relationships of the kidneys.

of the testis

of the spermatozoa in the seminiferous tubules contains interstitial cells which lie between the tubules and produce the male sex hormone (see page 114).

The *ovaries* lie just below the brim of the true pelvis between the external and internal iliac arteries and in close relation to the ureters. Each ovary is suspended by a short fold of peritoneum, the ovarian mesentery, from the back of the *broad ligament*, a much larger fold of peritoneum which attaches the uterus to the lateral wall of the pelvic cavity. The ovum discharged from one or other ovary during the middle of each uterine (menstrual) cycle passes into the adjacent opening of the oviduct which leads to the uterus. If fertilization takes place it usually does so in the oviduct (uterine tube) and the fertilized ovum begins to divide and differentiate before it enters the uterus. In the uterus it becomes embedded in the endometrium, which has become specially prepared for its reception. The ovaries are also responsible for the manufacture of the female sex hormones (see page 115).

The arteries to the testis or ovaries are given off about the middle of the aorta in the region where the sex glands developed before their later downward migration to the scrotal sacs or to the pelvic cavity. The right testicular or ovarian vein joins the inferior vena cava; the left vein joins the left renal vein.

The *uterus* or womb (Figs 3.53, 3.57) occupies the central portion of the female pelvis, lying in front of the rectum and above and behind the urinary bladder. It consists of a *body* limited above by the dome-shaped *fundus*. The fundus and upper part of the body are bent slightly

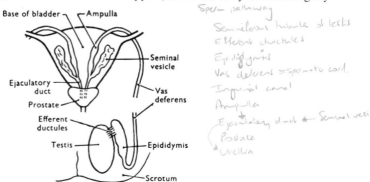

Fig. 3.63 Diagram of the male genital ducts. (By courtesy of Professor G. A. G. Mitchell and Dr E. L. Patterson.) *[see Fig 3.56]*

forward, or anteflexed, on the cervix. The body is continuous below and behind with the *cervix* which projects into the upper end of the vagina on its anterior aspect. The body and cervix lie on a plane which passes upwards and forwards with respect to the long axis of the vagina, in a normal position of anteversion. The uterus is a highly muscular structure lined by a specialized mucous membrane, the *endometrium*, containing many tubular glands and numerous blood vessels. Pelvic peritoneum covers the anterior and posterior surfaces of the uterus and is reflected laterally to form the broad ligaments, which assist in the stabilization of the uterine body.

The uterine body is somewhat mobile but the cervical region is held in place by the lateral pelvic ligaments, which are condensations of the pelvic fascia containing smooth muscle, and is supported by the levator ani muscles (pelvic diaphragm). The bilateral *round ligaments* of the uterus, which extend from the junction of the uterine body and tube to the deep inguinal ring, are attached to the upper aspect of the body on each side and add to its stability, especially when the organ is greatly enlarged during pregnancy. Various forms of prolapse of the uterus can occur when these restraining 'ligaments' become stretched and lax, as may be the case following multiple pregnancies.

The cervical region of the uterus is closely related to the ureters, as they pass forward to the bladder on each side, and in this region the major blood supply, in the form of the uterine arteries, pass medially across the pelvic floor in the base of the broad ligament, above the ureters, to reach the uterus. The uterine arteries are branches of the internal iliac vessels (page 109) and become greatly enlarged during pregnancy.

4. Regional Anatomy of the Head and Neck

THE SKULL

The skull, situated at the upper or cephalic end of the vertebral column (Fig. 1.2) consists of a number of bones united to one another at specialized joints known as sutures. Only a limited amount of movement takes place at the sutures but they play an important part in the growth of the skull (page 373). There is one bone, the mandible or lower jaw, which is united to each side of the base of the skull by a movable, diarthodial joint, the mandibular or temporomandibular joint. Movements of the mandible are produced by the muscles of mastication, and bring about the mastication or breaking up of food in the mouth cavity by the action of the upper and lower teeth.

The skull is divided into a postero-superior *cranial part* related to and protecting the brain, and an antero-inferior part, the *facial skeleton*, related to the orbital, nasal and oral cavities.

In examining a skull, it is best looked at in turn from the front (norma frontalis), the side (norma lateralis), from below (norma basalis), and from above (norma verticalis). Its interior can be examined after the removal of the skull cap, or calvarium.

Norma frontalis (Figs 1.2, 4.1)

When examined from the front the greater part of the skull is made up of the facial skeleton, the cranial part being limited to the forehead region above the orbital cavities. The facial skeleton consists of two parts: the *upper facial skeleton* related to the orbital and nasal cavities, the roof of the oral cavity (hard palate) and the upper teeth; and the *lower facial skeleton* consisting of the mandible (lower jaw) which carries the lower teeth.

The bony margin of the nasal cavities is roughly pear-shaped but shows considerable variation in size and form. It may be narrow or wide, and there is some correlation between nasal form and climatic conditions. In the living, the skeleton of the nose is completed by the alar cartilages laterally and in the midline by the septal cartilage.

The margins of the orbital cavities also vary in their form and size although to a lesser degree than the nasal cavities. There is, however, considerable variation in the width between the inner margins of the two orbital cavities across the bridge of the nose. The degree of depression at the bridge of the nose between the nasal and frontal bones is quite variable.

The degree of projection of the cheek bones (zygomatic or malar bones), and the extent to which they slope backwards shows a wide range of individual and racial variation.

In the forehead region the characters to be observed are the slope of the forehead (best observed in norma lateralis), and the degree of development of the superciliary bony ridges situated above the orbital margins. These are more marked in male skulls and in the skulls of primitive races.

In the lower part of the face the form and massiveness of the chin should be noted and also the size and regularity of the teeth. The height of the face is measured vertically from the nasion (where the nasal bones join the frontal bone in the middle line) to the menton (lower margin of the lower jaw in the midline). Facial width is the greatest distance between the outer surfaces of the zygomatic arches, and the *facial index* (width $\times$ 100 $\div$ height), gives an indication of the form of the face.

Optic nerve in optic foramen at same level as nasion

The bones seen from the front of the skull are the frontal zygomatic, maxilla, nasal and mandible. Within the orbital cavity (see page 186) are parts of the maxilla (in the floor); zygomatic and greater wing of the sphenoid (lateral wall); frontal and lesser wing of the sphenoid (roof); and of the lacrimal, ethmoid, and palatine (medial wall). At the back of the orbital cavity are the round optic foramen in the lesser wing of the sphenoid and the slit-like superior orbital fissure between the greater and lesser wings of the sphenoid. Both of these communicate with the interior of the cranial part of the skull. Between the lateral wall and floor of the orbital cavity is the slit-like inferior orbital fissure leading towards the back of the facial skeleton.

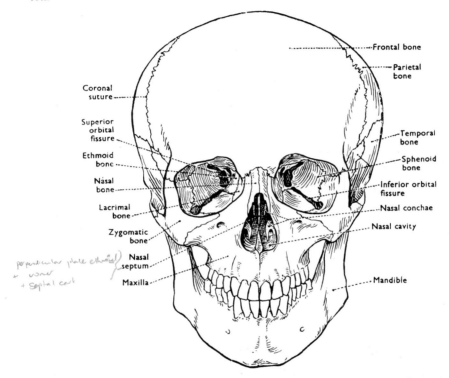

Fig. 4.1 The skull—viewed from the front (norma frontalis). (By courtesy of Professor G. A. G. Mitchell and Dr E. L. Patterson.)

The anterior opening of the nasal cavity is bounded by the maxillary bones below and at the sides, and the lower edges of the nasal bones above. Within the nasal cavity are the inferior turbinate bones (conchae) projecting from the lateral wall, and the middle and superior turbinate processes of the ethmoid bone at a higher level. In the middle line but often showing some deviation to the left or right are the vomer (below) and the perpendicular plate of the ethmoid (above), forming the bony part of the nasal septum. The notch between them which faces anteriorly is filled by the septal cartilage in the living subject. On each side of the nasal opening and below the lower orbital margin are the infra-orbital foramina, situated in the maxillary bones. The mandible shows a mental foramen on each side.

Norma lateralis (Fig. 4.2)
When the skull is examined from the side, the facial region occupies the front with the cranial part above and behind. When first examining the skull from the side it is better to remove

the lower jaw. Behind the orbital cavity a flying buttress, the zygomatic arch, unites the facial and cranial regions. Above the zygomatic arch and behind the outer margin of the orbital cavity is the temporal region (temporal fossa) of the cranium. It is limited above by the two parallel temporal lines. The height of the temporal lines at the side of the skull give an indication of the degree of development of the powerful temporal muscle, one of the muscles of mastication. Behind the position where the zygomatic arch joins the side of the cranium is the opening of the external auditory canal (external auditory meatus or ear hole). To its margin is attached the cartilaginous skeleton of the auricle (pinna or external ear). Behind and below the meatus is the cone-shaped mastoid process which is larger in male skulls.

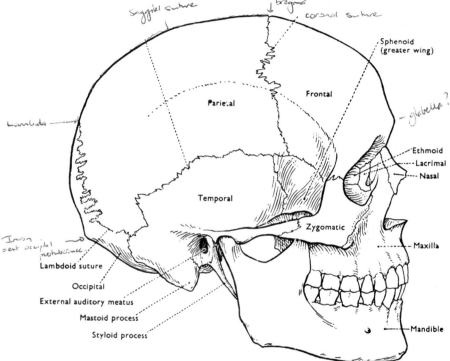

Fig. 4.2 The skull—viewed from the side (norma lateralis). (By courtesy of Professor G. A. G. Mitchell and Dr E. L. Patterson.)

On the inner side of the zygomatic arch, projecting downwards from the base of the skull and lying behind the maxilla, can be seen the outer surface of the lateral pterygoid plate of the sphenoid bone. The slit-like space between it and the maxilla in front is the pterygopalatine fossa (Fig. 5.18). When the mandible is in position its ascending portion behind the teeth (the ramus) lies on the outer side of these structures. At the upper end of the ramus is the transversely elongated condylar process which fits into (articulates with) a hollow fossa at the base of the skull in front of the external auditory meatus. This is the glenoid fossa. The shape of the cranial part of the skull is indicated by the *cranial* (*cephalic*) *index* which is the greatest width as measured above the ear holes × 100 and divided by the cranial length as measured from the glabella, the point in the middle line in front between the superciliary ridges (at the base of the forehead) and the most prominent point in the middle line at the back of the skull (occipital point). Long narrow skulls are termed dolicocephalic; short wide (round) skulls are brachycephalic.

The skull bones seen from the side of the skull are parts of the maxilla, palatine, zygomatic, nasal and mandible (facial skeleton); the frontal parietal, temporal, occipital and the greater wing and lateral pterygoid process of the sphenoid (cranial skeleton).

Norma basalis (Fig. 4.3)

The facial skeleton lies at the front of the skull and shows from this view the hard palate surrounded in front and at the sides by the upper dental arch. In the middle line towards the front of the hard palate is the incisive foramen while at the back of the palate on each side close to the dental arch is a palatine foramen. The shape of the palate and of the arch and narrow, or short and wide, and in many skulls there is often some degree of palatal deformity and dental irregularity. On each side of the hard palate and dental arch, the zygomatic process of the maxilla leads to the zygomatic arch. These join the cranial part of the

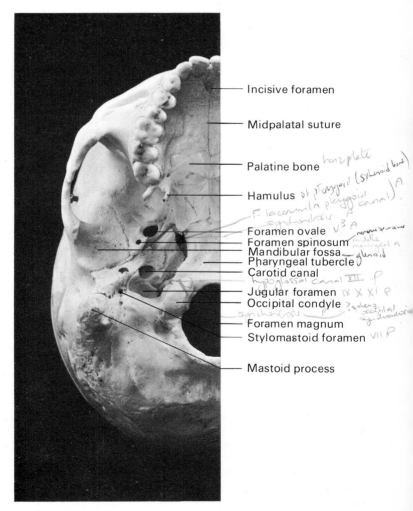

Incisive foramen

Midpalatal suture

Palatine bone

Hamulus

Foramen ovale
Foramen spinosum
Mandibular fossa
Pharyngeal tubercle
Carotid canal
Jugular foramen
Occipital condyle
Foramen magnum
Stylomastoid foramen

Mastoid process

Fig. 4.3 The adult human skull. Inferior aspect (norma basalis).

skull at the articular surfaces for the mandibular condyles. In the middle of the base of the cranial part of the skull is the large foramen magnum through which the spinal cord is continuous with the brain. On each side of the foramen magnum are the smooth occipital condyles where the skull articulates with the first cervical vertebra (the atlas) (Fig. 3.2). At the joints between the skull and atlas nodding movements take place. Rotary movements of the head take place chiefly between the first cervical vertebra and the second (axis) (Fig. 3.2). Lateral to the occipital condyles are the mastoid processes. In front of these the thin rod-like styloid processes project from the base of the skull. They vary considerably in length and are often broken. Between the mastoid and styloid of each side is the external opening of the canal of the facial nerve, the stylomastoid foramen.

On the inner side of the styloid processes are the jugular foramina, which are irregular in outline. In front of them are the rounded openings of the carotid canals. On the septum of bone between the jugular and carotid openings is the small tympanic foramen (canaliculus). On the inner side of each jugular foramen and above the occipital condyle is the anterior condylar foramen. Posterior condylar foramina are present in some skulls behind and above the occipital condyles.

On the inner side of each glenoid fossa is the spinous process of the greater wing of the sphenoid. In front of it is the small circular foramen spinosum and further forwards the important oval-shaped foramen ovale. On the inner side of these two foramina is the groove of the pharyngotympanic or Eustachian tube which is completed by cartilage; and on the inner side of the front part of this groove is situated the irregularly-shaped foramen lacerum, which shows considerable variation in regard to its size and shape. In front of the foramen lacerum the lateral and medial pterygoid plates of the sphenoid project downwards from the skull base towards the back of the dental arcade on each side. The hollowed-out area between the two plates is the pterygoid fossa. The medial pterygoid plate forms the side wall of the back part of each nasal cavity, and at its base is the small scaphoid fossa. The back of the nasal septum is formed by the vomer.

In front of the foramen magnum is a strong bar of bone in the middle line, the *mid-line cranial base*, continuous in front with the roof of the nasal cavities. Behind the foramen magnum the skull base shows markings for the attachments of the powerful post-vertebral neck muscles.

The bones seen from this aspect of the skull are the maxilla and palatine bones, forming the hard palate; the zygomatic bone and parts of the vomer, sphenoid, temporal and occipital bones.

Norma verticalis

When viewed from above the skull is generally ovoid in shape, although comparison of skulls show that there is a great deal of individual variation in the outline. Three sutures are evident: the coronal, sagittal and lambdoid sutures, at which the bones of the vault or calvarium, are strongly interlocked and bound by fibrous tissue during life.

The coronal suture (Latin, a crown) lies in a transverse direction between frontal and parietal bones. The sagittal suture lies between the two parietal bones and, in all but a few skulls where a metopic suture exists between the two halves of the frontal bone (page 221), it terminates anteriorly at the coronal suture. Posteriorly, the sagittal suture meets the inverted V-shaped lambdoid suture, which separates the parietal bones from the occipital bone. In the skull of an infant the anterior end of the sagittal suture terminates at the anterior fontanelle (Fig. 4.4), a diamond-shaped membrane-filled area which does not complete its ossification until about the end of the second year. The overall appearance of the fontanelle and the sutures behind it have probably given origin to the name of the sagittal suture, from the Latin word, sagitta, meaning an arrow.

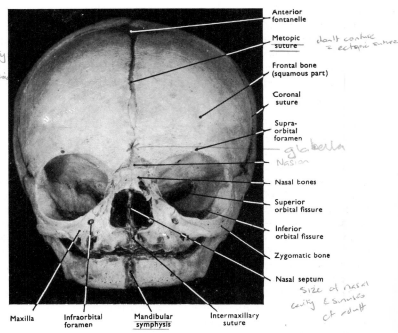

[handwritten:] Skull of newborn differs from that of an adult because it has a sy-ph-ysis menti T / a metopic suture T / a well developed tympanic plate F / a large nasal cavity / a deeply placed stylomastoid foramen F

[figure labels:] Anterior fontanelle — Metopic suture *[handwritten: don't confuse = ectopic suture]* — Frontal bone (squamous part) — Coronal suture — Supra-orbital foramen — *[handwritten: glabella / Nasion]* — Nasal bones — Superior orbital fissure — Inferior orbital fissure — Zygomatic bone — Nasal septum *[handwritten: size of nasal cavity & sinuses of adult]* — Maxilla — Infraorbital foramen — Mandibular symphysis — Intermaxillary suture

Fig. 4.4 Anterior view of the skull of a newborn infant.

The most prominent part of each parietal bone is the *parietal eminence* and close to the sagittal suture a distinct *parietal foramen* may be present on each side. The skull is rather flat here and the point of the sagittal suture which is on a level with the parietal foramina is termed the *obelion*. The point of junction of the sagittal and lambdoid sutures is termed the *bregma*, that of the sagittal and lambdoid sutures, the *lambda*.

On the front of the skull are the two frontal eminences and, in the mid-line between the superciliary arches the ridge-like prominence which is called the *glabella*.

Further details about individual skull bones and foramina are given in Chapter 000.

CHARACTERISTIC FEATURES OF THE HUMAN FACE

Certain features of the human craniofacial skeleton can best be appreciated by comparison with those of lower animal forms including the anthropoid apes. There are very obvious differences between the size of the cranium and the size of the facial skeleton when these comparisons are made, but the length of the cranial base is similar in both. In man the alveolar processes do not greatly exceed the size of the supporting skeleton, e.g. the body of the maxilla. The human dentition is small and unspecialized and, as a result, the facial buttress system is not strongly developed (page 236).

The cranial base in man is more flexed than in other animal forms and the upper facial skeleton does not normally project in front of the cranium. In the human skull, the relative enlargement of the cranium and reduction in size of the facial skeleton results in a less extensive development of the paranasal air sinuses. In man the foramen magnum remains beneath the skull, a good example of the persistence of foetal characteristics in human anatomy.

The growth of the nasal septal cartilage in the human is much less extensive than in other animals. It grows downwards more than forwards and the anterior segment of the cranial base maintains its foetal position, so that the human face lies beneath the cranium rather than in front of it.

DETERMINATION OF THE SEX AND AGE OF SKULLS

The chief criteria used in determining the sex of a skull are:

1. The size of the mastoid processes. These are larger and more robust in male skulls.

2. The upper orbital margins are sharper in the female skulls and the superciliary region flatter.

3. The forehead is more vertical in the female.

4. Muscle markings, especially at the angle of the mandible, along the lower border of the zygomatic arch and in the region of attachment of the post-vertebral muscles, are usually more marked in male skulls.

5. The posterior root of the zygoma usually forms a sharper ridge above the external auditory meatus in male skulls.

6. The cranial capacity is usually somewhat less in female skulls. This is correlated with body size.

The determination of sex from the skull is not as reliable as from the pelvic bones or the femur, and care must be taken, especially when skulls of different racial groups are being examined. A female Eskimo skull might show more 'male' markings than that of males in other races, especially in regard to those markings which depend on the degree of development of muscles.

The age of the skull is determined chiefly by the condition of the dentition and the closure of the cranial sutures. The three permanent molars, erupting at approximately 6, 12 and 18 years, are the most important teeth in determining age. The relative degree of wear of these teeth gives a clue to age after they have erupted, providing their opponents in the opposite jaw are in position. All three teeth show an approximately equal degree of wear by 35 to 40 years. The size of the maxillary tuberosity tends to enlarge for some time after the eruption of the third permanent molars, especially in dentitions which have been put to good use with a resultant wearing of the approximal or contact surfaces of adjacent teeth and a reduction in their mesiodistal diameters.

The sagittal and upper part of the coronal sutures begin to show areas of union between 25 and 30 years of age. The lambdoid suture commences to close a few years later and is not completely closed until at least the fortieth year. The sutures bounding the squamous part of the temporal bone show evidence of closure between 40 and 60 years, but complete closure of these sutures is very rare. All cranial sutures close, or fuse, first on the endocranial aspect.

The synchondrosis at the base of the skull between the occipital and sphenoid bones unites between 14 and 20 years, the process being completed earlier in female than in male skulls.

SUPERFICIAL STRUCTURES OF THE NECK

The neck is bounded below by the clavicles and the upper border of the sternum, and above by the lower border of the mandible and the base of the skull. It contains the seven cervical vertebrae of the vertebral column, the hyoid bone, the pharynx, larynx, the upper parts of the oesophagus and trachea, the great cervical vessels, parts of the ninth to the twelfth cranial nerves, the cervical plexus and the roots of the brachial plexus, the upper parts of the sympathetic nerve chain and the following muscles (Fig. 4.5):

1. The platysma muscle in the subcutaneous tissue of the front of the neck. It is supplied by the cervical branch of the seventh cranial nerve (facial nerve), which also supplies the stylohyoid and the posterior belly of the digastric muscle in the neck.

2. The trapezius and sternomastoid muscles, both supplied by the accessory nerve (eleventh cranial nerve) and by branches of the cervical plexus.

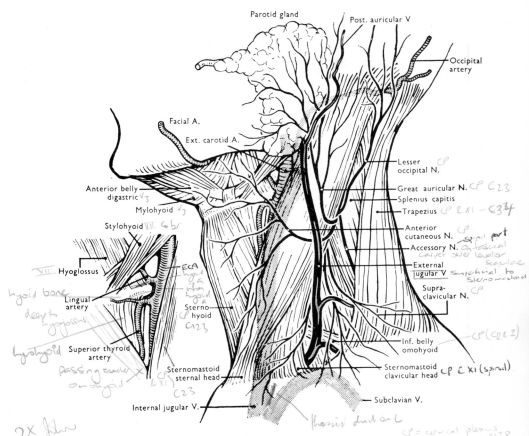

Fig. 4.5 The sternomastoid muscle and related structures.

3. The suprahyoid muscles: the digastric, stylohyoid, mylohyoid and geniohyoid muscles, and the muscles of the tongue. The group of muscles receives its nerve supply from various sources, but the tongue muscles are supplied by the hypoglossal (twelfth cranial) nerve.

4. The infrahyoid group of muscles, supplied by the anterior primary rami of the first three cervical nerves, through the cervical plexus (Figs 4.9, 4.11).

5. The laryngeal muscles, supplied by the vagus (tenth cranial) nerve (Fig. 4.53).

6. The pharyngeal muscles, supplied by the vagus and glossopharyngeal (ninth cranial) nerves (Fig. 4.48).

7. The anterior and lateral vertebral muscles (scalene muscles, longus cervicis and longus capitis, and levator scapulae) supplied by the cervical plexus.

8. The postvertebral muscles lying behind the vertebral column and supplied by the posterior primary rami of the cervical nerves (Fig. 4.10).

The *platysma* is a thin sheet of muscle which covers the side of the neck and arises from the deep fascia below the clavicle. It passes obliquely upwards on the side of the neck to the lower margin of the mandible, to which it is partly attached. Most of the anterior fibres of the muscle merge with the superficial muscles of the face, particularly those of the lower lip. The most anterior fibres reach the point of the chin and merge with the corresponding muscle fibres from the opposite side of the neck. The platysma is a remnant of a much more extensive superficial sheet of muscle in lower animals and it has limited function in man. It is seen best when it raises the skin of the neck into a series of vertical folds during grimacing, especially if the head is bent backwards. The platysma and other superficial muscles in the neck, such as the sternocleidomastoid and trapezius stand out prominently in exertions involving the upper limb, as in weight lifting and boxing.

A number of muscles which can be seen when the skin, subcutaneous tissue and the platysma are removed conveniently subdivide the neck into a number of muscular triangles (Fig. 4.6):

1. The *trapezius*. This large flat muscle covers the back of the neck from the skull above to the clavicle and shoulder blades (scapulae) below (Fig. 4.5). It arises from the medial third of the superior nuchal line of the occipital bone, the external occipital protuberance, the ligamental nuchae and the spines of all the thoracic vertebrae. It is inserted into the lateral third of the clavicle and the upper border of the spine of the scapula.

2. The *sternocleidomastoid* is a prominent strap-like muscle passing from the medial third of the clavicle and upper margin of the sternum upwards and backwards along the side of the neck to be attached to the mastoid process of the skull behind the ear and also to the lateral one-third of the superior nuchal line of the occipital bone (Fig. 4.6). Further details are given on page 135.

3. The *omohyoid muscle* consists of two bellies. The superior belly is a small strap-like muscle attached above to the lower border of the hyoid bone and, passing downwards and backwards, disappears beneath the anterior border of the sternomastoid and about the middle of its length (Fig. 4.6). The inferior belly of the omohyoid is continuous with the superior belly under cover of the sternomastoid. It appears at the posterior border of the sternomastoid at about the junction of its lower fourth and upper three-fourths. The omohyoid then passes laterally, deep to the trapezius and just above the clavicle to its attachment on the upper border of the scapula in the region of the suprascapular notch.

4. The *digastric* muscle (literally, two bellies). The anterior belly runs upwards and forwards from the upper surface of the hyoid bone to the lower border of the mandible below the chin (Fig. 4.6). The posterior belly is united to the anterior belly at the upper border of the hyoid bone by an intermediate tendon. It passes upwards and backwards to disappear under cover of the sternomastoid just behind the angle of the mandible.

These muscles demarcate anterior and posterior triangles of the neck (Fig. 4.6):

The posterior triangle

This is bounded behind by the anterior border of the trapezius, in front by the posterior border of the sternomastoid, and below by the upper border of the clavicle. It is subdivided by the inferior belly of the omohyoid into a larger occipital triangle above the omohyoid and a smaller subclavian triangle below the omohyoid. It contains (Fig. 4.5) the levator scapulae, scalenus medius and scalenus anterior muscles, which form its floor; the accessory nerve; branches of the cervical plexus and the roots of the brachial plexus; and below the omohyoid, the subclavian vessels and their branches.

The anterior triangle

This is bounded behind by the anterior border of the sternocleidomastoid, the middle line of the neck at the front, and the lower border of the mandible above. It is subdivided by the superior bellies of the omohyoid and digastric muscles into the following parts:

1. The *muscular*, or *thyroid*, *triangle* bounded by the lower part of the anterior edge of the sternocleidomastoid, the superior belly of the omohyoid and the middle line of the neck from the hyoid bone as far as the sternum (Fig. 4.6).

2. The *carotid triangle* bounded by the middle part of the anterior edge of the sternocleidomastoid, the superior belly of the omohyoid below and the posterior belly of the digastric above (Fig. 4.6).

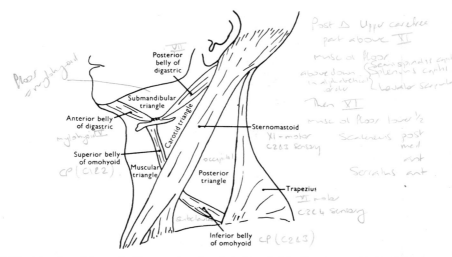

Fig. 4.6 The triangles of the neck. The anterior triangle is bounded by the sternomastoid, the lower border of the mandible and the mid-line of the neck anteriorly.

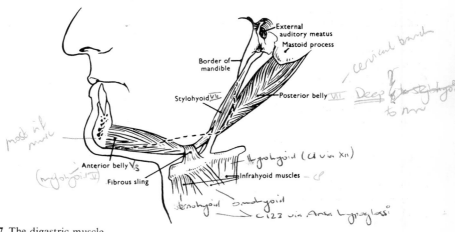

Fig. 4.7 The digastric muscle.

3. The *submandibular triangle* bounded by the anterior and posterior bellies of the digastric muscle and the lower border of the mandible (Fig. 4.6).

In the *muscular triangles* below the omohyoids and between the diverging sternocleidomastoid muscles of each side are the sternohyoid and underlying sternothyroid and thyrohyoid muscles. Between the edges of the sternohyoid muscles of the two sides there are the following structures, situated in the middle line of the neck from above downwards (Figs 4.8, 5.39):

1. The hyoid bone to which the omohyoid, sternohyoid and thyrohyoid muscles are attached.
2. The middle part of the thyrohyoid membrane.
3. The thyroid cartilage (Adam's apple).
4. The middle part of the cricothyroid membrane.
5. The cricoid cartilage.

6. The cartilaginous rings forming the skeleton of the upper part of the trachea. These are covered above (second, third and fourth rings) by the middle part (isthmus) of the thyroid endocrine gland and below by a plexus of veins.

The *carotid triangle* is so called because it contains the division of the common carotid artery into its two terminal branches, the external and internal carotid arteries. The external carotid artery gives off within the carotid triangle the ascending pharyngeal, superior thyroid, lingual, facial, and occipital arteries (Figs 4.14, 4.28).

The carotid triangle also contains the anterior facial vein as it runs across the submandibular and carotid triangles to form the common facial vein, by its union with the anterior division

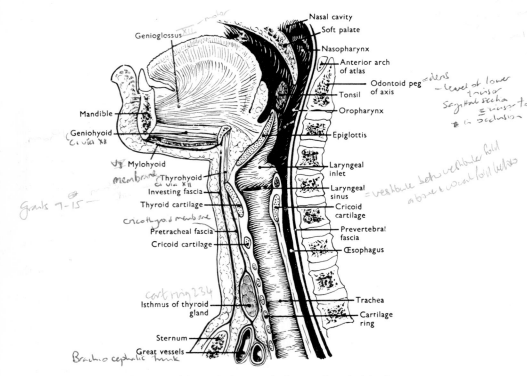

Fig. 4.8 Mid-line sagittal section of the neck showing the layers of cervical fascia.

of the posterior facial vein. The common facial vein passes deep to the sternomastoid to open into the internal jugular vein.

In the *submandibular triangle* the superficial part of the submandibular salivary gland lies beneath the lower border of the mandible between and overlapping the bellies of the digastric muscle. It is crossed by the anterior facial vein passing downwards towards the internal jugular vein at the anterior edge of the sternomastoid. The floor of the submandibular triangle is formed by the mylohoid muscle.

These various triangles of the neck serve as useful landmarks to the surgeon during operations in this region. For example, the submandibular triangle is important for access to the lower border and angle of the mandible when the jaw has been fractured or there is need for corrective surgery to adjust the length of the mandible. The deeper muscular compartments are important for surgical approaches to control infections spreading from the teeth.

The sternocleidomastoid muscle

This large muscle (Fig. 4.5) arises below from the upper surface of the medial third of the clavicle (clavicular head) and from the anterior surface of the upper end of the sternum (sternal head). There is usually a slight space between the two heads at their origin. The clavicular head is flattened and muscular, the sternal head rounded and tendinous. The muscle passes obliquely upwards on the side of the neck and is inserted into the anterior border and the outer surface of the mastoid process, and to a ridge extending backwards from the mastoid process (superior nuchal line) (Fig. 5.3). It is supplied by the accessory nerve (spinal part) and by branches of the cervical plexus (second and third cervical nerves).

The two muscles acting together help to rotate the head. The anterior parts of the muscles flex the head on the vertebral column, the posterior parts of the muscles extend the head, lifting the face upwards. One muscle acting alone (while the other muscle relaxes) turns the head upwards and to the opposite side. If the head is fixed, the muscles elevate the sternum and first rib, an action which is important during forced respiration. The outlines of the sterno-cleidomastoid muscles are easily seen beneath the skin and their form contributes to the grace-fulness of the neck and its movements.

STRUCTURES RELATED TO THE STERNOCLEIDOMASTOID MUSCLE

On its surface lie the external jugular vein, the anterior cutaneous and great auricular nerves (branches of the cervical plexus). Beneath it are found parts of the posterior belly of the digastric muscle, the stylohyoid muscle and the intermediate part of the omohyoid muscle. The accessory nerve runs deep to it and usually in part through it. The muscle covers the great vessels and nerves of the neck; the roots of the cervical plexus; the upper part of the phrenic nerve; the first part of the subclavian artery; and the subclavian vein.

Many important structures are seen when the sternocleidomastoid muscle is cut and re-flected, including:

1. The internal jugular vein and its common facial, lingual, and superior thyroid branches (Fig. 4.29).

2. The hypoglossal nerve crosses the internal and external carotid arteries just below the digastric muscle on its way to the tongue (Fig. 4.33).

3. The upper part of the common carotid and the lower parts of the internal and external carotid arteries (Fig. 4.14).

4. The superior thyroid artery arises from the anterior surface of the external carotid artery and passes downwards beneath the omohyoid to the upper border of the thyroid glands (Figs. 4.5, 4.28).

5. The lingual artery is given off the external carotid artery above the superior thyroid artery. It forms a characteristic upwardly directed loop which is crossed by the hypoglossal nerve. Both the nerve and the artery pass forwards deep to the digastric and mylohyoid muscles Fig. 6.7).

6. The facial artery arises from the external carotid artery above the lingual artery and soon passes deep to the digastric muscle.

7. The occipital artery leaves the posterior aspect of the external carotid artery opposite the origin of the facial artery and is crossed by the hypoglossal nerve. The artery runs along the lower border of the digastric muscle deep to the sternocleidomastoid towards the back of the skull where it supplies the scalp and the muscles at the back of the neck.

8. The vagus nerve lies between the internal jugular vein and the internal carotid and com-mon carotid arteries in the dense connective tissue sheath which surrounds them (carotid sheath). It descends through the neck and thorax to the abdominal cavity.

9. The posterior belly of the digastric muscle and the stylohyoid muscle lie beneath the upper part of the sternocleidomastoid and cross superficial to the external and internal carotid arteries, the internal jugular vein, the hypoglossal and vagus nerves.

10. The deep cervical lymph nodes form a chain along the course of the internal jugular vein (Fig. 6.14).

11. The anterior rami of the second to the fifth cervical nerves form a series of loops which make up the cervical plexus (Fig. 4.11). The superficial branches of the plexus, namely, the lesser occipital, great auricular, anterior cutaneous and supraclavicular nerves are closely related to the posterior border of the sternocleidomastoid. These are sensory nerves supplying the skin of the neck and scalp (Fig. 4.5).

12. The accessory nerve appears from beneath the posterior belly of the digastric muscle and crosses the middle of the posterior triangle to reach the trapezius (Fig. 4.5).

13. The lower part of the common carotid artery, the lower part of the internal jugular vein, and the vagus nerve lying between the artery and vein (Fig. 4.38).

14. The scalenus anterior muscle. This is crossed by the omohyoid muscle and by the phrenic nerve which run downwards on its surface. The phrenic nerve (C3.4.5) is the nerve supply to the diaphragm which it reaches after descending through the thoracic cavity (page 85). At the posterior or lateral margin of the scalenus anterior muscle are found the large roots of the brachial plexus and the third part of the subclavian artery.

The infrahyoid muscles (Figs 4.9, 4.38, 5.39)
This group of small strap-like muscles lies in front of the power part of the neck, between, and partly overlapped by the diverging sternocleidomastoid muscles.

The *sternohyoid* muscle on each side arises from the posterior surface of the medial end of the clavicle, from the posterior surface of the upper end of the sternum and from the capsule

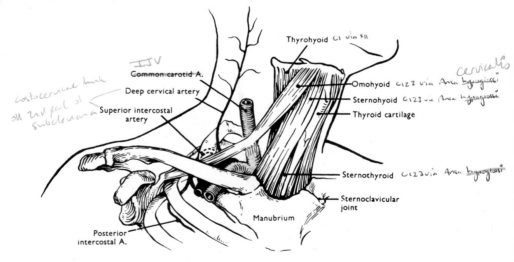

Fig. 4.9 The infrahyoid muscles and branches of the costocervical trunk of the subclavian artery.

of the sternoclavicular joint. It is inserted into the lower border of the body of the hyoid bone (Fig. 5.20).

The *sternothyroid* lies deep to the sternohyoid. It arises from the posterior surface of the upper end of the sternum and is inserted into the oblique line on the lateral surface of the thyroid cartilage. It lies on the surface of the lateral lobe of the thyroid gland of the same side.

The *thyrohyoid* lies in the same plane as the sternothyroid. It arises from the oblique line on the lateral surface of the thyroid cartilage and, passing somewhat laterally as it ascends, is attached to the lower border of the greater horn (cornu) of the hyoid bone. It covers the lateral part of the thyrohyoid membrane.

The *omohyoid* muscle is attached above to the lower border of the body of the hyoid. In passing backwards and downwards it crosses the upper part of the thyrohyoid and sterno-thyroid muscles, the great vessels of the neck (common carotid artery and internal jugular vein), the scalenus anterior and scalenus medius muscles, to be inserted under cover of the trapezius to the upper border of the scapula near the suprascapular notch. Its intermediate tendon, which connects the superior and inferior bellies of the muscle, lies under cover of the sterno-cleidomastoid muscle. Traction by the fascial layers below the omohyoid causes the two bellies to lie at an obtuse angle to one another.

All the infrahyoid muscles are supplied by the first, second and third cervical nerves through the cervical plexus and the ansa hypoglossi nerve loop (Fig. 4.11). They control the movements of the thyroid cartilage and hyoid bone from below and play a part in swallowing and mastica-tion. With the geniohyoid, which is also supplied by the cervical nerves, they represent the rectus musculature in the neck, and are the counterpart of the abdominal rectus group (page 87).

The cervical fascia

The spaces between the muscles of the neck are filled in by connective tissue. In life this tissue contains a great amount of tissue fluid and permits the easy movement of adjacent muscles in relation to one another. In certain places it becomes more dense, forming capsules for glands and sheath-like coverings for the muscles. Although this tissue is everywhere continuous it is usual to subdivide it for descriptive purposes into various layers. These include (Figs 4.8, 4.10):

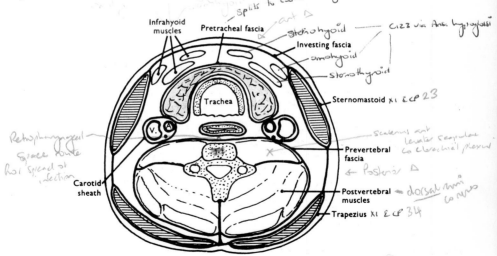

Fig. 4.10 Transverse section of the neck at the level of the isthmus of the thyroid gland showing the layers of cervical fascia.

The superficial or *general investing layer*. This layer lies beneath the subcutaneous tissue of the neck. Behind, it is attached to the spinous processes of the cervical vertebrae and the overlying ligamentum nuchae (neck ligament). It forms a capsule for the trapezius muscle; a roof (beneath the subcutaneous tissue) for the posterior triangle; a capsule for the sternomas-toid muscle; and then covers over the structures of the anterior triangle to the midline of the neck where it is firmly attached to the body of the hyoid bone. Below, it is attached to the clavicle and upper border of the sternum; above, to the base of the skull, the mastoid processes, the zygomatic arch, and the posterior and lower borders of the mandible. Below the hyoid bone, the fascia covers the infrahyoid group of muscles and splits just above the sternum to enclose the suprasternal space, in which is found the anterior jugular veins. As the fascia passes upwards above the hyoid bone it splits to form compartments for the submandibular and parotid glands. The deeper layer of the compartments is attached to the mylohyoid line of the mandible, the styloid process and the tympanic part of the temporal bone.

A *prevertebral layer* of fascia passes in front of the cervical part of the vertebral column, the adjacent prevertebral and lateral vertebral muscles (including the scaleni muscles and levator scapulae), the roots of the cervical and brachial nerve plexuses and the subclavian artery. It lies behind the pharynx and laterally it unites on each side with the investing layer beneath the trapezius.

The trunks of the brachial plexus (Fig. 4.35) and the subclavian artery (Fig. 4.28) are covered by a prolongation of the prevertebral fascia, the axillary sheath, which extends into the armpit. Between the posterior aspect of the pharynx and oesophagus and the anterior surface of the prevertebral fascia there is a potential space filled with loose connective tissue, the retropharyngeal space, which is a pathway for the spread of infections.

The *pretracheal fascia* lies in front of the trachea and encapsulates the infrahyoid muscles and the thyroid gland. Above, it is attached to the hyoid bone and the cricoid cartilage; below, it passes into the thoracic cavity where it becomes continuous with the fibrous layer of the pericardium. At the sides it merges with the superficial layer of the cervical fascia.

The *carotid sheath*. The common and internal carotid arteries and the internal jugular vein with the vagus nerve are surrounded by a layer of connecting tissue extending from the cranial base above into the thoracic cavity below.

The carotid sheath adjoins the deep layer of the capsule formed by the superficial layer of fascia around the sternomastoid muscle, is continuous medially with the pretracheal layer, and posteriorly with the prevertebral layer. The carotid sheath is attached to the base of the skull at the lower border of the tympanic plate and continues downwards along the vessels it contains to the aortic arch.

The *pharyngeal fascia* covers the wall of the pharynx. Between the pharynx, the subjacent prevertebral fascia and the more superficial investing layer laterally, there is on each side of the neck a region of less dense tissue forming a potential space for the spread of infective material and pus. This is the important parapharyngeal fascial space extending from the base of the skull to the thoracic cavity.

The connective tissue layers which make up the components of the cervical fascia form the series of natural planes which aid the surgeon, as well as directing the spread of haemorrhage and infection. The general investing layer, or superficial lamina, is important in surgical approaches to the lower border of the mandible because the parotid and submandibular glands, facial arteries and veins, the facial nerve and cervical cutaneous nerves lie within it or just deep to the fascia. Hence the surgeon takes care to cut through the investing layer close to its line of fusion with the mandible, so that these important structures are not damaged. For further information on the role of fascial planes and spaces in the spread of infections, see page 445.

The cervical plexus (Fig. 4.11)

This is the uppermost of a series of nerve plexuses formed by groups of spinal nerves and is derived from anterior primary rami of the first four cervical nerves. These form a series of loops with one another, lying deep to the internal jugular vein and on the transverse of the atlas (first loop) and the scalenus medius muscle. The superficial branches of the plexus are all sensory. They are the lesser occipital, the great auricular, the transverse cervical (or anterior cutaneous of the neck), and the supraclavicular nerves. Between them they supply the skin of the back of the auricle, part of the scalp, over the angle of the mandible (great auricular), the side and front of the neck (transverse cervical) and over the shoulder (supra-scapular). The skin of the back of the scalp and neck is supplied by the posterior rami of the upper cervical nerves which take no part in the formation of the cervical plexus.

The deep branches of the plexus are mainly motor nerves to various groups of muscles. These nerves include:

1. The *phrenic nerve* which, arising from the lower part of the plexus (third to fifth cervical nerves, mainly the fourth), turns over the lateral border of the scalenus anterior on to its anterior surface and there descends through the root of the neck deep to the subclavian vein into the thoracic cavity where it supplies the chief muscle of respiration, the diaphragm. Developmentally the diaphragm is a neck muscle which, in spite of its migration, maintains its original nerve supply from the cervical plexus. The phrenic nerve also contains some sensory fibres arising from the pleura covering the central part of the diaphragm and from the mediastinal structures of the thorax. In the thorax the phrenic nerves lie in the mediastinum between the pleura and the pericardium anterior to the roots of the lungs (Fig. 3.10). After crossing the subclavian artery each nerve passes medially in front of the internal thoracic or mammary artery (Fig. 4.34). The left phrenic nerve crosses lateral to the vagus nerve and the aortic arch whilst the right phrenic nerve accompanies the brachiocephalic vein and superior vena cava. On reaching the diaphragm most of the phrenic nerve fibres pierce the muscle and supply it from its abdominal aspect.

2. Branches to the scaleni muscles, the levator scapulae; the prevertebral muscles, the trapezius and sternocleidomastoid (Fig. 4.35).

3. The nerve supply to the rectus muscle group of the neck; the infrahyoid muscles and the geniohyoid. These muscles are supplied from a nerve loop, derived from C1.2.3, lying in front of the great cervical vessels, the ansa hypoglossi. The upper limb of the loop is from the first cervical nerve, which sends a branch to run within the perineurial sheath of the hypoglossal nerve, leaving it as an apparent branch of the hypoglossal, the descendens hypoglossi. From this part of the loop branches are given off to the geniohyoid and thyrohyoid muscles. The lower limb of the loop is formed by the second and third cervical nerves, forming the descendens cervicalis (Fig. 4.11).

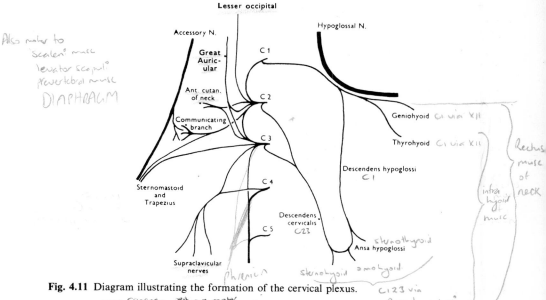

Fig. 4.11 Diagram illustrating the formation of the cervical plexus.

SUPERFICIAL STRUCTURES OF THE FACE

The general form of the face is determined by the underlying bony skeleton clothed by the overlying muscles and subcutaneous tissue. In certain regions such as the margins of the orbital cavities, the upper part of the nose, the cheekbones, over the zygomatic arches, and along the lower border of the mandible, there is a minimal amount of subcutaneous tissue. In these areas, therefore, the bone lies close to the covering skin and can be easily palpated. In other regions the soft tissues are not directly supported by bone, for example, the eyelids, lips and cheeks. These parts are particularly mobile and are responsible for the wide range of expression shown by the facial features of each individual.

Apart from the masseter muscle covering the ramus of the mandible, and the upper part of the temporal muscle covering the side walls of the cranial part of the skull, all the muscles of the face and scalp belong to the same muscle group. These are the subcutaneous *muscles of expression*. Their nerve supply is from the seventh (facial) cranial nerve. The skin of the face, except for a small area over the angle of the mandible and lower part of the auricle, is supplied by sensory branches of the fifth cranial nerve. This nerve has three parts or divisions (hence its alternative name of trigeminal nerve) each supplying a localized region of the skin of the face (see page 146).

The muscles of expression can be divided into groups as follows:

1. The muscles of the lips and cheeks.
2. The muscles of the nose.
3. The eyelid muscles.
4. The auricular muscles.
5. The muscles of the scalp.

There is also the sheet-like platysma, which extends downwards in the neck close beneath the skin (page 130).

Not all the muscles supplied by the facial nerve are muscles of facial expression. Specifically, the stylohyoid and the posterior belly of the digastric muscles do not have this rôle to play in the range of expressions. Other muscles which are not supplied by the facial nerve do take part in facial expression such as: elevation of the upper eyelid by the oculomotor nerve; rolling of the eyes by the oculomotor, abducent and trochlear nerves; and protruding of the tongue by the hypoglossal nerve. The association of the muscles of expression with only the facial nerve is therefore a simplification of the circumstances and the term is somewhat misleading, although widely stated.

THE MUSCLES OF THE LIPS AND CHEEKS (Figs 4.12, 4.14)

These can be divided into two groups, a deeper layer made up of the buccinator muscles in the cheeks and the sphincter-like orbicularis oris in the lips. The more superficial layer is made up of a series of small muscles arising from the maxillae, the zygomatic bones, the fascia covering the masseter muscle and from the mandible. These muscles converge upon the lips to enter them in a radial manner. Their names and origins are:

Levator labii superioris alaeque nasi
Levator labii superioris   from the facial surface of the maxilla.
Levator anguli oris
Zygomaticus major and minor: from the zygomatic bone.
Risorius: Arising from the fascia covering the masseter, as an upward extension of the platysma sheet.
Depressor labii inferioris
Depressor anguli oris   from the body of the mandible.
Mentalis

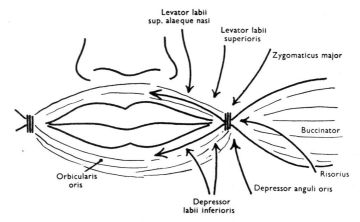

Fig. 4.12 Diagram representing the arrangement of the muscles of expression at the angle of the mouth.

Some of the fibres of the platysma muscle may enter the lower lip between the risorius and depressor labii inferioris.

The buccinator muscle (Figs 4.14, 4.25)
This important muscle of the cheek consists of a flat, thin, but strong set of fibres in contact with the mucous membrane of the oral vestibule. It is attached above and below to the outer surface of the maxilla and mandible in the region of the molar teeth close to the line of reflection of the gum mucoperiosteum from the alveolar processes; and by its posterior margin to the pterygomandibular raphe. Its fibres have three attachments:

1. To the deep surface of the mucous membrane of the cheek.
2. To a vertical musculotendinous septum (modiolus) situated about 10 mm lateral to the angle of the mouth.
3. Into the lips where they mingle with the fibres of the orbicularis oris.

At the angle of the mouth some of the lower fibres cross some of the upper fibres to enter the upper lip and *vice versa.*

Nerve supply
Seventh (facial) cranial nerve.

Relations
The deep surface is closely related to the mucous membrane of the cheek. Its superficial surface is related to the duct of the parotid gland which passes through it obliquely; the buccal pad of fat (Fig. 4.14); the facial artery and vein; the deep facial vein; the buccal artery and sensory nerve; the buccal branches of the facial nerve; the ~~sensory~~ buccal (or facial) lymph nodes; the zygomaticus major and minor and the risorius muscle.

The orbicularis oris muscle (Figs 4.12, 4.14, 4.19)
This complex muscle is made up of fibres derived from a number of sources:

1. The deeper fibres are derived partly from the buccinator muscle and partly from the vertical musculotendinous septum close to the angle of the mouth. The fibres decussate with one another or are inserted into the deep surface of the skin and mucous membrane of the lips. The fibre bundles cross the middle line and decussate with fibres from the opposite side. They insert into the skin of the opposite side and, with fibres of the levator superioris muscle,

contribute to the formation of the philtral ridge. In both the upper and lower lips small slips of muscle arise from the incisive fossae and contribute to the lip musculature (the superior and inferior incisive muscles).

2. The radial muscles enter the lips from above, below, and at the sides. The fibres of the levator anguli oris and the depressor anguli oris cross one another at the corners of the mouth; the former to enter the lower lip and the latter the upper lip. This decussation lies superficial to the decussation of the buccinator muscle. The radial muscles are inserted chiefly into the skin, but do not reach as close to the red margin of the lips as do the deeper fibres.

3. Small fibres pass from the mucous membrane to the skin between the main fibre bundles.

Nerve supply
The lower buccal and mandibular branches of the facial nerve.

Functional anatomy
The buccinator and orbicularis oris form a continuous sheet of muscle fibres passing around the face from one pterygomandibular raphe to the other. Each raphe is a fibrous band which extends from the tip of the hamulus of the medial pterygoid plate to the inner surface of the mandible, just above the posterior end of the mylohyoid line. The muscles converge on a common, node-like zone of attachment just lateral to and slightly above the corner of the mouth. This is the modiolus which can be fixed in various positions by the radial muscles of the lips, acting as 'guy ropes' or 'stays'.

When the modiolus is firmly fixed contraction of the buccinator applies force to the outer surfaces of the check teeth and contraction of the orbicularis oris results in pressure on the anterior teeth. If the contraction is maintained it will result in inversion of the red margins of the lips and a sealing of the opening between the lips, the oral commissure. This is dependent on the support provided by the underlying teeth and if these are absent or malpositioned, then the orbicularis muscle can pull backwards and inwards on contraction and there is inadequate sealing of the lips. The detailed arrangement of the radial lip muscles and the orbicularis oris, which are antagonist muscles, coupled with the delicate control of the position of the modiolus enables fine control of lip movements and lip pressures to take place during speech. The patterns of contraction during speech will be disrupted if the muscle sheets are inadequately supported by the teeth. This may be evident in severe malocclusion cases and can be corrected by orthodontic treatment. Lacerations that interrupt the angle of the mouth are quite debilitating and surgical repair techniques seek to minimize scarring and recreate the interlacing of muscle fibres at the modiolus.

The orbicularis oris and buccinator muscles are of particular importance in dentistry because they form the most important elements of a muscle sheet which lies on the outer aspect of the dental arches. In addition to their functions as muscles of expression and the part which they play in speech and the mastication of food, they are counter balanced by the forces of the tongue musculature on the inner aspect of the dental arches. If this balance is upset, say by mouth breathing, then some alteration in the position of teeth is likely to result. The protrusion of upper incisors which can result from abnormal positioning of the tongue is another example.

The attachments of the buccinator limit the depth of the vestibule of the mouth, specifically the sulcus between the gums and cheek. When the cheeks are stretched by pulling them outwards some ridges of mucous membrane will appear between the inside of the cheek and the gum margin. These ridges are produced by stretching of muscle fibre bundles of buccinator. It is important to register the position of these muscle slips during the taking of impressions for the construction of dentures, for the margins of the denture must be relieved at these sites. Otherwise, when it is worn, the denture will be moved each time the buccinator fibres contract. Similarly, with newly fitted dentures, the patient often experiences sore spots where small bundles of buccinator fibres lift the mucous membrane into tight contact with the edge of the denture.

Tender areas are produced by the constant rubbing of the mucous membrane against the denture base during chewing movements. This is easily corrected by trimming away a little of the denture base in the affected area and repolishing the denture margin. The contour of the anterior part of a denture is important because the orbicularis oris is a relatively powerful muscle which will dislodge the denture if the artificial teeth or the denture base are set too far forwards. The aesthetics of dentures depend not only on the correct colour and shape of the artificial teeth but also on the correct relationships between them and the facial muscles.

The radial muscles of the lips (Figs 4.12, 4.14)

1. *Levator labii superioris alaeque nasi*. Arises from the upper part of the frontal process of the maxilla at the inner margin of the orbital rim. As it descends it divides into a medial slip inserted into the lower margin of the alar cartilage of the nose and a lateral slip which enters the upper lip. ßVII

2. *Levator labii superioris*. Arises from the lower margin of the orbital rim above the infraorbital foramen on both sides of the zygomaticomaxillary suture. It descends to enter the upper lip. ßVII

3. *Zygomaticus major* and *minor*. Arise from the lateral surface of the zygomatic bones and pass downwards and somewhat medially to enter the upper lip. ßVII *levator labii superioris* below

4. *Levator anguli oris*. Arises from the canine fossa below the infraorbital foramen and passes downwards towards the angle of the mouth where some of its fibres enter the lower lip. ßVII

5. *Risorius*. Arises from the fascia covering the parotid gland and masseter muscles. Its fibres pass forward towards the angle of the mouth. ßVII

6. *Depressor labii inferioris*. Arises from the external oblique line of the mandible between the mental foramen and the region of the symphysis. Its fibres pass upwards to be inserted into the lower lip. mVII

7. *Depressor anguli oris*. Arises below the depressor labii inferioris. Its fibres pass towards the angle of the mouth and many of its fibres pass into the upper lip. mVII

Nerve supply

The radial muscles of the lips are innervated by the facial nerve. The depressor anguli oris and the depressor labii inferioris are supplied by the mandibular branch, and the remaining muscles by the buccal branches.

Functions

The radial muscles of the lips control expressive movements of the lip region, such as those concerned customarily with smiling, sorrow and scorn. These movements tend to overshadow the important dilator action of the muscles on the oral orifice, which takes place in association with the sphincteric action of the orbicularis muscle.

Gentle contraction of the fibres of the orbicularis oris draws the lips together; when more tightly contracted the muscle gathers the lips as in pouting or whistling. Parting of the lips is produced by the simultaneous contraction of the superior (maxillary) and inferior (mandibular) radiating muscles and the relaxation of the orbicularis oris sphincter. At rest the upper lip is supported by the upper incisor teeth. If these teeth are markedly protruded the upper lip may not be able to make contact with the lower lip in the resting position.

The mentalis muscle

This small conical shaped muscle is situated at the side of the frenulum of the lower lip and arises, one on each side, from the incisive fossa of the mandible which is close to the middle

just as inf oblique & sup oblique musc of eye lie below lev palps superioris & sup rectus & inf rectus line at the level of the apices of the incisor teeth. The fibres descend deep to the reflection of the oral mucous membrane from the alveolar bone and are inserted into the dermis of the skin covering the chin.

Nerve supply
Mandibular branch of the facial nerve.

Functions
The mentalis raises and protrudes the lower lip, dimples the skin over the chin and decreases the depth of the vestibule of the mouth behind the lower lip.

The muscles of the eyelids

Each eyelid contains a dense fibrous structure, the *tarsal plate*, attached to the upper and lower orbital margins respectively by a looser palpebral fascia and more firmly on the medial and lateral sides by band-like palpebral ligaments, of which the medial is the stronger. The medial palpebral ligament overlies the lacrimal sac, which occupies the lacrimal fossa on the medial bony wall of the orbital cavity (Figs 4.13, 4.44).

The chief muscle of the eyelids is the *orbicularis oculi* (Figs 4.13, 4.14), consisting of an orbital part overlying the bones surrounding the orbital cavities, and a palpebral part situated within eyelids. Both parts are attached at the medial corners of the orbital cavities to the medial palpebral ligament and adjacent bones (frontal and maxilla). From this position the orbital fibres radiate around the margins of the orbital cavities and gain partial attachment to the

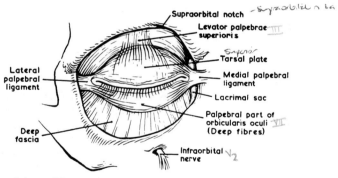

Fig. 4.13 Structure of the eyelids.

overlying skin; the palpebral fibres radiate within the eyelids and are there attached to the skin, the underlying tarsal plates and the lateral palpebral ligament. A bundle of fibres (the pars lacrimalis) passes deep to the lacrimal crest of the lacrimal bone.

The palpebral fibres of the orbicularis oculi are responsible for gentle closing of the eyelids as in sleep, while the orbital fibres draw the eyebrows forcibly downwards as the skin surrounding the orbital margin is drawn towards the medial angles of the orbits producing a series of skin wrinkles around the orbital margins. The pars lacrimalis regulates the size of the lacrimal sac from which tears pass via the nasolacrimal duct to the nasal cavity and is part of the mechanism concerned with the cleansing of the surface of the eyeball. The eyelids are opened by the relaxation of the palpebral fibres and the action of one of the orbital group of muscles, the levator palpebrae superioris (Fig. 4.13). The muscle arises at the back of the orbital cavity, passes over the upper surface of the eyeball and is inserted into the upper tarsal plate and the skin of the upper eyelid near its margin. When it contracts it elevates the upper eyelid (Fig. 4.45). It derives its nerve supply from the third cranial nerve.

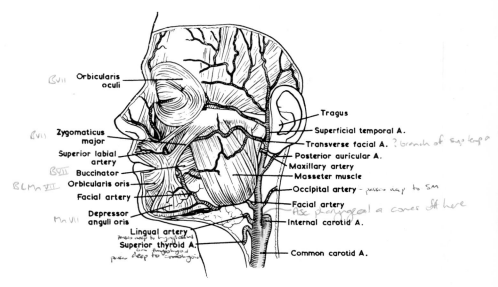

Fig. 4.14 Arteries and muscles of the facial region. *see 4-5*

The muscles of the nose

These are two small muscles on each side, the dilator and compressor naris. They arise from the facial surface of the maxilla at the side and lower border of the bony nasal aperture and are inserted into the skin and cartilages of the alae of the nose. Their actions are indicated by their names.

The muscles of the outer ear

These are a group of small intrinsic muscles within the external ear which are attached to parts of its cartilaginous skeleton, and three extrinsic muscles, the auriculares anterior, superior and posterior. The anterior and superior muscles arise from the epicranial aponeurosis and from the temporal fascia overlying the temporal muscle, while the auricularis posterior arises from the outer surface of the mastoid process above the insertion of the sterno-mastoid. The auricular muscles are inserted into the cartilage of the ear. In man they are rudimentary and usually non-functional; they are supplied by branches of the seventh cranial (facial) nerve. Some people can move their ears voluntarily.

The scalp and its muscles

The scalp consists from without inwards of *skin, superficial fascia, epicranial aponeurosis* (a thin dense sheet of connective tissue); a layer of loose *connective tissue*; and the periosteum covering the external surfaces of the cranial bones (Fig. 8.22). The skin is tightly bound to the underlying epicranial aponeurosis through the dense superficial fascia containing the hair follicles and in which run the larger blood vessels and nerves of the scalp. The epicranial aponeurosis is continuous in front with the frontalis muscle and behind with the occipitalis muscle. At the side of the skull it unites with the underlying temporal fascia. The fibres of the frontalis muscle gain a limited attachment to bone at the root of the nose (procerus muscle), and laterally to the anterior parts of the superior temporal lines. The intermediate fibres are attached to the skin of the eyebrows and forehead. The occipitalis muscle is attached at the back of the skull to the superior nuchal lines of the occipital bone. The frontalis and occipitalis acting together may move the scalp but the degree to which they can act varies considerably in different individuals. In the process of 'scalping' as performed in the past by American Indians the epicranial aponeurosis with the overlying skin was removed exposing the periosteum of the cranial bones.

Both the frontalis and occipitalis muscles are supplied by the seventh cranial (facial) nerve. The sensory nerve supply to the scalp is derived from branches of the ophthalmic and mandibular divisions of the trigeminal nerve in front of the ear and from the posterior rami of the upper cervical nerves behind the ear. Blood vessels enter the scalp from the upper margins of the orbital cavities (branches of the ophthalmic artery, itself a branch of the internal carotid artery); from the temporal region (the superficial temporal artery, a terminal branch of the external carotid artery); and from the back of the neck, the posterior auricular and occipital arteries, branches of the external carotid artery, given off in the carotid triangle) (pages 134, 162). Thus the scalp has a very good supply from its periphery and scalp wounds heal readily.

THE NERVE SUPPLY OF THE FACE

Branches of two cranial nerves supply most of the structures making up the superficial tissues of the face. These are the fifth (trigeminal) cranial nerve, the sensory nerve to the skin of the greater part of the face, and also to the mucous membrane lining the nasal and oral cavities; and the seventh (facial) cranial nerve, the motor nerve to the facial group of muscles. The fifth nerve also supplies the muscles of mastication and the seventh nerve supplies secretomotor fibres to the submandibular and sublingual salivary glands to the lacrimal glands (page 320).

The facial distribution of the fifth cranial nerve

While still within the cranial cavity the fifth nerve divides into three major divisions, the ophthalmic, maxillary and mandibular (Fig. 5.32). Each division gives branches to the deeper structures of the facial region and then the majority of its terminal branches appear on the face after passing through various foramina in the facial bones (Fig. 4.15).

The branches of the mandibular division supplying the skin of the lower and back parts of the face (Fig. 4.15) are the mental, the buccal and the auriculotemporal nerves. The mental nerve appears through the mental foramen in the body of the mandible and supplies the skin

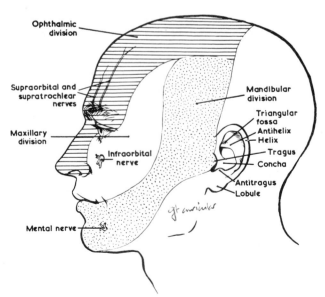

Fig. 4.15 The distribution of the trigeminal nerve on the face and the parts of the external ear.

of the lower lip and the skin covering the front of the mandible. The buccal nerve appears on the face at the anterior border of the masseter muscle, and lies on the outer surface of the buccinator muscle. It is the sensory nerve for the skin of the cheeks and deeper structures including the mucous membrane lining the cheek inside the mouth cavity (the oral vestibule). The auriculotemporal nerve passes upwards immediately in front of the ear and supplies the skin of its anterior part, particularly the tragus, and the scalp overlying the temporal muscle at the side of the skull.

The branches of the maxillary division appearing on the face supply the skin of its middle part. They are the infra-orbital nerve, the zygomaticofacial and the zygomaticotemporal nerves. The infra-orbital nerve appears through the infraorbital foramen in the maxilla at the side of the nose a short distance below the orbital margin and above the canine tooth. It gives off nasal branches to the skin of the side of the nose, a palpebral branch to the skin and conjunctiva of the lower eyelid and a descending labial branch to the skin and mucous membrane of the upper lip. The zygomaticofacial and zygomaticotemporal nerves emerge on the face through foramina of the same names on the facial and temporal surfaces of the zygomatic bone and supply the overlying skin.

The terminal branches of the ophthalmic division of the fifth nerve supply the skin of the upper part of the face and the forehead. They are the lacrimal, the supraorbital, the supratrochlear, the infratrochlear and the external nasal nerves. All these, except the latter, appear at the margin of the orbital cavity. The supraorbital is the longest branch and passes upwards to supply the skin of the forehead and the front of the scalp as far as the vertex of the head. The supratrochlear, infratrochlear and the lacrimal nerves supply the skin of the upper eyelid and at the lateral (lacrimal) and medial (infratrochlear) corners of the eye. The external nasal nerve appears between the nasal bone and alar cartilages of the nose from within the nasal cavity and supplies the skin over the lower part of the side of the nose.

It is important to note that the skin over the angle of the lower jaw is not supplied by the trigeminal nerve but receives its sensory innervation from the cervical plexus through the second cervical nerve. Loss of skin sensation in this region and even over the parotid gland in front of the ear has important clinical significance. It implies some disturbance of nerve pathways associated with the first cervical segment of the spinal cord rather than the fifth cranial nerve. Apparently during human evolution the increased size of the brain caused the skin of the face to be drawn upwards over the enlarging cranium and at the same time pulled the skin of the neck up over the angle of the jaw.

The areas of sensory distribution served by the trigeminal nerve are quite sharply defined and disturbances of sensation can give the dentist or the neurologist distinct clues about which part of the trigeminal system is involved in a disease process. Painful conditions involving the nerve, the most profound of which is *trigeminal neuralgia*, are frequently confined to the areas served by one of the divisions of the nerve. The onset of pain can be triggered by the slightest stimulation of an area of skin whose sensory supply comes from the same division of the nerve. These *trigger zones* frequently occur around the mouth and a draught of air or rubbing of the skin during shaving or while applying makeup are sufficient to stimulate the onset of pain. Viral infections of the trigeminal ganglion, which contains the cell bodies of the neurons whose peripheral processes make up the branches of the trigeminal nerve, can lead to an extremely painful condition called *herpes zoster* in which vesicular eruptions occur on the face, limited to the area of skin supplied by the affected division.

The loss of pain sensibility which follows either damage to the nerve branches or interruption of the pathway for nerve impulses by the administration of local anaesthetics for dental operations is well-known. For example, block anaesthesia of the mandibular nerve produces tingling and numbness of the skin of the lower lip and angle of the mouth which is a useful sign to the dentist that the anaesthesia of the lower teeth is effective.

The facial distribution of the seventh cranial nerve

The seventh cranial nerve after a complicated course through the temporal bone of the cranium appears at the base of the skull through the stylomastoid foramen, and gives off two main branches: an ascending branch which passes behind the auricle to supply the occipitalis and posterior auricular muscles, and a deep descending branch, which supplies the stylohyoid and posterior bellies of the digastric muscles in the neck. The nerve then enters the parotid gland where it further divides into a number of branches. These appear at the upper and anterior edges of the gland and pass forward over the face to supply the facial muscles; they are from above downwards (Fig. 4.16):

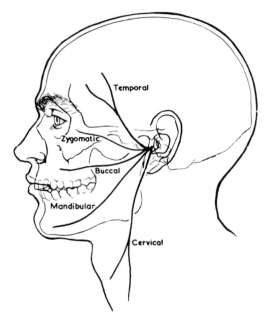

Fig. 4.16 Superficial branches of the facial nerve.

Temporal branches pass above the zygomatic arch to supply the anterior and superior auricular muscles, the upper fibres of the orbicularis oculi and the frontalis muscle.

Zygomatic branches pass horizontally forwards across the zygomatic arch to supply the lower upper part of the orbicularis oculi.

Buccal branches run above and below the level of the parotid duct to supply the buccinator, the muscles of the upper lip, the risorius, the muscles of the side of the nose and lower eyelid. The zygomatic and buccal branches freely communicate.

The *mandibular branch* runs on the masseter muscle superficial to the facial artery and outer surface of the mandible, to supply the muscles of the lower lip and the mentalis muscle. There may be two branches.

The *cervical branch* passes into the neck behind the angle of the mandible and supplies the platysma muscle. It may communicate with the mandibular branch.

Because the facial nerve passes through the parotid gland and some of its branches continue to be closely related to the superficial surface of the gland or to its duct, the nerve is susceptible to injury during surgical procedures involving the parotid gland. Surgeons take great care to identify the facial nerve as they dissect parotid tissue to avoid nerve damage with subsequent paralysis of facial muscles. Damage to the facial nerve, either by trauma, exposure, chill or viral infection, presents clinically as *Bell's* palsy which affects some or all of the muscles of expression on one side of the face. In a typical case the patient cannot close the eye or blink and the lips cannot be pursed as in whistling. The face is expressionless on the affected side and food tends to accumulate between the cheek and the molar teeth due to the paralysis of the buccinator muscle. Sometimes tears escape from the eye and run down over the upper part of the cheek due to the dropping of the lower eyelid, although there may be diminished lacrimal secretion depending on the site of injury to the nerve. Also depending on where the nerve has been injured, the mouth may be dry due to a reduced salivary secretion and the sensation of taste may be lost from the anterior part of the tongue. Bell's Palsy is a dramatic illustration of the function of the muscles of expression, the effects of muscle paralysis and loss of muscle tone.

THE BLOOD SUPPLY OF THE FACE

The arteries
The principal artery to superficial structures of the face is the *facial artery*, a branch of the external carotid, given off in the carotid triangle (Fig. 4.27). After giving branches to some of the deeper structures in the neck and the submandibular salivary gland, it passes over the lower border of the mandible, in front of the masseter muscle, to enter the face (Fig. 4.14). In the face the facial artery passes upwards and forwards towards the side of the nose, being located about half an inch from the corner of the mouth. It runs in a tortuous fashion among the facial muscles. In its course it lies in turn on the mandible, the buccinator muscle, on the maxilla and the levator anguli oris and deep to the platysma, risorius, zygomatic and levator labii superioris muscles. Its terminal part, sometimes known as the angular artery, may run within the substance of the levator labii superioris alaeque nasi muscle and ends by anastomosing with branches of the ophthalmic artery at the medial corner of the eye.

In its course the facial artery gives off a series of small branches which course backwards into the cheek, and larger branches which run forwards to enter the lips. These latter, the inferior and superior labial arteries, lie deep in the substance of each lip between the orbicularis oris muscle and the mucous membrane (Fig. 4.19). Commonly a lateral nasal branch is given off to supply the side of the nose. For further details see pages 164, 254.

The *superficial temporal artery* is one of the terminal branches of the external carotid (Figs 4.14, 4.41). It is given off within the substance of the parotid gland and it leaves the gland as its upper border along with the auriculotemporal nerve. After leaving the gland the artery crosses the zygomatic arch in front of the auricle where its pulsations can be felt, and runs upwards on the temporal fascia to supply the scalp through temporal and frontal branches. While still within the parotid gland the superficial temporal artery gives off the transverse facial artery. This runs forwards towards the upper part of the cheek between the zygomatic arch above and the parotid duct below and is accompanied by the zygomatic branches of the facial nerve. (The parotid duct lies a finger's breadth below the zygomatic arch.)

A number of small arteries appear on the face accompanying the terminal branches of the trigeminal nerve. These are:

1. The lacrimal, supra-orbital, supratrochlear and infratrochlear arteries, around the margin of the orbital cavity. These are terminal branches of the ophthalmic artery.

2. The infra-orbital artery, a branch of the maxillary artery.

3. The buccal and mental arteries, branches of the maxillary and inferior dental arteries respectively.

These anastomose with branches of the facial and transverse facial arteries in the tissues covering the facial skeleton.

The branches of the facial artery are loose and tortuous which is an advantage in their relationship to the constantly mobile facial muscles. Interruption of the facial artery by injury or during surgery is not usually serious because a collateral circulation develops rapidly from the transverse facial artery and the buccal, infraorbital and sphenopalatine branches of the maxillary artery.

The veins

The *anterior facial vein* commences at the medial corner of the eye by the union of the supra-orbital and supratrochlear veins and a branch which communicates with the ophthalmic veins within the orbital cavity (important in the spread of infections, see page 264). It runs down-wards in the angle between the nose and the cheek and then downwards and backwards through the facial muscles lying behind the facial artery (Fig. 4.17). In its course it receives veins from the side of the nose and lips and communicates with the mental and infraorbital

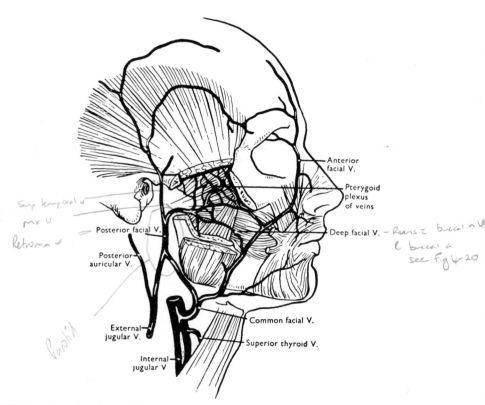

Fig. 4.17 The veins of the face and neck.

veins on the face and with the pterygoid venous plexus deep to the ramus of the mandible. Communication is through the deep facial vein, which runs with the buccal branch of the mandibular nerve. The pterygoid plexus in turn communicates with the cavernous sinus by emissary veins. After crossing the lower border of the mandible, the anterior facial vein is joined in the neck by the anterior branch of the posterior facial vein to form the common facial vein. This joins the internal jugular vein at the anterior border of the sternocleidomastoid muscle below the angle of the mandible.

The anterior facial vein drains a dangerous area of the face which includes the upper lip, the nasal septum and the skin around the sides of the nose. This region is prone to bacterial infection which may spread by the connections of the anterior facial vein to the cavernous sinus in the skull. It is of surgical significance that the vein has no valves; it passes between muscles which may displace blood clots in the vein by muscle contraction; and there is no deep fascia to act as a barrier to the rapid spread of inflammation. The cavernous sinus and its relations are described on page 264.

The *superficial temporal vein* runs with the superficial temporal artery and drains the scalp, the auricle, external auditory canal and the posterior part of the face (Fig. 4.17). It enters the upper aspect of the parotid gland where it is joined by the maxillary vein, draining the pterygoid plexus, to form the *posterior facial vein* (Fig. 4.17). This vein runs through the substance of the gland and then divides into two parts:

1. An anterior branch which runs downwards and forwards to pierce the fascial capsule of the parotid gland. Here it joins the anterior facial vein to form the common facial vein which enters the internal jugular vein.

2. A posterior branch which runs backwards, pierces the fascia surrounding the parotid gland and is joined on the surface of the sternocleidomastoid muscle by the posterior auricular vein draining the scalp behind the ear.

The union of these veins forms the *external jugular vein* which descends at the side of the neck to join the subclavian vein above the clavicle. For further details on facial veins see page 261.

Lymphatic drainage of the face   p306

The lymphatic drainage of the facial region is associated with blood vessels. The lymph nodes and systems of lymphatic channels are divided into superficial and deep groups. Lymphatic vessels that originate close to the mid-line often cross to the opposite side, particularly in the lips, tongue and palate. Lymphatic drainage patterns are of considerable surgical importance because infections and malignant tumours tend to spread along the lymphatic system.

Lymph vessels draining the lower lip pass under the chin to end in the *submental lymph nodes* two or three in number on each side of the mid-line. The vessels from the corner of the mouth, upper lip, side of the nose, cheek, and lower eyelid drain into the *submandibular nodes* about three to eight in number, in front of and below the angle of the mandible (Fig. 4.18).

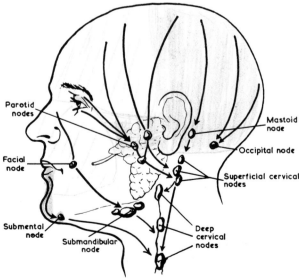

Fig. 4.18 Lymphatic drainage of the head.

The lymphatics of the lips communicate across the mid-line and cancers at or near the mid-line in either lip may spread to the lymph nodes on both sides of the neck. Radical surgery in these cases requires removal of the deep cervical nodes on both sides of the neck and nodes which are not enlarged may still be infected with cancer cells. Glands which are readily felt (palpable) may be enlarged only as a result of simple bacterial infections around the lip of cheek region. The upper lip has a more extensive lymphatic drainage than the lower and this is an important factor in the prognosis of lip cancer.

There are usually a few small nodes in the substance of the cheeks, the *buccal* or *facial nodes*. The vessels from the upper eyelid, the frontal and anterior temporal regions of the scalp, the skin of the front of the auricle and over the cheek bones drain into the *superficial parotid nodes* lying on the surface of the parotid gland within its fascial sheath. The skin of the back of the scalp and auricle is drained by vessels which enter the *mastoid* and *occipital nodes*. The submental, submandibular, parotid, mastoid and occipital nodes are known collectively as the pericervical group of nodes. They all communicate with the deep cervical nodes, which lie along the internal jugular vein close to the anterior border of the sternocleidomastoid muscle.

The structure of the lips and cheeks
The lips are covered by skin on their outer aspect and by mucous membrane on the oral surface. Between the skin and mucous membrane is the red margin of the lips (lipstick area) which consists of a modified skin without hair follicles or sebaceous glands and where the epithelial

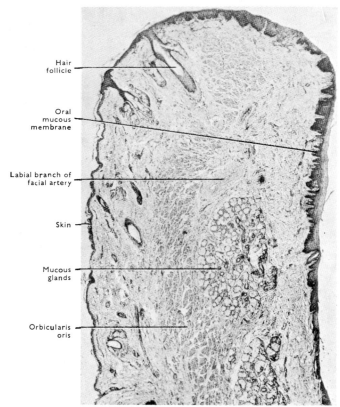

Hair follicle

Oral mucous membrane

Labial branch of facial artery

Skin

Mucous glands

Orbicularis oris

Fig. 4.19 Sagittal section through the lower lip × 18.

covering is thin and translucent. Beneath the skin lies subcutaneous tissue, in which hair follicles, sebaceous glands, deeper sensory nerve endings and blood vessels can be demonstrated. Deeper still lie the concentric muscle fibres of the orbicularis oris and the radiating muscles entering the lips from the adjacent parts of the face (Fig. 4.19). Some of the muscle fibres gain attachment to the subcutaneous connective tissue. Beneath the muscle layers is the submucosa containing the labial mucous glands whose ducts pierce the mucous membrane to enter the vestibule of the mouth. Running in the submucosa, close to the margin of the lips, are the labial arteries from which small branches pass to supply the mucous membrane, the muscles and the more superficial subcutaneous tissue. In the middle line, the mucous membrane lining the inner surface of each lip is connected to the adjacent gum by a vertical sagittally placed septum-like structure, the frenulum labii, of which the upper is the larger. When insufficiently reduced in size during development it passes on to the palate between the central incisor teeth and may be associated with an undesirable spacing of these teeth.

The cheeks form the side walls of the posterior part of the vestibule of the mouth cavity. Beneath the skin lie the facial muscles which radiate towards the corner of the mouth and the upper lip (zygomaticus minor, zygomaticus major and risorius). Deep to the superficial muscles lies the buccinator. Towards the back of the cheek the superficial muscles are separated from the buccinator by part of the *buccal pad of fat*. This is an encapsulated mass of fatty tissue, especially well developed in infants, extending backwards beneath the ramus of the mandible on the surface of the buccinator muscle.

The parotid duct passes around the front part of the buccal pad to pierce the buccinator muscle. Between the superficial muscles and the buccinator there is a network of nerve fibres forming a plexus, made up of the buccal branches of the seventh (facial) nerve and the buccal and infraorbital branches of the fifth (trigeminal) nerve. In this region the facial artery and vein pass between the superficial and deep muscles. There may be a few isolated *buccal or facial lymph nodes* in the cheeks. The inner surface of the buccinator is closely related to the mucous membrane of the vestibule of the mouth and many of its fibres are attached to the submucosa.

THE PAROTID GLAND AND ITS RELATIONS

The parotid is the largest of the salivary glands. It lies beneath the skin on the side of the face and is surrounded by fascia which is thickest laterally and inferiorly. Extensions of this fascia divide the gland into a series of lobules. The outline of the gland is irregular and adapted to adjacent structures—the mandible, temporal bone, the zygomatic arch and the cartilage of the external ear.

The parotid gland occupies a wedge-shaped space which is bounded in front by the ramus of the mandible and its related muscles, the masseter and medial pterygoid; behind by the sternomastoid and the posterior belly of the digastric (Figs 4.5, 4.18). Medially it is limited by the styloid process and its attached muscles, separating the deep part of the gland from the still more deeply placed internal jugular vein and the side wall of the upper part of the pharynx. The gland is surrounded by a sheath of cervical fascia, a modified thickened part of the general investing layer, and the structures which pass through the gland pierce this fascia.

The duct of the gland (Stensen's duct) appears at its anterior border and passes forwards across the masseter muscle about one inch below the upper border of the zygomatic arch. It can also be described as lying a finger's breadth below the zygomatic arch. At the anterior border of the muscle it turns sharply inwards, passing around the buccal pad of fat, to pierce the buccinator muscle and open into the mouth cavity. At birth the orifice of the duct is opposite the first deciduous molar and the duct opening gradually changes its position relative to the teeth during childhood due to growth of the dental arches. In the adult it is opposite to the upper second permanent molar tooth.

The parotid gland with its duct is an outgrowth from the mouth cavity. Growing backwards across the ramus of the mandible and the covering masseter muscle it invades the retromandibular space which has already been occupied by the facial nerve, external carotid artery and posterior facial vein. When this is remembered, it will be understood how these structures appear to lie within the substance of the gland in the adult.

The parotid gland can be described as having three surfaces, superficial, anteromedial (related to the mandible), posteromedial (related to the sternomastoid) and an upper and a lower pole. The upper pole of the gland is closely related to the lower surface of the external auditory meatus (glenoid lobe) and the lower pole lies between the angle of the mandible and the sternocleidomastoid muscle, superficial to the posterior belly of the digastric.

From the anteromedial surface of the gland a process of varying size passes forwards deep to the ramus of the mandible, between it and the medial pterygoid muscle. This is the pterygoid lobe. A ridge of gland substance between the anteromedial and posteromedial surfaces may pass deeply between the medial pterygoid muscle, the styloid process and its attached muscles to come into contact with the side wall of the pharynx.

The facial nerve enters the posteromedial surface of the gland close to the upper pole. Within the gland it lies more superficially than the other structures. It usually divides into two parts, an upper branch from which are given off the temporal, zygomatic and upper buccal branches and a lower branch from which arise the lower buccal, mandibular and cervical branches. These terminal branches of the nerve appear on the face along the anterior margin of the parotid gland as this overlaps the masseter muscle. Their distribution and clinical significance has already been described on page 148.

The external carotid artery pierces the fascial capsule of the gland at the upper border of the posterior belly of the digastric muscle and enters the deep surface of the lower pole. Just before entering the gland substance it gives off the posterior auricular artery which, leaving the capsule of the gland, runs upwards and backwards along the upper border of the digastric muscle. One of the terminal branches of this artery enters the stylomastoid foramen from which the facial nerve emerges. Within the parotid gland the external carotid artery gives off small twigs to supply the gland substance and divides into its terminal branches, the superficial temporal artery and the maxillary artery. The superficial temporal artery emerges from the upper pole of the gland with the auriculotemporal nerve lying immediately behind it, and crosses the root of the zygomatic arch in front of the external auditory meatus. The maxillary artery leaves the anteromedial surface of the gland, usually running for a short distance within the pterygoid lobe. It passes into the pterygoid region deep to the ramus of the mandible (Fig. 4.20).

The posterior facial (retromandibular) vein is formed within the capsule of the gland and sometimes within the gland substance, by the union of the superficial temporal and maxillary veins. The posterior facial vein then passes through the parotid gland between the superficial branches of the facial nerve and the more deeply placed external carotid artery. It divides near the lower border of the gland to appear at the deep surface of the lower pole as two branches, the anterior and posterior divisions of the vein. The former pierces the gland capsule to enter the submandibular space, where it is joined by the anterior facial vein to form the common facial vein. The latter leaves the posterior part of the gland capsule and, on the surface of the sternocleidomastoid, is joined by the posterior auricular vein to form the external jugular vein (see page 262).

Lymphatic drainage from the region of the parotid gland is into the deep cervical lymph nodes, which commonly become involved with extensions of parotid tumours. The pre-auricular nodes lie superficial to the capsule of the parotid gland and are seldom involved in pathologic conditions of the gland. Subfascial nodes are found in the gland and inferior to it.

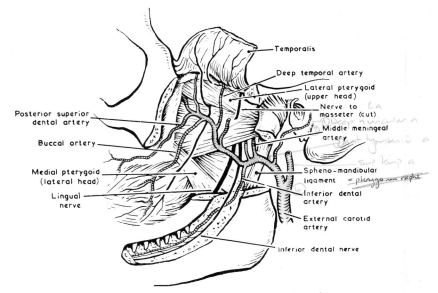

Fig. 4.20 Distribution of the maxillary artery - *below ramus & med pterygoid m*

The auriculotemporal nerve has a very short but important course through the upper pole of the parotid gland (glenoid lobe) for, while passing through, it gives off secretomotor fibres to the gland. These are derived from the ninth cranial nerve (glossopharyngeal) and pass via the tympanic plexus, lesser superficial petrosal nerve, and otic ganglion, to join the auriculotemporal nerve. The sympathetic nerve supply reaches the gland through the perivascular nerve plexus around the external carotid artery as postganglionic nerve fibres whose cell bodies are located in the superior cervical ganglion of the sympathetic nerve chain (see page 174).

STRUCTURE OF THE SALIVARY GLANDS

Parotid gland

The parotid gland is classified as a compound tubulo-alveolar type and is surrounded by a dense fibrous tissue capsule from which septa subdivide the gland substance into lobes and lobules. Nerves, arteries, veins and lymphatics run in the areolar tissue septa in close relation to the secreting units. The gland is a pure serous secreting gland and produces a special enzyme—ptyalin. The secretory elements are the terminal cul-de-sacs or *acini* of the duct system and are composed of pyramidal secretory cells, resting on a basal lamina. Within each acinus individual cells are in different secretory phases as they form secretion (zymogen) granules which are discharged into the lumen of the acinus.

On the parotid gland the ducts leading from each acinus are known as *intercalated ducts*. These are often absent in mucous secreting glands. They are lined by low cuboidal epithelial cells and in turn open into *striated ducts*, characterized by a tall columnar epithelium resting on a distinct basal lamina. The vertically striated appearance of the cells, which gives the duct its name, is due to the presence of radially arranged mitochondria and extensive infoldings of the basal cell membrane. These features are only seen using the electron microscope. Functionally, the importance of the striated duct is its role in the transfer of water and electrolytes to and from the saliva produced by the acinar cells.

The striated ducts open in turn into the interlobular *excretory ducts*, usually lined by a pseudostratified columnar epithelium. The excretory ducts open into the final common duct (Stensen's duct), which is lined by stratified squamous epithelium where it becomes continuous with the oral epithelium, as it enters the mouth cavity.

Submandibular gland
As in the parotid gland, the glandular cells are subdivided by fibrous tissue septa which, however, are less dense and compact. The gland contains both mucous and serous elements. The mucous acini open into less prominent intercalated ducts and the usual sequence of striated and excretory ducts. The striated ducts are longer and more conspicuous than in the parotid gland. The majority of the acini are serous and the mucous variety are usually mixed, for they have serous crescents (or demilunes) at their periphery.

Sublingual gland
The capsule of this collection of glands is less well developed than in the parotid and submandibular gland. The gland is predominantly a mucous type gland with some mixed units and a few serous elements. Intercalated ducts and the striated ducts are short and less numerous than in other glands.

STRUCTURES RELATED TO THE MANDIBULAR RAMUS

The muscles of mastication
This group of muscles, which move the lower jaw at the mandibular joint, consists of the temporalis, the masseter, the lateral and medial pterygoid, and the digastric muscles.

The masseter muscles and the anterior (vertical) parts of the temporalis muscles close the jaw. The lateral pterygoid muscles acting together protrude the lower jaw and, assisted by the digastric muscles, open the jaw by rotating the mandible so that the condyles come forward on to the articular eminences and the angle of the mandible moves backwards. The posterior (horizontal) fibres of the temporalis muscle draw the jaw backwards. The medial pterygoid muscles assist the masseters in closing the jaws and also assist the lateral pterygoids in its protrusion. A more detailed analysis of the action of the muscles of mastication is given later on page 316.

The masseter, temporalis and pterygoid muscles are all supplied by branches of the mandibular division of the fifth cranial nerve. The digastric has a double nerve supply, its anterior belly being supplied by the fifth nerve, its posterior belly by the seventh nerve. The two parts of the muscle have a different developmental origin hence the double nerve supply. However in opening the jaw they act together from the origin of the posterior belly at the base of the skull to the insertion of the anterior belly to the chin region of the mandible, the intermediate tendon moving within the fibrous sheath which attaches it to the hyoid bone and acts as a pulley.

The maxillary artery
This is one of the two terminal branches of the external carotid artery given off within the substance of the parotid salivary gland (Figs 4.20, 6.24). It runs forward deep to the ramus of the mandible to reach the lower border of the lateral pterygoid muscle, and in this part of its course it lies on the sphenomandibular ligament and medial pterygoid muscle. This is described as the first part of the artery. On reaching the lateral pterygoid it may pass either superficial or deep to the muscle. If it passes deep to the muscle it usually appears for a short distance between the two heads to give off its buccal, anterior deep temporal and posterior superior dental branches. The part of the vessel related to the lateral pterygoid is the second part. After giving off the posterior superior dental artery, the maxillary artery enters the pterygopalatine fossa where it divides into its terminal branches (third part).

In its *first part* the maxillary artery gives off the deep auricular, anterior tympanic and middle meningeal branches from its upper surface and the inferior dental artery from its lower surface. The ascending branches pass deep to the lateral pterygoid muscle. The auricular branch enters the external auditory canal between its bony and cartilaginous parts, and supplies its lining mucous membrane; the tympanic branch enters the middle ear through the squamotympanic fissure. The middle meningeal artery enters the cranial cavity through the foramen spinosum to supply the dura mater covering the brain.

The inferior dental artery runs downwards and somewhat forwards on the lateral surface of the sphenomandibular ligament and the medial pterygoid muscle towards the mandibular foramen where it enters the mandibular canal, accompanied by the inferior dental nerve. Just before it enters the foramen it gives off its mylohyoid branch which, with the mylohyoid nerve, runs along the under surface of the mylohyoid diaphragm on the floor of the mouth. Within

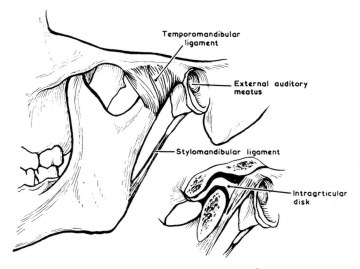

Fig. 4.21 Ligamentous structures of the temporomandibular joint.

the mandibular canal the inferior dental artery supplies the lower teeth and the bony tissue of the mandible. At the mental foramen branches pass on to the surface of the bone to supply the adjacent skin and mucous membrane and anastomose with branches of the facial artery.

The *second part* of the maxillary artery gives off a number of branches to supply the muscles of mastication, including the artery to the masseter and the two deep temporal arteries. From the second part of the maxillary artery there is also given off the buccal artery. This branch runs with the buccal nerve to reach the cheek on the surface of the buccinator muscle which it helps to supply.

The posterior superior dental artery leaves the parent trunk as the maxillary artery enters the pterygopalatine fossa. It descends on the posterior (temporal) surface of the maxilla to enter one or more foramina above the tuberosity. It is accompanied by the posterior superior dental nerve. Other branches of the posterior dental artery continue on the surface of the bone to supply the gum related to the upper molar teeth. The artery and nerve run forwards in neurovascular canals within the bone to supply the mucous membrane lining the back of the maxillary air sinus and the upper molar teeth. For further details about the maxillary artery see page 257.

The maxillary vein and pterygoid venous plexus

The maxillary vein has a less extensive course than the artery. It usually consists of a large vessel corresponding to the first part of the artery. The *pterygoid plexus* consists of a number of veins related to the pterygoid muscles. It communicates with the cavernous sinus within the cranial cavity through the foramen ovale and the foramen of Vesalius, if this is present. It also communicates with the ophthalmic veins through the pterygopalatine fossa and inferior ophthalmic fissure (Fig. 4.17). Into the pterygoid plexus drain veins corresponding to the branches of the second and third parts of the maxillary artery. The plexus is drained from behind the maxillary vein, which joins the superficial temporal vein within the capsule of the parotid gland to form the posterior facial vein. The plexus also drains forwards into the anterior facial vein along the deep facial vein which runs with the buccal artery and nerve on the surface of the buccinator muscle. Below, the pterygoid plexus communicated with the venous plexus on the wide wall of the pharynx (pharyngeal plexus). All these venous connections are of importance in the spread of infection (see page 264).

The mandibular joint

The mandibular joint is situated between the condyle of the mandible and the articular surface of the temporal bone. The *capsule* of the joint is attached below to the neck of the condyle. Above it is attached to the margins of the articular area of the temporal bone, extending to the anterior edge of the articular eminence in front and behind to the squamotympanic fissure. The capsule is strengthened on its outer side to form the lateral ligament, a dense fan-shaped band of fibres passing from the outer surface of the articular eminence (zygomatic tubercle) downwards and somewhat backwards to be attached to the condylar head below its lateral pole. This ligament prevents backward and forward dislocation of the condyle and, with the corresponding ligament of the opposite side, it prevents lateral dislocation of the jaw. The capsule of the joint is weak anteriorly (Fig. 4.21) so that dislocation frequently occurs in this direction.

The *articular disk* or meniscus is a dense sheet of fibrous tissue dividing the joint cavity into a large upper and a smaller lower compartment (Figs 4.22, 4.23, 4.41). In front the disk is attached to the anterior margin of the articular eminence. At the sides it blends with the capsule of the joint and both are firmly attached to the medial and lateral poles of the condyle. Behind, the disk divides into an upper layer which is attached in front of the squamotympanic fissure, and a lower layer which is attached to the back of the head of the condyle. The back

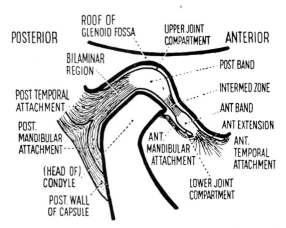

Fig. 4.22 Section of the articular disk, glenoid fossa and the head of the mandibular condyle. (By courtesy of the *British Dental Journal*.)

of the capsule is united with these temporal and mandibular layers of the disk. The disk is not of the uniform thickness, but shows an anterior and a posterior thickened region in which the connective tissue is particularly dense. Centrally, the disk is thinner and may sometimes be perforated, so that the two joint compartments communicate with one another. This seems to have a negligible effect on joint function. The back of the disk is of looser texture and contains elastic fibres, especially the upper part, blood vessels, sensory nerve endings and many large vascular spaces.

The lateral pterygoid muscle is inserted partly into the front of the disk (upper head) and partly to the fossa at the front surface of the neck of the condyle below the articular surface (Fig. 5.15). When the muscle contracts disk and condyle are pulled forward together.

The *condyle* is shaped like a gabled roof with a transverse ridge running lateromedially, and anterior and posterior surfaces sloping inferiorly from it. The lateral and medial poles of the condyle are usually marked by distinct bony tubercles for the attachment of the capsule and disk. The articular surface consists of a surface layer of dense fibrous tissue. In children and young adults this is a perichondrium lying upon the surface of the condylar growth cartilage (Fig. 4.23). In adults it is a periosteum lying upon bone which has replaced the cartilage. For further details see page 309.

The mandibular nerve

The sensory and motor roots of the mandibular division of the fifth cranial nerve leave the cranial cavity through the foramen ovale in the great wing of the sphenoid bone. Below the foramen the two roots unite to form the trunk of the mandibular nerve deep to the lateral pterygoid muscles on the surface of the tensor palati muscle and in front of the middle meningeal artery (Fig. 4.24). From the trunk are given off two small branches, the nerve to the medial pterygoid muscle and the nervus spinosus. The latter passes with the middle meningeal artery into the cranial cavity through the foramen spinosum and is a sensory nerve to a large part of the dura mater, the outermost of the meningeal coverings of the brain. Between the trunk of the mandibular nerve and the tensor palati muscle lies the small otic parasympathetic ganglion from which nerve fibres pass along the auriculotemporal nerve to the parotid gland (Fig. 4.24).

From the trunk of the mandibular nerve four large branches are given off. These are (Fig. 4.24):

1. The anterior division.
2. The auriculotemporal nerve.
3. The inferior dental nerve.
4. The lingual nerve.

⎫
⎬ Together making up the posterior division.
⎭

1. The *anterior division* of the nerve passes downwards and forwards to appear between the two heads of the lateral pterygoid muscle as the buccal nerve. While lying deep to the lateral pterygoid, to both heads of which it gives branches, it also sends two branches to the temporal muscle, which appear at the upper border of the lateral pterygoid, and a nerve to the masseter muscle. This branch emerges from between the heads of the lateral pterygoid to enter the deep surface of the masseter by passing below the zygomatic arch and above the mandibular notch. The buccal nerve, a sensory nerve, passes forwards deep to the base of the coronoid process, sometimes through the lower part of the temporal muscle, to appear on the face of the surface of the buccinator muscle. It supplies the skin and mucous membrane of the cheek as far forward as the angle of the mouth (Fig. 4.20).

2. The *auriculotemporal nerve* arises by two roots which embrace the middle meningeal artery just below the foramen spinosum and deep to the lateral pterygoid muscle (Fig. 4.24). It crosses lateral to the sphenomandibular ligament immediately below the spine of the sphenoid and lies close to the inner side of the capsule of the mandibular joint. It then winds around the back of the neck of the condyle below the attachment of the joint capsule and

Temporal bone
Upper joint cavity
Elastic part of disk
Intra-articular disk
Lower joint cavity
Articular cartilage
Vascular part of disk
Condylar head

Fig. 4.23 Sagittal section of the mandibular condyle and temporomandibular joint in a child.

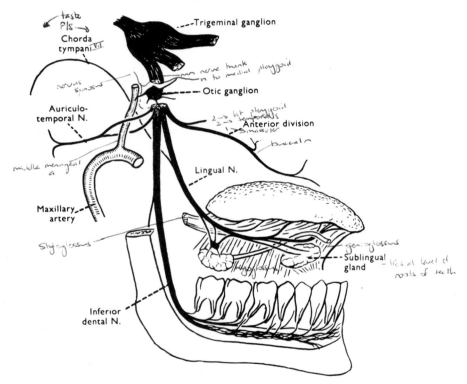

taste
P/s →
Trigeminal ganglion
Chorda tympani VII
nerve trunk, pterygoid
n to medial pterygoid
nervus spinosus
Otic ganglion
Auriculo-temporal N.
2 → lat pterygoid
2 → temporalis
3 masseter
Anterior division
buccal n
middle meningeal a
Lingual N.
Maxillary artery
Styloglossus
genioglossus
Sublingual gland — lies at level of roots of teeth
hyoglossus
Inferior dental N.

Fig. 4.24 The mandibular division of the trigeminal nerve.

passes upwards to enter the small glenoid lobe of the parotid gland. It leaves the gland by crossing the root of the zygomatic arch in front of the external auditory meatus (Fig. 6.32). Branches of the auriculotemporal nerve (sensory fibres) supply the front of the auricle, the mucous membrane of the external auditory canal, part of the outer surface of the tympanic membrane (ear-drum), the mandibular joint structures and the skin of the temporal region.

3. The *inferior dental nerve* at first lies deep to the lateral pterygoid muscle, appearing at its lower border in front of the inferior dental artery on the outer surface of the medial ptery-

goid muscle (Fig. 4.20). At the mandibular foramen, on the deep surface of the mandibular ramus, it gives off its mylohyoid branch and enters the mandibular canal (Fig. 4.25).

The mylohyoid nerve is the only muscular branch of the posterior division of the mandibular nerve. It supplies the mylohyoid muscle and the anterior belly of the digastric. Soon after entering the mandibular canal the nerve gives off a branch to supply the teeth posterior to the canine. At the level of the mental foramen it divides into an incisive branch or plexus and the mental nerve which passes through the mental foramen and divides into several filaments supplying the skin and mucous membrane of the lower lip. The incisor teeth, and in some cases the canine, are supplied from the incisive plexus. In other cases the canines are supplied by the posterior dental branch.

The site of entry of the inferior dental nerve into the mandibular foramen is important for it is the landmark for block anaesthesia of the nerve and its identification and retraction

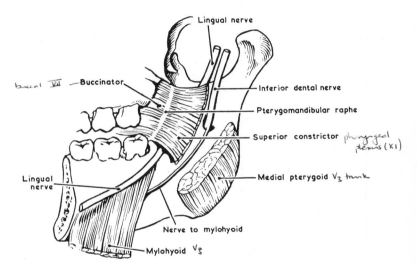

Fig. 4.25 The retromolar region of the mandible (after Last).

during surgical procedures involving the ramus of the mandible. The location of the mandibular foramen is variable but usually it is approximately 1 cm above the occlusal plane. With severe loss of the alveolar ridge that is typical of many edentulous mandibles the mental nerve and artery lie close to the crest of the bony ridge where they are likely to be traumatized by pressure from dentures. Surgical procedures are used to transpose the neurovascular bundle to a more inferior position in the mandible where it is then free from compression.

4. The *lingual nerve* lies somewhat in front of the inferior dental nerve (Fig. 4.24). While under cover of the lateral pterygoid muscle it is joined by the chorda tympani nerve and communicates with the inferior dental nerve. The chorda tympani is a branch of the seventh cranial (facial) nerve within the petrous part of the temporal bone (Fig. 4.26). It appears through the squamotypanic fissure. The chorda tympani carries taste fibres from the lingual nerve to the facial nerve and secretomotor (parasympathetic) fibres from the facial to the lingual nerve. The latter are destined for the submandibular and sublingual salivary glands (page 281). The lingual nerve lies below the lateral pterygoid on the surface of the medial pterygoid muscle deep to the mandibular ramus. It then enters the floor of the mouth between the styloglossus medially and the body of the mandible laterally. It passes below the last molar tooth and the lower border of the superior constrictor muscle, where this is attached to the mandible at the inner edge of the retromolar triangle (Fig. 4.25). The lingual nerve is the sensory nerve

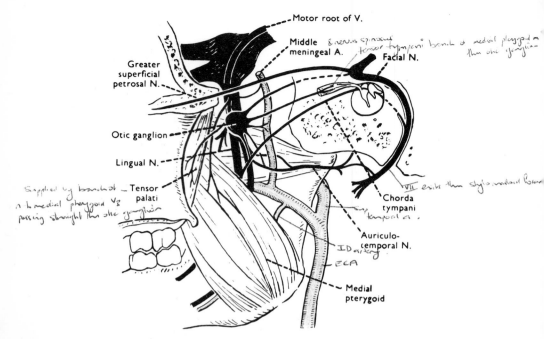

Fig. 4.26 The otic ganglion and its connections seen from the medial aspect (after Gray).

for the anterior two-thirds or oral part of the tongue, the floor of the mouth and the gum on the inner (lingual) side of the lower teeth.

5. The *otic ganglion* (Fig. 4.26) receives fibres from the ninth cranial nerve via the tympanic plexus situated on the inner wall of the middle ear, and the lesser superficial petrosal nerve. The latter leaves the cranial cavity by the foramen ovale or through the suture between the sphenoid and temporal bones. From the ganglion secretomotor fibres reach the parotid gland via the auriculotemporal nerve. Two small twigs from the nerve to the internal pterygoid muscle pass through the ganglion without interruption to supply the tensor tympani muscle of the middle ear and the tensor palati muscle of the soft palate. These are small muscles which developmentally are muscles of mastication. They develop in relation to Meckel's cartilage, but during foetal life they take on other functions. For further details on the mandibular nerve see page 248.

DEEP STRUCTURES OF THE NECK

The external carotid artery

The external carotid artery can be divided into three parts:

1. That below the level of the digastric muscle in the carotid triangle where it gives off its ascending pharyngeal, superior thyroid, lingual, facial, and an occipital branch which passes posteriorly.

2. A part deep to the digastric and stylohyoid muscles. In this part it is separated from the internal carotid artery by the stylopharyngeus muscle, the styloglossus muscle, the glosso-pharyngeal nerve and the pharyngeal branch of the vagus.

3. Above the digastric and stylohyoid muscles it pierces the fascial capsule of the parotid gland, gives off the posterior auricular artery and enters the gland substance where it divides

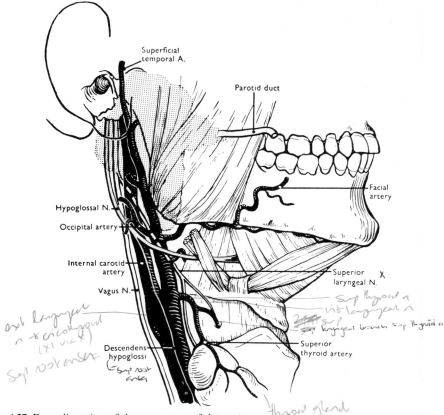

Fig. 4.27 Deep dissection of the upper part of the neck.

into its terminal branches, the maxillary artery and the superficial temporal artery (Fig. 4.27). This part provides small branches which supply the gland.

1. The *superior thyroid artery* runs downwards in the carotid triangle, lying close to the upper edge of the inferior constrictor and behind the superior cornu of the hyoid bone. It lies on the lateral part of the thyrohyoid membrane at the lateral border of the thyrohyoid muscle. It then passes deep to the omohyoid, sternohyoid and sternothyroid muscles and enters the upper pole of the thyroid gland (Figs 4.10, 4.27). Deep to the artery and its superior laryngeal branch runs the superior laryngeal branch of the vagus nerve. The superior thyroid artery, as it passes behind the hyoid bone, gives off a small branch (infrahyoid artery) which runs forwards along the lower border of the bone. A larger laryngeal branch (Fig. 4.27) enters the larynx with the internal laryngeal nerve by piercing the lateral part of the thyrohyoid membrane. Just before passing deep to the omohyoid muscle, the superior thyroid artery gives a branch to the sternocleidomastoid muscle. This vessel runs downwards and backwards along the upper border of the omohyoid muscle and crosses superficial to the carotid sheath to reach the sternocleidomastoid. When it reaches the thyroid gland, the superior artery pierces the fascial sheath surrounding the gland and divides into three main branches which run on the surface of the capsule of the gland, within the fascial sheath, on its anterior and posterior surfaces and along the anterior margin of the gland. The branch to the posterior surface of the gland anastomoses with an ascending branch of the inferior thyroid artery (page 270).

2. The *lingual artery* will be described on page 254.
3. The *facial artery* can be divided into four parts:
   a. In the carotid triangle,
   b. Deep to the digastric and stylohyoid muscles on the wall of the pharynx,
   c. In the submandibular triangle,
   d. On the face (page 149).

In the carotid triangle it lies on the middle constrictor of the pharynx. Deep to the digastric and stylohyoid muscles it lies on the stylopharyngeus and the superior constrictor muscles which separate it from the lower pole of the tonsil. Here it gives off tonsillar branches which pierce the superior constrictor muscle. An ascending palatine branch passes upwards on the surface of the superior constrictor and over its upper border, to enter the soft palate with the tensor and levator-palati muscles. The facial artery then turns forwards over the upper border of the stylohyoid muscle to enter the groove in the submandibular gland where it gives off its submental branch and branches to the gland, before crossing the lower border of the mandible to enter the face (page 149).

4 and 5. The *occipital artery* (page 135) and *posterior auricular artery* (page 154) require little further description. They are both distributed to the muscles at the back of the neck and supply the tissues of the post-auricular region of the scalp. The occipital artery is in contact with the occipital bone in a groove medial to that occupied by the attachment of the posterior belly of the digastric muscle (Fig. 4.3).

6. The *ascending pharyngeal artery* arises from the medial (deep) surface of the external carotid artery as its first and smallest branch (Fig. 4.28). It ascends on the middle and superior constrictor muscles of the pharynx and deep to the stylopharyngeus muscle. It gives off branches to the muscles and side wall of the upper pharynx and to the soft palate. It terminates as a small meningeal branch which enters the skull through the foramen lacerum and as small branches to the auditory (Eustachian) tube.

7 and 8. The *maxillary artery* and the *superficial temporal artery* have been described already (pages 149, 156).

The internal carotid artery

The internal carotid artery, like the external carotid, is a terminal branch of the common carotid artery. It commences in the carotid triangle at the level of the upper border of the thyroid cartilage and terminates within the cranial cavity by dividing into the anterior and middle cerebral arteries (Fig. 4.28).

At its origin the wall of the artery is somewhat swollen to form this carotid sinus. Sensory nerves, belonging to the autonomic nervous system and concerned with the regulation of blood pressure, pass from the sinus to join the glossopharyngeal and vagus nerves and later the brain stem (pages 250, 395). The internal carotid artery can be divided into three parts, a cervical part, a part running through the carotid canal, and the terminal intracranial part, which is closely associated with the cavernous venous sinus at the side of the body of the sphenoid bone.

The long cervical part of the artery runs in the carotid sheath together with the upper part of the internal jugular vein and the vagus and hypoglossal nerves. It lies in the angle between the wall of the pharynx (the middle and superior constrictors) medially, and the prevertebral muscles and fascia behind. It is separated from the more superficial external carotid artery by the styloglossus and stylopharyngeus muscles; the lower end of the styloid process, if this is well developed; the pharyngeal branch of the vagus and the glossopharyngeal nerve (Fig. 4.26). Deep to the internal carotid artery the superior laryngeal branch of the vagus passes downwards and forwards. The artery gives off no large branches in the neck. The internal carotid artery enters the cranial cavity by passing along a 'S'-shaped carotid canal in the temporal bone (Fig. 4.41). Within the canal it is closely related to the anterior wall of the middle

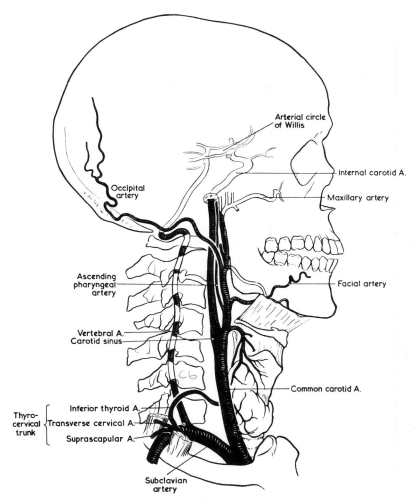

Fig. 4.28 Arteries of the head and neck drawn from a radiograph. The arteries had been injected with a radio-opaque substance. Note the formation of the circle of Willis.

ear and gives off a number of small branches which help to supply its lining mucous membrane. The intracranial part of the artery runs through the cavernous sinus (Fig. 5.9) and terminates by dividing into the anterior and middle cerebral arteries. From this part of the artery the large ophthalmic artery and small branches are given off to supply the closely related pituitary gland.

The internal jugular vein

This great vein commences at the jugular bulb which is attached to the margins of the jugular foramen. Into the bulb drain the sigmoid and inferior petrosal sinuses; between them these sinuses drain most of the other cranial sinuses. The internal jugular vein runs through the neck in the carotid sheath to terminate behind the sternoclavicular joint where it joins with the subclavian vein to form the brachiocephalic (innominate) vein. As it runs through the neck its chief tributaries are (Fig. 4.29):

1. The *pharyngeal vein*, draining the pharyngeal venous plexus, situated on the side wall of the pharynx.

2. The *lingual vein*, draining the tongue and the structures in the floor of the mouth (page 263).

3. The *common facial vein*, draining many of the superficial and deep structures of the face (page 150).

4. The *superior and middle thyroid veins*, draining the thyroid gland, the laryngeal region and the upper pretracheal structures.

In its upper part the internal jugular vein is accompanied by the internal carotid artery and in its lower part by the common carotid artery. The vagus nerve passes with the vein through the jugular foramen and descends with it through the neck in the carotid sheath. At the commencement of the vein the glossopharyngeal (ninth), accessory (eleventh), and hypoglossal (twelfth) cranial nerves lie between the vein and the internal carotid artery. The accessory nerve (spinal part) soon leaves the great vessels by inclining backwards either super-

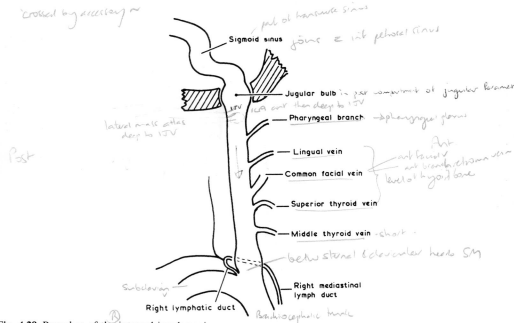

Fig. 4.29 Branches of the internal jugular vein.

ficial or deep to the vein to enter the substance of the upper part of the sternomastoid muscle. The glossopharyngeal nerve passes forwards between the internal and external carotid arteries and in close relationship with the stylopharyngeus muscle, which it supplies. It enters the pharynx to supply its mucous membrane and that covering the posterior one-third of the dorsum of the tongue. The hypoglossal nerve runs downwards between the internal jugular vein and internal carotid artery until it reaches the carotid triangle. It turns forwards across the internal and external carotid arteries to enter the floor of the mouth deep to the hypoglossus and mylohyoid muscles to supply the muscles of the tongue.

The internal jugular vein lies anterolateral to the rectus capitis lateralis muscle, the arch of the atlas, the levator scapulae muscle, the scalenus medius and·scalenus anterior muscles, the roots of the cervical plexus and the phrenic nerve. On the left side the terminal part of the thoracic duct lies behind its lower part. The vein lies under cover of the anterior border of the sternocleidomastoid muscle for the whole of its course and is crossed from above downwards by the accessory nerve; the facial nerve (between the stylomastoid foramen and the parotid gland), the styloid process and its muscles; the posterior belly of the digastric; the posterior auricular and occipital arteries; the sternocleidomastoid branch of the superior thyroid artery; the omohyoid muscle; and the anterior jugular vein (Figs 4.5, 4.27). For further details see page 263.

The *last four cranial nerves*. These emerge from the cranial cavity close together and run for part of their course in the upper part of the carotid sheath.

The glossopharyngeal nerve

This nerve leaves the cranial cavity through the jugular foramen in its own sheath of dura mater. It lies at first between the internal jugular vein and internal carotid artery (Fig. 4.30). It then crosses the artery lying close to the posterior border of the stylopharyngeus muscle as it passes deep to the external carotid artery. The nerve passes with the muscle into the wall of the pharynx at the upper border of the middle constrictor and deep to the hyoglossus muscle. Here it divides into branches which are distributed to:

1. The mucous membrane of the posterior third of the dorsal (pharyngeal) part of the tongue (lingual branches).

2. The mucous membrane covering the tonsils (tonsillar branches).

3. The mucous membrane covering the upper part of the pharynx, the lining of the auditory tubes and the middle ear (pharyngeal branches). These are all sensory branches.

The only muscle supplied by the glossopharyngeal nerve is the stylopharyngeus. It receives fibres related to visceral sensation from the carotid sinus and carotid body (the sinus nerve).

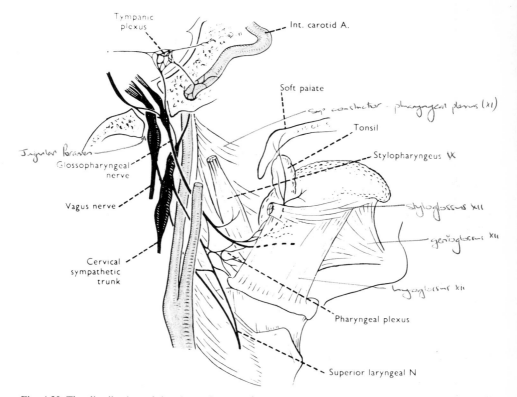

Fig. 4.30 The distribution of the glossopharyngeal nerve.

This is part of the mechanism regulating the control of blood pressure. Taste fibres from the back of the tongue and the circumvallate papillae run with the lingual branches of the nerve. Secretomotor fibres are distributed to the parotid gland via the tympanic nerve, the tympanic plexus, the lesser superficial petrosal nerve, the otic ganglion and the auriculotemporal nerve. For further details see page 320.

The vagus nerve (Figs 4.31, 4.38)
This nerve leaves the cranial cavity with the accessory nerve through the jugular foramen and descends through the neck within the carotid sheath. Just below the jugular foramen it shows two swellings, the superior or jugular ganglion and the inferior ganglion or ganglion nodosum, containing the cell bodies of sensory nerve fibres belonging to the nerve. In the region of the smaller superior ganglion the vagus is joined by the cerebral part of the accessory nerve.

The branches of the vagus nerve in the neck are:

A *recurrent sensory meningeal branch* returning to the cranial cavity via the jugular foramen to supply part of the dura mater.

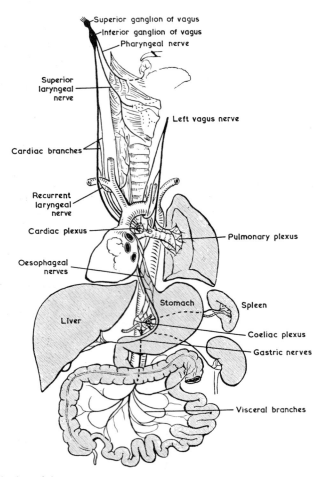

Fig. 4.31 The distribution of the vagus nerve.

The *auricular branch* passes through the temporal bone and is a sensory nerve to the back part of the external surface of the tympanic membrane and the external auditory meatus (Arnold's nerve). Within the brain stem the fibres of this nerve join the group of cells forming the sensory nucleus of the fifth cranial nerve.

The *pharyngeal branch* forms a plexus on the wall of the pharynx (with branches of the glossopharyngeal nerve and the superior cervical ganglion). It supplies the muscles of the pharynx, except the stylopharyngeus muscle, and the muscles of the soft palate except the tensor palati supplied by the mandibular division of the fifth cranial nerve. — seq'Al.' — V₃ motor

The *superior laryngeal nerve* passes deep to the internal carotid artery and divides on the wall of the pharynx into a. an upper sensory *internal laryngeal branch* which pierces the lateral part of the thyrohyoid membrane accompanied by the superior laryngeal branch of the superior thyroid artery. It is distributed to the mucous membrane of the larynx above the vocal cords, the lower part of the pharynx and the epiglottis; and the lower, motor branch, the *external laryngeal nerve* which passes deep to the infrahyoid muscles on the surface of the inferior constrictor to supply it and the cricothyroid muscle (Fig. 4.27).

Cardiac branches (usually two in number) arise from the vagus in the neck and run with the main nerve into the thoracic cavity where they join the cardiac plexus and help to regulate the rate of the heart beat.

The *recurrent laryngeal nerves.* On the right side the recurrent laryngeal nerve is given off as the vagus crosses the first part of the subclavian artery at the medial (anterior) border of the scalenus anterior muscle (Fig. 4.31). It passes below and behind the artery and then runs upwards between the trachea and oesophagus to enter the larynx from below. On the left side the nerve is given off at a lower level, in the thoracic cavity, and forms a loop around the arch of the aorta. Behind the aorta it reaches the groove between the trachea and oesophagus and returns to the neck. Both nerves in the larynx supply all the internal laryngeal muscles except the cricothyroid and contain sensory fibres which are distributed to the larynx *below* the vocal cords and to the trachea. For further details see page 198.

The accessory nerve

This nerve consists of two parts (Fig. 4.32). The *spinal part* arises from the side of the upper end of the spinal cord and enters the cranial cavity through the foramen magnum. Within the cranial cavity it is joined by the *cerebral* (*cranial*) *part* arising from the brain stem below the pons. The united nerve trunk leaves the cranial cavity through the jugular foramen in the same sheath of dura mater as the vagus nerve. The trunk then divides so that the cerebral part of the nerve joins the vagus and the spinal part runs backwards superficial (or sometimes deep to) the internal jugular vein to enter the sternocleidomastoid muscle. The cerebral part of the nerve is distributed to the muscles of the pharynx, soft palate, and larynx; these fibres accompanying the pharyngeal and recurrent laryngeal branches of the vagus. The spinal part of the nerve helps to supply the sternocleidomastoid and trapezius muscles and has already been described (page 136). For further details see page 252.

The hypoglossal nerve

The nerve leaves the skull through the anterior condylar foramen but soon comes into close relation with the nerves emerging from the jugular foramen between the internal jugular vein behind and the internal carotid artery in front. Below the posterior belly of the digastric, in the carotid triangle, it hooks around the occipital artery and crosses the internal and external arteries and the loop of the lingual artery (Fig. 4.33). It leaves the carotid triangle by passing deep to the posterior belly of the digastric and the posterior edge of the mylohyoid, to enter the floor of the mouth.

As the hypoglossal nerve enters the carotid sheath at the base of the skull it is joined by a branch of the first cervical nerve (anterior primary ramus). These cervical fibres run with

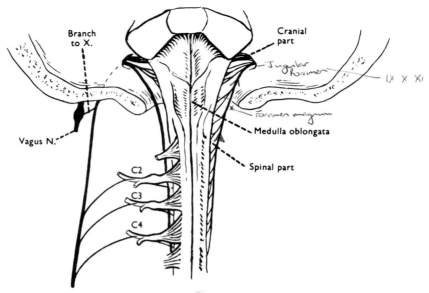

Fig. 4.32 The formation of the accessory nerve. XI

Branch to X.

Cranial part

Jugular foramen

Vagus N.

IX X XI

Foramen magnum

Medulla oblongata

Spinal part

C2

C3

C4

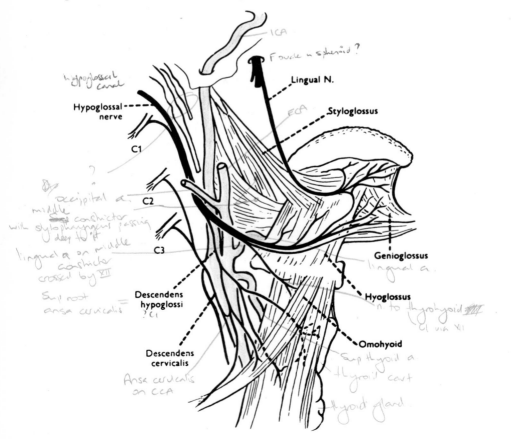

Fig. 4.33 The distribution of the hypoglossal nerve and the formation of the ansa hypoglossi.

ICA

Fovale n sphenoid ?

Lingual N.

hypoglossal canal

Hypoglossal nerve

ECA

Styloglossus

C1

occipital a.

middle constrictor with stylopharyngeus passing deep to it

lingual a on middle constrictor crossed by XII

Sup root ansa cervicalis

C2

C3

Genioglossus

lingual a.

Descendens hypoglossi ?C1

Hyoglossus

n to thyrohyoid C1 via XII

Descendens cervicalis

Omohyoid

Sup thyroid a

thyroid cart

Ansa cervicalis on CCA

thyroid gland

the hypoglossal ~ a) supplies (all) the muscles of the tongue T
b) arises from the pons F medulla
c) is related in its course to the superomedial surface of mylohyoid T
d) conveys fibres from the 3rd cervical n F – C1
e) contributes to the n supply of the pharyngeal muscle F (cranial XI vagus)

the hypoglossal nerve and are distributed as meningeal and muscular branches. The latter are distributed to the geniohyoid, thyrohyoid and, through the descendens hypoglossi, to the other infrahyoid muscles (Fig. 4.33). The fibres proper of the hypoglossal nerve are entirely distributed to the muscles of the tongue. For further details see page 284.

The subclavian artery and related structures

The subclavian artery arises on the right side from the brachiocephalic artery (the first branch of the arch of the aorta) behind the sternoclavicular joint (Fig. 4.28). On the left side the subclavian artery usually arises directly from the aortic arch (Fig. 3.23). The thoracic part of the left artery is referred to elsewhere (page 81). The cervical part of the subclavian artery is divided into three sections by its relation to the scalenus anterior muscle, which crosses the artery at the base of the neck (Fig. 4.34).

The artery rests on the *suprapleural membrane* (Sibson's fascia), which separates it from the apex of the lung above the level of the first rib; the scalenus medius muscle; and the outer surface of the first rib. Also lying behind the artery on the surface of the scalenus medius are the roots of the lower cervical (eighth) and the first thoracic nerves which unite to form the lower trunk of the brachial plexus. At the outer (lower) border of the first rib the subclavian artery continues as the axillary artery and enters the upper limb.

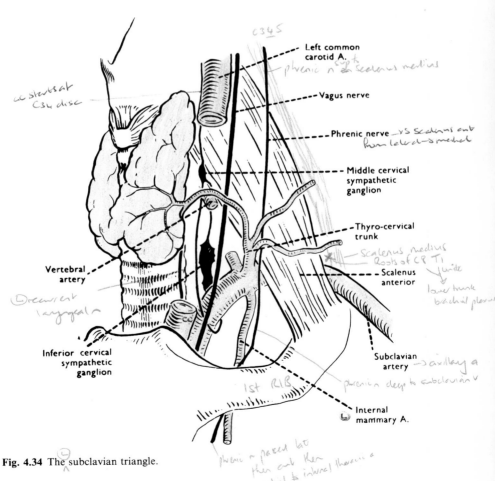

Fig. 4.34 The subclavian triangle.

The terminal part of the internal jugular vein lies superficial to the first part of the subclavian artery (on the medial side of the scalenus anterior) and the vertebral vein; the sternohyoid and sternothyroid muscles; and the sternocleidomastoid muscle. The artery is crossed here by the vagus nerve, its cardiac branches, on the right side by the recurrent laryngeal nerve which hooks around it, and by branches of the sympathetic nerve trunk.

Superficial to the second part of the artery are the scalenus anterior muscle, the phrenic nerve and the sternocleidomastoid muscle.

The third part is less deeply placed and lies beneath the deep cervical fascia lateral to the scalenus anterior. It is crossed by the terminal part of the external jugular vein.

The branches of the first part of the subclavian artery are the vertebral artery, the thyrocervical trunk and the internal thoracic artery (Fig. 4.34).

The *vertebral artery* (Fig. 4.28) is given off the upper surface of the subclavian artery and runs upwards in a deeply placed triangular cleft (the vertebral triangle) bounded by the scalenus anterior on the lateral side and the longus cervicis, lying on the bodies of the lower cervical vertebrae, on the medial side. Deep to the artery in this space lie the roots of the seventh and eighth cervical nerves, the inferior cervical sympathetic ganglion, the transverse process of the seventh cervical vertebra, and the neck of the first rib. In front lie the vertebral vein, the inferior thyroid artery, and the contents of the carotid sheath (internal jugular vein, common carotid artery and vagus nerve). At the apex of the triangle the artery passes through the foramen in the transverse process of the sixth cervical vertebra. In its further course the vertebral artery passes through similar foramina in the upper five cervical vertebrae, winds round in a groove on the upper surface of the posterior arch of the atlas and enters the cranial cavity through the foramen magnum. In the neck the vertebral artery gives spinal branches which pass between the vertebrae to supply the spinal cord and its meninges. Within the cranial cavity the two vertebral arteries unite to form the basilar artery which contributes to the formation of the *circle of Willis* at the base of the brain (Fig. 4.28). This is also joined by the middle cerebral branches of the internal carotid arteries. From this arterial ring branches supply various parts of the cerebral cortex, brain stem, cerebellum and internal structures of the brain.

The *thyrocervical trunk* is a short artery which divides at the anterior (medial) edge of the scalenus anterior into: a. two muscular branches (the transverse cervical and suprascapular arteries) crossing the surface of the scalenus anterior and the phrenic nerve to supply the muscles of the lower part of the neck and shoulder region; and b. the inferior thyroid artery which ascends along the anterior (medial) edge of the scalenus anterior to the apex of the vertebral triangle where it turns sharply medially across the vertebral vessels (Fig. 4.34). Here it lies closely related to the middle cervical sympathetic ganglion and lies behind the carotid sheath and its contents. As the artery enters the lower pole of the lateral lobe of the thyroid gland it is closely related to the recurrent laryngeal nerve. These are important surgical relations.

The *internal thoracic artery* is given off the lower surface of the subclavian artery close to the anterior (medial) border of the scalenus anterior muscle near its insertion. It enters the thoracic cavity by passing deep to the front of the first rib where it is crossed by the phrenic nerve (Fig. 4.34). It lies close to the edge of the sternum and gives branches to the mammary glands (page 82).

The second part of the subclavian artery gives off the *costocervical trunk* deep to the scalenus anterior muscle (Fig. 4.9). This branch passes backwards over the dome of the cervical part of the pleura to reach the anterior aspect of the neck of the first rib. Here it divides into a descending branch, the superior intercostal artery, which enters the thoracic cavity to supply the first two intercostal spaces, and a deep cervical branch which reaches the postvertebral muscles by passing backwards between the transverse process of the seventh cervical vertebra and the first rib. It anastomoses with a descending branch of the occipital artery.

The third part of the subclavian artery seldom gives off any branches, but sometimes one of the muscular branches of the thyrocervical axis arises here (usually the transverse cervical).

The *subclavian vein* lies at a lower level than the artery but passes superficial to, i.e. in front of, the muscle and is therefore a more superficial structure in the subclavian triangle (Figs 4.28, 4.34). It joins the internal jugular vein to form the brachiocephalic (innominate) vein which passes into the thoracic cavity behind the first rib. As the subclavian vein lies on the scalenus anterior muscle it is joined by the external jugular vein which enters the subclavian triangle from above by crossing superficial to the omohyoid muscle. Other veins join the subclavian in the region either separately or as a common trunk with the external jugular. One of these, the anterior jugular vein descending close to the middle line of the neck in the muscular triangle, turns laterally beneath the lower part of the sternocleidomastoid muscle to reach the subclavian vein.

The brachial plexus (Fig. 4.35)

This is one of the largest nerve plexuses and is formed from the anterior primary rami of the last four cervical and first thoracic nerves. The plexus serves to associate the nerves supply-

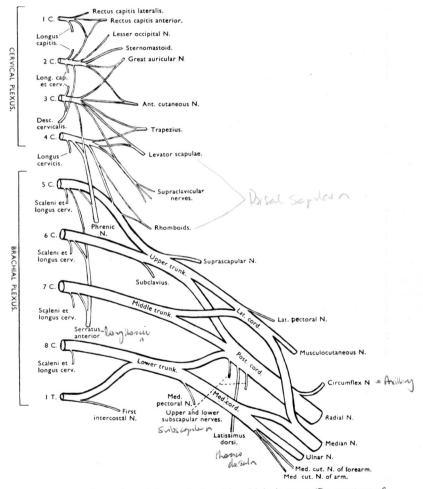

Fig. 4.35 Diagram showing formation of the cervical and brachial plexuses. (By courtesy of Professor G. A. G. Mitchell and Dr E. L. Patterson.)

ing the upper limb by a regrouping of nerve fibres from the various segments of the spinal cord to form peripheral nerves whose constituent fibres are thus best arranged for muscular innervation and muscle group action.

The nerves forming the brachial plexus are in series with those comprising the cervical plexus (page 138) and lie in the posterior triangle of the neck between the scalenus anterior and the scalenus medius muscles. In the triangle the plexus is covered by the deep fascia, and crossed by the external jugular vein, the inferior belly of the omohyoid muscle and the transverse cervical artery. At the root of the neck of the plexus behind the clavicle into the axilla where its branches surround the axillary artery, a continuation of the subclavian artery.

The plexus is divided into roots, trunks, divisions, cords and branches.

The *roots of the plexus* which are short and are the individual anterior primary rami of spinal nerves C5–8 and T1.

The *trunks of the plexus* are formed in the posterior triangle behind the clavicle:

1. The upper trunk from the union of C5 and C6,
2. The middle trunk derived from C7, and
3. The lower trunk from C8 and T1.

The lower trunk often forms a groove on the upper surface of the first rib behind the groove for the subclavian artery (Fig. 3.5).

The *anterior and posterior divisions of the plexus* formed by the splitting of the trunks, indicating the fibre bundles destined for the anterior and posterior aspects of the upper extremity.

The *cords of the plexus* are three in number:

1. The lateral cord—from the anterior divisions of the upper and middle trunks.
2. The medial cord formed by the anterior division of the lower trunk.
3. The posterior cord—from the union of the posterior divisions of all three trunks.

The nerve branches of the cord are indicated in Figure 4.35. The median and ulnar nerves supply the anterior (or ventral) surface of the forearm and hand, whilst the radial nerve supplies the musculature on the posterior (or dorsal) aspect.

The cervical sympathetic trunk

This is made up of white rami communicantes leaving the spinal cord with the upper thoracic nerves; three cervical ganglia, superior, middle and inferior; and various branches of distribution (grey rami communicantes) (Fig. 4.36). The white rami contain preganglionic myelinated nerve fibres while the grey rami consist of postganglionic nonmyelinated nerve fibres. The trunk which connects the ganglia leaves the thorax by passing in front of the neck of the first rib. In the neck it lies on the prevertebral muscles (longus cervicis and longus capitis) and behind the prevertebral fascia. In front of it is the carotid sheath and its contents. On its posterolateral side are the roots of the cervical nerves to which it gives branches of distribution.

The large superior cervical ganglion lies at the level of the second and third cervical vertebrae; the small middle ganglion at the level of the sixth vertebrae, where it is closely related to the loop of the inferior thyroid artery; while the inferior cervical ganglion lies between the transverse process of the seventh cervical vertebra and the first rib in close relation with the vertebral artery (Fig. 4.34).

Each ganglion gives off the following branches of distribution:

1. Communicating branches to the cervical nerves.
2. Vascular branches which run along the larger blood vessels.
3. Cardiac branches which pass downwards to the thoracic cavity where they unite with branches of the vagus nerve to form the cardiac plexus.

The *superior cervical ganglion* (Fig. 4.30) gives off vascular branches along the internal carotid artery, which are distributed to the cranial cavity and the contents of the orbital cavity including the lacrimal gland. Branches are also distributed along the external carotid artery and its branches (maxillary, facial and lingual), to the superficial and deep structures of the face, including the parotid, submandibular and sublingual salivary glands, and the mucous glands of the nasal and oral cavities. In addition branches join the upper four cervical nerves and the glossopharyngeal, vagus and hypoglossal nerves. A branch is given to the pharyngeal plexus (to which the vagus and glossopharyngeal nerves also contribute). Finally, the superior ganglion gives off the superior cervical cardiac nerve which descends through the neck to the thoracic cavity to join the cardiac plexus (Fig. 4.36).

The *middle cervical ganglion* gives off vascular branches distributed along the inferior thyroid

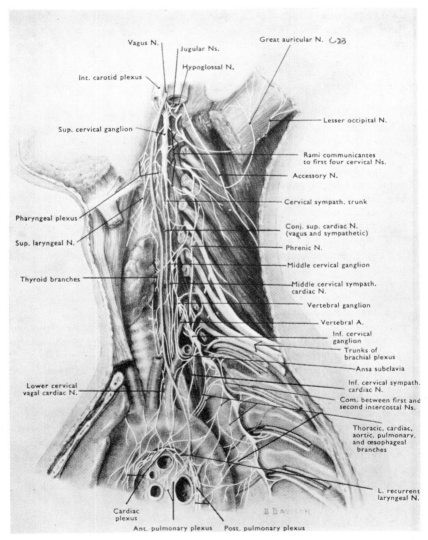

Fig. 4.36 Dissection of the neck and upper part of the thorax from the left side showing cardiac and pulmonary branches from the sympathetic trunk and vagus nerve converging on the region of the aortic arch and root of the lung to form the cardiac and pulmonary plexuses. (By courtesy of G. A. G. Mitchell and Dr E. L. Patterson.)

artery; communicating branches to the fifth and sixth cervical nerves; and the middle cervical cardiac nerves which enter the thorax to join the cardiac plexus. Some of the communicating branches between the middle and inferior ganglia form a loop (ansa subclavia) around the first part of the subclavian artery (Figs 4.34, 4.36).

The *inferior cervical ganglion* (Fig. 4.34) distributes vascular branches along the subclavian artery and vertebral arteries; communicating branches to the seventh and eighth cervical nerves and the inferior cervical cardiac nerve.

The deep lateral and prevertebral muscles of the neck (Fig. 4.10)
These muscles cover the anterolateral surfaces of the cervical vertebrae and have a close relationship to the pharynx and oesophagus, the main vessels and large nerves of the neck.
The *scalenus anterior* arises from the anterior tubercles of the transverse processes of the third and sixth cervical vertebrae. It is inserted into the scalene tubercle on the upper surface of the first rib (Figs 3.5, 4.34). The phrenic nerve runs downwards on its anterior surface.

The *scalenus medius* arises from the posterior tubercles of the second to the sixth cervical vertebrae and is inserted into a roughened area on the upper surface of the first rib behind the groove for the subclavian artery.

The *scalenus posterior* arises from the posterior tubercles of the transverse process of the fourth to the sixth cervical vertebrae and is inserted into a roughened impression on the outer surface of the second rib.

The *longus capitis* arises from the anterior tubercles of the transverse processes of the third to the sixth cervical vertebrae and is inserted into the inferior surface of the basilar part of the occipital bone (Fig. 5.3) anterolateral to the pharyngeal tubercle.

The *longus cervicis* spans the region from the third thoracic to the atlas vertebra, as a series of muscular slips between the bodies or transverse processes of the vertebrae.

The nerve supply to all these deep muscles is from the anterior primary rami of cervical nerves. The muscles act with the postvertebral musculature in movements of the head and neck. They are flexors and rotators of the head and neck and are able to function as important accessory muscles of respiration when the head and neck are fixed. This is because they can raise the first two ribs and thus help to expand the thoracic cage.

The postvertebral muscles
These muscles form a complex mass in the floor of the posterior triangle of the neck and behind the vertebral column and act in conjunction with the prevertebral musculature. The most important are (Fig. 4.10):

Semispinalis capitis arises from the transverse processes of the upper six thoracic vertebrae and the articular processes of the last four cervical vertebrae. It is inserted between the superior and inferior nuchal lines on the occipital bone near the mid-line (page 205).

Splenius capitis and cervicis arise from the mid-line ligamentum nuchae and the spinous processes of the lower cervical and upper four thoracic vertebrae. The splenius capitis is inserted into the mastoid process of the temporal bone and the adjoining part of the superior nuchal line on the occipital bone. The splenius cervicis is inserted into the posterior tubercles of the transverse processes of the fourth cervical vertebrae.

Levator scapulae arises from the posterior tubercles of the transverse processes of the first four cervical vertebrae and passes downwards to be inserted into the upper part of the medial border of the scapulae.

The suboccipital muscles
These are four in number:
The *obliquus capitis inferior* extends from the spine of the axis to the transverse process of the atlas.

The *obliquus capitis superior* extends from the transverse process of the atlas to the occipital bone above the inferior nuchal line.

The *rectus capitis posterior major* passes from the spine of the axis to the occipital bone below the inferior nuchal line.

The *rectus capitis posterior minor* extends from the posterior tubercle of the atlas to the occipital bone medial and deep to the rectus major muscle.

The suboccipital muscles are innervated by the posterior primary ramus of the first cervical nerve. They form the boundaries of the *suboccipital triangle* which is filled in by dense connective tissue containing the suboccipital plexus of veins. This venous plexus communicates with emissary veins in the occipital region of the skull. Emissary veins provide connections between veins on the inside and those on the outside of the skull bones. In the depths of the triangle the vertebral artery passes horizontally along the upper border of the posterior arch of the atlas on its way to the foramen magnum. The posterior ramus of C1. appears in the triangle between the arch of the atlas and the vertebral artery. The suboccipital muscles are concerned with movements of the head on the neck.

THE HEAD AND NECK IN SAGITTAL SECTION

This view helps to relate the major features of the oral cavity and upper parts of the respiratory and alimentary tracts to one another (Figs 4.8, 4.37, 5.19):

The *bony mid-line cranial base* extends from the foramen magnum to the root of the nose (nasion). From behind forwards it is made up of the body of the occipital and sphenoid, the ethmoid and the frontal bones. Note the pituitary gland in its fossa in the body of the sphenoid at the base of the cranial cavity; the sphenoid air sinus and the frontal air sinus. Further details of the cranial base are given on page 381.

The *cervical part of the vertebral column*. From above downwards it is made up of:

1. The arches of the atlas.
2. The body and odontoid process of the axis, which is the second cervical vertebra. The atlas rotates around the odontoid process;
3. The bodies of the third to seventh cervical vertebrae, united by the fibrocartilaginous intervertebral disks, lie in front of the spinal canal.

Note that when the mouth is closed in a resting position the hard palate lies at about the same level as the arch of the atlas; the lower incisors are opposite the body of the second cervical vertebra; the hyoid bone is opposite the body of the third cervical vertebra; the lower border of the mandible is at about the same level as the body of the fourth cervical vertebra, and the cricoid cartilage lies opposite the body of the fifth or sixth cervical vertebra.

The *nasal cavity*. The mid-line saggital plane passes through the nasal septum which is covered by mucus membrane and consists of the vomer, the perpendicular plate of the ethmoid and the septal cartilage.

Parasagittal planes of section demonstrate surface features of the lateral wall of the nose (Fig. 4.37) including the turbinate processes or nasal conchae, the opening of the pharyngotympanic tube, the curvature of the hard and soft palates and the relationship of the sphenoidal air sinus to the posterior part of the roof of the nasal cavity.

In the *mouth cavity* the muscles of the tongue and floor of the mouth fill the space between the palate, mandible and hyoid bone. The oropharynx passes vertically downwards between the posterior aspect of the tongue and the prevertebral region. The tonsil lies on the lateral wall at the entrance to oropharynx.

The *laryngeal apparatus* lies below and behind the tongue and at the upper end of the trachea. Note the sectioned cricoid, thyroid and epiglottic cartilages and in the side wall of the larynx

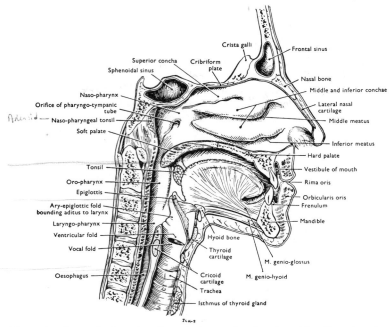

Fig. 4.37 Paramedian sagittal section through the nasal and oral cavities. Note that the cricoid cartilage lies one vertebra above its normal level. (By courtesy of Professor G. A. G. Mitchell and Dr E. L. Patterson.)

the vocal and vestibular folds with the opening of the laryngeal sinus between them. The orifice of the larynx faces upwards and backwards. For further details see page 198.

The *pharynx* extends from the base of the skull to about the level of the sixth cervical vertebra behind the nasal, oral and laryngeal cavities. The upper part behind the nasal cavities is the *nasopharynx* and into its side wall on each side opens the auditory or pharyngotympanic tube. The intermediate part behind the tongue and the mouth is the *oropharynx*.

The lower part of the pharynx lies behind the larynx and leads inferiorly into the oesophagus. For further information on the pharynx see page 191.

The *isthmus of the thyroid* lies in front of the second, third and fourth cartilaginous rings of the trachea below the level of the larynx, and superficial to this are some of the infrahyoid muscles (see page 136).

THE NECK IN TRANSVERSE SECTION

The more important features seen in transverse (cross) sections of the neck (Figs 4.10, 6.4) are:

1. The bodies of the cervical vertebrae or their intervertebral disks, the spinal cord and spinal nerves passing forwards to the cervical and brachial plexuses (Fig. 4.35).

2. The pharynx, at the highest levels of the neck, leading to the oesophagus, larynx and trachea. The pharynx and oesophagus have thin muscular walls, the larynx and trachea have a cartilaginous skeleton (pages 68, 92). The trachea is always patent, the oesophagus opens only during the passage of a bolus of food.

3. The great vessels and nerves within the carotid sheath, the common carotid artery or

its terminal branches; the internal jugular vein; the vagus, hypoglossal and glossopharyngeal nerves.

4. Muscles or muscle groups, of which the most relevant are the:
   a. Sternocleidomastoid
   b. The prevertebral muscles *— scalene & longus*
   c. The infrahyoid and suprahyoid muscles *— attached to hyoid bone from below & above.*
   d. The styloid musculature *styloglossus XII   stylohyoid VII   stylopharyngeus IX .*

Other important structures, depending on the cervical level, include the hyoid bone, the thyroid and parathyroid glands, the cervical and brachial plexuses, the cervical sympathetic trunk and its ganglia, cervical lymph nodes, the cervical deep fascia and its compartments.

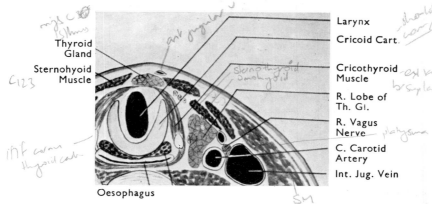

Thyroid Gland

Sternohyoid Muscle

Oesophagus

Larynx

Cricoid Cart.

Cricothyroid Muscle

R. Lobe of Th. Gl.

R. Vagus Nerve

C. Carotid Artery

Int. Jug. Vein

Fig. 4.38 Diagram representing a transverse section of the neck at the level of the entrance into the larynx. (By courtesy of the Anatomical Society of Great Britain and Ireland.)

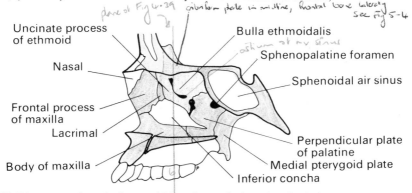

Uncinate process of ethmoid

Nasal

Frontal process of maxilla

Lacrimal

Body of maxilla

Bulla ethmoidalis

Sphenopalatine foramen

Sphenoidal air sinus

Perpendicular plate of palatine

Medial pterygoid plate

Inferior concha

Fig. 4.39 Diagram to show the bones which make up the lateral wall of the nasal cavity.

THE NASAL CAVITIES

The nasal cavities occupy the middle of the facial skeleton. They are bounded above from before backwards by the nasal, frontal, ethmoid (cribriform plates) and sphenoid bones (Fig. 4.39); below by the hard palate, which consists of the palatal processes of the maxillary and palatine bones (Fig. 4.37). Each cavity is separated from the other by the nasal septum consisting of the perpendicular plate of the ethmoid and the vomer. The persisting cartilaginous part of the septum is in front between the two bones (Fig. 4.40). The lateral wall of each nasal cavity contains parts of the maxillary, nasal, lacrimal, ethmoid, inferior turbinate and palatine

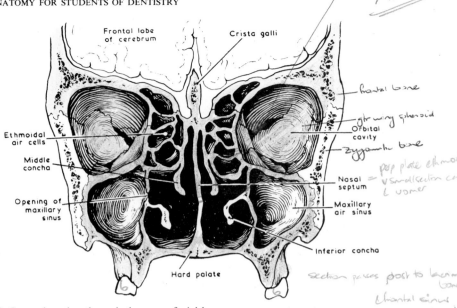

? is this maxilla

frontal bone

gt wing sphenoid

zygomatic bone

pep plate ethmoid
v small section cart
& vomer

Fig. 4.40 Coronal section through the upper facial bones.

Section passes post to lacrimal bone
Ethmoid sinus
Back to sup concha sphenoid
palatine bone &
medial pteryg

bones and the medial pterygoid plate (Fig. 4.39). Opening into each cavity are the frontal, maxillary, ethmoidal and sphenoidal air sinuses and the nasolacrimal duct.

Each cavity can be divided into a narrow, upper, interorbital *olfactory region*, bounded chiefly by the parts of the ethmoid bone (cribriform plate, perpendicular plate, and the upper turbinate process of the bone); and a lower *respiratory region* (Fig. 4.40). The upper region of the nasal cavity is lined by the olfactory mucous membrane containing the sensory endings of the olfactory (first cranial) nerve (page 424). The wider, respiratory part of each nasal cavity lies below the level of the orbital cavities. It is bounded on the lateral side by the maxillary bone, the inferior turbinate, the palatine bone and the medial pterygoid plate of the sphenoid. Its mucous membrane is thicker, more vascular, more glandular than the olfactory mucosa. It is supplied by branches of the ophthalmic and maxillary divisions of the fifth cranial nerve. The surface area of the respiratory mucous membrane is increased by the middle and inferior turbinate processes (conchae). The main passage for inspired air is beneath the inferior turbinate (the inferior meatus) into which opens the lacrimal duct from the lacrimal sac. Into the middle meatus (beneath the middle turbinate process of the ethmoid) are the openings for the frontal, maxillary, the anterior and middle ethmoid air sinuses. Into the superior meatus (beneath the superior turbinate process of the ethmoid) open the posterior ethmoidal air sinuses.

The part of the nose projecting on the face (external nose) has as its skeleton the nasal bones, forming the bridge of the nose, and a group of cartilaginous elements made up of the front part of the mid-line septal cartilage and bilateral alar cartilages. The latter consist of two major elements and a number of smaller units.

The lateral wall of the external nose and nasal cavities can be divided into three areas (Fig. 4.42).

A *flattened, anterior, region* consisting of the vestibule of the nostril (lined by skin and carrying hairs) and the atrium (lined by respiratory mucous membrane).

The *middle, or turbinate, region*, divided by the three turbinate processes (conchae) into four slit-like grooves. These are:

1. The inferior meatus beneath the inferior turbinate bone. contains lacrimal duct opening

2. The middle meatus beneath the middle turbinate process of the ethmoid bone.

– contains openings for maxillary anterior & middle ethmoid air sinuses

3. The superior meatus beneath the superior turbinate process of the ethmoid bone.

4. The olfactory sulcus running beneath the cribriform plate of the ethmoid above the superior turbinate process. Behind, it leads to the spheno-ethmoidal recess into which opens the sphenoidal air sinus.

3. The *posterior region*. This is a short segment of the lateral wall behind the turbinate processes bounded by the medial pterygoid plate of the sphenoid.

Function

The hairs of the vestibule clean the inspired air of gross particles. Within the nasal cavity most of the inspired air passes along the inferior meatal channel above the hard palate, but some is deflected along the middle and superior meatal channels and along the olfactory sulcus. In the nasal cavity the air is cleaned by the mucous secreted by the numerous glands, and warmed by the highly vascular mucous membrane. The nasal mucous membrane is lined by a pseudostratified columnar ciliated epithelium in its respiratory part. The cilia act in such a manner as to direct mucus produced by goblet cells in the epithelium downwards and backwards towards the nasopharynx. The upper olfactory region is lined by a non-ciliated epithelium which contains special bipolar olfactory nerve cells (page 424).

The human nasal cavity is much less complicated in its structure than that of the majority of mammals. Complexity of the turbinate processes appears to be related to two distinct functions:

1. Increasing the area covered by olfactory mucosa in animals with a highly developed sense of smell.

2. Increasing the area covered by the respiratory (vascular) mucous membrane in animals in which the nasal cavity plays an important part in the regulation of heat loss by the body.

The Paranasal air sinuses

Certain evaginations, diverticula or outgrowths from the nasal cavity are a constant feature and are called the *paranasal air sinuses*. They are paired bilateral structures frequently of unequal size, and are lined by a continuation of the nasal mucous membrane in the form of a ciliated columnar epithelium. Some of them cause a considerable lightening of the bones in which they lie, notably the maxillary air sinus, by a hollowing out process which occurs as they develop. They also impart some degree of resonance to the voice, probably quite an incidental function in man.

The epithelial lining of the air sinus helps to clean the nasal passages by secreting a sticky fluid, mucin, which is moved by ciliary action into the meatuses of the nose. The sinuses develop as outpouchings of the embryonic nasal cavity and are less than one-tenth of their adult size at the time of birth. With increasing age in the adult, the cavities expand into adjacent bony areas such as the alveolar ridge of the maxilla, the palatine or the frontal bones. Thus the middle and upper parts of the facial skeleton are more likely to shatter, or comminute, from trauma in older persons and diseases involving the alveolus are more likely to extend into the maxillary sinuses in old age.

The *maxillary air sinus*. The maxillary sinus (or antrum of Highmore) occupies the body of the maxillary bone (Figs 4.40, 4.41). If it is large it may extend into the zygomatic bone. The roof of the sinus is separated from the orbital cavity by the relatively thin orbital plate of the maxilla through which runs the infra-orbital groove and canal containing the infra-orbital vessels and nerve. The floor of the sinus is closely related to the apices of the permanent molar teeth in the adult and usually also to the second premolar. If the sinus is large it may even be related to the apical region of the first premolar. The dimensions of an average sinus are: width 25 mm, length (front to back) 30 mm, and height 35 mm. It is present at birth but is at that stage little more than a slit-like outpouching of the nasal cavity. It grows rapidly during the eruption of the deciduous teeth and has reached about half its adult size by three

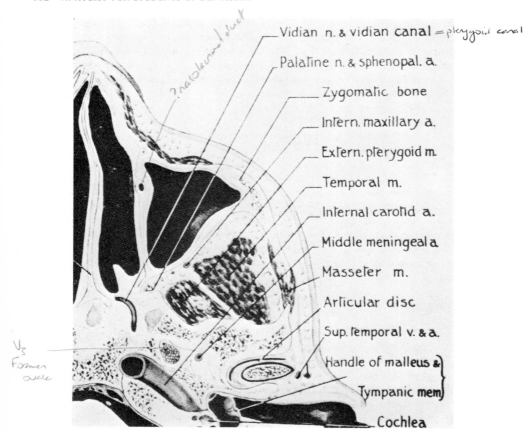

Vidian n. & vidian canal = pterygoid canal

Palatine n. & sphenopal. a.

Zygomatic bone

Intern. maxillary a.

Extern. pterygoid m.

Temporal m.

Internal carotid a.

Middle meningeal a

Masseter m.

Articular disc

Sup. temporal v. & a.

Handle of malleus &

Tympanic mem

Cochlea

? nasolacrimal duct

V₃
Former
ovale

Fig. 4.41 Drawing of a horizontal section through the upper part of the face to show the nasal cavities, the maxillary air sinus, the mandibular joint and related structures. (By courtesy of the Anatomical Society of Great Britain and Ireland.)

years of age. It is important to be aware of these dental relationships when extracting these teeth for a root or part of a root of a tooth may be forced easily through the thin floor of the sinus into the sinus cavity, from which it can only be removed with difficulty. The floor of the antrum is sometimes partly subdivided into a number of compartments by one or more incomplete septa.

The anterior wall of the sinus is bounded by the facial surface of the maxilla. The posterior wall is related to the pterygopalatine fossa. The middle and anterior dental nerves run in canals in the lateral and anterior walls of the sinus. Before its eruption the upper canine is closely related to the anterior wall, and the front of a large sinus is related to the nasolacrimal duct. The opening into the antrum from the nasal cavity (middle meatus) is close to the roof of the antrum high up on its medial, or nasal, wall (Figs 4.40, 4.42). The opening lies below the bulla ethmoidalis and is hidden by the uncinate process of the ethmoid bone. Because of the high position of the opening foreign matter, such as purulent material or pus, does not drain easily from the sinus, for gravity tends to cause retention of the fluid in the most dependent part of the cavity, i.e. along its floor above the roots of the teeth. Furthermore, as both the frontal air sinus and the maxillary sinus can drain into the infundibulum, fluid from the frontal sinus tends to gravitate towards the maxillary sinus opening. Infections of the frontal air sinus are therefore often followed by a secondary infection involving the maxillary sinus. In other instances, due to the position of the uncinate process over the maxillary opening, infected materials are directed backwards towards the nasopharynx without necessary involvement of the maxillary sinus. The maxillary sinus, like the other sinuses, is lined by a mucoperiosteum supporting a layer of columnar or cubical cells which are ciliated in certain regions. The medial (nasal) wall of the antrum is bounded by the nasal surface of

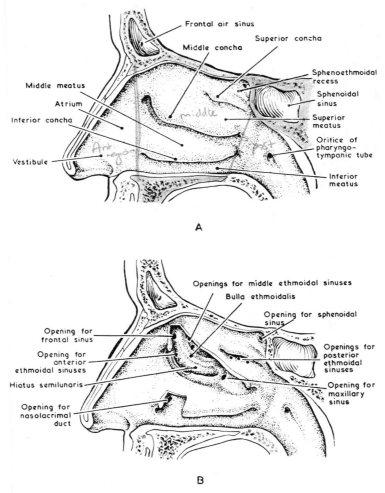

Fig. 4.42 The lateral wall of the nasal cavity: A, as in the living state; B, showing the openings of the paranasal air sinuses.

the body of the maxilla and by parts of the palatine, lacrimal, ethmoid and inferior turbinate bones (Fig. 4.39). The presence of these bones reduces considerably the size of the opening between the antrum and the nasal cavity during life. The nerve supply of lining mucous membrane is from branches of the superior dental and infra-orbital nerves.

The *frontal sinuses*. These occupy the frontal bone above the superior orbital margin. They extend upwards in the forehead region and backwards in the roof of the orbital cavities to a varying degree. Each sinus is separated from its neighbour by a septum which is often oblique. One sinus may be considerably larger than the other (usually the left). In 50 per cent of subjects the frontal sinus communicates directly with the front of the semilunar groove at the side wall of the middle meatus, the *infundibulum*, via the frontonasal duct. One or more of the anterior ethmoidal air cells open into the front part of this curved passage. The infundibulum is reached from the nasal cavity through a curved fissure, the *hiatus semilunaris*, which is bounded above by the bulla ethmoidalis and below by the sharp edge of the uncinate process of the ethmoid bone. The nerve supply to the mucous membrane of the frontal sinus is from the supraorbital nerve.

The *ethmoidal sinuses*. These are a group of small cavities lying within the facial part of the ethmoid bone (Fig. 4.40) or between it and adjacent bones such as the frontal, lacrimal, maxilla and palatine. They are closely related to the medial wall of the orbital cavity and the floor of the anterior cranial fossa (on either side of the cribriform plate). They are divided into three groups. The *anterior group* opens into the hiatus semilunaris of the middle meatus or into the terminal part of the infundibulum of the frontal sinus. The *middle group* opens on the surface of an elevation (bulla ethmoidalis) on the side wall of the middle meatus situated immediately above the hiatus semilunaris. The *posterior* or *superior group* opens into the superior meatus (Fig. 4.42).

The *sphenoidal sinuses*. These occupy the body of the sphenoid bone (Fig. 4.42). They may, if large, extend for some distance into the great wings. They are bilateral but vary considerably in size, and the septum between the sinus of each side is usually asymmetrical (often deflected to the right side). The sinuses are related to the pituitary fossa, the cavernous venous sinus and its contents, and the structures passing through the optic foramen. Each sinus opens into the spheno-ethmoidal recess at the upper posterior part of the respective nasal cavity through an opening in its anterior wall (Fig. 4.42). The sphenoidal and ethmoidal sinuses are supplied by the ethmoidal branches of the ophthalmic division of the trigeminal nerve.

Because the ethmoidal sinuses vary considerably in size and number and extend into adjacent parts of the ethmoid, sphenoid, lacrimal and palatine bones as age increases, surgical reconstruction of the interorbital skeleton may be very difficult. The inner table of the frontal bone is quite thin where it forms the inner wall of the frontal sinus and easily punctured or eroded by chronic sinus infections. This may result in a pathway for the spread of micro-organisms into the anterior cranial fossa. The floor of the maxillary sinus may be partitioned by bony septa, especially in the posterior part of the sinus. This may hinder the removal of foreign objects such as the tips of the roots of teeth that may have been dislodged into the sinus cavity during tooth extraction. Because the opening of the maxillary sinus is high on its medial wall drainage of the sinus cavity does not occur in the erect position. It may be necessary to provide artificial drainage for collections of blood or pus, by piercing the wall of the sinus through the inferior meatus or through the canine fossa. This is done using an intraoral approach, the Caldwell–Luc procedure.

Blood vessels of the nasal cavities and air sinuses

The *sphenopalatine (or nasal) artery*. This is one of the terminal branches of the third part of the maxillary artery given off in the pterygopalatine fossa (Fig. 5.18). It enters the nasal cavity through the sphenopalatine foramen and supplies the mucous membrane covering the greater part of the lateral wall of the cavity. Over the turbinate processes (conchae) the mucosa is especially thick and vascular containing large sinus-like vascular spaces. The veins corresponding to the branches of the sphenopalatine artery leave the nasal cavity through the sphenopalatine foramen and in the pterygopalatine fossa communicate with veins from the orbital cavities (ophthalmic veins) and with the pterygoid venous plexus.

The sphenopalatine artery sends a long septal (nasopalatine) branch across the roof of the nasal cavity to run downwards and forwards on the nasal septum and anastomose at the incisive foramen with terminal branches of the greater palatine arteries. Branches of the sphenopalatine artery also assist in supplying the paranasal sinuses (maxillary and ethmoidal).

The *posterior and anterior ethmoidal branches of the ophthalmic artery* (page 259) supply the sphenoidal, ethmoidal and frontal air sinuses. The anterior ethmoidal artery also supplies the mucous membrane of the lateral wall and septum at the front of the nasal cavity.

Branches of the posterior, middle and anterior superior dental arteries (page 298) help to supply the mucous membrane of the maxillary sinus.

The mucoperiosteum of the nasal cavities is thick and highly vascular over the turbinate processes. Control of haemorrhage from this part of the mucosa is quite difficult because the arterial supply comes from different sources, namely, the ophthalmic, maxillary, pharyngeal and palatal arteries.

The nerve supply of the nasal cavities (Fig. 4.43)

The *olfactory nerves*. These nerve fibres are extremely short and arise from the olfactory cells of the nasal mucous membrane lining the roof of the nasal cavity (cribriform plate region), the superior nasal concha and the upper part of the nasal septum. They pierce the cribriform plate in about fifteen to twenty branches and terminate in the olfactory bulbs. The olfactory nerve fibres are concerned with the special sense of smell (see page 427).

The *anterior ethmoidal nerve* is a branch of the nasociliary nerve given off in the orbital

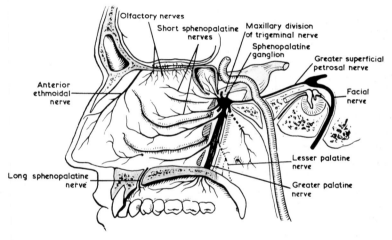

Fig. 4.43 The nerve supply of the lateral wall of the nasal cavity. (Part of the vertical plate of the palatine bone has been cut away.) cf Fig 4.37

cavity. It enters the nasal cavity towards the front of the cribriform plate and is distributed to the nasal mucous membrane covering the upper anterior region of the lateral nasal cavity (atrium), the adjacent nasal septum and the anterior ethmoidal sinuses. It terminates as the external nasal nerve which supplies an area of skin at the side of the nose (page 247). The *posterior ethmoidal nerve* supplies the ethmoidal and sphenoidal sinuses.

Sphenopalatine nerves. These branches of the maxillary division of the trigeminal nerve (page 247) enter the nasal cavity through the sphenopalatine foramen (page 229). The lateral branches (*short sphenopalatine nerves*) are distributed over the greater part of the lateral wall of the nasal cavity (superior, middle and inferior conchae and the superior, middle and inferior meatus). A branch (*long sphenopalatine* or *nasopalatine*) crosses the roof of the nasal cavity and descends on the nasal septum supplying the covering mucous membrane. After passing through the incisive canal terminal branches communicate with the greater palatine nerves and small branches help to supply the supporting tissues of the central and lateral incisor teeth.

The *nasal branch of the anterior superior dental nerve* (page 302) supplies an area of mucous membrane over the lower anterior region of the lateral nasal wall.

The *greater and lesser palatine nerves* give off small branches, as they descend through the palatine canal in the lateral wall of the nose, to supply the mucous membrane covering the lower posterior part of the nasal wall.

Nasal lymphatics

The lymphatics of the nasal cavities and air sinuses drain backwards towards the free edge of the medial pterygoid plate where they pierce, or pass above, the superior constrictor muscle to join the upper deep cervical lymph nodes. Nasal lymphatics also communicate with the subarachnoid space along the coverings of the olfactory nerves. Tumours of the nose can spread therefore into the cranial cavity as well as into the neck.

THE ORBITAL CAVITIES

The orbital cavities are situated between the cranial and facial regions of the skull and are separated from one another by the upper orbital part of the nasal cavities (Figs 4.40, 4.41). The rim of each orbital cavity is subcutaneous and is bounded by the frontal bone above; the zygomatic bone laterally and below; the maxilla and its frontal process below and on the medial side. At the upper margin of the orbital rim, at about the junction of the outer two-thirds and inner third, is the supra-orbital notch or foramen. There is great variation in the form of the margin of the orbital cavity in different individuals (Fig. 4.44).

The *roof* of each orbital cavity, separating it from the anterior cranial fossa and the frontal lobes of the brain, is formed by the orbital plate of the frontal bone and, at the back of the cavity, by the lesser wing of the sphenoid.

The *lateral wall* of each cavity is formed by the orbital plate of the zygomatic bone separating it from the temporal fossa, and the orbital plate of the great wing of the sphenoid situated between the orbital cavity and the middle cranial fossa.

The *floor* of the orbital cavity is formed by the orbital surface of the maxilla separating the orbital cavity from the maxillary sinus. It is grooved by the infra-orbital groove leading forward to the infra-orbital canal and foramen.

The *medial wall* of the orbital cavity is made up of a number of bones (Fig. 4.44). These are from before backwards: the frontal process of the maxilla, the lacrimal, the orbital plate of the ethmoid, the orbital process of the palatine and part of the body of the sphenoid. The medial wall of each orbital cavity is closely related to the upper part of the nasal cavity and the ethmoidal and sphenoidal air sinuses. In relation to the frontal process of the maxilla and the lacrimal bone is the lacrimal fossa leading to the nasolacrimal canal which opens into the inferior meatus of the nasal cavity (page 180).

At the back of each orbital cavity are two openings communicating with the cranial cavity (Fig. 4.44). These are:

1. The rounded optic foramen between the lesser wing and body of the sphenoid bone.

2. The slit-like superior orbital fissure, situated between the lesser and greater wings of the sphenoid bone.

Between the floor and lateral wall of the back part of the orbital cavity is the slit-like inferior orbital fissure communicating with the infratemporal fossa and the pterygopalatine fossa.

The eyeball is well protected by the bones of the orbit and fractures of the bones can be quite extensive without necessarily affecting sight. Crush injuries of the upper face commonly involve the maxilla and the bones of the orbit, for instance in automobile accidents where the head strikes the steering wheel or instrument panel. Fractures of the orbital bones may involve the maxillary, frontal or ethmoidal air sinuses as well as extend to the temporal fossa. Injury to the orbital muscles produce abnormal positions of the eyeball relative to the orbital

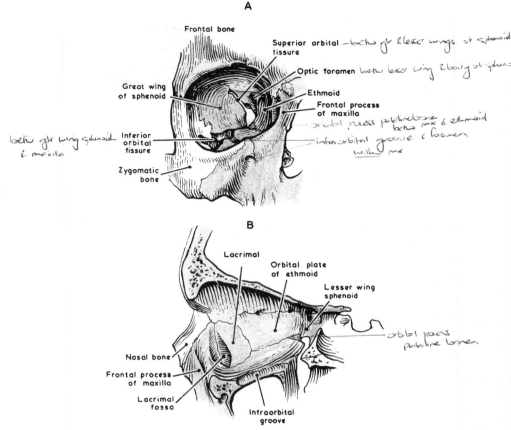

Fig. 4.44 The bones of the orbit. A, anterior view; B, medial wall.

cavity and trauma to the appropriate nerve branches may cause malfunction of the periorbital muscles, with drooping of the upper eyelid. Venous communications between the orbit, the face and the interior of the skull make the spread of infection after injury a potential clinical problem and of course damage to the front of the eyeball is of particular concern.

The eyeball and its associated structures are surrounded by loose orbital fat and the eyeball is suspended and separated from the fatty tissue by a fascial sheath, the *fascia bulbi* or Tenon's fascia. The volume of the orbital fat changes under a variety of conditions and alters the appearance of the eyes. When there is extensive weight and fat loss, as in debilitating diseases such as cancer, the loss of orbital fat causes the eyes to appear sunken. Conversely, a form of thyroid disease, thyroid goitre, causes an increase in the amount of fat behind the eyeball and the eyes appear to protrude much farther than usual.

The *contents of the orbital cavity* are (Fig. 4.45):

1. The eyeball and the optic nerve which are described on pages 425 and 428.
2. The vessels and nerves supplying the eyeball and its muscles.
3. The orbital muscles which are responsible for the movements of the eyeball and the eyelids.
4. The lacrimal apparatus.

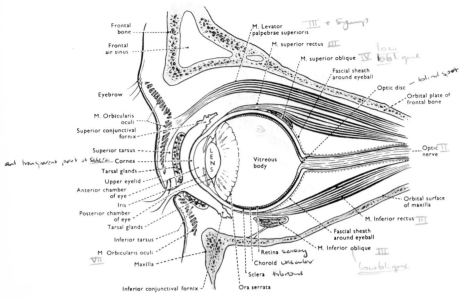

Fig. 4.45 Sagittal section of the eye and orbit. (By courtesy of Professor G. A. G. Mitchell and Dr E. L. Patterson.)

The orbital vessels and nerves

The ophthalmic artery is a branch of the internal carotid artery commencing within the cranial cavity and enters the orbital cavity through the optic foramen with the optic nerve. Within the orbital cavity it supplies the eyeball and its muscles, the optic nerve, and the lacrimal gland. Terminal branches of the artery appear on the face at the orbital margin; these are the lacrimal, supra-orbital, supratrochlear, and dorsal nasal arteries (Fig. 5.33), while other branches (anterior and posterior ethmoidal) supply parts of the nasal cavity and the air sinuses. Further details are given on page 259.

The ophthalmic veins drain through the superior orbital fissure into the cavernous sinus (page 264). They communicate with the anterior facial vein at the medial margin of the orbital rim and with the pterygoid and pharyngeal venous plexuses through the inferior orbital fissure. This is an important pathway for the spread of infection to the cranial cavity.

The *nerves of the orbital cavity are*:
1. The optic nerve.
2. The branches of the ophthalmic division of the trigeminal nerve.
3. The nerves to the muscles of the eyeball.
4. Branches of the parasympathetic component of the autonomic system, the ciliary para-sympathetic ganglion, and branches of the sympathetic component derived from the superior cervical ganglion (page 174).

The orbital muscles (Figs 4.45, 4.46):
1. *Levator palpebrae superioris;* which arises above the optic foramen from the back part of the orbital cavity, and passes forwards to be inserted into the skin of the upper eyelid (Fig. 4.45), and also into the tarsal plate and conjunctiva.
2. The *four rectus* (straight) *muscles*, superior, medial, inferior and lateral. They arise at the back of the orbital cavity from a common tendinous ring surrounding the optic foramen

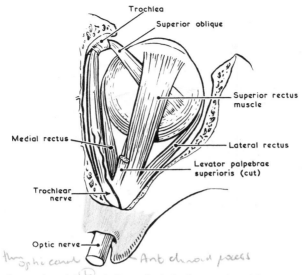

Fig. 4.46 Extrinsic muscles of the eyeball, particularly the superior oblique.

and part of the superior orbital fissue. Each muscle is inserted into the outer coat of the eyeball (the sclera) a short distance behind the corneoscleral junction.

    3. The *two oblique muscles;* the superior and inferior oblique.

        a. The superior oblique rises from the back of the orbital cavity and above and on the medial side of the optic foramen (Fig. 4.47). It runs forward close to the medial wall of the orbital cavity. Near the medial orbital margin it becomes tendinous and passes through a fibrous pulley (the trochlea) attached to the frontal bone, and turns laterally to be inserted into the upper and lateral surfaces of the eyeball just behind its equator.

        b. The inferior oblique muscle arises from a slight depression in the orbital surface of the maxilla close to the junction of the floor and medial walls of the cavity and close to the orbital margin. It passes laterally below the eyeball and is inserted into the lower and lateral surfaces of the eyeball just behind the equator.

    The rectus and oblique muscles, acting together and with the eye muscles of the other side, are responsible for the delicate complicated movements of the eyeballs, essential for true stereoscopic vision.

    The primary movements of the eyeball are:

    1. Upward movement produced by the superior rectus assisted by the inferior oblique to correct inward deviation.

    2. Downward movement produced by the inferior rectus assisted by the superior oblique to correct inward deviation.

    3. Outward movement produced by the lateral rectus muscle.

    4. Inward movement produced by the medial rectus.

    5. Upward and outward movement is produced by the inferior oblique.

    6. Downward and outward movement is produced by the superior oblique.

    The nerve supply to the orbital muscles is as follows:

    1. The oculomotor (third cranial) nerve supplies the levator palpebrae superioris, the superior, medial and inferior rectus muscles and the inferior oblique muscle.

    2. The trochlear (fourth cranial) nerve supplies the superior oblique muscle.

    3. The abducent (sixth cranial) nerve supplies the lateral rectus muscle.

p428

These nerves enter the orbital cavity through the superior orbital fissure.

The ciliary parasympathetic ganglion lies at the back of the orbital cavity behind the eyeball and close to the lateral side of the optic nerve. Parasympathetic preganglionic fibres reach it in branches of the oculomotor nerve, and postganglionic fibres (short ciliary nerves) pass to the eyeball from the ganglion to supply the smooth muscle of the ciliary process and iris (the sphincter pupillae), which control the size of the pupil of the eyeball.

Trauma or surgical lesions to either the cervical sympathetic trunks or to branches from the internal carotid sympathetic plexus that enter the orbit results in Horner's syndrome, the signs of which include constriction of the pupil, drooping of the eyelid and vasodilation on the affected side.

The lacrimal apparatus

This consists of the lacrimal gland, the lacrimal ducts, the conjunctival sac, the puncta lacrimalia, the lacrimal canaliculi, the lacrimal sac and the nasolacrimal duct (Figs 4.41, 4.47). The lacrimal gland occupies the upper lateral part of the orbital cavity just within the orbital rim. The ducts of the gland, up to twelve in number, open into the adjacent part of the conjunctival sac, formed by the capillary space between the surface of the eyeball and the inner surface of the eyelids. The secretion of the lacrimal gland (the tears) is carried medially by the blinking movements of the eyelids, produced by periodic involuntary contraction of the palpebral part

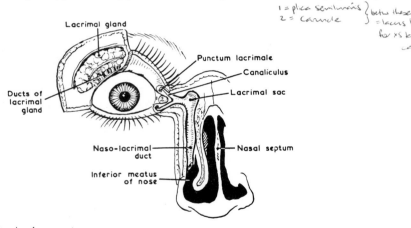

Fig. 4.47 The lacrimal apparatus.

of the orbicularis oculi, to the inner corner of the eye where it enters the puncta lacrimalia. These are the minute papillary openings of the lacrimal canaliculi, located on the margin of each eyelid. Each lacrimal canaliculus runs close to the margin of the corresponding eyelid and then passes either above or below the medial palpebral ligament to end in the lacrimal sac, which occupies the lacrimal fossa situated towards the front of the medial wall of the orbital cavity.

From the lacrimal sac the nasolacrimal duct runs in the bony nasolacrimal canal to open at the lateral wall of the nasal cavity beneath the inferior turbinate bone (Fig. 4.47). The lacrimal secretion is drawn into the lacrimal canals and lacrimal sac by the negative pressure created within the nasolacrimal duct during respiration. If the secretion of the lacrimal glands is excessive, or if some part of the drainage system is blocked, tears run on to the face. The nasal orifice of the nasolacrimal duct is equipped with a flap-like valve of mucous membrane so that a sudden rise in intranasal pressure does not cause fluid to ascend through the canal to the lacrimal sac.

Secretomotor fibres reach the lacrimal glands from the seventh nerve along the great superficial petrosal nerve to the sphenopalatine parasympathetic ganglion, situated in the pterygopalatine fossa. From the ganglion, postganglionic fibres run first with the zygomatic branch of the maxillary nerve, then, by means of a nerve loop, they join the lacrimal nerve for the last part of their course (Figs 5.27, 6.33).

THE PHARYNX, OESOPHAGUS, TONSILS AND SOFT PALATE

The pharynx

The pharynx is a tube-shaped cavity extending from the base of the skull above to the level of the sixth cervical vertebra below, where it becomes continuous with the oesophagus. It lies in front of the cervical part of the vertebral column and behind the nasal and oral cavities and the larynx. It is about 6 inches (15 cm) in length and is divided into three parts for descriptive purposes (Fig. 4.37).

The *nasopharynx* lies behind the nasal cavities and contains the openings of the auditory tubes on its lateral ways and has the pharyngeal tonsil (adenoids) on its posterior wall. Its mucous membrane is lined by ciliated columnar epithelium in its upper part and supplied by sensory branches of the fifth and ninth cranial nerves. It becomes closed off from the oral pharynx when the soft palate is raised during swallowing. Its walls are firmly attached to the base of the skull (basi-occipital and basisphenoid bones) and to the margins of the posterior nares (nasal cavities).

The *oropharynx* (oral pharynx) lies behind the pharyngeal surface of the tongue. The oral cavity communicates with the oral pharynx between the anterior pillars of the fauces (palatoglossal folds). The mouth is shut off from the pharynx when the soft palate lies against the dorsum of the tongue and is in communication with the pharynx when the soft palate is raised. The mucous membrane of the oral part of the pharynx is lined by stratified squamous epithelium continuous with that lining the oral cavity. Its sensory nerve supply is derived from branches of the ninth cranial nerve. On the side wall of the oral pharynx, behind the anterior pillars of the fauces and in front of the posterior pillars or palatopharyngeal folds, are situated the tonsils (Fig. 6.9). The dorsum of the tongue forms the greater part of the anterior wall of this part of the pharynx and contains numerous nodes of lymphoid tissue. The oral region of the pharynx is surrounded by the superior and middle constrictor muscles. The muscous membrane is closely attached to the inner surface of these muscles with the vertical fibres of the palatopharyngeus (extending from the soft palate to the thyroid cartilage) and the salpingopharyngeus (extending from the auditory or Eustachian tube to the thyroid cartilage), intervening.

A ring of lymphoid tissue encircles the entrance to the oropharynx, consisting of the tonsils, the lingual lymphoid tissue and lymph nodules in the soft palate. This lymphoid tissue collectively forms an important defence mechanism against the spread of infections from the mouth into the throat and body tissues. At one time the tonsils were not thought to have great clinical significance and they were removed routinely when inflamed. Nowadays tonsillectomy is carried out much more selectively. The pharyngeal tonsil or adenoids also plays a role in the prevention of infection and, like the tonsils, may become infected and swollen. It may be desirable to remove the tonsils and adenoids surgically under such circumstances, not only to remove the source of infection but also to correct the tendency towards mouth breathing and restore the normal quality of speech, which becomes quite nasal in this condition.

The *laryngeal part of the pharynx* lies behind the epiglottis, the laryngeal opening and the back of the cricoid cartilage. At the sides it partly surrounds the tube-like laryngeal apparatus forming two pouch-like extensions, the *piriform recesses*. Below it is continuous with the oesophagus. The laryngeal part of the pharynx is in direct continuity with the oral part. Its sensory nerve supply is from the ninth and tenth cranial nerves. The middle and inferior constrictor muscles, which surround it and to which its mucous membrane is united, are attached to the hyoid bone and to the thyroid and cricoid cartilages. The *hyoid bone* is a most important element in the pharyngeal apparatus. It has a number of muscles attached to it and its movements play a vital part in the mechanism of swallowing (see page 321).

The side walls of the pharynx are formed for the greater part by the three *constrictor muscles* (Fig. 4.48). Above the upper edge of the superior muscle there is a space in which the muscle

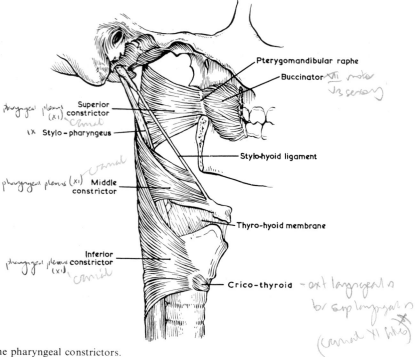

Fig. 4.48 The pharyngeal constrictors.

wall is incomplete on each side. Here the mucous membrane is attached to a fibrous sheet made up of the union of the *pharyngeal aponeurosis* (which elsewhere covers the inner surface of the muscles) and the *buccopharyngeal fascia* (which elsewhere covers the outer surface of the pharyngeal muscles and is continued on to the buccinator muscle). Above, this sheet (*pharyngobasilar fascia*) is attached to the cranial base. Between the two fibrous layers the levator palati and tensor palati muscles run downwards for part of their course before entering the soft palate.

At the upper edge of the superior constrictor the ascending pharyngeal and ascending palatine arteries (the latter a branch of the facial artery) enter the pharyngeal mucous membrane and the soft palate.

Between the middle and superior constrictors, which overlap one another in the posterior pharyngeal wall, the glossopharyngeal nerve and stylopharyngeus muscle pass from the outer surface of the superior constrictor to the inner surface of the middle constrictor.

The stylopharyngeus muscle (Figs 4.48, 5.26)
This arises from the inner (medial) side of the root of the styloid process. It enters the pharynx at the upper edge of the middle constrictor muscle. It is inserted into the superior and posterior

borders of the thyroid cartilage. It is supplied by the ninth (glossopharyngeal) cranial nerve and it elevates the pharyngeal wall.

A space between the lower border of the middle constrictor and the upper border of the inferior constrictor is filled in by the thyrohyoid membrane, which is pierced by the internal branches of the superior laryngeal vessels and nerve.

Below the inferior border of the inferior constrictor muscle, in the angle between the oesophagus and trachea, the recurrent laryngeal nerve and vessels pass into the larynx.

The superior constrictor muscle of the pharynx (Figs 4.48, 6.4)

This muscle arises from the lower part of the posterior edge of the medial pterygoid plate, from the pterygoid hamulus, from the back of the pterygomandibular raphe, from the back of the mylohyoid ridge of the mandible and the mucous membrane of the side of the tongue. The upper fibres pass upwards and backwards and are inserted to the pharyngeal tubercle at the under surface of the basi-occipital bone in front of the foramen magnum. The remainder of the muscle is attached to the posterior mid-line tendinous raphe which gives insertion to the fibres of all three pharyngeal constrictors. Some of the upper fibres are usually thickened to form a nasopharyngeal ridge (Passavant's ridge), which, with the soft palate, helps to close off the nasopharynx during swallowing.   *? = velopharyngeal sphincter*

Nerve supply

The pharyngeal plexus (cranial part of the eleventh or accessory cranial nerve).

Functions

Draws the posterior pharyngeal wall forwards; the upper fibres (Passavant's bundle) constrict the lower end of the nasopharynx.

The middle constrictor of the pharynx (Fig. 4.48)

This originates from both the greater and lesser cornu of the hyoid bone and from the lower end of the stylohyoid ligament. Its fibres pass backwards to gain attachment to the posterior mid-line tendinous raphe.

Nerve supply

The pharyngeal plexus (cranial part of the eleventh or accessory cranial nerve).

Functions

Constricts the oral pharynx from above downwards during swallowing.

The inferior constrictor of the pharynx (Fig. 4.48)

This arises from:

1. The side of the cricoid cartilage.
2. The inferior cornu and oblique line of the thyroid cartilage.
3. From a fibrous arch passing between the thyroid and cricoid cartilages superficial to the cricothyroid muscle. The lower fibres are horizontal; the upper fibres ascend towards the common insertion to the posterior mid-line tendinous raphe. Some of the lower fibres of the muscle are continuous with the circular fibres of the oesophagus.

The inferior constrictor muscle of the pharynx has two parts. The *thyropharyngeus* part of the muscle has an upper border that encloses the middle and superior constrictor muscles as it arches upwards to the level of Passavant's ridge. The lower fibres are thinned out and do not have the other constrictor muscles to support them below the level of the vocal cord. This is a weak area of the pharyngeal wall.

The *cricopharyngeus* muscle is rounded and thicker than the other constrictors and extends without interruption from one side of the cricoid arch to the other, without formation of a posterior mid-line raphe. The muscle acts as a sphincter at the lower part of the pharynx and becomes continuous with the circular muscle of the oesophagus.

Nerve supply

The pharyngeal plexus (cranial part of the eleventh or accessory cranial nerve).

Function

Constricts the laryngeal region of the pharynx from above downwards during swallowing.

On the wall of the pharynx and in the covering connective tissue is the pharyngeal venous plexus, and the pharyngeal plexus of nerves (Fig. 4.36). The venous plexus drains the wall of the pharynx and the pharyngeal mucous membrane, and also the soft palate. The tributaries of the plexus communicate above with the pterygoid plexus and below drain into the internal jugular vein. The nerve plexus is made up of branches from the vagus, glossopharyngeal, and superior cervical sympathetic ganglion. The branches from the glossopharyngeal nerve are sensory; those of the vagus (from the accessory) are motor fibres to the pharyngeal muscles (except for the stylopharyngeus, supplied by the glossopharyngeal nerve) and the muscles of the soft palate (except for the tensor palati, supplied by the fifth cranial nerve). The sympathetic fibres pass to the smooth muscle of the pharyngeal blood vessels and to its mucous glands. The lymphatics draining the pharynx enter the deep cervical chain of lymph nodes.

Closely related to the wall of the pharynx on each side are the common, internal and external carotid arteries, parts of the superior thyroid lingual and facial arteries, the ascending pharyngeal artery, the ninth, tenth, eleventh and twelfth cranial nerves and the sympathetic nerve trunk. The internal jugular vein and the cervical chain of lymph nodes are less intimately related. The deep part of the parotid gland may come in contact with it. The styloid muscles, the pterygoid muscles, the posterior belly of the digastric and the omohyoid muscles are all closely related to the pharyngeal wall. The mandibular nerve lies on its outer side as it emerges from the foramen ovale, the superior and inferior laryngeal nerves also are in contact with the pharyngeal musculature in the neck (Figs 6.7, 4.30).

The cervical part of the oesophagus

The oesophagus begins in the neck at the lower border of the cricoid cartilage opposite to the sixth cervical vertebra, where it is continuous with the laryngeal part of the pharynx (Fig. 4.37). The oesophagus descends in front of the vertebral column to enter the superior mediastinum of the thorax (Figs 3.10, 4.10) and the thoracic part of its course is described on page 83. In the neck the oesophagus is related in front to the trachea and, because it inclines towards the left side in the lower part of the neck, to the left lobe of the thyroid gland. Posteriorly it lies in front of the vertebral column and prevertebral muscles, particularly the longus cervicis. Laterally the oesophagus is related to the carotid sheath, its contents and to the lobes of the thyroid gland. The recurrent laryngeal nerves ascend between it and the trachea and the thoracic duct ascends for a short distance along its left border.

The palatine tonsils

These are masses of lymphoid tissue which project to a varying degree into the isthmus between the mouth and oral pharynx. They are actually in the pharynx, lying immediately behind the anterior pillars of the fauces (Figs 4.37, 6.4, 6.10). Their medial free surfaces are covered by a stratified squamous epithelium and show a number of clefts or *crypts*. Their lateral attached surfaces are separated from the superior constrictor muscle by a fibrous layer which is part of the pharyngeal aponeurosis. The upper pole of each tonsil reaches the soft palate and lies beneath a fold of mucous membrane (*plica semilunaris*). In the upper part of the tonsil there is usually a deep intratonsillar cleft; the lower pole is in contact with the tongue. The tonsil on each side is at the level of the angle of the mandible, being separated from the bone by the medial pterygoid muscle, the styloglossus muscle and the superior constrictor. The latter is pierced by branches of the facial artery, which enter the tonsil from its deep surface and by lymph vessels which drain from the tonsil to the upper deep cervical lymph nodes, especially into a node below the angle of the mandible and closely related to the intermediate tendon of the digastric muscle (*jugulodigastric lymph node*). Other blood vessels reach the tonsil from the lingual artery (dorsal lingual branches), the ascending pharyngeal and descending (greater) palatine arteries. The veins pierce the superior constrictor muscle and enter the pharyngeal venous plexus.

The soft palate
This structure plays an important part in swallowing, in speech, and in mastication. It consists of a central aponeurotic sheet, the *palatal aponeurosis* attached to the posterior edge of the hard palate and contains the following muscles (Fig 4.49):

The levator palati muscle (Fig. 4.50)
This arises from the lower surface of the apex of the petrous temporal bone and the medial side of the cartilage of the auditory tube. Within the soft palate it is attached to the palatal aponeurosis and to the corresponding muscle of the opposite side.

Nerve supply
From the pharyngeal plexus (cranial part of the eleventh or accessory cranial nerve).

Function
Elevates the soft palate.

The tensor palati muscle (Figs 4.49, 4.50, 6.29)
This muscle arises from the scaphoid fossa at the base of the medial pterygoid plate and from the lateral side of the cartilage of the auditory tube. It descends on the lateral side of the medial pterygoid plate in close relation to the origin of the medial pterygoid muscle and ends in a tendon which enters the soft palate above the superior constrictor by hooking around

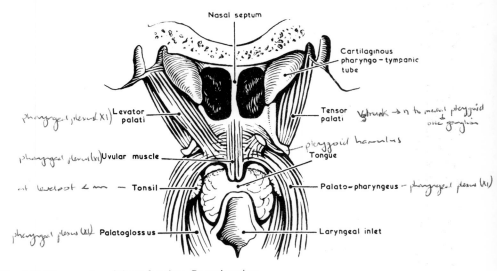

Fig. 4.49 The muscles of the soft palate. Posterior view.

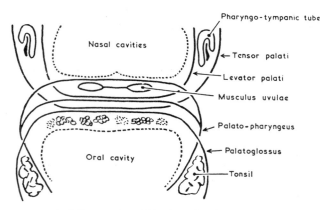

Fig. 4.50 Details of the arrangement of the muscles within the soft palate.

the pterygoid hamulus. There is a small bursa between the muscle tendon and the bone. Within the palate the tendinous fibres spread out to become attached to the palatal aponeurosis and the posterior edge of the hard palate.

Nerve supply
The mandibular division of the fifth (trigeminal) cranial nerve, through the otic ganglion and the branch to the medial pterygoid.

Function
Tightens the soft palate over the dorsum of the tongue.

The palatoglossus muscle (Figs 4.50, 6.9)
This arises within the soft palate from the under surface of the palatal aponeurosis and is continuous across the middle line with the muscle of the opposite side. It runs in the palatoglossal fold (anterior pillar of the fauces), and is inserted into the side of the back part of the tongue.

Nerve supply
From the pharyngeal plexus (derived from the cranial part of the eleventh or accessory cranial nerve).

Functions
Elevates the tongue to the palate and depresses the soft palate to the dorsum of the tongue; the two muscles acting together approximate the palatoglossal arches.

The relations of the tongue muscles are described in the section on the floor of the mouth (page 279).

The palatopharyngeus muscle (Figs 4.49, 4.50, 6.9)
This originates within the soft palate in two layers.
1. The upper posterior layer lies beneath the nasal mucous membrane and is continuous with the fibres of the muscle of the opposite side.
2. The lower anterior layer arises from the palatal aponeurosis and the posterior margin of the hard palate. At the side of the soft palate the two layers unite and the fibres descend downwards and backwards in the palatopharyngeal fold on the side wall of the pharynx and behind the tonsil. The muscle is inserted into the posterior border of the thyroid cartilage

and the mucous membrane of the pharynx. It merges with fibres descending from the cartilage of the auditory tube (salpingopharyngeus muscle).

Nerve supply
From the pharyngeal plexus (cranial part of the eleventh or accessory cranial nerve).

Function
Draws the wall of the pharynx upwards, forwards and medially; the two muscles acting together approximate the palatopharyngeal arches (posterior pillars of the fauces).

The uvular muscle (Fig. 4.49)
This arises from the posterior nasal spine at the back of the hard palate and passes backwards through the soft palate to the tip of the uvula.

Nerve supply
The pharyngeal plexus (accessory nerve).

Function
Shortens the uvula.

Immediately above the oral mucous membrane covering the under surface of the soft palate is a mass of lymphoid and mucous gland tissue which extends forwards for some distance on the under surface of the hard palate. The mucous glands secrete mucus on to the dorsum of the tongue. Their parasympathetic secretomotor fibres are derived from the sphenopalatine ganglion (page 320).

The blood supply of the soft palate is derived from a number of sources. These include:

1. The ascending palatine branch of the facial artery (page 253).

2. The dorsal lingual branches of the lingual artery, which reach the palate through the palatoglossal folds (page 282).

3. The lesser palatine branches of the palatine artery, a branch of the third part of the maxillary artery, which reach the soft palate through the lesser palatine foramina.

The veins of the soft palate drain into the pharyngeal plexus; the lymphatics into the upper deep cervical lymph glands, which they reach after piercing the superior constrictor muscle.

The tensor palati muscle is supplied by the mandibular division of the trigeminal nerve; the remaining muscles by the cranial part of the accessory nerve via the vagus nerves and the pharyngeal plexus. The sensory nerve supply is from the maxillary division of the fifth cranial nerve through the lesser palatine branches. The glossopharyngeal nerve also supplies the lateral part of the soft palate with sensory fibres. The mucous membrane lining the under (oral) surface of the soft palate is covered by stratified squamous epithelium; on the upper nasal surface it is covered by a ciliated columnar epithelium.

Developmental clefts of the soft palate produce disturbances of speech and swallowing by interrupting the continuity of the palatal aponeurosis and the actions of the tensor and levator palati muscles on this sheet of tissue. The horizontally arranged fibres of the tensor palati cannot form the firm barrier of tissue characteristic of the normal soft palate. The levator palati cannot lift this tissue to make contact with the posterior wall of the pharynx or create the velopharyngeal sphincter with the pharyngeal constrictors of the pharynx to control the flow of air and passage of food in the soft palate region.

Surgical procedures to correct clefts of the soft palate seek to reestablish the continuity of the tensor palati across the mid-line of the palate and enable velopharyngeal contact to occur. The surgical procedure sometimes involves fracturing the pterygoid hamuli to allow the tensor palati muscles to be drawn together at the midline more easily. Other procedures involve lifting the mucoperiosteum over the hard palate to permit the soft tissue to be moved posteriorly and thus create the needed 'soft palate' tissue. The part played by the soft palate in swallowing is referred to on page 322.

Oropharyngeal lymphatic tissue—clinical significance

Lymphatic tissue masses, namely, the tonsils, the adenoids, the lymphoid nodules of the posterior part of the dorsum of the tongue and of the soft palate, form an important zone of protective tissue encircling the entrance from the mouth and nose into the pharynx. As in other parts of the body the main function of this tissue is to prevent the spread of infections, in this case into the throat. There is reason to believe from clinical observations that this tissue has an important part to play, not only in the prevention of common inflammatory infections, but also in the prevention of diseases such as poliomyelitis. Consequently the tonsils and adenoids are removed only for essential reasons.

THE LARYNX

The larynx lies opposite the lower end of the pharynx above the upper part of the trachea. Its functions are respiratory, sphincteric, protective and phonation. It allows the passage of air from the pharynx into the trachea and lungs. Sphincteric action by laryngeal muscles prevents the entry of food into the larynx during swallowing and the aspiration of food or liquid during vomiting. Afferent nerve terminations in the laryngeal mucosa are the beginning of reflex pathways associated with coughing, breathing, and swallowing. The production of speech depends on the accurate and smooth coordination of a large number of laryngeal muscles which are easily disordered under a variety of circumstances.

The larynx consists of a number of cartilages making up its skeleton (with the hyoid bone), a special group of muscles, and the vocal folds, which by their movement regulate the size of the entrance to the trachea. The larynx is lined by a mucous membrane continuous with that of the trachea and the pharynx. Its position alters to a considerable extent during the act of swallowing.

The opening of the larynx faces backwards and slightly upwards. It is bounded by the epiglottis in front, the mucous membrane covering the aryepiglottic muscles at the sides, and the arytenoid cartilages behind. Within the aryepiglottic folds are two small nodules of cartilage (the cuneiform and corniculate cartilages). Between the laryngeal inlet and the vestibular fold lies the region of the larynx known as the *vestibule*. Each vestibular fold of the laryngeal mucous membrane overhangs the entrance to a laterally placed cul-de-sac, the *laryngeal sinus* (Fig. 4.51), which continues upwards in the side wall of the larynx as the *saccule*. The mucous membrane of the sinus and the saccule contains numerous mucous glands which lubricate the vocal cords lying immediately below the opening of each laryngeal sinus. Between the vocal cords the laryngeal canal is narrowed to form the *rima glottidis* which varies in size during respiration and voice production. Beneath the vocal cords the larynx leads to the trachea.

In the upper part of the laryngeal vestibule and over the vocal folds the epithelium is of the stratified squamous type. The rest of the vestibule and the lower parts of the larynx are covered by ciliated columnar epithelium. The mucous membrane is loosely attached to the walls of the larynx above the vocal folds and can swell rapidly to the extent that it may obstruct the airway, as in diphtheria. Because the laryngeal lining below the vocal folds is firmly bound to underlying tissues it is not subject to the same degree of swelling. For this reason tracheostomy is a successful way to allow air to enter the lungs, even though the upper part of the larynx may be obstructed by oedematous swelling. The sensory nerve supply to the mucous membrane is from the internal laryngeal nerves above the vocal cords and from the recurrent laryngeal nerves below them. These nerves are branches of the tenth cranial nerve (vagus).

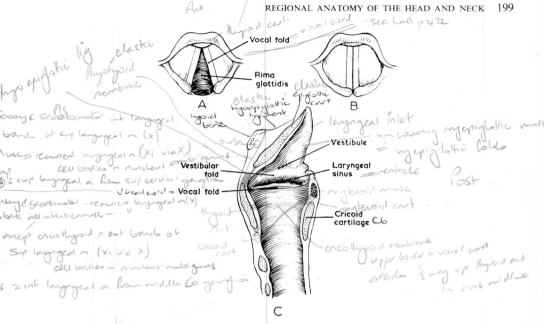

Fig. 4.51 Soft tissue features of the larynx. A. The rima glottidis from above with the vocal folds abducted. B. Adducted vocal folds. C. Median sagittal section of the larynx.

The lymphatics drain forwards to pierce the thyrohyoid and cricothyroid membranes to enter the pretracheal lymph nodes.

The laryngeal cartilages (Figs 4.37, 4.38, 4.52)

The cartilages of the larynx are the thyroid, cricoid, the epiglottis and the arytenoid cartilages.

The *thyroid cartilage* is the largest of the laryngeal cartilages and consists of two laminae, or plates, united in the middle line in front so as to form a projecting keel-like eminence (the Adam's apple). Above this projection there is a deep V-shaped thyroid notch. The posterior border of each lamina is prolonged upwards as a superior horn or cornu attached by ligamentous tissue to the greater cornu of the hyoid bone, and downwards as an inferior horn which articulates at the side of cricoid cartilage. The outer surface of the lamina is marked by an oblique line along which are attached the thyrohyoid, sternothyroid and inferior constrictor muscles.

The *cricoid cartilage* is shaped like a signet ring and is the lowest of the laryngeal cartilages. It is attached below to the uppermost cartilaginous ring of the trachea. It is narrow at the front and sides but is much deeper posteriorly where it gives attachment to the arytenoid cartilages. The inferior horn of the thyroid cartilage articulates with its lateral aspect. The *epiglottic cartilage* is a thin leaf-like structure made of elastic fibrocartilage (Figs 4.8, 4.52). It projects upwards behind the tongue and the body of the hyoid bone, to which it is attached by the elastic hyo-epiglottic ligament. Its tapering lower end is attached to the inner (posterior) surface of the thyroid cartilage by the thyro-epiglottic ligament, below the level of the thyroid notch.

The *arytenoid cartilages* are pyramidal in shape. Each articulates at its base with the cricoid cartilage. The lateral angle of the base forms a muscular process and a vocal process projects forwards to give attachment to the vocal ligament. The apex of each arytenoid cartilage projects upwards and slightly backwards and articulates with the cricoid cartilage by means of a synovial joint.

The thyroid cartilage is united to the hyoid bone by the *thyrohyoid membrane* (Fig. 4.52). This is attached below to the upper margin of the thyroid cartilage and the anterior edge and summit of its superior cornu. Above it is attached to the upper margin of the hyoid bone passing behind this structure (a bursa intervening) and to the inner surface of its greater cornu. The posterior edge of the membrane on each side is somewhat thickened, cord-like and contains a small mass of cartilage (cartilago triticea). The membrane contains a large amount of elastic tissue and is pierced on each side by the internal laryngeal nerve and vessels.

The central part of the *cricothyroid membrane* extends from the upper border of the cricoid cartilage to the middle of the lower border of the thyroid cartilage (Fig. 4.52). The lateral

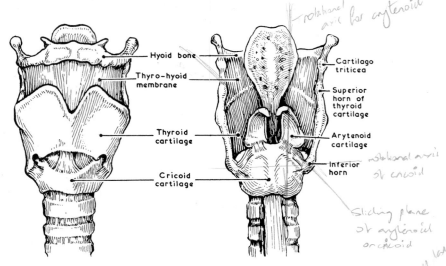

Fig. 4.52 The laryngeal cartilages. Anterior view on the left, posterior view on the right.

part of the membrane, however, is attached to the vocal process of an arytenoid cartilage behind and to the inner surface of the lamina of the thyroid close to the middle line in front (Fig. 4.51). Between these attachments the membrane forms a somewhat thickened free margin, the vocal cord, to which the overlying mucous membrane is closely attached.

The cricoid cartilage can rotate backwards or forwards relative to the thyroid cartilage on a transverse axis passing through the joints between the inferior cornua of the thyroid cartilage and the cricoid cartilage. Each arytenoid cartilage can rotate around a vertical axis running from its apex to its base. It can also slide towards or away from its fellow on the upper edge of the cricoid cartilage. The thyroid and cricoid cartilages can be elevated relative to the hyoid bone which itself has a large range of movement.

The movements of the larynx fall into two groups:

1. Movement of the whole larynx, as occurs during the act of *swallowing*.

2. Movements between the laryngeal cartilages (cricoid, arytenoid and thyroid), as occurs during *voice production*.

The muscles used in swallowing are described on page 322.

The laryngeal muscles (Figs 4.38, 4.48, 4.53)
Functionally the laryngeal muscles can be divided into three groups:

1. Those which regulate the size of the laryngeal inlet; the thyro-epiglottic and aryepiglottic muscles.

2. Those which open and close the space between the vocal cords; the cricoarytenoids (lateral and posterior) and the arytenoid muscle.

3. Those which regulate the degree of tension of the vocal cords; the cricothyroid and vocalis muscles.

The *cricothyroid muscles*. Each of these arises from the front and lateral surface of the cricoid cartilage and is inserted into the inferior cornu and lower border of the posterior part of the thyroid lamina. The posterior fibres of the muscle when they contract cause the thyroid cartilage to slide forwards relative to the cricoid cartilage. The distance between the thyroid and arytenoid cartilages thus becomes greater and the tension of the vocal folds is increased. These muscles elevate the anterior part of the cricoid cartilage and depress the posterior part with the arytenoid cartilages and so tense the vocal cords.

The *lateral crico-arytenoid muscles*. Each of these arises from the upper border of the lateral

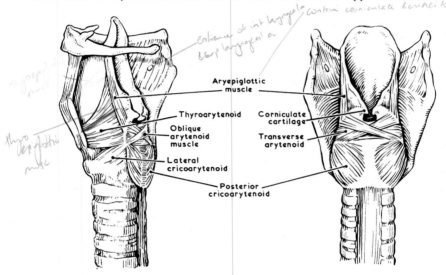

Aryepiglottic muscle
Thyroarytenoid
Oblique arytenoid muscle
Lateral cricoarytenoid
Corniculate cartilage
Transverse arytenoid
Posterior cricoarytenoid

Fig. 4.53 The intrinsic muscles of the larynx. Lateral view on the left, posterior view on the right.

surface of the cricoid cartilage and is attached behind to the muscular process of an arytenoid cartilage. Acting together they rotate the arytenoid cartilages so that the vocal cords are approximated (adducted) (Fig. 4.51B).

The *posterior crico-arytenoid muscles*. Each of these arises from a shallow depression on the posterior surface of the cricoid cartilage and is inserted into the muscular process of an arytenoid cartilage. Acting together they rotate the arytenoid cartilages so that the vocal cords are separated (abducted) (Fig. 4.51A).

The *arytenoid muscle*. This passes from one arytenoid cartilage to the other. It is made up of transverse and oblique fibres. The action of this muscle is to approximate (adduct) the arytenoid cartilages and thus help to close the rima glottidis. The oblique fibres can also help to close the laryngeal inlet.

The *thyro-arytenoid muscles*. Each muscle arises from the inner surface of the thyroid lamina close to the middle line. Some of the fibres (thyro-epiglottic) pass upwards to become attached to the side of the epiglottis; others pass in the wall of the larynx to be attached to the base and lateral surface of the arytenoid cartilage; other fibres (the vocalis muscle) run alongside the vocal cords and are attached to the vocal process of the arytenoid cartilage. The latter part of the muscle regulates the tension of the vocal cords.

The *aryepiglottic muscles*. Each of these muscles passes from the apical region of an arytenoid cartilage to the epiglottis. They regulate the size of the laryngeal orifice.

All the laryngeal muscles, except the cricothyroid, are supplied by the recurrent laryngeal branch of the vagus. The cricothyroid muscles are supplied by the vagus through its superior and external laryngeal branches.

Clinical anatomy. The recurrent laryngeal branch of the vagus supplies most of the intrinsic muscles of the larynx and therefore is a major factor in phonation. It ascends to the larynx in close relation to the inferior thyroid artery, and thus may be damaged during operations on the thyroid gland, with resultant serious disturbance or loss of speech. Inflammatory conditions affecting the lining of the larynx, *laryngitis*, are characterized by disturbances of phonation in the form of hoarseness or temporary loss of the voice. Laryngeal tumours may obstruct the airway and cause progressive respiratory embarrassment and difficulty in swallowing. Small nodular tumours on the vocal folds, which are quite common and often benign, give a characteristic hoarse quality to the voice. Tumours of the deep structures of the neck, for example the oesophagus or pharynx, may invade laryngeal structures and removal of parts of the recurrent laryngeal nerves with permanent loss of phonation may be inevitable during corrective surgery. In such cases the patient may compensate by developing oesophageal speech.

The cervical part of the trachea

The trachea is continuous with the lower end of the larynx and descends from the level of the sixth cervical vertebra to the upper border of the fifth thoracic vertebrae, where it divides into the main bronchi which enter the lungs (page 62). In the neck the trachea is covered anteriorly by the skin, superficial and deep fascia, investing and pretracheal layers, and is overlapped by the sternohyoid and sternothyroid muscles (Fig. 5.39). The isthmus of the thyroid gland lies in front of the second, third and fourth rings of the trachea. Above the isthmus the trachea is crossed by an arterial anastomosis between the two superior thyroid arteries. Below the isthmus it is related to the inferior thyroid veins, the thyroidea ima artery, the remains of the thymus gland and, in the child, is crossed by the brachiocephalic artery a short distance above the manubrium of the sternum. Laterally the trachea is related to the lobes of the thyroid gland, which extend downwards to the level of the sixth tracheal ring, and to the common carotid and inferior thyroid arteries. The recurrent laryngeal nerve ascends towards the larynx on the posterolateral aspect of the trachea in a groove between it and the oesophagus.

THE ROOT OF THE NECK

This important region (Figs 4.8, 4.9, 4.28, 4.36), situated at the junction of the thoracic cavity and the neck, is bounded posteriorly by the first thoracic vertebra, laterally by the first rib and anteriorly by the upper border of the manubrium of the sternum. Through this *thoracic inlet*, in the middle line in front of the vertebral column, the oesophagus, trachea and the thoracic duct pass into or out of the thoracic cavity. Lateral to these structures are the recurrent laryngeal nerves which lie in the groove between the trachea and oesophagus; the brachiocephalic veins; the brachiocephalic artery; the left common carotid and subclavian arteries; the vagus and phrenic nerves; the cervical cardiac nerves derived from the vagus nerves and the cervical sympathetic trunks. On each side, immediately in front of the neck of the first rib, there is the sympathetic trunk often with its stellate ganglion; the anterior primary ramus of the first thoracic spinal nerve which passes over the first rib to take part in the formation of the brachial plexus, and the superior intercostal artery, a branch of the costocervical trunk of the subclavian artery, descending to supply the tissues in the first two intercostal spaces.

Closely associated with the phrenic nerves and crossing in front of or behind them are the internal thoracic arteries. Each is a branch of the first part of the subclavian artery and descends into the thorax over the lung apex. Arising from the upper convexity of the first part of the subclavian artery, the vertebral artery passes upwards to enter the foramen transversarium of the sixth cervical vertebra. Lateral to the origin of the vertebral artery is the thyrocervical trunk (also a branch of the subclavian) which divides almost immediately into transverse cervical, suprascapular and inferior thyroid arteries. *the 1st 2 bend the phrenic n onto scalenus ant*

In front of the trachea are the inferior thyroid veins which drain into the left brachiocephalic vein and the smaller midline thyroidea ima artery, most often a branch of the brachiocephalic artery but occasionally arising directly from the aortic arch. The apices of the lungs ascend into the root of the neck through the thoracic inlet rising one to two inches above the level of the first ribs in front and occupying on each side of the vertebral column a tent-like space beneath the scalenus anterior and scalenus medius muscles. Each lung apex lies immediately behind the subclavian artery from which it is separated by a thickened portion of the parietal pleura—the *suprapleural membrane* of Sibson's fascia. This membrane is a derivative of a small muscle slip which arises from the transverse process of the seventh cervical vertebra. *scalenus minimus*

It will be appreciated that a tumour growing in the constricted space of the root of the neck could produce serious effects by exerting pressure on these important structures. Similar effects could result from the fracture of neighbouring bones.

5. Systematic Anatomy of the Head and Neck

THE BONES OF THE SKULL

The skull is described (page 124) as consisting of two parts:

1. The cranium which surrounds, supports and protects the brain;
2. The facial skeleton which is related to the orbital cavities, the nasal cavities and the mouth (Figs 5.1, 5.2).

The facial skeleton consists of an upper part which is firmly united to the cranium, and a lower part, the mandible, attached to the cranium on each side by the ligaments of the mandibular joints.

The individual bones which make up the parts of the skull are:

CRANIAL BONES

The occipital bone
The temporal bones
The sphenoid bone
The cribriform and perpendicular plates of the ethmoid bone
The frontal bone
The parietal bones

FACIAL BONES

The maxillae
The mandible
The palatine bones
The zygomatic bones
The nasal bones
The vomer
The facial ethmoid bones
The inferior turbinate bones
The lacrimal bones
The hyoid bone (which is situated in the neck)

CRANIAL BONES

The occipital bone (Figs 5.2, 5.3, 5.4)

Position
Base and back of the cranial part of the skull, posterior cranial fossa.

Parts
Basi-occipital, squamous, condylar.

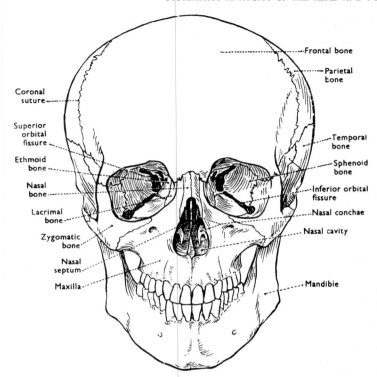

Fig. 5.1 The skull—viewed from the front (norma frontalis). (By courtesy of Professor G. A. G. Mitchell and Dr E. L. Patterson.)

Articulations
Parietals, temporals, sphenoids, atlas.

Foramina and other features
Foramen magnum, anterior condylar canal, posterior condylar canal, jugular foramen (between occipital and temporal).

The occipital is a large saucer-like bone situated at the back of the cranium with its concavity facing upwards and forwards, containing in its lower part the foramen magnum through which the spinal cord is continuous with the brain stem. The outer surface of its squamous part, behind and above the foramen magnum, gives attachment to some of the muscles of the back of the neck. These muscles play an important role in supporting the skull on the vertebral column. They include:

1. Elements of the great sacrospinalis muscle mass on the posterior aspect of the vertebral column—semispinalis and longissimus capitis.
2. Muscles which have attachment to elements of the pectoral girdle—sternomastoid (sternum and clavicle) and trapezius (clavicle and scapula).
3. Muscles attached to the atlas or axis vertebra involved in nodding or side to side movements of the head—rectus capitis posterior major and minor, obliquus capitis superior and inferior (which is not attached to the skull) form the boundaries of the suboccipital muscular triangle and are described on page 176.

In the middle line at the upper end of the region of muscle attachment is the external occipital

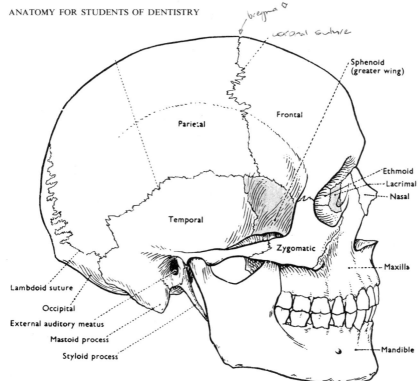

Fig. 5.2 The skull—viewed from the side (norma lateralis). (By courtesy of Professor G. A. G. Mitchell and Dr E. L. Patterson.)

protuberance, the central point of which is called the inion. The most projecting part of the skull, however, is at a varying distance above the protuberance. There is great individual variation in the shape of this part of the bone forming the back of the skull.

The inner surface of the squamous region is related to the cerebellum immediately above the foramen magnum and shows on each side of the middle line two large shallow cerebellar fossae. These are limited above by the grooves related to the transverse dural venous sinuses, to the lips of which is attached the tentorium cerebelli. Above the sinus grooves the bone is related to the occipital lobes of the cerebral hemispheres of the brain. These are separated in the middle line by the posterior part of the superior sagittal venous sinus and the attachment of the falx cerebri.

The condylar part of the bone lies at the sides of the foramen magnum and carries the convex occipital condyles on its under surface. The condyles articulate with the concave articular fossae on the upper surface of the atlas vertebra. Nodding movements of the head on the vertebral column occur at these joints. Rotatory movements occur at the plane joints between the atlas (first cervical vertebra) and the axis (second cervical vertebra).

On the cranial surface of the occipital bone just above the foramen magnum on each side is the internal opening of the anterior condylar canal. Its external opening lies immediately above and in front of the condyles. It transmits the hypoglossal (twelfth) cranial nerve. The posterior condylar canal, when present, opens on the exterior of the cranium behind the condyle; it transmits an emissary vein from the terminal part of the transverse (sigmoid) venous sinus. Between the outer margin of the condylar part of the occipital bone and the base of the petrous part of the temporal bone is the jugular foramen, the meeting place of the sigmoid

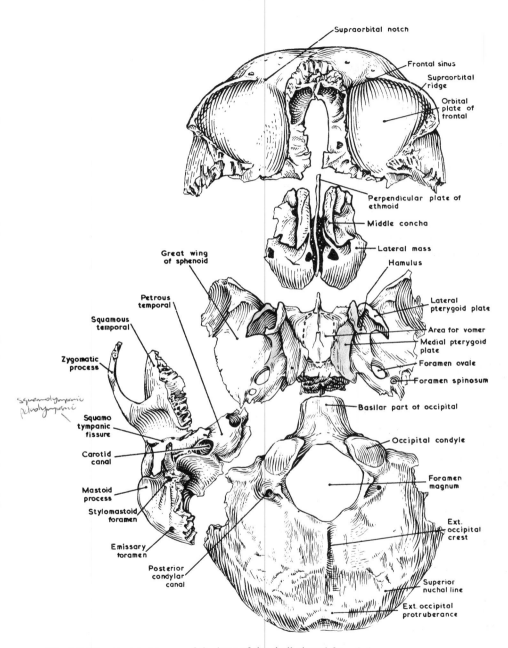

Fig. 5.3 Disarticulated bones of the base of the skull viewed from below.

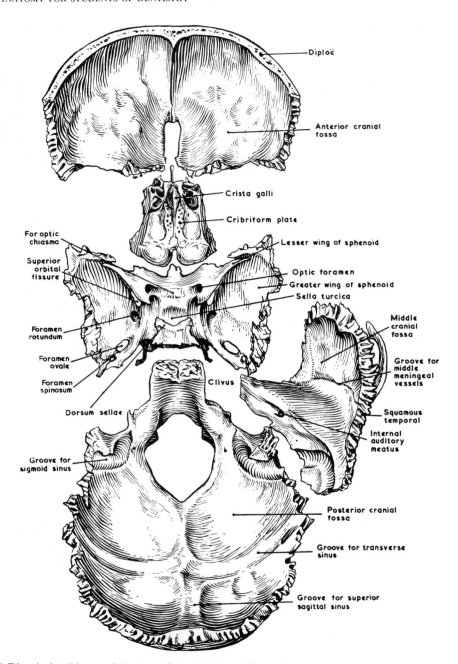

Fig. 5.4 Disarticulated bones of the base of the skull viewed from above.

and inferior petrosal venous sinuses. At the foramen these unite to form the internal jugular vein (page 165). The glossopharyngeal, vagus and accessory (ninth, tenth and eleventh cranial) nerves also pass through the jugular foramen.

The basilar part of the bone lies in front of the foramen magnum and is united in the adult skull with the body of the sphenoid bone. Up to about 16 years of age, however, the two bones are separated by a plate of cartilage spheno-occipital synchondrosis), which plays an important part in the growth of the skull (Fig. 2.14). Between the side of the basi-occipital and petrous temporal bones on the inner (cranial) surface are the grooves for the inferior petrosal venous sinuses running backwards to the jugular foramina. The upper surface (clivus) of the basi-occipital is related to the medulla of the brain stem and the vertebral and basilar arteries. The whole of the inner surface of the occipital bone forms the base of the posterior cranial fossa. The inferior (external) surface of the basilar part shows a small mid-line pharyngeal tubercle, to which the upper fibres of the superior constrictor muscles of the pharynx and the mid-line pharyngeal raphe are attached. The most anterior point on the margin of the foramen magnum is called the *basion*, the posterior limit of the mid-line cranial base.

Development

The squamous part of the bone above the muscle markings develops in membrane from two bilateral centres of ossification appearing in the eighth week of intra-uterine life. The remainder of the bone develops in the back part of the chondrocranium (page 367) within which centres appear for the lower squamous region (two centres), the condylar region (two centres, one on each side), and the basilar part (one centre).

Between three and four years of age the squamous part unites with the condylar part; union between the basilar and condylar parts takes place about the sixth year.

The temporal bone (Figs 5.3, 5.4, 5.5)

Position

Side and base of the cranial part of the skull, temporal fossa, middle cranial fossa, posterior cranial fossa.

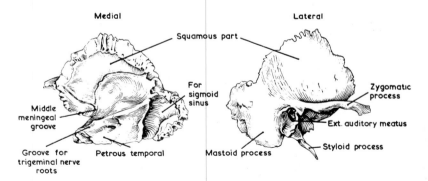

Fig. 5.5 Medial and lateral aspects of the temporal bone.

Parts

Squamous, with zygomatic process and glenoid fossa; tympanic plate; petromastoid; styloid process.

Articulations

Occipital, sphenoid, parietal, sometimes frontal, zygomatic, mandible.

Foramina and other features

Stylomastoid canal, canal for tympanic nerve, canal for auricular nerve, jugular foramen (between temporary and occipital), foramen lacerum (between temporal and sphenoid), external auditory canal, internal auditory canal, bony part of auditory tube, middle ear cavity, tympanic antrum and mastoid air cells.

The squamous part of the temporal bone forms the lower side wall of the cranial part of the skull. The zygomatic process projects from its lateral surface and passes anteriorly to articulate with the zygomatic bone, to form the zygomatic arch. Above the zygomatic arch the temporal muscle is attached. Below the zygomatic arch the squamous region of the bone forms the glenoid fossa and articular eminence of the mandibular joint (page 307). The squamous part of the bone also forms the upper rim and roof of the external auditory meatus. Its inner surface is related to the lateral surface of the temporal lobe of the cerebral hemisphere and shows well-marked grooves for the branches of the middle meningeal artery and their accompanying veins.

The tympanic plate forms the anterior wall and floor of the external auditory meatus and the floor of the middle ear cavity (tympanic cavity). It is separated from the glenoid fossa by the squamotympanic (petrotympanic) fissure.

The petromastoid part of the temporal bone extends from the mastoid process, behind the external auditory meatus, forwards and inwards on the base of the skull to the foramen lacerum, forming a wedge of bone between the great wing of the sphenoid on its outer side and the basilar part of the occipital and body of the sphenoid on its inner side (Figs 5.1, 6.29).

The mastoid process is especially well-developed in male skulls and gives attachment to the sternomastoid muscle of the neck. On its medial side is a deep groove for the origin of the posterior belly of the digastric muscle. At the front of this groove is the opening of the stylomastoid foramen. Through this emerges the facial (seventh) cranial nerve and into it passes a branch of the posterior auricular artery. The foramen is the termination of the facial canal which begins on the interior of the skull at the internal auditory meatus. In front of the stylomastoid foramen and embraced behind by the tympanic plate is the styloid process, which varies considerably in length.

On the under surface of the wedge-like petrous process in front of the styloid process are the openings of the jugular foramen and the carotid canal. The latter is the passage by which the internal carotid artery and the carotid nerve plexus (from the superior cervical sympathetic ganglion) enter the cranial cavity.

At the medial side of the under surface of the petrous process there is a deep groove running from the foramen lacerum in front to the jugular foramen behind. It separates the temporal and occipital bones. At the lateral side of the under surface of the petrous process is another groove separating the temporal and sphenoid bones. Behind it leads to the opening of the bony part of the auditory tube. In the living the groove is converted into a canal by a floor of cartilage forming the terminal cartilaginous part of the tube that opens in front into the nasopharynx close to the base of the medial pterygoid plate.

Between the carotid canal and the jugular foramen is the small opening of the tympanic canal along which the tympanic branch of the glossopharyngeal nerve reaches the middle ear.

As seen from the interior of the cranial cavity the petrous part of the temporal bone shows a superior surface forming part of the floor of the middle cranial fossa, and a posteromedial surface forming the side wall of the posterior cranial fossa (Fig. 5.6).

On the superior surface a small groove runs from a foramen (hiatus for the greater superficial petrosal nerve) to the foramen lacerum. It contains the greater superficial petrosal nerve, a branch of the facial nerve given off in the facial canal.

On the posteromedial surface is the large opening of the internal auditory meatus. It transmits the facial nerve, which emerges later from the stylomastoid foramen and the vestibulo-cochlear (eighth cranial) nerve, which commences as a nerve of special sense in the inner ear.

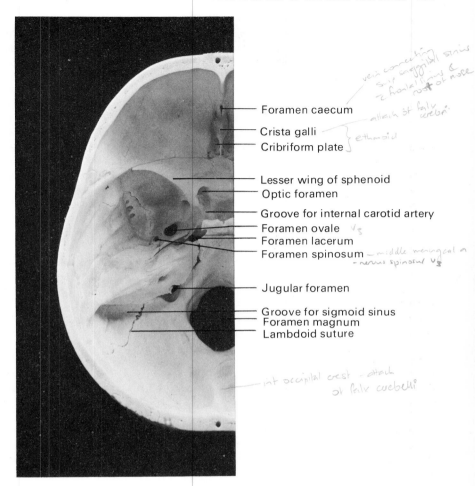

Fig. 5.6 The left side of the base of the skull viewed from above after removal of the vault.

Between the superior and posteromedial surfaces of the petrous temporal, the groove for the superior petrosal venous sinus runs from the region of the foramen lacerum backwards and laterally to the groove for the transverse venous sinus. Near the apex of the petrous temporal is the trigeminal impression for the trigeminal nerve and its ganglion (Fig. 5.10).

The middle ear (tympanic cavity) (Figs 4.41, 5.6, 5.7)
This is an elongated slit-like cavity situated in the temporal bone. Its medial wall is formed by the petrous temporal bone; its roof by the tegmen tympani; its floor by the tympanic plate, and its lateral wall contains the tympanic membrane. Into its front end opens the auditory (pharyngotympanic) tube; from its back end opens a cul-de-sac, the tympanic antrum, with which the mastoid air cells communicate.

Between the *tympanic membrane* in the lateral wall of the cavity and the oval *vestibular foramen* in the medial wall, the cavity is traversed by a chain of bony ossicles, the *malleus*, the *incus* and the *stapes*. The footplate of the stapes is attached around the margin of the vestibular foramen. Vibrations of the tympanic membrane move the bony ossicles so that the footplate of the stapes moves like a piston at the commencement of the vestibular canal of the inner ear. The complicated membranous canals of the inner ear contain fluid (endolymph) which is set in motion by the movement of the stapes at the vestibular foramen (see page 429).

The auditory tube, the tympanic cavity, the inner surface of the tympanic membrane, the tympanic antrum and the mastoid air cells are all lined by a mucous membrane continuous with that of the nasopharynx. The mucous membrane also covers the bony ossicles. The sensory nerve supply is derived from the glossopharyngeal nerve, and its blood supply from a variety of sources including branches of the posterior auricular artery (page 164), the tympanic branch of the maxillary artery (page 156), some branches of the internal carotid given off in the carotid canal, and branches of the ascending pharyngeal and ascending palatine arteries.

The tympanic cavity contains two small muscles, the tensor tympani, which arises within

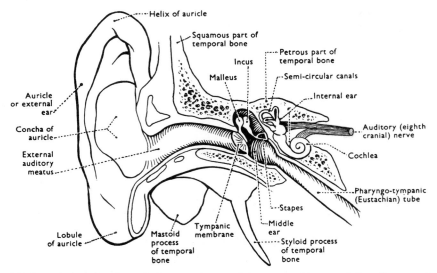

Fig. 5.7 A vertical section through the petrous part of the temporal bone and the adjacent soft parts to show the external, middle and internal ears. (By courtesy of Professor G. A. G. Mitchell and Dr E. L. Patterson.)

a bony canal above the auditory tube and is attached to the malleus, and the stapedius muscle, which arises within a pyramidal process at the back wall of the cavity and is inserted into the stapes. These muscles regulate the movements of the ossicles. The tensor tympani is supplied by the mandibular division of the fifth nerve, as the malleus develops from Meckel's cartilage in the first pharyngeal arch; and the stapedius muscle by a branch of the seventh (facial) nerve (page 249) because the stapes develops from the second or hyoid arch.

The *tegmen tympani*, a thin shelf of bone projecting laterally from the petrous temporal and articulating with the squamous temporal, forms the roof of the tympanic cavity and separates the middle ear from the temporal lobe of the brain. The floor of the middle ear is separated from the jugular bulb by the tympanic plate. On the medial wall of the cavity as well as the vestibular foramen is the cochlear foramen, closed by the secondary tympanic membrane, a part of the inner ear mechanism, and a rounded promontory grooved by the nerves making up the tympanic plexus (page 250). Along the angle between the roof (tegmen tympani) and the medial wall runs the intermediate part of the canal for the facial nerve (page 248). Into the front of the middle ear cavity opens the auditory tube; above it is the canal for the tensor tympani muscle and below it a plate of bone separating the middle ear from the carotid canal.

Development
The squamous and tympanic parts each develop in membrane from single ossification centres. The styloid process develops from two centres in the back of the cartilaginous skeleton of the second (hyoid) pharyngeal arch. The petromastoid part of the bone develops from four centres of ossification in the cartilaginous capsule of the inner ear. At birth the squamous, tympanic and petromastoid are separate bony elements and the styloid process is still largely cartilaginous. The tympanic ring unites with the squamous part about the time of birth; the petromastoid unites with the squamous part during the first year.

The malleus, meaning a hammer (one of the middle ear bones), develops from the most distal portion of Meckel's cartilage (the skeleton of the first, or mandibular, pharyngeal arch). The incus (anvil) develops from the back of the pterygoquadrate bar, in the posterior part of the maxillary process, and the stapes (stirrup) from the most distal part of the cartilaginous skeleton of the second, or hyoid, pharyngeal arch. All three ear ossicles are ossified at birth and form a chain of bony elements across the middle ear from the tympanic membrane to the opening of the inner ear canal (fenestra vestibuli), which is filled by the footplate of the stapes (see page 211).

The sphenoid bone (Figs 5.3, 5.4, 5.6, 5.8)

Position
Base of cranial part of skull, part of upper facial skeleton, temporal fossa, orbital cavity, anterior and middle cranial fossa.

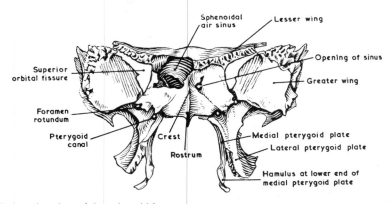

Fig. 5.8 Anterior view of the sphenoid bone.

Parts
Body (presphenoid and postsphenoid), greater and lesser wings, medial and lateral pterygoid plates, sphenoid conchae.

Articulations
Occipital, temporal, frontal, ethmoid, palatine, vomer, zygomatic.

Foramina and other features
Optic foramen, foramen rotundum, ovale and spinosum, pterygoid canal, superior orbital fissure, foramen lacerum (with temporal bone) and sphenoidal air sinuses. These are all bilateral.

The sphenoid bone consists of a body which occupies the middle of the cranial base between the basilar part of the occipital bone behind and the ethmoid (cribriform plate) in front; a greater and lesser wing attached on each side to the body; and medial and lateral pterygoid plates which project from its lower surface.

The upper surface of the body of the sphenoid is hollowed out for an endocrine gland, the *hypophysis cerebri* or *pituitary* gland, to form the hypophysial or pituitary fossa. The fossa is bounded behind by the transverse, vertical, shelf-like crest, the dorsum sellae; in front by a low transverse ridge, the tuberculum sellae. Between the tuberculum sellae and the upper edge of the dorsum in the living stretches a double fold or diaphragm of dura mater, forming the roof of the pituitary fossa. This is pierced in the middle by an opening for the stalk (infundibulum) which unites the gland to the base of the brain. On each side of the pituitary fossa the dura mater splits into two layers to contain the cavernous venous sinus (Fig. 5.9). Small sinuses, the intercavernous sinuses, connect the cavernous sinuses of the two sides with one another.

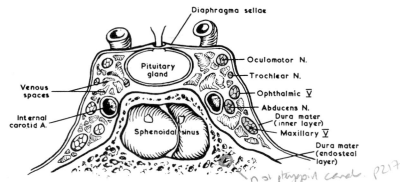

Fig. 5.9 Coronal section through the cavernous venous sinuses and body of the sphenoid.

In front of the tuberculum sellae is a shallow groove running between the optic foramen on each side. In spite of its name this optic groove does not contain the optic chiasma, uniting the two optic nerves which pass through the optic foramina with the ophthalmic arteries. The chiasma is in fact related to the tuberculum sellae. Between the anterior edge of the optic chiasma groove and the back of the cribriform plate of the ethmoid, the upper surface of the body of the sphenoid forms part of the floor of the anterior cranial fossa (planum sphenoidale). The pituitary fossa forms the middle (mid-cranial) part of the middle cranial fossa, while the body of the sphenoid behind the dorsum sellae forms part of the posterior cranial fossa (Fig. 5.6).

At the side of the body of the sphenoid within the cranial cavity a shallow groove, bounded laterally by a sharp bony margin called the lingula, runs from the foramen lacerum towards the optic foramen. It indicates the intracranial course of the internal carotid artery. At the back of the optic foramen the artery gives off the ophthalmic branch, which enters the orbital cavity through the optic foramen, and ends as the middle and anterior cerebral arteries which with the corresponding arteries of the opposite side and the basilar artery (derived from the two vertebral arteries entering through the foramen magnum) form an arterial circle (circle of Willis) at the base of the brain. From this circle of arteries the brain derives its blood supply (Fig. 5.10).

The under surface of the body of the sphenoid, together with the basilar part of the occipital bone, forms the roof of the nasopharynx from the pharyngeal tubercle on the occipital bone, as far forward as the back of the nasal septum. In front of this body of the sphenoid forms the roof of the back part of each nasal cavity and supports the upper end of the vomer.

The lesser wing of the sphenoid bone on each side is attached to the upper surface of the

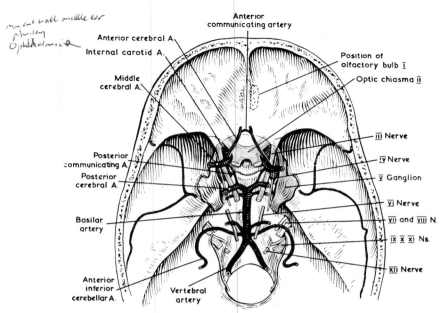

Fig. 5.10 Base of the cranium viewed from above showing the cranial nerves and the formation of the arterial circle of Willis.

anterior part of its body (presphenoid element) by two roots which form the roof and base of the optic foramen (Fig. 5.11). The upper surface of each lesser wing (orbitosphenoid) forms part of the floor of the anterior cranial fossa; the under surface forms part of the roof of the back of the orbital cavity. Its posterior margin is free and separates the anterior and middle cranial fossae. The anterior edge of the lesser wing articulates with the orbital plate of the frontal bone in the floor of the anterior cranial fossa. On either side of each optic foramen the lesser wings project backwards as the anterior clinoid processes to give attachment to the tentorium cerebelli.

The greater wings of the sphenoid (alisphenoids) are attached to the sides of the posterior part of the body (postsphenoid element) between the inner end of the superior orbital fissure in front and the foramen lacerum behind. Each greater wing has three surfaces:

An *upper cerebral surface* forming part of the anterior wall, the floor and side wall of the middle cranial fossa (Figs 5.6, 5.11).

A *lower and outer surface* forming the roof of the infratemporal fossa and part of the side wall of the temporal fossa (Figs 5.3, 6.18).

An *anterior of orbital surface* forms the back part of the lateral wall of the orbital cavity and the posterior wall of the upper part of the pterygopalatine fossa (Figs 4.44, 5.8).

The upper cerebral surface of the greater wings of the sphenoid on each side show three foramina (Fig. 5.11).

The *foramen rotundum* passes forwards to open into the pterygopalatine fossa of the articulated skull and transmits into the maxillary division of the trigeminal nerve.

The *foramen ovale* opens on the roof of the infratemporal fossa and transmits the motor and sensory roots of the mandibular nerve, the accessory meningeal artery, an emissary vein connecting the cavernous sinus with the pterygoid venous plexus and sometimes the lesser superficial petrosal nerve.

The *foramen spinosum* also opens on the roof of the infratemporal fossa, immediately in front of the spine of the sphenoid. It transmits the middle meningeal branch of the mandibular nerve (nervus spinosus) and the middle meningeal artery (Fig. 5.6).

Between the anterior edge of the cerebral surface of the greater wing and the lesser wing is the superior ophthalmic (or orbital) fissure. Through it pass the third, fourth and sixth cranial nerves (motor nerves to the muscles of the eyeball), the ophthalmic division of the fifth nerve (lacrimal, frontal and nasociliary branches), two ophthalmic veins, a small ophthalmic branch of the middle meningeal artery and a meningeal branch of the lacrimal artery.

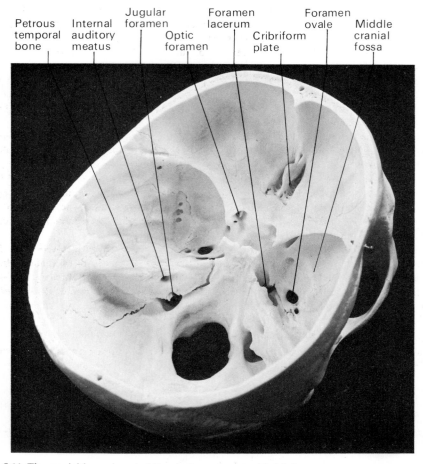

Fig. 5.11 The cranial base viewed obliquely from above and behind.

On the outer surface of the greater wing, the infratemporal crest separates the infratemporal fossa, which faces downwards, from the temporal surface of the bone, which faces outwards. The temporal fossa at the side of the skull is made up of the greater wing of the sphenoid, the squamous part of the temporal, the parietal and part of the frontal bone (Figs 4.2, 6.18). It gives origin to the temporal muscle.

To the infratemporal fossa on the under surface of the great wing of the sphenoid is attached the upper head of the lateral pterygoid muscle.

The pterygoid plates descend from the under surface of the greater wings at the region of the attachment to the body of the sphenoid. In front the two plates are united except in the lower part where a notch separates them. In the articulated skull the notch is filled in by the tuberosity of the palatine bone. Between the medial and lateral pterygoid plates on each side is the large pterygoid fossa which faces backwards. The lateral pterygoid plate gives attachments to the lower head of the lateral pterygoid muscle on its outer surface and to the upper (deep) head of the medial pterygoid on its inner surface.

The medial pterygoid plate forms the side wall of the back of the nasal cavity. On its lateral side the tensor palati muscle descends towards the soft palate. At the lower end of the medial plate there projects the pterygoid hamulus to which is attached the pterygomandibular raphe (Fig. 5.8). At the base (upper end) of the medial pterygoid plate is the small boat-shaped scaphoid fossa from which the tensor palati muscle takes origin. On the inner side of the scaphoid fossa, in the living, the auditory tube opens at the junction of the roof and side wall of the nasopharynx.

From the front of the foramen lacerum the pterygoid canal runs forwards through the sphenoid bone to open at the back of the pterygopalatine fossa in the articulated skull (Figs 5.8, 6.18). It transmits the nerve of the pterygoid canal (Vidian nerve), and a branch of the third part of the maxillary artery (Fig. 4.41).

The nerve of the pterygoid canal is formed by the union of the greater superficial petrosal nerve, containing parasympathetic fibres from the facial nerve, with the carotid nerves (deep petrosal) composed of sympathetic nerve fibres ascending from the cervical sympathetic trunk.

The front surface of the body of the sphenoid in a disarticulated bone shows a mid-line crest or rostrum which articulates with the perpendicular plate of the ethmoid (in the upper part of the nasal septum). On either side of the crest are the openings of the sphenoidal air sinuses. In the articulated skull these openings are partly closed by the sphenoidal conchae which in early life are separate bones.

The sphenoidal air sinuses occupy the front part of the body of the sphenoid (Fig. 5.8) and are closely related to the pituitary gland, the internal carotid arteries, the cavernous venous sinuses and the optic nerves, as the latter traverse the optic foramina. If the sinuses are large they may extend into the roots of the greater wings and the back of the body of the sphenoid. The sinuses are usually asymmetrical in size and shape.

Development

Centres of ossification appear in the chondrocranium for the following parts of the sphenoid bone beginning in the eighth week of intra-uterine life:

1. A centre for each of the lesser wings.
2. Two centres for the presphenoid part of the body.
3. A centre for each sphenoidal concha and lingula.
4. A centre for the root of each of the greater wings (adjacent to the foramina rotundum and ovale).
5. Two centres for the postsphenoid part of the body.
6. A centre for the hamulus of each medial pterygoid plate.

Centres appear in membrane for:

1. The squamous part of the great wing, and the lateral pterygoid plate.
2. The medial pterygoid plates.

The presphenoid and postsphenoid parts of the body unite in man late in foetal life (eighth month). The place of union is immediately in front of the pituitary fossa. In most animals they are separated by a plate of cartilage until after birth.

At birth the sphenoid consists of three parts: the body with the lesser wings, and, on each side, the greater wings with the pterygoid plates. During the first year the greater wings commence to unite with the side of the body around the margins of the pterygoid canal.

The ethmoid bone (Figs 5.3, 5.4, 5.14)

Position
Floor of anterior cranial fossa, lateral wall of nasal cavity, nasal septum, medial wall of orbital cavity.

Parts
Cribriform plate, crista galli, perpendicular plate, facial part labyrinth).

Articulations
Sphenoid, frontal, lacrimal, maxilla, palatine, vomer, inferior turbinate.

Foramina and other features
Foramina in cribriform plate for olfactory nerve filaments, anterior and posterior ethmoidal foramina, foramen caecum (between ethmoid and frontal bone) and the ethmoidal air cells.

The ethmoid bone contributes both to the cranial and facial regions of the skull. Developmentally it consists of a mid-line cranial element (mesethmoid) from which develops the *crista galli* and the *perpendicular plate*; and two bilateral box-like *facial parts*. The two parts of the bone are united after birth by the ossification of the *cribriform plate*.

As seen from the interior of the cranial cavity (Fig. 5.6), the cribriform plates, divided by the mid-line projection of the crista galli, occupy the middle of the anterior cranial fossa between the orbital plates of the frontal bone and the planum sphenoidale of the anterior part of the body of the sphenoid. The cribriform plates separate the anterior cranial fossa from the nasal cavities and through them the filaments of the olfactory nerves pass to the olfactory bulbs of the brain from the olfactory mucous membrane. The crista galli gives attachment to the front of the falx cerebri, the large vertical septum of dura mater separating the cerebral hemispheres. At the front of the crista galli is situated the foramen caecum. Through it a small vein may connect the superior sagittal sinus with the veins of the roof of the nasal cavity.

As seen from the orbital cavity the ethmoid forms a smooth quadrilateral-shaped plate in the medial wall between the frontal bone above and the maxilla and palatine bones below, with the lacrimal in front and the sphenoid behind (Fig. 4.44). At the suture separating the frontal and ethmoid are two small openings, the anterior and posterior ethmoidal foramina. They open in the anterior cranial fossa at the side of the cribriform plate. They transmit the ethmoidal nerves and arteries which are branches of the nasociliary nerve and the ophthalmic artery.

In the nasal cavities the ethmoid forms:

The *upper part of the lateral wall*. This is the facial part or labyrinth of the bone. It contains the ethmoidal air cells, anterior, middle, and posterior groups, and sends two projecting processes, the superior and middle (ethmoidal) turbinate processes (conchae) and a curved uncinate process into the nasal cavity (Figs 4.39, 4.40).

The *roof of the middle region of each nasal cavity* formed by the cribriform plate.

The *upper part of the nasal septum* formed by the perpendicular plate of the ethmoid (mesethmoid) (Fig. 5.14).

The facial portion of the ethmoid helps to reduce the size of the opening of the maxillary antrum in the articulated skull, and sends a process (the uncinate process) downwards to articulate with the inferior turbinate bone (Fig. 4.39).

Development
The ethmoid develops in the front part of the cranial basal plate of the chondrocranium and in the nasal capsule. A centre of ossification for each facial part (labyrinth) appears during the fifth month of foetal life and this part of the bone is ossified at birth. During the first year a centre appears at the base of the crista galli and gradually spreads downwards as the perpendicular plate in the cartilage of the nasal septum. The cranial and facial parts unite at the cribriform plate between two and five years.

The frontal bone (Figs 5.1, 5.3, 5.4)

Position
Anterior part of cranium, floor of anterior cranial fossa, roof of orbital and nasal cavities.

Parts
Squamous, orbital plates, nasal and zygomatic processes.

Articulations
Parietal, sphenoid, ethmoid, maxilla, nasal, lacrimal, zygomatic, sometimes the temporal (at pterion).

Foramina and other features
Foramen caecum (between frontal and ethmoid), supra-orbital notch or foramen and the frontal air sinus.

The *squamous* (leaf-like) part of the frontal bone forms the forehead region of the skull. Its posterior margin articulates with the parietal bones along the upper part of the coronal suture above the pterion. Its inner, or cerebral surface of the squamous plate, lies in front of the frontal lobes of the brain and shows in the middle line the crest and groove for the superior sagittal venous sinus.

The outer surface of the squamous part is covered by the tissues of the scalp. It is continuous with the horizontal orbital plates at the upper margins of each orbital cavity. The upper orbital margin shows a notch or foramen (the supra-orbital notch or foramen), at about its middle, which transmits the supra-orbital vessels and nerves. Between the two orbital cavities the frontal descends to a lower level (nasal process) and articulates with the nasal, frontal process of the maxilla, and lacrimal bone on each side. Beneath the upper part of each nasal bone lies a deeper shelf-like process of the frontal bone forming part of the roof of each nasal cavity. The nasal spine supports the nasal bones and gives attachment in the middle line to part of the cartilaginous nasal septum. The point in the mid-line of the surface of the skull where the frontal and nasal bones meet is called the *nasion* (Fig. 5.20).

Above each orbital margin the frontal bone shows a slight convex bulging, the *superciliary ridge or arch*. These are continuous across and above the root of the nose as a smooth elevation, the *glabella*. The superciliary arches and glabella are usually better developed in males and were very massive in certain forms of prehistoric man, forming the supra-orbital torus.

On the outer side of the orbital cavity each frontal bone descends as the *zygomatic process* to articulate with the zygomatic bone thus forming the lateral orbital margin. Behind this the frontal articulates with the orbital plate of the zygomatic (separating the lateral wall of the orbital cavity from the front of the temporal fossa), and with the greater wing of the sphenoid.

The *orbital plates* make up the triangular roof of each orbital cavity as far back as the lesser wing of the sphenoid. Medially each orbital plate lies above the upper surface of the facial ethmoidal labyrinth and forms the roof for some of the ethmoid air cells. Between each orbital plate is the deep ethmoid notch filled in, in the articulated skull, by the cribriform plate of the ethmoid (Figs 5.4, 5.11).

The *frontal air sinuses* in the adult occupy the glabellar and superciliary regions and may extend backwards into the orbital plates. They are of variable size and are often asymmetrical (Figs 4.42, 5.3).

Development

The frontal bone develops in membrane. An ossification centre appears above each orbital margin during foetal life. There may be accessory centres for the nasal and zygomatic pro-

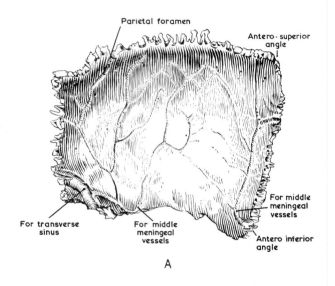

A

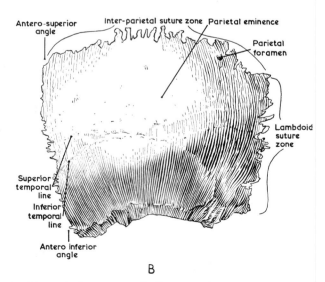

B

Fig. 5.12 The parietal bone. A. Inner aspect. B. Outer aspect.

cesses. At birth the bone is in two halves separated by the mid-line *metopic suture*. In the majority of individuals union between the two halves commences in the first or second year and is complete by the sixth year; the metopic suture may, however, persist into adult life, in some 6–8 per cent of individuals.

The parietal bone (Figs 5.2, 5.12)

Position
Side of cranial part of skull.

Articulations
Parietal of opposite side, frontal, greater wing of sphenoid, temporal, occipital.

Foramina
Parietal.

The parietal is a quadrilaterally-shaped bone situated at the side and roof of the cranial vault. Its convex outer surface shows the inferior and superior temporal lines. The former marks the extent of attachment of the temporal muscle to the bone, and the latter the line of attachment of the overlying temporal fascia. Above the superior temporal line the outer surface of the parietal is directly covered by the tissues of the scalp. The posterior superior part of the bone may show a small foramen, the parietal foramen, through which an emissary vein connects the sagittal venous sinus with the veins of the scalp. A line joining the parietal foramina intersects the sagittal suture at the *obelion*.

The inner surface of the bone shows two or more well marked grooves for the anterior and posterior branches of the middle meningeal vessels, and at the postero-inferior angle of the bone is part of the groove for the transverse dural sinus (sigmoid part).

Development
Each parietal bone ossifies in membrane. Two centres appear for each bone about the seventh week of intra-uterine life and soon unite to form a single centre from which ossification extends in a radial manner towards the margins of the bone.

FACIAL BONES

The maxilla (Figs 5.13, 5.14)
The two maxillary bones make up the greater part of the upper facial skeleton.

Position
Side and front of the upper part of the face, roof of oral cavity, side wall and floor of the nasal cavity, floor of orbital cavity.

Parts
Body; zygomatic, palatal, frontal and alveolar processes.

Articulations
Maxilla of opposite side, zygomatic, palatine, ethmoid, lacrimal, frontal, nasal, vomer, mandible (through occlusion of the teeth).

Foramina and other features
Posterior superior dental, incisive, palatine canal (with palatine bone), nasolacrimal canal (with lacrimal and inferior turbinate bones), infra-orbital groove, canal and foramen and the maxillary antrum.

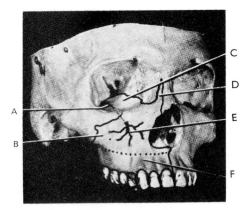

Fig. 5.13 Photograph of an adult skull. The maxilla is heavily outlined and the parts of the bone are shown. The developmental subdivision into neural and alveolar areas is indicated by the interrupted line—below this line is the alveolar area. A, infra-orbital groove; B, zygomatic process; C, floor of the orbit; D, frontal process; E, terminal branches of the infraorbital nerve; E, outer alveolar wall. (By courtesy of *The Dental Practitioner*.)

The body of each maxillary bone has a facial, posterior (infratemporal), nasal, and orbital surface. It contains the large maxillary air sinus (antrum) which communicates with the nasal cavity (Fig. 4.40).

The *facial surface* of the bone is bounded above by the lower orbital margin, below by the alveolar border, medially by the nasal aperture and the mid-line intermaxillary suture, and laterally by the zygomatic process. To this surface are attached certain of the muscles of expression, and opening on to it below the middle of the orbital margin is the infra-orbital foramen which transmits the terminal branches of the infra-orbital nerve and blood vessels.

The facial surface usually shows two shallow depressions:

The *incisive fossa* above the sockets of the incisor teeth and below the lower border of the nasal opening and bounded laterally by the canine eminence formed by the outer wall of the socket of the canine tooth.

The *canine fossa* lying behind the canine eminence between the infra-orbital foramen above and the alveolar margin below, and in front of the zygomatic process.

The degree to which these fossae are developed varies to a considerable extent in different individuals.

The *posterior* (*infratemporal*) *surface* faces the infratemporal fossa. It is separated from the lateral pterygoid plate by the pterygomaxillary fissure and curves medially to form the anterior boundary of the pterygopalatine fossa of the articulated skull. The posterior surface is separated from the anterior surface by the zygomatic process of the bone, and the bony ridge (key-ridge) which descends from it towards the alveolar border. Above, the posterior surface becomes continuous with the orbital surface at the inferior ophthalmic fissure. The buccinator muscle is attached to the posterior surface (above the line of reflection of the gum) and extending forward below the zygomatic process to the premolar region. This surface also shows two or more small foramina, the openings of the posterior superior dental canals along which the posterior superior dental nerves and vessels reach the upper teeth and their supporting structures.

The lower medial part of the posterior surface of the maxilla is separated from the pterygoid plates in the articulated skull by the tuberosity of the palatine bone.

The smooth *orbital surface* forms the floor of the orbital cavity and the roof of the maxillary air sinus. In its posterior part it is grooved by the infra-orbital groove leading to the infra-orbital canal and infra-orbital foramen which opens on the facial surface.

In its middle part the *nasal surface*, in the disarticulated bone, shows a large irregular opening leading to the maxillary air sinus.

Behind this opening the nasal surface is roughened and in the articulated skull is overlapped from behind by the vertical plate of the palatine bone (Fig. 4.39). It shows a groove running obliquely downwards and forwards, which is converted into the palatine canal when the palatine bone is in position.

Below the opening of the antrum and as far forward as the anterior nasal margin, the bone

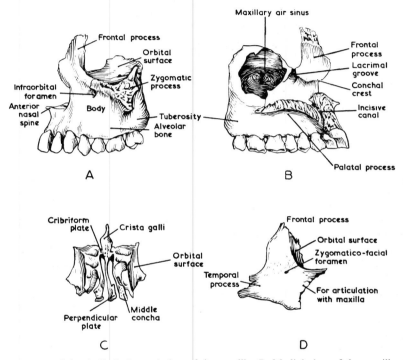

Fig. 5.14 Bones of the skull. A. Lateral view of the maxilla. B. Medial view of the maxilla. C. Anterior view of the ethmoid. D. Lateral view of the zygomatic bone.

is smooth and is covered by the mucous membrane of the inferior meatus of the nasal cavity in the living.

In front of the opening the nasal surface shows a deep groove which in the articulated skull is converted into a canal (the *nasolacrimal canal*) by the lacrimal and inferior turbinate bones.

Above the opening of the antrum, at the junction of the nasal surface and orbital surface, the bone may show one or more excavations which in the articulated skull form part of the wall of one or more ethmoidal air cells. In front of the nasolacrimal groove the maxillary bone reaches a higher level as the frontal process of the bone. A *horizontal crest* divides the nasal surface of the frontal process into a lower part related to the inferior meatus and an upper part in front of the middle and superior meatuses. To this conchal crest is attached the anterior end of the inferior turbinate bone (inferior concha).

In the articulated skull the opening to the maxillary antrum is greatly reduced by parts of other bones which contribute to the side wall of the nasal cavity. These include: the vertical plate of the palatine bone behind, the inferior turbinate bone below, the facial part (labyrinth) of the ethmoid above, and the lacrimal bone in front (Fig. 4.39).

Each maxillary bone has four processes. These are:

The *zygomatic process* separating the facial and infratemporal (posterior) surfaces. It articulates with the zygomatic bone; the articulating surface faces outwards (Fig. 5.14).

The *frontal process* (Figs 5.1, 5.13, 5.14) articulates with the nasal process of the frontal bone between the nasal bone in front and the lacrimal bone behind. Its nasal surface has already been described. Its outer surface shows a vertical ridge, the medial orbital margin, dividing the process into a posterior part at the front of the medial wall of the orbital cavity, forming part of the fossa for the lacrimal sac, and an anterior part which lies outside the orbital cavity.

The *alveolar process* carries the upper teeth and ends behind at the alveolar bulb in young skulls, or the maxillary tuberosity in adult skulls.

The *palatal process* projects inwards from the inner surface of the bone between the alveolar process below and the nasal surface above. With the corresponding process of the opposite side it forms the greater part of the hard palate that separates the oral and nasal cavities. The floor of the nasal cavity (upper surface of the hard palate) is smooth, but in the middle line it shows an anteroposterior crest with which the vomer bone articulates, to form the lower part of the nasal septum. The under surface of the maxillary palatal process is roughened by the attachment of the palatal mucoperiosteum. In the angle between the alveolar and palatal processes a groove runs backward to the greater palatine foramen which in the articulated skull opens on the hard palate between the maxillary and palatine bones. This foramen is at the lower end of the palatine canal and through it and along the groove pass the greater palatine vessels and nerves.

The *incisive canals* commence towards the front of the floor of each nasal cavity. Each canal opens into the mid-line incisive foramen, which is complete when the two maxillary bones are articulated. The incisive canals transmit the terminal branches of the nasopalatine (septal) nerve and vessels, branches of the maxillary division of the trigeminal nerve and the maxillary artery (Figs 4.3, 4.39).

In the maxilla of a young person a suture runs outwards on the palate from the incisive foramen to between the lateral incisor and canine teeth. Within the nasal cavity it may ascend a short distance on the frontal process. It does not, however, appear on the facial surface of the bone. This suture separates the premaxillary element from the main part of the maxilla. In animals other than man the *premaxilla* is a separate bone before and after birth.

Development
The greater part of the maxilla develops in membrane on the outer side of the nasal capsule. There is also a premaxillary centre from which develops the premaxillary element of the bone. The two parts unite on the facial surface early in foetal life and this early union is characteristic of the human face. For further information see page 362.

The mandible (lower jaw) (Figs 5.1, 5.2, 5.15, 7.8)

Position
Lower facial skeleton.

Parts
Body, alveolar process, ramus, coronoid process, angle, condyle.

Articulations
Temporal (at mandibular joint), maxilla (through occlusion of teeth).

Foramina

Mandibular (inferior dental), mental.

The mandible at birth consists of a right and left half united by cartilage and fibrous tissue at the symphysis in the middle line. The two parts unite during the first year to form a single bone.

The *body* of the mandible is curved like a horseshoe and presents a lower border which is subcutaneous. The upper border, the alveolar process, carries the lower teeth. The outer, facial surface, gives origin to some of the facial muscles and shows on each side the opening of the mental foramen facing upwards and backwards. In about 50 per cent of cases the foramen lies below the apex of the second premolar (see page 435). The inner, lingual surface

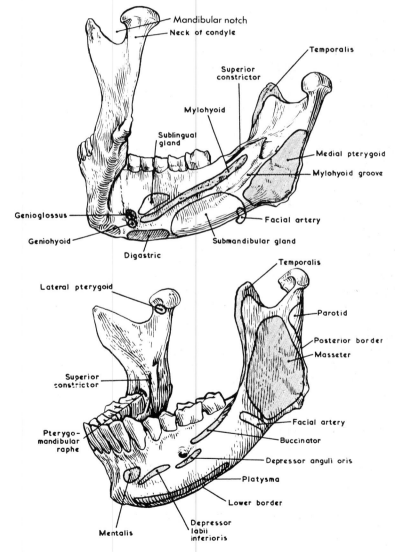

Fig. 5.15 The mandible showing its muscle attachments and relationships.

is related to the floor of the mouth cavity, the tongue, the submandibular and sublingual salivary glands, and to a number of muscles, including the mylohyoid or oral diaphragm.

The *ramus of the mandible* on each side forms a vertical plate of bone with an outer and inner surface, a posterior border, an anterior border, and carrying on its upper part two processes. To the anterior of these, the *coronoid process*, is attached the temporal muscle. The posterior process, the *condyle*, articulates with the temporal bone at the mandibular joint. To the outer surface of the mandibular ramus is attached the masseter muscle. The inner surface shows at its centre the mandibular or *inferior dental foramen*, the posterior opening of the inferior dental canal which runs through the bone to terminate in front at the mental foramen. Behind and below the mandibular foramen the inner surface of the ramus is roughened for the insertion of the medial pterygoid muscle. This region is known as the angle of the mandible.

From the tip of the coronoid process the anterior edge of the ramus descends towards the body of the bone where it is continuous with the *external oblique ridge* on the outer surface of the body. This ridge becomes less marked as it is traced towards the mental foramen. It separates the alveolar process above from the basal element of the bone below. Above the level of the ridge the alveolar process is covered by gum tissue. To the ridge, as far forward as the mental foramen, is attached the buccinator muscle. Below and in front of the foramen as far as the middle line are attached the depressor anguli oris, the depressor labii inferioris and the mentalis muscle. Along the lower border of the body of the bone some fibres of the platysma muscle are attached. At the anterior border of the masseter muscle, at the region of continuity between the ramus and body of the bone, the facial artery and vein lie on the bone as they enter the face from the neck.

From the tip of the coronoid process a ridge descends to become continuous on the inner side of the body of the bone with the *mylohyoid ridge* to which is attached the mylohyoid muscle. The back of the mylohyoid ridge is close to the alveolar margin in the region of the last molar tooth; the front of the ridge is continuous with that of the opposite side at the middle line and is close to the lower border of the bone. In the middle line immediately above the mylohyoid attachment are two small tubercles, the *genial tubercles*, which may, however, be united to form a vertical flange of bone. To these are attached the geniohyoid and genioglossus muscles. On either side of the middle line between the mylohyoid ridge above and the lower border of the body of the bone below, are the shallow *digastric fossae* for the attachment of the anterior bellies of the digastric muscles.

The inner surface of the back of the body below the level of the mylohyoid attachment is in close relation to the superficial part of the submandibular salivary gland.

The inner surface of the front of the mandibular body above the level of the mylohyoid ridge is closely related to the outer surface of the sublingual salivary gland.

On the inner surface of the ramus a groove descends downwards and forwards from the opening of the inferior dental (mandibular) canal. In this groove lie the mylohyoid nerve and vessels.

Between the *mylohyoid groove* and the alveolar margin the lingual nerve lies in contact with the bone for a short distance before it comes to lie on the upper (deep) surface of the mylohyoid muscle.

On the inner side of the mandibular foramen a tongue-like projection of bone, the *lingula*, gives attachment to the sphenomandibular ligament.

From the tip of the coronoid process to the condyle the upper edge of the ramus forms the *mandibular notch* (or coronoid notch). The bony margin passes on the outer side of the neck of the condyle as a ridge of bone which ends at the outer pole of the head of the condyle (Fig. 5.15). This bony ridge forms the outer boundary of the pterygoid fossa on the front surface of the neck of the condyle; the lower head of the lateral pterygoid muscle is attached to the fossa. For further details in regard to certain parts of the mandible see page 439.

Development
The mandible develops from a single centre of ossification on each side. This centre appears early in foetal life in the membranous condensation which develops on the outer side of Meckel's cartilage. The greater part of the bone develops in membrane, but a mass of secondary cartilage forms in its back part and persists after birth as the cartilage of the condyle which plays an important part in the growth of the lower jaw. For further details see page 356.

Palatine bone (Figs 4.39, 5.16)

Position
Back of hard palate, side wall of nasal cavity, orbital cavity, pterygopalatine fossa.

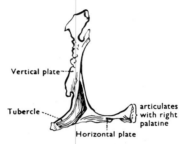

Vertical plate

Tubercle

articulates with right palatine

Horizontal plate

Fig. 5.16 The left palatine bone viewed from behind. (By courtesy of Professor G. A. G. Mitchell and Dr E. L. Patterson.)

Parts
Palatine (horizontal) plate, nasal (vertical) plate, tuberosity.

Articulations
Maxilla, ethmoid, sphenoid, vomer, inferior turbinate.

Foramina
Sphenopalatine (with sphenoid), palatine canal (with maxilla), greater and lesser palatine foramina.

The *horizontal plate* of each palatine bone contributes to the formation of the back part of the hard palate (Fig. 5.17). Its posterior margin limits the hard palate and to it is attached the palatal aponeurosis and some of the muscles of the soft palate. The angle between the horizontal and the vertical plates forms the medial side of the greater palatine foramen.

The *vertical plate* of the palatine bone forms by its inner surface part of the side wall of the nasal cavity. It articulates behind with the medial pterygoid plate and in front overlaps the back part of the nasal wall of the maxilla. It helps to fill in the opening to the maxillary antrum (Fig. 4.39). Between the vertical plate and the maxilla in the articulated skull the *palatine canal* descends from the pterygopalatine fossa to the palatine foramina. Between its articulation with the surface of the maxilla in front and the medial pterygoid plate behind, the vertical plate forms a bony partition separating the nasal cavity from the pterygopalatine fossa (Fig. 4.39). It forms the medial wall of the fossa. The nasal surface of the vertical plate is divided into a lower and upper part by a horizontal crest to which the back of the inferior turbinate bone is attached. The area below the crest forms part of the side wall of the inferior meatus; the area above the crest forms part of the side wall of the middle and superior meatuses. At its upper end the vertical plate of the palatine shows a deep notch, the *sphenopalatine notch* (Fig. 5.16). The notch is converted into the sphenopalatine foramen in the articulated skull.

Through this foramen the nasal nerves and blood vessels enter or leave the nasal cavity from the pterygopalatine fossa (Figs 4.39, 5.18).

In front of the sphenopalatine notch the *orbital process* of the vertical plate passes upwards and forwards to form part of the medial wall of the orbital cavity, articulating with the sphenoid, ethmoid and maxillary bones (Fig. 4.39). It usually contains a small air sinus. Behind the notch the *sphenoidal process* of the vertical plate passes upwards to articulate with the sphenoid bone at the base of the medial pterygoid plate.

The *tuberosity of the palatine bone* passes backwards from the posterior edge of the angle of union between the horizontal and vertical plates. It articulates with the posterior surface of the maxilla above its tuberosity (Fig. 5.17) and fills in the notch between the medial and lateral pterygoid plates (Fig. 5.18). It gives attachment to some fibres of the medial pterygoid muscle.

The neck of the tuberosity shows on its under (palatal) surface one or two small foramina which transmit the lesser palatine nerve and vessels that supply the soft palate.

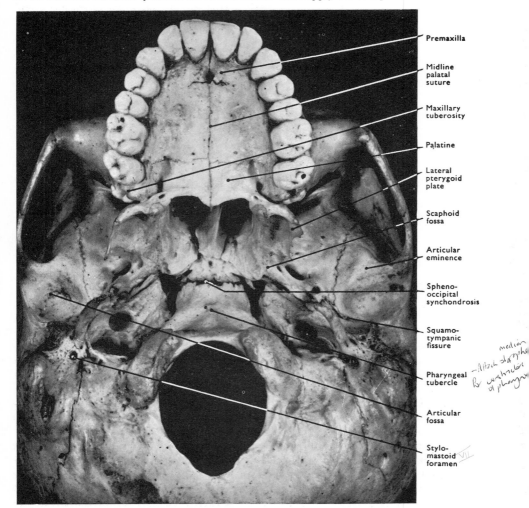

Fig. 5.17 Inferior aspect of the base of the skull.

Development

The palatine bone develops during the foetal life in membrane in close relation to the inner surface of the cartilage of the nasal capsule from a single centre of ossification at the site of the vertical plate of the bone close to the tuberosity (see page 365 for further details).

sphenopalatine foramen (p185) is an upper part of medial wall of fossa

The pterygopalatine (sphenopalatine) fossa (Fig. 5.18)

Between the maxilla and the lateral pterygoid plate is the pterygomaxillary fissure which leads medially into the pterygopalatine fossa. This lies at the back of the facial skeleton beneath the posterior part of the orbital cavity. It is bounded in front by the posterior surface of the maxilla; behind by the anterior surface of the pterygoid plates of the sphenoid and medially by the vertical (nasal) plate of the palatine bone separating the fossa from the nasal cavity. Above it opens into the orbital cavity through the medial part of the inferior orbital fissure. Below it leads to the palatine canal and the greater and lesser palatine foramina on the hard palate. Into the back of the pterygopalatine fossa opens the foramen rotundum from the middle cranial fossa, and the pterygoid canal (Vidian's canal) from the front of the foramen lacerum.

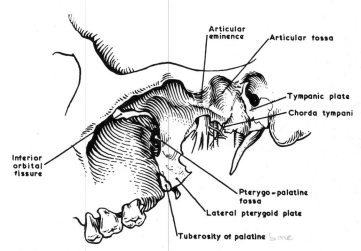

Fig. 5.18 The (sphenopalatine) pterygopalatine fossa.

the ... of the pterygopalatine fossa,

At the upper part of its medial wall the sphenopalatine foramen communicates with the nasal cavity. Laterally the fossa opens into the front of the infratemporal fossa at the anterior edge of the lateral pterygoid plate.

The maxillary artery enters from the infratemporal region and its terminal branches pass through the pterygopalatine fossa to the orbital cavity, nasal cavity, roof of the mouth and the nasopharynx (page 257). The maxillary division of the fifth cranial nerve enters through the foramen rotundum. Its branches correspond to those of the third part of the maxillary artery (page 302). The fossa also contains the sphenopalatine parasympathetic ganglion (page 320). & p185

During childhood the back part of the alveolar bulb of the superior alveolar process projects into the lower part of the fossa and may overlap the outer surface of the lateral pterygoid plate (page 224, Fig. 7.46). With the downward growth of the maxillary alveolar process, the pterygoid plates and the hard palate, the palatine canal is elongated between the lower end of the fossa and the palatine foramina. The upper part of the fossa, however, is a relatively stable region after the first decade.

The zygomatic (malar) bone (Figs 4.41, 5.2, 5.14)

Position
Upper part of face, orbital cavity, temporal fossa, zygomatic arch.

Parts
Facial surface, orbital plate, maxillary process, temporal process, frontal process.

Articulations
Maxilla, great wing of sphenoid, frontal, temporal.

Foramina
Foramina for temporal and facial branches of zygomatic nerve (from the maxillary division of the trigeminal).

As seen from its facial surface the zygomatic bone has a triradiate form. The *maxillary process* passes forwards and inwards to articulate with the zygomatic process of the maxilla. It forms the lateral half of the lower orbital margin. The *frontal process* passes upwards to articulate with the frontal bone and forms the lateral margin of the orbital cavity; the *temporal process* passes backwards to articulate with the zygomatic process of the temporal bone and completes the zygomatic arch. The facial surface gives attachment to the zygomaticus major and minor muscles and shows a small foramen through which the zygomaticofacial sensory nerve appears on the face.

From the facial part of the bone the orbital plate passes inwards as a vertical shelf of bone contributing to the side wall of the orbital cavity and separating the orbital cavity from the temporal fossa. This closing of the orbital cavity on the lateral side is a characteristic feature of the primate skull. The orbital plate articulates behind with the great wing of the sphenoid. It shows on its temporal surface a small foramen through which the zygomaticotemporal nerve enters the temporal fossa.

The zygomatic bone is an important element in the facial buttress system whereby the forces of mastication are transmitted from the alveolar process to the cranial base (see page 236).

Development
The zygomatic bone develops in membrane from a single centre of ossification which appears early in foetal life.

The nasal bone (Figs 5.1, 7.49)

Position
Face and nasal cavity.

Articulations
Maxilla, frontal, ethmoid, nasal bone of the opposite side.

Each nasal bone forms the anterior part of the roof of the nasal cavity. It articulates above with the nasal process of the frontal bone, laterally with the anterior margin of the frontal process of the maxilla, and medially with its fellow of the opposite side. The point in the middle line where the internasal suture meets the frontal bone is known as the *nasion*. This is an important landmark in measurements of the cranial base and analysis of facial growth. To the lower margin is attached the lateral (alar) cartilage of the nose. The outer surface is smooth and covered by skin and some fibres of the nasal muscles. The inner surface is slightly concave and forms the roof of the anterior part of the nasal cavity. It is traversed from above downwards by a groove which is occupied by a branch of the anterior ethmoidal nerve and

Table 5.1 Skull foramina. A. Seen on exterior surfaces of face and cranium

Foramen	Position	Contents
Supra-orbital foramen or notch	Frontal bone, upper margin of orbit.	Supra-orbital vessels and nerve; nerve a branch of ophthalmic division of trigeminal.
Infra-orbital foramen	Facial surface of maxilla.	Infra-orbital vessels and nerve; nerve a branch of maxillary division of trigeminal.
Mental foramen	Facial (buccal) surface of the body of the mandible, usually below premolars.	Mental vessels and nerve; nerve a branch of mandibular division of trigeminal.
Zygomaticofacial foramen	Facial surface of zygomatic bone.	Zygomaticofacial vessels and nerve; nerve a branch of maxillary division of trigeminal.
Zygomaticotemporal foramen	Temporal surface of zygomatic bone.	Zygomaticotemporal vessels and nerve; as above.
Mandibular foramen	Inner surface of ramus of mandible.	Inferior dental vessels and nerve; artery a branch of maxillary artery.
Incisive foramen	Front of hard palate behind upper central incisor teeth.	Terminal branches of sphenopalatine vessels and nerve; nerve a branch of maxillary division of trigeminal.
Greater palatine foramen	Back of hard palate at level of last molar tooth, between maxillary and palatine bones, termination of palatine canal from pterygopalatine fossa.	Greater palatine vessels and nerve; nerve a branch of maxillary division of trigeminal; artery a terminal branch of maxillary artery.
Lesser palatine foramina	In tuberosity of palatine bone, behind greater palatine foramen; 2–3 in number.	Lesser palatine vessels and nerve, origins as above.
Foramen ovale	In great wing of sphenoid bone, on lateral side of auditory tube.	(1) Mandibular division of trigeminal nerve (2) Emissary vein (3) Meningeal artery.
Foramen spinosum	In great wing of sphenoid bone in front of spine of sphenoid.	(1) Middle meningeal artery (2) Nervus spinosus; a branch of mandibular division.
Foramen lacerum	Between sphenoid and petrous temporal bones: filled in by fibrocartilage in living.	Emissary vein; meningeal artery, a branch of ascending pharyngeal artery.
Carotid canal	Petrous temporal bone.	Internal carotid artery and sympathetic nerves.
Jugular foramen	Between occipital and petrous temporal bones.	(1) Jugular bulb (2) 9th, 10th, 11th cranial nerves (3) Meningeal branch of ascending pharyngeal artery.
Tympanic canaliculus	Between jugular foramen and carotid canal.	Tympanic branch of 9th nerve.
Stylomastoid foramen	Temporal bone between mastoid and styloid processes.	(1) Facial nerve (2) Stylomastoid branch of occipital or posterior auricular artery.
Squamotympanic (petrotympanic) fissure	Between squamous and tympanic parts of temporal bone behind the glenoid fossa.	(1) Chorda tympani nerve (2) Tympanic branch 1st part maxillary artery.
External auditory meatus	Closed in living by tympanic membrane.	None.

Table 5.1 (*cont.*)

Foramen	Position	Contents
Anterior condylar foramen = Hypoglossal canal	Occipital bone, above the condyle.	12th cranial nerve; meningeal branch ascending pharyngeal artery.
Posterior condylar foramen	Inconstant; behind condyle of occipital bone.	Emissary vein; meningeal branch occipital artery.
Mastoid foramen	Behind mastoid process; sometimes multiple.	Emissary veins.
Foramen magnum	Occipital bone.	(1) Spinal cord and covering meninges (2) Spinal root 11th nerve (3) Vertebral arteries.

B. Seen on inner surface of cranial base

Foramen	Position	Contents
Foramen caecum	In front of crista gali.	Emissary vein.
Cribriform plate	Ethmoid.	Olfactory nerves.
Optic foramen	Between upper and lower roots lesser wing of sphenoid.	(1) Optic nerve (2) Ophthalmic artery.
Superior orbital (ophthalmic) fissure	Between lesser and greater wings of sphenoid.	(1) Ophthalmic division of trigeminal nerve (2) 3rd, 4th and 6th cranial nerves (3) Ophthalmic veins (4) Meningeal branch lacrimal artery.
Foramen rotundum	Great wing of sphenoid; floor of middle cranial fossa.	Maxillary division of trigeminal nerve.
Foramen ovale	As on exterior surface.	
Foramen spinosum	As on exterior surface.	
Foramen lacerum	As on exterior surface.	
Carotid canal	As on exterior surface.	
Hiatus canalis facialis	Anterior surface petrous temporal.	Great superficial petrosal branch of 7th nerve.
Internal auditory meatus	Posterior surface petrous temporal.	7th and 8th cranial nerves; internal auditory artery.
Jugular foramen	As on exterior surface.	
Anterior condylar foramen	As on exterior surface.	
Mastoid foramen	As on exterior surface.	
Foramen magnum	As on exterior surface.	

C. In pterygopalatine fossa

Foramen	Position	Contents
Foramen rotundum	As on interior cranial surface.	
Pterygoid canal	Runs forward from foramen lacerum.	Nerve of pterygoid canal; artery branch of 3rd part maxillary.
Sphenopalatine foramen	Medial wall pterygopalatine fossa between vertical plate palatine and body of sphenoid.	Nasal nerves (maxillary division trigeminal) and nasal branches of the 3rd part of maxillary artery.
Palatine canal	Leading to greater and lesser palatine foramina.	See palatine foramina.

Table 5.1 (*cont.*)

Foramen	Position	Contents
Pharyngeal canal	Between ala of vomer and under surface of body of sphenoid.	Nerve and artery branches of maxillary division and maxillary artery.

D. In orbital cavities

Foramen	Position	Contents
Inferior orbital (ophthalmic) fissure	Between maxilla and great wing of sphenoid leading to infra-orbital groove in floor of orbital cavity.	Infra-orbital vessels and nerve. Vein to pterygoid plexus.
Superior orbital fissure	As on interior of cranium.	
Optic foramen	As on interior of cranium.	
Anterior and posterior ethmoidal foramina	Medial wall orbital cavity between frontal and ethmoid.	Ethmoidal nerves and vessels, branches of ophthalmic division trigeminal and ophthalmic artery.
Nasolacrimal canal	Between maxillary, lacrimal and inferior turbinate bones.	Nasolacrimal duct.

artery. This nerve supplies the tissues of the nose in the region of the alar cartilage and is then known as the external nasal nerve.

Where the bones of each side articulate they rest from above downwards on the nasal spine of the frontal bone, the perpendicular plate of the ethmoid and the septal cartilage.

Development
Each nasal bone develops in membrane in close relation to the nasal capsule from a single centre of ossification which appears early in foetal life.

The vomer (Figs 5.3, 5.19)

Position
Nasal septum.

Articulations
Body of sphenoid, perpendicular plate of ethmoid, maxilla, septal cartilage.

The vomer articulates behind with the lower surface of the body of the sphenoid (Figs 5.18, 5.19). Here two wing-like processes (alae) project on each side beneath the sphenoid to articulate laterally with the medial pterygoid plate and the sphenoid process of the palatine to form the floor of the pharyngeal canal. The posterior margin of the vomer forms the posterior edge of the nasal septum, separating the posterior nasal apertures. The lower border of the bone articulates with the mid-line crest on the upper surface of the hard palate as far forward as the incisive foramen. The upper border of the vomer articulates with the sphenoid and the perpendicular plate of the ethmoid and in front is grooved to receive the cartilage of the nasal septum (Fig. 5.19). The surfaces of the vomer are covered by nasal mucous membrane and grooved by the nasopalatine (septal vessels) and nerve. In the membrane above the opening of the incisive canal is the rudimentary vomeronasal organ.

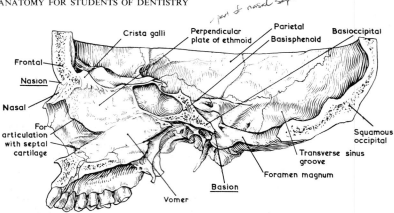

Fig. 5.19 Mid-line sagittal section of the bones of the cranial base.

Development

The vomer develops in foetal life from two centres of ossification in the perichondrium covering the posterior-inferior edge of the septal plate of the nasal capsule. The centres soon unite forming a Y-shaped trough of bone along the lower edge of the septal cartilage. At about seven years of age the ossification of the perpendicular plate of the ethmoid in the back of the septal cartilage extends to the vomer and the two bones unite.

The ethmoid bone (Fig. 5.14)

The facial part of the ethmoid bone is described with the cranial bones (page 218).

The inferior turbinate bone or inferior concha

Position

Lateral wall of the nasal cavity.

Articulations

Maxilla, lacrimal, ethmoid, palatine.

   The inferior turbinate is a curved plate of bone attached to the side wall of the nasal cavity and projects inwards to form the roof of the inferior meatus and the floor of the middle meatus (Figs 4.40, 4.41). It is covered by nasal mucous membrane, which is thick and vascular in this respiratory region of the nasal cavity.

   In front the inferior turbinate is attached to the horizontal crest on the frontal process of the maxilla; behind to a similar crest on the vertical plate of the palatine bone. Between these attachments the turbinate fills in the lower part of the opening of the maxillary sinus (Fig. 4.39), and shows two vertical processes. The anterior of these articulates with the lower end of the lacrimal bone at the inner side of the nasolacrimal groove of the maxilla converting this into the nasolacrimal canal, which opens beneath the inferior turbinate into the inferior meatus. The posterior vertical process articulates with the descending (uncinate) process of the facial ethmoid behind the definitive opening to the sinus, when all the bones on the lateral side of the nasal cavity are in articulation.

Development

The inferior turbinate develops in the inturned lower end of the lateral wall of the cartilaginous nasal capsule from a single centre of ossification.

The lacrimal bone (Figs 4.39, 4.44, 5.1, 5.2)

Position
Orbital cavity, nasal cavity, nasolacrimal canal.

Articulations
Maxilla, inferior turbinate, ethmoid, frontal.

The lateral or orbital surface of the lacrimal bone forms part of the medial wall of the orbital cavity (Fig. 4.44). It articulates above with the frontal bone; behind with the facial ethmoid, and in front with the frontal process of the maxilla. It is divided by a vertical crest into an anterior part, which, with the back part of the frontal process of the maxilla, forms a groove-like depression on the medial wall of the orbital cavity just inside the orbital margin. This is the *lacrimal fossa* for the lacrimal sac. From the lower end of the lacrimal bone, a process descends into the nasal cavity forming the medial wall of the nasolacrimal canal or duct (Fig. 4.39). It articulates below with the ascending lacrimal process of the inferior turbinate.

The inner or nasal surface of the lacrimal bone forms part of the side wall of the nasal cavity and it helps to fill in the opening in the maxillary bone leading to the maxillary sinus.

Development
The lacrimal bone develops, during foetal life, from a single centre appearing in membrane in close relation to the cartilage of the nasal capsule.

The hyoid bone (Fig. 5.20)
This is a U-shaped bone situated in the upper part of the neck. It consists of a central *body* which is continued backwards on each side as the *greater cornua* (horn). On its upper surface at the junction of the body and greater cornua on each side are the conical-shaped *lesser cornua*. They are usually connected to the rest of the bone by a synovial joint which disappears in many adults and is replaced by fibrous tissue. The upper part of the *anterior surface* of the

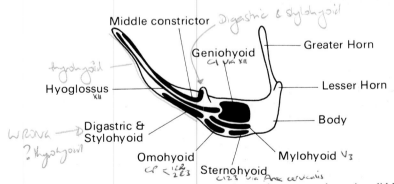

Fig. 5.20 Anterolateral view of the hyoid bone. Muscle attachments are shown in solid black on the right half of the bone.

body of the bone shows two shallow fossae separated by a mid-line vertical ridge. They give attachment to the geniohyoid and the medial part of the hyoglossus muscles. Below the fossae the anterior surface of the bone is smooth and gives attachment to the mylohyoid, sternohyoid and omohyoid muscles. To the greater cornua are attached the middle constrictor, which continues its attachment on to the lesser cornua, the lateral part of the hyoglossus, the thyrohyoid and the stylohyoid. To the tip of the lesser cornua is attached the stylohyoid ligament; to the tip of the greater cornua the rounded posterior part of the thyrohyoid membrane. The *posterior surface* of the body of the bone is smooth and is separated from the middle portion of the thyrohyoid membrane, which is attached to the upper edge, by some areolar tissue and a small bursa.

Development
The hyoid bone develops from the cartilages of the second and third visceral arches. In these a pair of centres appear for the body (about the time of birth), in the greater cornua towards the end of foetal life, and in the lesser cornua during the first or second year. The second visceral arch gives rise to the lesser horn and probably the upper part of the body of the bone, the greater horn and the lower part of the body being derived from the cartilage of the third arch.

Facial buttress system (Fig. 5.21)
With the eruption of the teeth certain regions of the facial skeleton become strengthened so as to withstand the forces of mastication. In the body of mandible the cortical bone of the body becomes thickened to support the alveolar process. The thickening is especially marked along the lower border and in the region of the external and internal oblique lines. In the ramus the lower and posterior borders are thickened and also the internal coronoid ridge

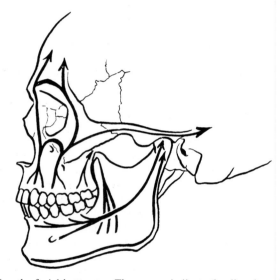

Fig. 5.21 Diagram showing the facial buttresses. The arrows indicate the direction of thrust from the teeth to the cranial base. The teeth are in post-normal occlusion. (After Weinmann and Sicher.)

running upwards on the inner side of the coronoid process from the back of the alveolus and the mylohyoid (internal oblique) ridge.

These thickenings, or trajectories, are formed in response to forces exerted by the muscles of mastication. Their presence and alignment can be confirmed in radiographs, or by the study of sections through the mandibular body and ramus. Analyses of stress patterns in the human mandible using photoelastic models of the jaws confirms this correlation between structure and function. The behaviour of the models confirms that a bone with the basic shape and dimensions of the mandible requires reinforcement in exactly those areas where bony reinforcement is known to exist.

In the upper facial skeleton the bone is especially adapted in three places to carry the forces of mastication from the alveolar process and the palatal vault. These are:

1. The *frontomaxillary buttress* passes on each side of the nasal orifice to the frontomaxillary sutures.

2. The *zygomatic buttress* commences in the adult above the first and second permanent molar teeth as the zygomatic process of the maxilla and divides into two limbs: a vertical limb forming the lateral rim of the orbital cavity, and a horizontal limb, the zygomatic arch. During facial growth the upper teeth move forward relative to the lower edge of the zygomatic buttress.

3. The *pterygopalatine buttress* supports the maxilla from behind (Fig. 5.18).

Muscular processes

The important muscular processes of the skull associated with the buttress system are:

1. The zygomatic arches, lateral pterygoid plates, and the coronoid and angular processes of the mandible; related to the muscles of mastication.

2. The nuchal region of the occipital bone and the mastoid processes related to the neck muscles.

3. The styloid processes.

Space for the growing muscles of mastication is provided for between the face and the vertebral column (infratemporal region) by the growth of the cranial base. Growth of the muscles to adult size is closely related to the completion of the dentition and the development of the facial buttress system. A period of rapid growth in all these regions takes place during adolescence. Experiments have shown that certain structures, such as the coronoid process of the mandible, depend entirely on the presence and activity of the attached muscles. The zygomatic arch, on the other hand, although depending on muscle action for its full development, does not entirely disappear if the masseter is cut or put out of action. As well as being a process for muscular attachment it is part of the facial buttress system.

The alveolopalatal region (Figs 5.17, 8.9)

This consists of the upper and lower alveolar processes and the bones of the hard palate. During foetal life and the first year after birth when the interfrontal midpalatal, and mandibular (symphysial) sutures make up a *sagittal suture system*, growth in width of the oral cavity is partly the result of separation of the bones bounding this suture system. By the end of the first year the frontal and mandibular sutures are uniting, and although the midpalatal suture remains, it is unlukely to be an important side of growth after this time.

Further growth of the oral cavity is the result of bone deposition along the alveolar margins, on the under-surface of the palate and on the facial surface of the maxillary and mandibular bones. There may be a certain amount of resorption on the lingual surfaces of the alveolar processes, but increase in the width of the palate is largely the result of the downward and outward direction of growth of the alveolar process.

The palatal vault, as well as bounding the oral cavity, is part of the facial buttress system whereby the alveolar processes carrying the teeth are supported, and the stresses of mastication distributed to the cranium and the cranial base.

MUSCLES OF MASTICATION

The masseter muscle (Figs 4.14, 5.15, 6.4)

This muscle consists of superficial and deep parts or heads. The *superficial head* arises from the lower border of the zygomatic arch as far back as the zygomaticotemporal suture. That part of the superficial head which takes origin from the front of the zygomatic bone is tendinous. The general direction of the fibres is downwards and slightly backwards, and they are attached to the lower border of the ramus of the mandible especially in the region of the angle. The *deep head* of the muscle arises from the whole length of the deep surface of the zygomatic arch and may be subdivided into a superficial and deeper portion. The latter is attached to the outer surface of the coronoid process, the former to the outer surface of the ramus between the coronoid process and the angle. The fibres of the deep head are vertical in direction. Some of the posterior fibres are attached to the anterolateral edge of the articular disk of the mandibular joint (page 311). The superficial and deep heads (both portions) become continuous in front, forming one or two intramuscular pockets which open backwards and through which nerves and vessels pass to the muscle tissue.

Nerve supply
The mandibular division of the fifth (trigeminal) nerve, via a branch which passes through the coronoid notch of the mandible.

Functions
The masseter muscles elevate the lower jaw and draw it slightly forwards. With the medial pterygoid of the same side they regulate the position of the angle of the mandible in the vertical plane.

Relations
Superficial to the masseter lies part of the parotid gland, its duct, the risorius muscle, the transverse facial artery and the terminal (facial) branches of the facial nerve (except for the cervical branch). Deep to it lies the ramus of the mandible, the insertion of the temporal muscle, and the masseteric vessels and nerve which approach its deep surface. The anterior part of the muscle is superficial to the posterior part of the buccinator muscle, the buccal nerve and the buccal pad of fat. The terminal portion of the parotid duct turns around the anterior edge of the muscle prior to penetrating the buccinator muscle and the oral mucous membrane.

The temporalis muscle (Figs 4.41, 5.22)
This arises by fleshy fibres from the whole of the temporal fossa, with the exception of that part formed by the zygomatic bone at the back of the orbital cavity, and from the deep surface of the covering temporal fascia to within a short distance of the zygomatic arch. The muscle fibres are inserted for the greater part into a flat intramuscular aponeurosis which is in turn attached to the tip and anterior margin of the coronoid process of the mandible. Some of

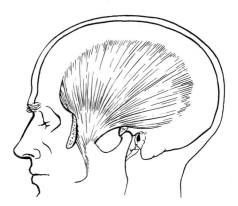

Fig. 5.22 The left temporal muscle. The zygomatic arch has been removed.

the deeper fibres are attached to the anterior coronoid fossa as far down as the attachment of the oral mucous membrane (Fig. 5.15). For purposes of functional analysis the temporalis muscle can be divided into an anterior part in which the fibres are vertical in direction and elevate the mandible as they contract; and a posterior part in which the fibres are more horizontal in direction and retract the mandible when they shorten.

Nerve supply
The mandibular division of the fifth (trigeminal) cranial nerve.

Functions
The anterior fibres of the temporal muscles elevate the lower jaw; the posterior fibres draw the condyle backwards into the glenoid fossa and help to remove pressure from the head of the condyle as the teeth are clenched.

Relations
Related to the superficial surface of the muscle is the scalp, the anterior and superior auricular muscles, the temporal fascia, the superior temporal vessels and auriculotemporal nerve, the zygomatic arch and upper part of the masseter. The deep surface is related to the temporal fossa, the lateral pterygoid and part of the medial pterygoid and buccinator, the deep temporal vessels and nerves, the maxillary artery, the buccal artery and nerve, the pterygoid venous plexus. The nerve and artery to the masseter are related to its posterior border.

The lateral pterygoid muscle (Figs 4.20, 5.15, 5.23)
This arises by two parts or heads:

An *upper* (*superior*) *head* from the infratemporal surface of the great wing of the sphenoid between the foramen ovale and the infratemporal crest. The fibres arising from the crest are tendinous.

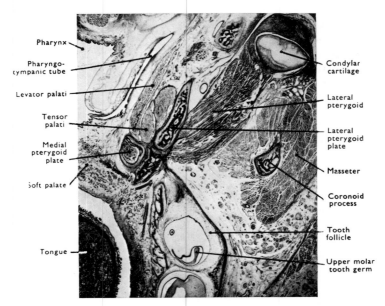

Fig. 5.23 Horizontal section through the facial region and upper jaw of a six months human foetus.

A *lower* (*inferior*) *head* from the outer or lateral surface of the lateral pterygoid plate of the sphenoid bone.

At their origins the two heads are separated by a slight space through which passes the buccal nerve and the maxillary artery (page 156). The fibres of the upper head are horizontal in direction, pass beneath the articular eminence and are attached to the front of the articular disk of the mandibular joint (page 311). The fibres of the lower head run upwards, backwards and slightly outwards and are attached to the fossa on the anterior surface of the neck of the condyle of the mandible.

The mandibular articular disk is developmentally a part of the lateral pterygoid muscle which in foetal life passes through the petrotympanic fissure to be attached to the back (malleolar) portion of Meckel's cartilage (Fig. 7.34).

Nerve supply
The mandibular division of the fifth (trigeminal) cranial nerve.

Functions
The upper head of the lateral pterygoid muscle draws the articular disk forwards with the condyle as the condyle is drawn forwards by the lower head. The two lateral pterygoid muscles acting together protrude the lower jaw. They are also used in opening the mouth by producing a rotatory movement of the mandible around a horizontal axis (see page 312). The lateral pterygoid of one side acting with the medial pterygoid moves the point of the chin across the middle line towards the opposite side of the face, as in grinding movements of the jaws.

Relations
Deep to the lateral pterygoid lies part of the deep head of the medial pterygoid; sometimes the second part of the maxillary artery; the mandibular division of the trigeminal nerve and the otic ganglion; the middle meningeal, deep auricular and tympanic branches of the first part of the maxillary artery; the auriculotemporal nerve; the upper part of the sphenomandibular ligament; the lingual and inferior dental nerves and the chorda tympani; the deep temporal nerves; the tensor palati and levator palati muscles.

Superficial to the lateral pterygoid muscle are the lower end of the temporal muscle and the coronoid process of the mandible; the buccal nerve (related to the lower head of the muscle); the deep facial vein which drains into the pterygoid venous plexus; in about 50 per cent of cases the second part of the maxillary artery; finally, the superficial head of the medial pterygoid at its origin.

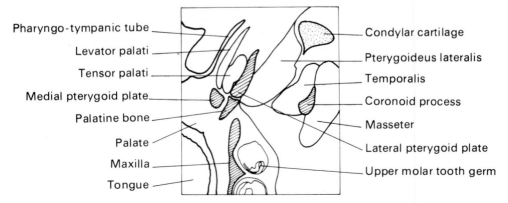

Fig. 5.24 Diagram to show the main features seen in Fig. 5.23 in outline form.

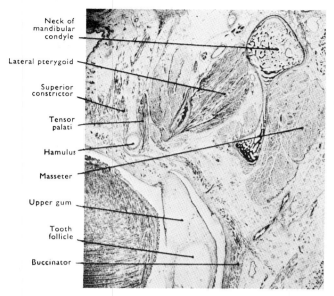

Neck of mandibular condyle

Lateral pterygoid

Superior constrictor

Tensor palati

Hamulus

Masseter

Upper gum

Tooth follicle

Buccinator

Fig. 5.25 Horizontal section through the facial region of the same foetus as in the previous photograph but at a lower level.

At the upper border of the upper head appear the deep temporal nerves with their accompanying vessels and the nerve and artery to the masseter. Between the two heads appears the buccal nerve, which curves outwards towards the cheek and the maxillary artery, on its way to the pterygopalatine fossa. At the lower border of the lower head are found the inferior dental artery and nerve, the lingual nerve and the sphenomandibular ligament. The pterygoid venous plexus surrounds the muscle.

The medial pterygoid muscle (Figs 4.20, 5.15)
This arises by two heads:

1. A larger *deep head* from the medial surface of the lateral pterygoid plate and that part of the tuberosity of the palatine bone which forms the lower anterior boundary of the pterygoid fossa. The interior surface of this head is tendinous and closely related at its origin to the tensor palati muscle as this lies on the outer surface of the medial pterygoid plate.

2. The smaller *superficial head* arises from the tuberosity of the maxilla and that part of the tuberosity of the palatine bone which appears between the maxilla and the lateral pterygoid plate at the lower end of the pterygopalatine fossa (Fig. 5.18).

The muscle is inserted by tendinous fibres to the inner surface of the angle of the mandible behind and below the inferior dental (mandibular) foramen (Fig. 5.15).

Nerve supply
The mandibular division of the fifth (trigeminal) cranial nerve.

Functions
The medial pterygoid muscles elevate the lower jaw and protrude it. They also tilt the angle of the mandible medially.

Relations
Deep to the upper part of the muscle lies the side wall of the pharynx and the tensor palati muscle, the styloglossus and stylopharyngeus. In the region of its insertion the medial pterygoid

muscle is related to the submandibular gland, the digastric and stylohyoid muscles, the facial artery, and the superior constrictor of the pharynx.

The superficial (lateral) relations of the muscle are the lower head of the lateral pterygoid muscle, the inferior dental vessels and nerve, the lingual nerve, the sphenomandibular ligament, the pterygoid venous plexus, the mandibular foramen and the ramus of the mandible.

THE SUPRAHYOID MUSCLES

The digastric muscle (Figs 4.7, 5.15)
The *anterior belly* of the muscle is inserted into a fossa on the deep surface of the body of the mandible just above the lower border and close to the middle line. The anterior belly is sometimes divided into a lateral and medial part. The *posterior belly* arises from the digastric notch on the inner (medial) side of the mastoid process of the temporal bone. The intermediate tendon lies above the greater cornu of the hyoid bone and is bound to it by a fibrous tissue sheath lined by a small bursa, which permits easy movement of the digastric relative to the bone.

Nerve supply
To the anterior belly—the mandibular division of the trigeminal nerve; to the posterior belly—the facial nerve.

Functions
It is an accessory muscle of mastication. The two parts of the digastric acting together depress the chin and assist in opening the mouth when the hyoid bone is fixed by the infrahyoid group of muscles. They elevate the hyoid bone when the mandible is fixed in position by the primary muscles of mastication, as during swallowing.

Relations
The posterior belly of the digastric muscle, with the stylohyoid at its upper border, passes superficial to the internal jugular vein, the internal and external carotid arteries and the vagus, accessory and hypoglossal nerves. The occipital artery runs close to its lower border and the posterior auricular artery to its upper border. This part of the muscle lies deep to the sternoclei-

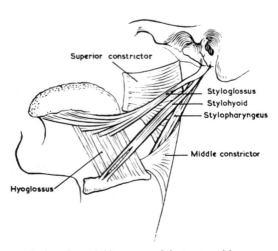

Fig. 5.26 The muscles attached to the styloid process of the temporal bone.

domastoid muscle. Its tendon lies superficial to the posterior lower part of the hyoglossus muscle and is here embraced by the divided tendon of the stylohyoid. The anterior belly lies beneath the mylohyoid muscle deep to the platysma and is related to submental and submandibular lymph nodes, the submental artery and the nerve and artery to the mylohyoid.

The stylohyoid muscle (Fig. 8.26)
This arises from the back of the styloid process near its root and is inserted to the hyoid bone at the junction of the greater cornu with the body. The tendon of insertion is divided and embraces the intermediate tendon of the digastric (Fig. 4.7).

Nerve supply
The seventh (facial) cranial nerve.

Function
The stylohyoid muscle draws the hyoid bone upwards and backwards.

The mylohyoid muscle (Figs 6.3, 6.4, 6.5)
This arises from the mylohyoid ridge on the inner surface of the body of the mandible extending from the middle line in front to opposite the last molar tooth. The fibres run slightly downwards and backwards, the more posterior fibres reach the body of the hyoid bone, the others end in a mid-line tendinous raphe which extends from the symphysis of the jaw to the hyoid bone. The two muscles together make up the *oral diaphragm* (Fig. 6.5).

Nerve supply
The mandibular division of the fifth (trigeminal) cranial nerve (nerve to mylohyoid).

Functions
The two mylohyoid muscles acting together elevate the tongue against the palate and play an important role in the initiation of swallowing (see pages 280, 321). They also draw the hyoid bone upwards and forwards.

Relations
Related to the inferior or superficial surface is the anterior belly of the digastric muscle, the superficial part of the submandibular gland, the submental artery, the mylohyoid vessels and nerve, the submental and submandibular lymph nodes and the platysma muscle.

Deep to or above the muscle lies the deep portion of the submandibular gland, the sublingual gland, the lingual and hypoglossal nerves, the submandibular ganglion, the duct of the submandibular gland, the hyoglossus, genioglossus, styloglossus and geniohyoid muscles (Fig. 6.7).

The geniohyoid muscle (Figs 4.8, 5.15)
This arises by tendinous fibres from the inferior genial tubercle on the inner surface of the symphysial region of the mandible. It lies above the mylohyoid diaphragm. Its fibres run backwards and slightly downwards to be inserted into the anterior surface of the body of the hyoid bone. Developmentally the geniohyoid muscles belong to the same group as the infrahyoid muscles and derive their nerve supply from the same source.

Nerve supply
The first and second cervical nerves through a branch of the hypoglossal nerve (see page 252).

Functions
When acting from the mandible it draws the hyoid bone forwards and upwards; acting from a fixed hyoid it assists in opening the mouth.

THE EXTRINSIC TONGUE MUSCLES

The genioglossus muscle (Figs 4.8, 5.15)
This arises by tendinous fibres from the superior genial tubercle on the inner surface of the body of the symphysial region of the mandible. It enters the tongue where it forms a large fan-shaped muscle, its fibres of attachment radiating from the hyoid bone behind to the tip of the tongue in front. A few of the lower fibres are attached to the upper surface of the hyoid bone; the remainder to the mucous membrane of the pharyngeal and oral parts of the tongue. Within the tongue substance its fibres intermingle with those of the intrinsic muscles.

Nerve supply
The twelfth (hypoglossal) cranial nerve.

Function
The posterior fibres protrude the tongue; the anterior fibres retract the tongue. The whole muscle can depress the tongue and form a concavity on its dorsal surface.

The hyoglossus muscle (Figs 5.26, 5.20, 6.7)
This arises from the upper surface of the greater cornu and from the lateral part of the body of the hyoid bone. A few deeper fibres may arise from the base of the lesser cornu (chondroglossus). The lingual artery passes between the two parts of the muscle. The hyoglossus with the chondroglossus is inserted into the mucous membrane at the side and dorsum of the tongue.

Nerve supply
The twelfth (hypoglossal) cranial nerve.

Function
Depresses and assists retraction of the tongue.

The styloglossus muscle (Figs 4.33, 5.26)
This originates from the front of the styloid process near its tip and from the stylomandibular ligament. It is inserted into the side of the tongue decussating with the fibres of the hyoglossus muscle.

Nerve supply
The twelfth (hypoglossal) cranial nerve.

Function
Draws the tongue backwards and upwards.

The palatoglossus muscle (Figs 4.49, 4.50)
This muscle was described on page 196 as one of the muscles of the soft palate. It must be referred to here also, for it is a member of the extrinsic tongue muscle group. When the soft palate is fixed by the action of the levator and tensor palati muscles, contraction of the palatoglossus elevates the tongue to the under surface of the palate. This is important during swallowing. When both palatoglossal muscles contract the space between the anterior pillars of the fauces is narrowed, so that the entrance into the oropharynx becomes smaller.

MUSCLE GROUPS, THEIR DEVELOPMENT AND GROWTH

The muscles of the head and neck can be grouped as follows:
 1. Muscles of mastication (pages 156, 279).
 2. Muscles of the face and scalp (pages 140, 145).
 3. Muscles of the tongue (pages 244, 284).
 4. Muscles of the soft palate (page 195).
 5. Muscles of the orbital cavity (page 188).
 6. Muscles of the middle ear (page 211).
 7. Prevertebral muscles (page 176).
 8. Postvertebral muscles (page 176).
 9. Suprahyoid muscle group (pages 136, 242).
 10. Infrahyoid muscle group (page 136).
 11. Pharyngeal muscles (pages 193, 343).
 12. Laryngeal muscles (page 200).
 13. Sternomastoid and trapezius (pages 132, 135).

The actions of these muscles have been discussed in various parts of the text. Certain head and neck muscles as well as producing actual movements at various joints play an important part in the maintenance of body posture. They are sometimes called *postural muscles* or '*antigravity*' muscles. They include the postvertebral muscles responsible for maintaining the poise of the head on the cervical vertebral column, and the muscles of mastication, especially the masseter, temporal and internal pterygoid muscles which maintain the mandible in its normal resting position so that the distance between the opposing teeth is about 3–5 mm. From this position the teeth can be brought into occlusion of the various movements of mastication can be initiated. The facial muscles are unique in that in their movements they are used to express various emotional states, although the regulation of the orbital, nasal and oral orifices is their prime function.

The postvertebral and prevertebral muscles, the infrahyoid muscles and the geniohyoid muscle are supplied by branches of the cervical nerves; the others by the various cranial nerves; the sternocleidomastoid and trapezius are supplied by both the cervical plexus and accessory nerve.

The muscles of mastication, the facial muscles, the muscles of the soft palate, pharynx and larynx are developed from the pharyngeal (visceral) arches (page 341). The vertebral muscles, the orbital muscles and the muscles of the tongue and the infrahyoid muscles develop from the upper (cervical) somites (page 330).

It should be remembered that the hypoglossal nerve is in its development a member of the spinal series which has been taken up into the cranial cavity by the inclusion of vertebral elements into the skull in evolution and that the part of the accessory nerve (spinal part) which supplies the sternomastoid and trapezius muscles arises from the upper part of the spinal cord.

Muscles develop from groups of primitive muscle cells (myoblasts) which continue to divide until about the middle of foetal life. After that time further growth is by increase in size (hypertrophy) of individual fibres. Nerve connection with the developing fibres occurs early in development but reflex movements and normal functional activity occur later in foetal life. Spontaneous gasping movements involving the muscles of mastication and swallowing, the facial muscles, and the diaphragm, begin in foetal life and from these the complex activities of suckling, swallowing, chewing and breathing differentiate by the time of birth. Muscles gain their skeletal attachments sometime after differentiation of the myoblasts commences.

With use muscles increase in size. This is partly due to hypertrophy of individual fibres above the normal size with an increase in the thickness of each fibre. There is probably also a development of reserve fibres which in normal activity have not reached their full development. With disuse muscles become reduced in size (atrophy) and may become partly replaced by fibrous tissue. Hypertrophy of muscle is closely associated with hypertrophy of motor-nerve fibres and with a thickening and increased density on the part of the correlated skeletal elements. The efficiency of a muscle depends on the number of motor units which can be brought into action at a given time; the delicacy of muscle movement depends on the degree of development of the proprioceptive mechanism (pages 317, 422).

At birth the muscles of the cheeks and lips are relatively better developed than the muscles of mastication. Between birth and adult life the facial muscles increase four times in weight, while the muscles of mastication increase seven times. Of the latter, the masseter and medial pterygoid are better developed at birth than the temporal and lateral pterygoid muscles. The mandible moves up and down and backwards and forwards, but there is little side to side movement. The tongue musculature is well developed and the tongue shows a wide range of movement. Full control of muscle action depends upon the completion of neuromuscular connections, the establishment of spinal and higher level reflexes, the myelination of the nerve fibres, training, and a certain amount of practice.

During growth of the craniofacial skeleton certain muscles undergo a continual process of reattachment to the adjacent bone. For example, the lower head of the lateral pterygoid is continually being detached below and re-attached above in order to maintain a constant relationship to the backward and upward growth of the mandibular condyle.

IMPORTANT NERVES OF THE HEAD AND NECK

The trigeminal (fifth) cranial nerve

This is the largest of the cranial nerves and its peripheral connections resemble a spinal nerve in that it arises by separate *sensory and motor roots* and that the sensory root has a large ganglion. Its sensory fibres are associated with nerve endings subserving common sensation in the face, the fore part of the head, as well as the eye, the nose, part of the external ear and tympanic membrane, the mouth including the oral part of the tongue, the teeth and their supporting structures, and the dura mater. The motor root supplies the muscles of mastication, a muscle of the soft palate (tensor palati) and a muscle of the middle ear (tensor tympani). The two roots emerge at the side of the pons and enter a recess in the dura mater (Meckel's cave) situated over the tip of the petrous temporal bone and the foramen lacerum (Fig. 8.23). In Meckel's cave the ganglion of the sensory root lies above the motor root and each of the *three divisions* which arise from the ganglion leaves the base of the cave with a covering of dura mater reflected upon it. The motor root runs with the mandibular division. The maxillary and ophthalmic divisions have no somatic motor fibres associated with them.

Ophthalmic division (Fig. 5.27)
After leaving Meckel's cave the ophthalmic division runs through the cavernous sinus in close relationship to the third, fourth and sixth cranial nerves and the internal carotid artery. Within the front part of the sinus it divides into three large branches; the lacrimal, frontal and nasociliary, which enter the orbital cavity and leave the cavernous sinus by passing through the superior ophthalmic (or orbital) fissure between the lesser and greater wings of the sphenoid bone.

The *lacrimal nerve* runs along the lateral wall of the orbital cavity and is joined by communicating fibres (carrying secretomotor parasympathetic fibres) from the zygomatic branch of the maxillary division (Fig. 5.27). It gives branches to the lacrimal gland at the lateral side of the orbital cavity and terminal branches to the skin and conjunctiva of the lateral sides of the eyelids.

The *frontal nerve* passes forward beneath the roof of the orbital cavity dividing at about the middle of the cavity into a larger supra-orbital and a smaller supratrochlear branch. The former continues straight forward and leaves the orbital cavity through the supra-orbital notch or foramen at about the middle of the upper edge of the orbital rim. The nerve then passes upwards to supply the tissues of the forehead and scalp. It gives a branch to the frontal sinus. The supratrochlear branch passes more medially and leaves the orbital cavity at its medial angle above the fibrous tissue sling or pulley for the superior oblique muscle (hence its name). It supplies the upper eyelid and skin of the forehead above the root of the nose.

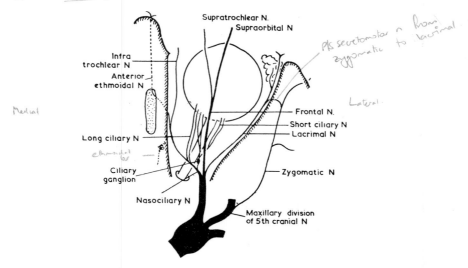

Fig. 5.27 Distribution of the ophthalmic division of the trigeminal nerve.

The *nasociliary nerve* runs on the inner side of the orbital cavity and there gives off two ethmoidal branches which supply the mucous membrane of the sphenoidal and ethmoidal air sinuses. The larger anterior ethmoidal branch enters the front part of the nasal cavity at the front of the cribriform plate supplying the mucous membrane of the roof of the nasal cavity and small areas of the lateral and medial (septal) walls. A terminal branch leaves the nasal cavity between the bony nasal margin and the alar cartilage to supply an area of skin over the lower half of the nose. The terminal branch of the nasociliary nerve is called the infratrochlear nerve and passes below the superior oblique pulley (trochlea) to supply skin and conjunctiva at the medial angle of the eye, the lacrimal sac, the upper part of the nasolacrimal duct, and the skin over the upper half of the nose.

Maxillary division (Fig. 5.32)
After leaving Meckel's cave the middle or maxillary division of the trigeminal nerve runs for a short distance in the base of the cavernous sinus, giving off in this part of its course a small meningeal branch to the dura mater at the front of the middle cranial fossa. It leaves the cranial cavity through the foramen rotundum in the great wing of the sphenoid bone below and behind the superior ophthalmic (orbital) fissure to enter the upper part of the pterygopalatine fossa. It continues across the fossa to leave it in front through the inferior ophthalmic fissure as the infra-orbital nerve, which runs on the floor of the orbital cavity, first in the infra-orbital groove and then in the infra-orbital canal, and terminates on the facial surface of the maxilla below the middle of the lower margin of the orbital rim at the infra-orbital foramen.

The branches of the maxillary division in the pterygopalatine fossa are:
1. Posterior superior dental nerve (page 302).

2. Palatine nerves (page 305).
3. Nasal nerves (sphenopalatine nerves) (page 185).
4. Pharyngeal nerves to the mucous membrane of the roof of the pharynx.
5. Zygomatic nerve.

The *zygomatic nerve* enters the orbit by the inferior ophthalmic fissure and runs on the lateral wall, where it divides into two branches which pierce the zygomatic bone. The posterior (temporal) branch enters the front of the temporal fossa behind the orbital cavity and then pierces the temporal fascia at the anterior margin of the temporal muscle to supply the skin between the eye and the auricle. The anterior (facial) branch appears through one or two small foramina on the facial surface of the zygomatic bone and supplies the overlying area of skin. Parasympathetic fibres derived from the pterygopalatine ganglion run with the zygomatic nerve to the lacrimal nerve via the communicating branch between them. These fibres are secretomotor to the lacrimal gland.

The other branches of the maxillary divisions have already been described in sufficient detail.

The branches of the infra-orbital nerve are the middle and anterior superior dental nerves (page 302), and its three terminal branches, the palpebral, nasal and labial (page 147).

The mandibular division (Fig. 4.24)
Almost immediately after leaving Meckel's cave, the sensory and motor roots of the mandibular division pass through the foramen ovale in the great wing of the sphenoid bone. Outside the skull the roots unite to form a short trunk which gives off small branches to the tensor palati and tensor tympani muscles, to the otic ganglion, and the medial pterygoid muscle. A recurrent sensory branch (nervus spinosus) which returns to the cranial cavity through the foramen spinosum supplies a large area of dura mater in the middle cranial fossa in relation to the temporal lobe of the brain. The trunk of the mandibular division lies on the outer surface of the tensor palati muscle with the otic ganglion between the nerve trunk and the muscle. Behind it lies the middle meningeal artery and laterally is the upper head of the lateral pterygoid muscle. The auditory tube lies close to the nerve trunk as it emerges from the foramen ovale.

After a short course the mandibular nerve trunk divides into a larger posterior division, giving off in turn the auriculotemporal (page 313), inferior dental (page 300), and lingual branches (page 304), and a smaller anterior division, which provides branches for the temporal, lateral, pterygoid and masseter muscles (page 159), and continues on to the buccinator muscle as a sensory branch, the buccal nerve (page 304). The course and distribution of the terminal branches of the mandibular division have been considered in sufficient detail in the text where indicated.

The trigeminal nerve contains sensory and motor somatic fibres; the latter are associated with the mandibular division and its branches. Fibres of special sense (taste) and secretomotor fibres (parasympathetic) are associated with certain branches of the nerve (lingual, palatine, lacrimal and nasal), but do not reach the brain stem with the fifth nerve.

The facial (seventh) cranial nerve (Figs 4.26, 8.23)
This nerve is attached to the brain stem at the upper end of the medulla in close relation to the lower border at the pons by a motor and sensory root. The two roots cross the subarachnoid space above the vestibulocochlear nerve (eighth cranial) and enter the internal auditory meatus. Within the meatus the two roots of the nerve unite and at the bottom of the internal auditory meatus the facial nerve enters the facial canal and takes a winding course through the temporal bone to emerge through the stylomastoid foramen at the base of the skull. At first the facial canal passes outwards and slightly forwards through the petrous part of the temporal bone between the semicircular canals of the inner ear behind, and the cochlea in

front, to reach the medial wall of the middle ear. Here it takes a sharp turn backward (the genu) and runs along the angle between the roof and medial wall of the middle ear cavity. At the back of the cavity it takes a second more gentle turn downwards to the stylomastoid foramen.

At the genu is situated the sensory nucleus of the nerve and from this point a branch of the nerve, containing secretomotor sympathetic fibres, passes forwards in a bony canal, to open behind the foramen lacerum. This branch of the facial nerve is the greater superficial petrosal nerve (Fig. 8.31) which joins the internal carotid or deep petrosal nerve (sympathetic fibres from the superior cervical ganglion) to form the nerve of the pterygoid canal (Vidian nerve). In the pterygopalatine fossa this nerve joins the back of the pterygopalatine (sphenopalatine) parasympathetic ganglion (page 320). In its course through the tympanic cavity the facial nerve gives off some small branches to the tympanic plexus to which the glossopharyngeal nerve also contributes (page 250).

In the terminal descending limb of its intracranial course the facial nerve gives off motor fibres to the stapedius muscle of the middle ear and the chorda tympani branch. The latter enters the middle ear cavity from behind, crosses the tympanic membrane and leaves the tympanic cavity, through the squamotympanic fissure (Fig. 5.17) to join the lingual nerve. The chorda tympani nerve contains taste fibres (sensory) from the anterior two-thirds of the tongue (page 287), and secretomotor parasympathetic fibres destined for the submandibular parasympathetic ganglion (page 282).

Immediately outside the stylomastoid foramen the facial nerve gives off three motor branches:

1. The posterior auricular nerve to the posterior auricular muscle and to the occipitalis muscle of the scalp.

2. A nerve to the posterior belly of the digastric.

3. A nerve to the stylohyoid muscle.

The facial nerve then pierces the fascial sheath of the parotid gland and enters the gland substance, where it usually divides into two large branches. From the upper branch arise the temporal, zygomatic and upper buccal branches; from the lower the lower buccal, the mandibular and cervical branches. The terminal branches appear at the anterior border of the parotid gland and radiate across the face forming a nerve plexus with one another and with the terminal branches of the trigeminal nerve.

The *temporal branch* supplies the anterior and superior auricular muscles, the frontalis muscle, the upper part of the orbicularis oculi and the corrugator supercilii.

The *zygomatic branch* supplies the outer fibres of the orbicularis oculi.

The *buccal branches* supply the buccinator, the muscles of the upper lip, the risorius and the muscles of the nose.

The *mandibular branch* supplies the muscles of the lower lip and the mentalis muscle.

The *cervical branch* supplies the platysma muscle in the neck and may send a communicating branch to join the mandibular branch.

The facial nerve contains the following types of nerve fibres:

1. A large number of somatic motor fibres which supply the muscles of the scalp, auricle, eyelids, nose, lips and cheek (muscles of expression), the platysma, posterior belly of digastric and stylohyoid muscles in the neck, and the stapedius muscle of the middle ear.

2. A smaller number of parasympathetic motor fibres (secretomotor), which supply the pterygopalatine and submandibular parasympathetic ganglia, the submandibular and sublingual salivary glands, the lacrimal gland, the mucous glands of the palate and the nasal cavity.

3. Taste fibres (special sensory) from the oral part of the tongue (anterior two-thirds) and the palate. These fibres have their central connections in the upper part of the tractus solitarius in the medulla oblongata (Fig. 8.10).

4. There is some clinical evidence that the facial branches of the nerve may contain sensory fibres, as well as the motor fibres to the muscles of expression.

The glossopharyngeal (ninth) cranial nerve (Fig. 4.30)

This is attached to the side of the upper part of the medulla below the pons by three or four filaments containing both sensory and motor fibres. The nerve passes through the front part of the jugular foramen in a sheath of dura mater. During its course through the foramen it shows two small gangliform swellings, the jugular and petrosal ganglia. Outside the skull it runs in the upper part of the carotid sheath between the internal carotid artery and the internal jugular vein. At the lower border of the stylopharyngeus muscle it leaves the sheath and runs on the surface of the muscle deep to the external carotid and ascending palatine arteries to the upper border of the middle constrictor. Here it passes deep to the constrictor muscle and enters the oral pharynx.

The chief branches of the glossopharyngeal nerve are:

The *tympanic branch* given off just below the jugular foramen. It enters the tympanic cavity (middle ear) through a small canal, the opening of which lies between the jugular foramen and the carotid canal. In the middle ear it joins with a branch of the facial nerve to form the tympanic plexus from which arise sensory branches to the mucous membrane of the middle ear, the tympanic antrum and the auditory tube; and the lesser superficial petrosal nerve which contains secretomotor fibres derived from the glossopharyngeal nerve and which supply the otic ganglion and the parotid salivary gland (Fig. 8.27).

The *carotid branch* descends on the internal carotid artery to the carotid sinus and carotid body (Fig. 4.28). It carries sensory (autonomic) fibres and takes part in the regulation of blood pressure.

A *motor branch* to the stylopharyngeus muscle.

Pharyngeal (*sensory*) branches to the pharyngeal plexus on the side wall of the pharynx (page 251). From the plexus sensory nerves, derived from the glossopharyngeal nerve, pass to the pharyngeal mucous membrane.

Tonsillar branches ascend deep to the hyoglossus muscle to form a plexus around the tonsil. From the plexus sensory branches are distributed to the upper part of the pharynx and the commencement of the auditory tube.

Lingual branches pass deep to the hyoglossus muscle and constitute the sensory nerve supply to the mucous membrane of the pharyngeal part (posterior third) of the tongue.

Communicating branches with the superior cervical sympathetic ganglion, the vagus and facial nerves.

The glossopharyngeal nerve contains:

1. Sensory somatic fibres subserving common sensation from nerve endings in the mucous membrane of the middle ear and its extensions, the auditory tube, the nasal and oral regions of the pharynx and the pharyngeal part of the tongue.

2. Taste fibres for the posterior third of the tongue and circumvallate papillae. These fibres have their central connections in the tractus solitarius below the upper part of the tract, which is devoted to the taste fibres travelling via the facial nerve (Figs 8.10, 8.11).

3. Motor somatic fibres to the stylopharyngeus muscle.

4. Sensory autonomic fibres from the carotid body and sinus.

5. Motor autonomic (parasympathetic or secretomotor) fibres which pass via the tympanic nerve and otic ganglion to the parotid salivary gland and probably via the buccal and mental nerves to the mucous glands of the cheek and lower lip (page 320).

The vagus (tenth) cranial nerve (Figs 4.30, 4.31)

This is attached by a series of filaments to the brain stem at the side of the medulla below and in series with those of the glossopharyngeal nerve. The nerve trunk passes through the jugular foramen in the same sheath of dura mater as the accessory nerve. Within the foramen it shows a gangliform swelling, the superior ganglion, and at a lower level, outside the skull, a larger inferior (nodose) ganglion. The cranial part of the accessory nerve joins the vagus at the inferior ganglion and its fibres are distributed with the pharyngeal and laryngeal branches of the vagus. The ganglia also communicate with the facial, glossopharyngeal, hypoglossal and sympathetic nerves.

Outside the jugular foramen the vagus enters the carotid sheath between the internal jugular vein and the internal carotid artery and descends within the sheath through the neck. On the right side it enters the thoracic cavity after crossing the first part of the subclavian artery; on the left side after descending between the left common carotid and subclavian arteries it crosses in front of the aortic arch. On the right side the recurrent laryngeal branch hooks around the subclavian artery; on the left side it passes around the arch of the aorta.

In the thorax the vagus passes behind the hilum of the lung to the anterior surface of the oesophagus where the nerves of each side unite to form the oesophageal plexus. From the plexus two trunks pass through the diaphragm in front of (left vagus) and behind (right vagus) the oesophagus. Within the abdomen the left vagus is distributed chiefly to the anterior surface and lesser curvature of the stomach, and to the liver and gall bladder. The right nerve is distributed chiefly to the back of the stomach, its greater curvature, and along branches of the superior mesenteric artery to the midgut derivatives (page 343). It sends branches to the coeliac, splenic and renal plexuses.

The branches of the vagus nerve in the neck are:

A *recurrent meningeal branch* given off the superior (jugular) ganglion passes back through the jugular foramen to supply the dura mater of the posterior cranial fossa.

The *auricular nerve (of Arnold)* given off the superior ganglion enters a canal through a small foramen between the jugular foramen and the styloid process. The canal passes close to the canal of the facial nerve and terminates between the external auditory meatus and the mastoid process. Here the nerve divides into a branch joining the posterior auricular nerve (a branch of the facial nerve) and a branch which supplies the mucous membrane of the external auditory canal and the outer surface of the tympanic membrane (with a branch of the auriculotemporal nerve).

A *small branch distributed to the carotid sinus and carotid body.*

The *pharyngeal branch* comes off the vagus in the region of the inferior ganglion. It consists chiefly of fibres derived from the cranial part of the accessory nerve. It crosses the internal carotid artery and passes deep to the external carotid to reach the pharyngeal wall where it contributes to the pharyngeal plexus. It contains motor fibres to the constrictor muscles of the pharynx, and the palatopharyngeus, levator palati, palatoglossus and uvular muscles of the soft palate.

The *superior laryngeal nerve* comes off the vagus in the region of the inferior ganglion and like the pharyngeal nerve consists chiefly of fibres derived from the cranial part of the accessory nerve. It passes deep to the internal carotid artery and the superior thyroid artery where it divides into a larger internal laryngeal (sensory) branch which accompanies the internal laryngeal artery through the thyrohyoid membrane to the laryngeal part of the pharynx and upper part of the larynx. The smaller motor branch (the external laryngeal) nerve passes deep to the infrahyoid muscles to reach the cricothyroid and inferior constrictor muscles (Fig. 4.48).

In the neck the vagus gives off an *upper and lower cardiac branch* which descends with the main trunk to the thoracic cavity where they join the cardiac plexus. They are accompanied by cardiac branches derived from the superior and middle sympathetic ganglia.

As already stated, the *recurrent laryngeal nerves* have a somewhat different course on each side: the right nerve passing around the subclavian artery, the left around the aortic arch. The nerves ascend between the oesophagus and trachea to enter the larynx from below. They contain motor fibres to the muscles of the larynx (except for the cricothyroid) and sensory branches to the laryngeal mucous membrane below the vocal folds and the mucous membrane of the trachea. Some motor fibres to the lower part of the inferior pharyngeal constrictor may run in the recurrent laryngeal nerves. The majority of the fibres of the recurrent nerves are derived from the cranial part of the accessory nerve. In the thoracic cavity the vagus gives off branches to the cardiac and pulmonary autonomic plexuses. The distribution of the nerves in the abdominal cavity has been described.

The vagus nerve contains the following fibres:

1. Somatic sensory fibres derived from nerve endings in the dura mater, auricle and tympanic membrane (meningeal and auricular branches). These, however, properly belong to the trigeminal nerve.

2. Somatic (pharyngeal) motor fibres derived from the accessory nerve and distributed to certain muscles of the pharynx, soft palate and larynx.

3. Sensory and motor autonomic (parasympathetic) fibres to the smooth muscle, glands and mucous membrane of the alimentary canal from the oesophagus to the transverse colon, to the liver, gall bladder, pancreas, spleen and kidneys, to the muscle of the heart, to the smooth muscle, glands, and mucous membrane of the respiratory system, from the larynx to the terminal bronchioles.

These fibres together with fibres derived from the sympathetic system regulate the rate of the heart beat, blood pressure, peristaltic gut movements, glandular secretions, the cough reflex and the amount of air entering the lungs.

The accessory (eleventh) cranial nerve (Fig. 4.32)
This is a motor nerve and consists of two parts which differ both in their origin and distribution. The cranial part arises from the side of the medulla below the vagus nerve and the spinal part from the side of the spinal cord as far down as the attachment of the fifth cervical nerve. The spinal portion enters the cranial cavity through the foramen magnum behind the vertebral artery and within the cranial cavity unites with the cranial part to form a common nerve trunk. This passes through the jugular foramen in the same sheath of dura mater as the vagus nerve and outside the skull lies between the internal jugular vein and the internal carotid artery. Here the trunk divides, and the cranial part of the nerve joins the vagus in the region of the inferior ganglion and is distributed through the pharyngeal and laryngeal branches of the vagus to certain of the pharyngeal, soft palate and laryngeal muscles. The spinal part passes backwards either superficial or deep to the internal jugular vein and through the substance of the sternocleidomastoid muscle, which it supplies, to reach the trapezius muscle after crossing the posterior triangle of the neck (Fig. 4.5).

The hypoglossal (twelfth) cranial nerve (Fig. 4.33)
This is attached by a series of filaments to the side of the medulla. These pass behind the vertebral artery to form a common nerve trunk in the anterior condylar canal through which the nerve leaves the cranial cavity. At the base of the skull the nerve enters the upper part of the carotid sheath and winds around the vagus nerve giving off a communicating branch. In this region it is joined by branches from the first and second cervical nerves and from the superior cervical sympathetic ganglion. Within the carotid sheath it descends between the internal jugular vein and the internal carotid artery to where the lower border of the digastric muscle crosses the vessels. Here the nerve turns forwards hooking around the origin of the occipital artery and lying superficial to the external carotid, facial and lingual arteries in the carotid triangle. It leaves the carotid triangle by passing deep to the digastric and stylohyoid muscles and enters the floor of the mouth between the mylohyoid and hyoglossus muscles. On the hyoglossus it communicates with the lingual nerve and then passes forwards to lie on and penetrate the genioglossus muscle and enter the substance of the tongue below the

sublingual gland. Within the tongue it gives branches to its intrinsic and extrinsic muscles, except the palatoglossus.

The other branches of the hypoglossal nerve are derived from the fibres which join it from the first cervical nerve. They are:

A *meningeal branch* supplying the dura mater of the posterior cranial fossa via the anterior condylar canal.

A *descending branch* (descendens hypoglossi) given off as the hypoglossal nerve passes around the occipital artery. It is in reality composed of first cervical nerve fibres and is joined by a loop from the second and third cervical nerves (descendens cervicalis) to form the ansa hypoglossi from which branches are distributed to the omohyoid, sternohyoid and sterno-thyroid muscles (page 139).

Muscular branches (derived from C1, 2) given off the hypoglossal nerve to the thyrohyoid and geniohyoid muscles.

IMPORTANT ARTERIES OF THE HEAD AND NECK

The external and internal carotid arteries and their branches have been described already in the text but in this section some of the more important vessels, particularly those to the oral and facial tissues, are described in greater detail.

The facial artery (Fig. 5.28)

This is given off the external carotid artery in the carotid triangle a short distance above the lingual artery; sometimes the two arteries arise from a common stem. At its origin it lies on the outer surface of the middle constrictor muscle of the pharynx. It ascends in the neck, lying on the middle and superior constrictor muscles deep to the posterior belly of the digastric and the stylohyoid muscles. At the upper border of the latter it passes forwards and enters a deep groove in the substance of the submandibular salivary gland below the mylohyoid muscle. It leaves the gland to reach the face at the lower border of the mandible at the anterior

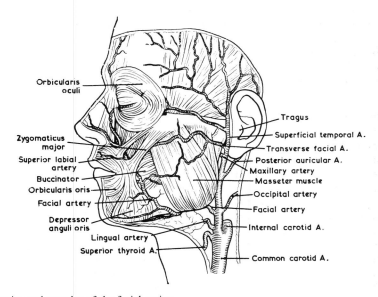

Fig. 5.28 Arteries and muscles of the facial region.

edge of the masseter muscle, piercing the investing layer of the deep cervical fascia which is attached to the lower border of the jaw. As it lies in contact with the bone it is covered by fibres of the platysma muscle and is crossed here or at a lower level by the mandibular branch of the facial nerve. In the face it lies on the buccinator and levator anguli oris beneath the risorius, zygomaticus major and minor and levator labii superioris muscles. At the side of the nose (angular artery) it runs through the fibres of the levator anguli superioris alaeque nasi. It can be divided into three parts for descriptive purposes:

Cervical part, on the side wall of the pharynx.

Submandibular part, in relation to the superficial portion of the submandibular gland.

Facial part.

The branches given off the first part are:

1. Ascending palatine artery, which continues to ascend on the side wall of the pharynx passing between the styloglossus and stylopharyngeus muscles. At the upper border of the superior constrictor it enters the soft palate with the levator palati muscle.

2. Tonsillar branches, given off close to the summit of the loop of the facial artery. They pierce the superior constrictor muscle to enter the deep surface of the tonsil.

The branches given off the second part are glandular branches to the submandibular gland and the submental artery. The latter run on the under surface of the mylohyoid muscle close to the lower border of the jaw. Close to the middle line it sends branches to the skin and muscles of the chin region.

The branches given off the third part are:

1. The inferior labial artery arises below the angle of the mouth. It enters the lower lip deep to the depressor anguli oris. In the lip it penetrates the orbicularis oris and runs towards the middle line beneath the mucous membrane (Fig. 4.19). It may divide into two branches in the lip.

2. The superior labial artery is given off at the level of the angle of the mouth and runs towards the middle line between the orbicularis oris and the mucous membrane. Close to the middle line it gives off an ascending branch to the lower edge of the nasal septum.

3. The lateral nasal branch to the side of the nose.

4. Small posterior branches to the tissues of the cheek.

The terminal part of the facial artery at the side of the nose is usually called the angular artery. It anastomoses with terminal branches of the dorsal nasal branch of the ophthalmic artery at the inner side of the rim of the orbital cavity. On the face the facial artery and its branches anastomose with branches of the mental, infra-orbital, transverse facial and buccal arteries. From its origin to the angular artery the facial artery is highly tortuous to permit its adaptation to the movements of the tissues and parts with which it is related.

The lingual artery (Figs 4.28, 5.29)

This is given off the external carotid artery in the carotid triangle, between the facial artery above and the superior thyroid artery below. It lies on the outer surface of the middle pharyngeal constrictor and passes upwards behind the greater cornu of the hyoid bone. It then descends, forming a loop which is crossed by the hypoglossal (twelfth) cranial nerve and passes deep to the hyoglossus muscle and superficial to the chondroglossus and genioglossus. It ascends on the side of the tongue along the anterior border of the hyoglossus, passing deep to the hypoglossal and lingual nerves, the duct of the submandibular gland, and the sublingual gland, to enter the substance of the tongue. The relationship of the artery to the hyoglossus muscle enables it to be subdivided into three parts for descriptive purposes:

The first part of the artery, from its origin to the posterior border of the hyoglossus muscle, gives off a small hyoid branch which runs along the upper border of the greater cornu and body of the hyoid bone superficial to the hyoglossus muscle.

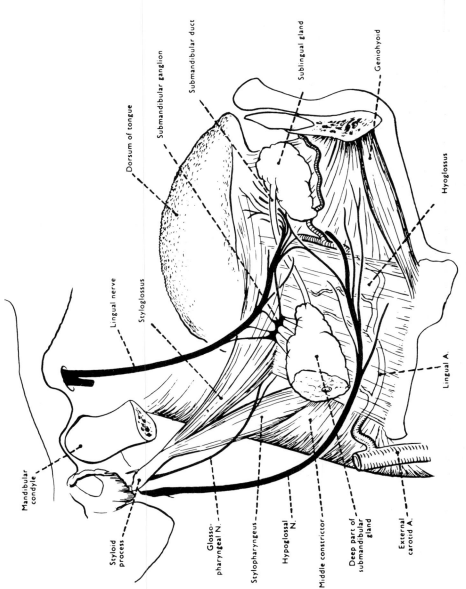

Fig. 5.29 The submandibular region.

The second part of the artery lies deep to the hyoglossus muscle, gives off one or more dorsal arteries of the tongue (dorsalis linguae) which ascend on the side of the tongue between the genioglossus and hyoglossus and give branches to the tongue, tonsil, the back part of the floor of the mouth, and to the soft palate via the anterior pillar of the fauces (palatoglossal fold).

The third part of the artery enters the substance of the tongue along the anterior border of the hyoglossus muscle. It gives off a large sublingual branch which runs forwards beneath the mucous membrane of the floor of the mouth between the genioglossus on its inner side and sublingual gland, the mucous membrane of the front part of the floor of the mouth and the adjacent gum on the inner (lingual) side of the lower teeth. The lingual artery in its course is highly tortuous to permit its adaptation to the movements of the tongue and hyoid bone.

The superior thyroid artery (Figs 4.10, 5.30)
This is given off the external carotid artery in the carotid triangle. It may share a common stem with the lingual artery. It lies superficial to the middle constrictor muscle and descends downwards and forwards along the posterior border of the thyrohyoid muscle, deep to the omohyoid, sternohyoid and sternothyroid muscles, to reach the apex of the lateral lobe of the thyroid gland. Here it perforates the fascial sheath of the gland and divides into three main branches. One descends along the anterior margin of the lateral lobe and upper border of the isthmus; the second branch descends on the anterior surface of the gland; and the third

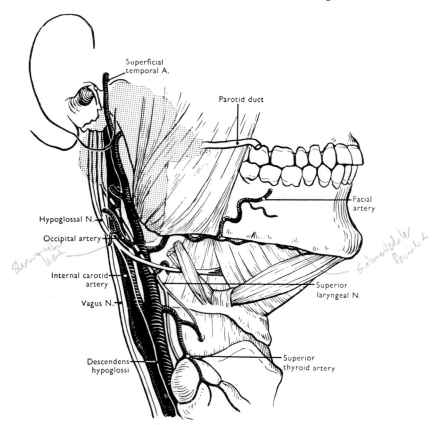

Fig. 5.30 Deep dissection of the upper part of the neck.

along the posterior surface to anastomose with an ascending branch of the inferior thyroid artery. From these vessels, which anastomose freely with one another, branches pierce the capsule of the gland and supply it.

Other branches of the superior thyroid artery are:

1. A small infrahyoid branch which passes along the lower border of the hyoid bone.

2. A large superior laryngeal artery which runs with the internal laryngeal branch of the vagus nerve to enter the pharynx and larynx by piercing the thyrohyoid membrane.

3. A sternomastoid branch passing downwards and backwards along the upper border of the omohyoid muscle to the sternocleidomastoid muscle.

4. A cricothyroid branch which runs towards the middle line of the neck on the surface of the cricothyroid membrane (see also page 163).

The maxillary artery (Figs 4.28, 5.31, 5.32)
This commences as one of the terminal branches of the external carotid artery within the substance of the parotid gland. From its origin it pierces the fascial sheath of the parotid compartment, runs forwards deep to the neck of the mandible, and comes into relation to the lower border of the lateral pterygoid muscle. It then passes on to either the deep or superficial surface of the lower head of the lateral pterygoid. If it passes deep to the muscle it reappears between the two heads close to their origin. It continues on the lateral (outer) side of the lateral pterygoid plate to enter the pterygopalatine fossa where it gives off its terminal branches.

For descriptive purposes the artery is divided into three parts by its relationship to the lateral pterygoid muscle:

The *first part* of the artery, from its origin to the lower border of the muscle, enters the infratemporal region between the neck of the condylar process of the mandible on its outer side and the sphenomandibular ligament and inferior dental nerve on its inner side. It lies at a variable distance below the line of attachment of the capsule of the mandibular joint and the auriculotemporal nerve. It gives off the following branches.

1. The auricular branch which ascends behind the capsule of the mandibular joint to enter the external auditory meatus between its bony and cartilaginous parts to supply the mucous membrane and outer surface of the tympanic membrane.

2. The tympanic branch which ascends on the inner side of the capsule of the mandibular joint to enter the middle ear via the squamotympanic fissure. Within the ear it helps to supply the mucous membrane and the inner surface of the tympanic membrane.

3. The middle meningeal artery is a vessel of moderate size (Figs 4.41, 5.31). It ascends deep to the lateral pterygoid muscle, between the two roots of the auriculotemporal nerve, to enter the cranial cavity through the foramen spinosum. It lies behind the mandibular nerve and the otic ganglion and usually gives off a small branch (accessory meningeal artery) which enters the cranial cavity through the foramen ovale. Within the cranium these meningeal arteries supply the dura mater covering the brain in the middle cranial fossa and temporal region. The middle meningeal artery usually gives a small branch to the orbital cavity through the superior ophthalmic fissure; sometimes this branch takes the place of the lacrimal branch of the ophthalmic artery.

4. The inferior dental artery descends immediately behind the inferior dental nerve, lying between the ramus of the mandible on its outer side and the sphenomandibular ligament and medial pterygoid muscle on its inner side, to reach the foramen of the inferior dental canal giving off its mylohyoid branch before entering the canal. Within the mandibular canal the inferior dental artery gives off branches to the teeth and their supporting structures, the medullary cavity of the body of the mandible, to the cancellous bone of the ramus and to the condylar growth cartilage. It terminates by dividing into mental and incisive branches. The former emerges through the mental foramen to supply the skin and mucous membrane and muscles

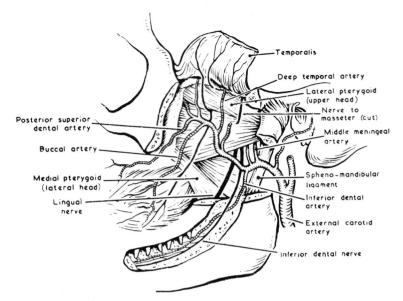

Fig. 5.31 Distribution of the maxillary artery.

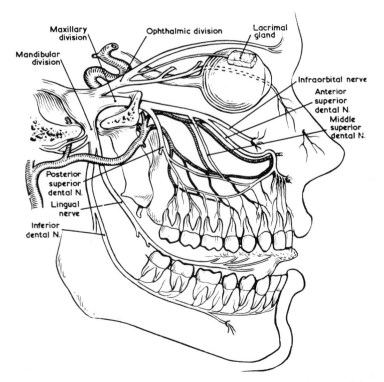

Fig. 5.32 The nerve supply and blood supply to the upper teeth. The main features of the ophthalmic and mandibular divisions of the trigeminal nerve are also shown.

of the lower lip and cheek, and anastomoses with branches of the facial artery. The incisive artery supplies the canine and incisor teeth and their supporting structures (Fig. 5.31).

The *second part* of the maxillary artery lies either superficial or deep to the lower head of the lateral pterygoid muscle. In the superficial position the artery lies between the lateral pterygoid on its inner side and the insertion of the temporal muscle on its outer side. If it passes deep to the lateral pterygoid (in 40 to 50 per cent of cases) it lies on the upper part of the medial pterygoid muscle and in close relationship to the mandibular nerve as this emerges from the foramen ovale.

The branches of this part of the artery are distributed to the muscles of mastication (pterygoids, temporal and masseter) and to the buccinator through the buccal artery which runs with the nerve of the same name. There are usually two deep temporal branches.

The *third part* of the artery passes on the outer side of the lateral pterygoid plate to enter the pterygopalatine fossa. If the second part of the artery passes deep to the lateral pterygoid muscle it appears between the two heads at the posterior edge of the pterygoid plate to become continuous with the third part. The branches of the third part of the maxillary artery are:

1. The posterior superior dental artery, given off before the main vessel enters the pterygopalatine fossa, descends on the back of the maxilla to supply the upper molar teeth, the gums in relation to their buccal surfaces, and the cheek (page 298). *helps to supply mucous membrane max sinus*

2. The greater palatine artery descends through the pterygopalatine fossa and palatine canal to appear on the hard palate through the greater palatine foramen (Fig. 4.3). It gives off a number of small branches which reach the soft palate through the lesser palatine foramina, and nasal branches, which enter the lower posterior part of the nasal cavity by piercing the vertical plate of the palatine bone. On the hard palate the artery runs forwards to the incisive foramen where it anastomoses with the terminal branches of the artery of the opposite side and the long sphenopalatine (nasopalatine) artery.

3. The sphenopalatine artery passes from the pterygopalatine fossa to the nasal cavity (spheno-ethmoidal recess) through the sphenopalatine foramen. Within the nasal cavity it supplies the mucous membrane covering the greater part of the lateral wall of the cavity, the sphenoid air sinus, the maxillary antrum, the ethmoidal air cells and the infundulum of the frontal sinus. A branch crosses the roof of the back part of the nasal cavity (behind the cribriform plate) to reach the nasal septum on which it descends as the nasopalatine or septal branch to the incisive foramen. It supplies the mucous membrane covering the nasal septum. Terminal branches anastomose with the greater (descending) palatine arteries, *as they ascend* through the incisive foramen and the septal branch of the facial artery (page 254).

4. The infra-orbital artery enters the orbital cavity through the inferior *orbital* ophthalmic fissure and runs forwards at first in the infra-orbital groove, then through the infra-orbital canal to emerge on the facial surface of the maxilla at the infra-orbital foramen. It gives off orbital, middle and anterior superior dental branches (page 298); and terminal branches on the face to the lower eyelid, the side of the nose and to the upper lip. These anastomose with branches of the facial artery.

5. The pharyngeal artery passes backwards beneath the body of the sphenoid bone or through the pharyngeal canal between the sphenoid and vomer. It supplies the mucous membrane on the roof of the back part of the nasal cavity and nasopharynx and gives branches to the sphenoid air sinus.

6. The artery of the pterygoid (Vidian) canal passes backwards along the bony canal at the base of the pterygoid plates to the foramen lacerum. It gives branches to the auditory tube, and to the mucous membrane of the side wall and roof of the nasopharynx.

The ophthalmic artery (Fig. 5.33)
This is a branch from the internal carotid artery after it emerges from the cavernous sinus. It enters the orbital cavity through the optic foramen in close relationship to the optic nerve.

It lies on the lateral side of the nerve before crossing above it to reach the medial wall of the orbit. It terminates by dividing into a supratrochlear and dorsal nasal branch; the latter anastomoses with the facial artery (page 253).

Other branches of the ophthalmic artery are:

1. The central artery of the retina which runs forwards in the substance of the optic nerve to reach the eyeball. This is its most important branch.

2. The lacrimal artery, which is given off before the main artery crosses the optic nerve and continues forwards on the lateral side of the eyeball to reach the lacrimal gland. It gives terminal (lateral palpebral) branches to the eyelids and a recurrent meningeal branch which

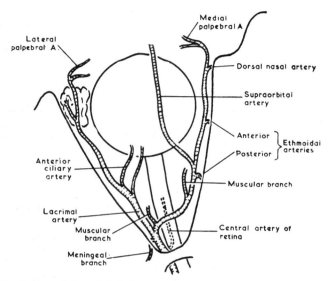

Fig. 5.33 The distribution of the ophthalmic artery.

passes backwards through the superior orbital fissure to supply the meninges of the middle cranial fossa.

3. Muscular branches to the muscles in the orbital cavity.

4. A group of ciliary arteries which pierce the eyeball and are distributed chiefly to its middle (choroidal) coat.

5. The supra-orbital artery passing with the corresponding nerve through the supra-orbital notch to supply the skin and tissues covering the frontal bone.

6. Anterior and posterior ethmoidal branches leave the medial side of the orbit with the corresponding nerves and are distributed to the anterior cranial fossa (meningeal branches), the ethmoidal, sphenoidal and frontal air sinuses, and the nasal mucosa.

7. Medial palpebral branches supply the medial sides of the upper and lower eyelids, and anastomose with the lateral palpebral branches of the lacrimal artery.

ARTERIAL ANASTOMOSES OF THE FACE

The chief blood vessels to the facial region are the lingual, facial and maxillary branches of the external carotid artery; the transverse facial branch of the superficial temporal artery; and the terminal branches (lacrimal, supra-orbital, supratrochlear and dorsal-nasal branches) of the ophthalmic artery, itself a branch of the internal carotid artery.

The most important sites of anastomosis between these vessels are as follows:

Between the lingual artery and other vessels.

1. The dorsal lingual artery anastomoses with the ascending palatine, descending (greater) palatine, and tonsillar branches of the facial artery in the region of the soft palate and the anterior pillar of the fauces.

2. The sublingual artery anastomoses with the submental branch of the facial artery and the artery to the mylohyoid in the floor of the mouth.

Between the facial artery and other vessels.

1. The ascending palatine branch anastomoses with the descending palatine and dorsal lingual vessels.

2. The tonsillar branches anastomose with the dorsal lingual vessels.

3. The submental branch anastomoses with the sublingual, mylohyoid and inferior labial vessels.

4. Branches on the face anastomose with the mental, infra-orbital, buccal, transverse facial, and dorsal nasal vessels.

Between the maxillary artery, its branches and other vessels.

1. The inferior dental branch anastomoses with the submental and sublingual vessels via the mylohyoid branch.

2. The inferior dental branch anastomoses with the facial artery via the mental artery.

3. The middle meningeal anastomoses with the ophthalmic artery via the ophthalmic branch of the former.

Sometimes the ophthalmic artery is a branch of a large middle meningeal artery.

4. The buccal artery anastomoses with branches of the posterior superior dental and facial artery in the cheek. Sometimes the buccal artery is enlarged to replace part of the facial artery.

5. The greater palatine artery anastomoses with other vessels to the soft palate and at the incisive foramen with the long sphenopalatine branch of the nasal artery.

6. The nasal artery and its branches anastomose with branches of the ophthalmic artery (ethmoidal vessels), the pharyngeal artery, the anterior superior dental artery, the superior labial artery (septal branches) and the (greater) descending palatine artery.

THE VEINS OF THE HEAD AND NECK

The more important veins of the head and neck have been described elsewhere but here a general account will be given of the venous drainage as a whole. It should be remembered that the distribution of veins is much more variable than that of arteries.

The anterior facial vein (Fig. 5.34)

This, the main venous drainage of the superficial tissues of the face, commences at the inner corner of the orbital cavity as the angular vein, anastomosing with terminal branches of the ophthalmic veins and the supratrochlear vein. It runs downwards and backwards, lying superficial to the facial muscles and behind the facial artery, to the lower border of the mandible at the anterior edge of the masseter muscle, where it pierces the investing layer of the deep cervical fascia (page 137). It then crosses the surface of the superficial part of the submandibular gland to join with the anterior branch of the posterior facial vein, forming the short common facial vein. This enters the internal jugular vein in the carotid triangle a short distance below the posterior belly of the digastric muscle.

On the face the chief tributaries of the anterior facial vein are the lateral nasal from the side of the nose, the superior and sometimes inferior labial veins from the lips, draining into its anterior aspect; and the deep facial vein, which joins it from behind. The anterior facial vein anastomoses with the cavernous sinus of the cranial cavity via the ophthalmic veins and with the pterygoid venous plexus via the buccal vein (deep facial vein). This is important in the spread of infections from the face to deeper tissues.

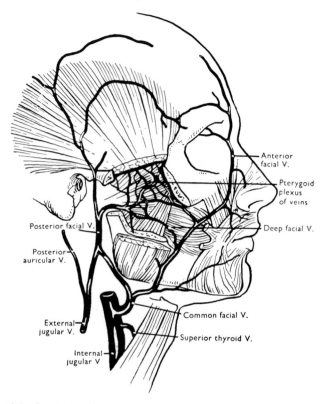

Fig. 5.34 The veins of the face and neck.

The posterior facial or retromandibular vein (Figs 5.34, 5.35)
This commences in the substance of the parotid gland and within the parotid fascial space by the union of the superficial temporal vein (draining the temporal region of the scalp) and the maxillary vein. Both these vessels pierce the fascial sheath to enter the parotid compartment.

Close to the lower pole of the parotid gland the posterior facial vein divides into:

1. An *anterior branch* passing forwards from the parotid to the submandibular fascial compartment (pages 447, 449), where it forms the common facial vein after joining the anterior facial vein.

2. A *posterior branch* piercing the back of the parotid compartment to lie on the surface of the sternomastoid muscle where it is joined by the posterior auricular vein draining the scalp behind the auricle, to form the external jugular vein.

The external jugular vein (Figs 4.5, 5.34)
This descends across the sternomastoid muscle to its posterior border, beneath the superficial layer of the investing fascia. Here it is joined by veins draining the subclavian region at the

base of the neck and enters the subclavian vein. It is usually joined here by the terminal part of the anterior jugular vein.

The anterior jugular vein

This commences below the chin in the submental region and descends close to the middle line of the neck on the surface of the infrahyoid muscles, where it may be joined by an inferior labial vein. There is often a communicating vein joining the common facial vein and anterior jugular vein. Above the suprasternal notch the veins on either side are usually united by a communicating branch (jugular arch). The anterior jugular vein then turns backwards, running deep to the sternomastoid muscle across the carotid sheath and its contents, to enter the subclavian triangle, where it usually joins the terminal part of the external vein. It may, however, enter the subclavian vein directly.

The lingual vein

Drains the tongue and floor of the mouth cavity. It usually passes superficial to the hyoglossus muscle, but branches also run with the lingual artery deep to the muscle. The lingual vein drains into the internal jugular vein.

The internal jugular vein (Figs 4.5, 4.10, 4.29, 5.34)

This commences as the jugular bulb at the jugular foramen with the union of the transverse (sigmoid part) and inferior petrosal venous sinuses. It descends through the neck within the carotid sheath with the vagus nerve and the internal and common carotid arteries, close to the anterior margin of the sternomastoid muscle, which usually overlaps it. At the base of the neck the internal jugular and subclavian veins unite to form the brachiocephalic vein. The tributaries of the internal jugular vein are the pharyngeal plexus, the common facial vein, the lingual, superior thyroid, and middle thyroid veins. The latter drain directly from the thyroid gland. An inferior thyroid vein or veins descends in front of the trachea to join the left brachiocephalic vein in the thorax. (See page 269 for further details.)

The cranial venous sinuses (Figs 8.22, 5.9)

1. The *superior sagittal sinus* runs between the outer and inner layers of dura mater, above the greater intercerebral fissure, in the base of the falx cerebri. It commences in front at the foramen caecum and ends behind on the inner side of the occipital protuberance, where it usually turns to the right and continues as the right transverse sinus. The left transverse sinus usually commences in the same region as a continuation of the straight sinus which runs backwards along the line of union of the falx cerebri and falx cerebelli. The two transverse sinuses frequently communicate with one another at their commencement (the confluence of the sinuses).

2. The *straight sinus* is formed by the union of the small inferior sagittal sinus, which runs backwards along the lower free edge of the falx cerebri, and the great cerebral vein (of Galen) which drains the depths and base of the cerebral hemispheres. The straight sinus is inclined backwards and downwards towards the internal occipital protuberance in the region of union between the falx cerebri and the tentorium cerebelli.

3. The *transverse sinuses* run along the line of attachment of the falx cerebelli to the occipital bone as far as the petrous temporal bone, where each sinus turns downwards to run on the side wall and floor of the posterior cranial fossa to reach the jugular foramen (Fig. 5.19). The part of the transverse sinus below the tentorium cerebelli is usually called the sigmoid sinus. It receives the superior petrosal sinus from the cavernous sinus at its commencement.

4. The *inferior petrosal sinus* runs in the angle between the petrosal temporal and occipital bones from the cavernous sinus in front to the jugular foramen behind, where it unites with the sigmoid (transverse) sinus at the jugular bulb.

5. The *cavernous venous sinus* lies at the side of the pituitary fossa, communicating freely via intercavernous sinuses with the sinus of the opposite side behind and in front of the stalk (infundibulum) of the gland. It drains backwards to the sigmoid sinus (via the superior petrosal sinus) and to the jugular bulb (via the inferior petrosal sinus) and forwards towards the face along the ophthalmic veins which enter it from the orbital cavity through the superior ophthalmic fissure.

The cavernous sinus is about 1 cm wide and 2 cm long. It transmits along its length the intercranial parts of the third, fourth and sixth cranial nerves passing to the orbital muscles and the ophthalmic and maxillary divisions of the fifth (trigeminal) cranial nerves (Figs 5.9, 5.35). The internal carotid artery runs forward through the sinus, with the sixth cranial nerve on its lateral aspect, and both structures are covered with endothelium. The other cranial nerves lie in the lateral wall of the sinus between the outer layer of dura mater and the endothelium lining the sinus. The cavernous sinus has a number of important relations, including the pituitary gland medially, the sphenoidal air sinus inferiorly, the optic chiasma superiorly and the middle cranial fossa on its lateral aspect. The close relation of the oculomotor, trochlear and abducent nerves to the cavernous sinus makes them susceptible to damage in thrombosis of the sinus or by aneurysm (swelling) of the internal carotid artery. Inflammation of the meninges (meningitis) may damage these nerves; they lie close together at the anterior end of the sinus near the superior orbital fissure, where they may be compressed by meningeal tumours. Tumours of the pituitary gland tend to occlude the internal carotid artery and press on the abducent nerve before affecting the other cranial nerves in the sinus, so that ocular signs are first to appear, due to interruption of the motor fibres to the lateral rectus muscle. The connections between the facial vein and the cavernous sinus through the ophthalmic veins makes the sinus a potential target for the spread of infections from the face, particularly from around the nose.

6. The various *venous sinuses* receive blood vessels from the brain, its covering meninges and the diploic vascular channels of the skull vault. They communicate with extracranial veins (other than the internal jugular and ophthalmic veins which are the chief drainage channels) through many of the basal foramina via numerous *emissary veins*. These include veins passing through the foramina ovale, rotundum, lacerum and the posterior condylar canal. In addition important emissary veins pass through the parietal and mastoid foramina.

A series of emissary veins connect the face with the cranial venous sinuses in the region of the pituitary fossa and the middle cranial fossa. Veins from the pterygoid plexus of veins communicate with those inside the skull near the opening of the facial canal, through the foramen ovale, foramen lacerum and less frequently through the foramen spinosum. Direct venous communication also occurs between the ophthalmic veins and the cavernous sinuses through the superior orbital fissure. Because they are devoid of valves, emissary veins are a potential pathway for the spread of bacteria carried in the blood stream from the face and jaws to the cavernous sinus. Infections of the side of the nose, palate and canine fossa may gain access to the inside of the skull by spreading from the angular facial vein to the ophthalmic veins and hence through the superior orbital fissure to the cavernous sinus. Skin infections around the angle of the nose are potentially dangerous in this respect. Haemorrhage from emissary veins in the base of the skull frequently rupture when the sphenoid bone is fractured, as often occurs in automobile accidents. Progressive neurological signs and symptoms may be produced by pressure on neighbouring cranial nerves.

Venous plexus of the face and neck

1. The pterygoid venous plexus is closely related to the lateral and medial pterygoid muscles (Fig. 5.34). It drains backwards into the posterior facial vein along the maxillary vein; and forwards into the anterior facial vein via the buccal (deep fascial) vein. It communicates with the inferior ophthalmic veins and the veins of the pterygopalatine fossa, which correspond

to the branches of the third part of the maxillary artery, at the back of the orbit; with the pharyngeal plexus deep to the medial pterygoid muscle on the surface of the superior constrictor muscle; and with the cavernous sinus via the emissary veins passing through the foramina ovale and lacerum.

The veins draining the upper and lower teeth and their supporting structures enter the pterygoid venous plexus (page 298).

2. The pharyngeal venous plexus lies on the side wall of the pharynx in relation to the constrictor muscles. It drains into the internal jugular vein and communicates with the pterygoid venous plexus above and with the pharyngeal plexus of the opposite side behind the pharynx it is joined by veins draining the tonsils, the soft palate and the pharyngeal mucous membrane.

Nerves of the head and neck

The cranial nerves are described on page 418, the cervical plexus on page 138, and the sympathetic trunk on pages 85 and 174.

ENDOCRINE GLANDS OF THE HEAD AND NECK

The pituitary gland or hypophysis cerebri (Figs 5.35, 5.36, 5.37)

The pituitary gland is a small ovoid body, the major part of which is contained in the hypophysial fossa of the sphenoid bone. It measures approximately 1 cm both in its anteroposterior and transverse diameters. It is connected to the base of the brain by the stalk or *infundibulum* which passes through an aperture in the centre of the diaphragm sellae, the fold of dura mater which covers over the sella turcica.

The gland has important relationships with:

1. The optic chiasma formed by the junction of the two optic nerves and which lies above the gland on the front part of the diaphragma sellae;

2. The internal carotid artery in the cavernous venous sinus laterally;

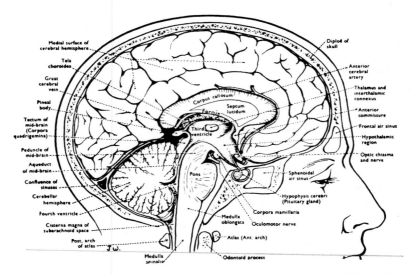

Fig. 5.35 Median sagittal section through skull and brain. The *falx cerebri* which lies in the midline in the longitudinal fissure of the brain has been removed to show the medial surface of the left cerebral hemisphere. The *tentorium cerebelli* overlies the cerebellum and is shown as a double line between the cerebral and cerebellar hemispheres. (By courtesy of Professor G. A. G. Mitchell and Dr E. L. Patterson.)

3. The sphenoidal air sinus which lies below it; the superior orbital fissures which lie antero-lateral to the gland.

Tumours of the gland often extend upwards, for there is little resistance to expansion in this direction, and exert pressure on the optic chiasma producing characteristic disturbances of vision. Tumours which spread laterally invade the cavernous sinus and may occlude the internal carotid artery which passes forwards through it, as well as pressing on the cranial

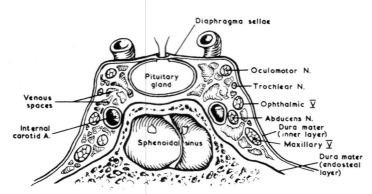

Fig. 5.36 Coronal section through the cavernous venous sinuses and body of the sphenoid.

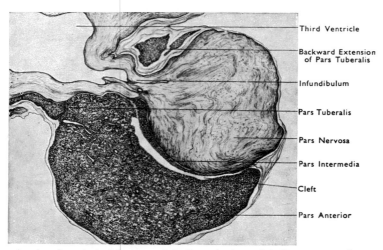

Fig. 5.37 Diagram showing the principal features of the hypophysis cerebri. (By courtesy of Professor G. A. G. Mitchell and Dr E. L. Patterson.)

nerves within the sinus. These are the abducent, oculomotor, trochlear and the ophthalmic and maxillary divisions of the trigeminal nerve. The pituitary gland is relatively inaccessible but it can be approached surgically through the nose, nasopharynx and the sphenoidal sinus.

The pituitary gland is well supplied by hypophysial arteries which enter its upper or lower surfaces from branches which originate from the internal carotid artery, near the origin of the ophthalmic arteries, or from the internal carotid as it passes through the cavernous sinus. The blood supply of the pituitary drains into a plexus of small venous sinuses around the organ and into the neighbouring cavernous sinuses.

Functional anatomy

The gland consists of two developmentally distinct lobes, anterior and posterior in position. The anterior lobe or *adenohypophysis* develops from a diverticulum (Rathke's pouch) in the roof of the primitive mouth cavity. After initial development this connection with the mouth is obliterated. The adenohypophysis can be subdivided further into several parts. These are:

1. The *pars distalis*, which makes up the bulk of the adenohypophysis;

2. The *pars tuberalis*, which is an upward extension of the distal part along the infundibulum; and

3. The *pars intermedia* which lies between the adenohypophysis and the posterior part of the pituitary gland. This portion of the pituitary is rudimentary in man.

The posterior lobe or *neurohypophysis* is developed as a hollow downgrowth from the floor of the forebrain, in the region of the third ventricle, and becomes a slender, solid cord of cells in the region of the infundibulum (stalk) which connects the hypophysis to the tuber cinereum in the floor of the third ventricle of the brain (Fig. 5.35).

The adenohypophysis is composed of anastomosing cords of cells, separated by sinusoidal capillaries which fall into two main types: chromophobes and chromophils. Based on their staining reaction the chromophil cells are further subdivided into acidophils and basophils, so that three types of cells are readily distinguished in routine histologic preparations: *chromophobes* (about 50 per cent), *acidophils* (35 per cent) and *basophils* (15 per cent). The cells are concerned with the secretion of at least six different hormones. Amongst these the following can be mentioned:

1. *Somatotrophin* or growth hormone, which stimulates body growth by promoting endochondral ossification. It seems probable that the acidophils are responsible for the production of the growth hormone and tumours of these cells produce a condition in adults called acromegaly, in which there is enlargement of many bones in the body, including the mandible and the calvarium.

2. *Thyrotrophin* or thyroid stimulating hormone, which influences the activities of the thyroid gland.

3. *Adrenocorticotrophin*, ACTH, which is concerned with the activities of the adrenal cortex.

4. Various hormones which have activities concerned with the female reproductive system.

The posterior lobe consists of neural tissue, chiefly neuroglial cells called *pituicytes*. This part of the gland is believed to store secretions produced by the hypothalamus, i.e. an antidiuretic hormone which suppresses overactivity by the kidneys and *oxytocin*, which has an effect on uterine musculature. These neurosecretions are thought to be carried from nerve cell bodies in the hypothalamus along their axons to the posterior lobe.

The pineal body (Fig. 5.35)

The pineal body, or epiphysis cerebri, is a small cone-shaped structure which projects backwards from the region of the third ventricle of the brain (Fig. 5.35). It is attached to the upper part of the posterior wall of the third ventricle by a stalk which is hollowed out by a narrow extension from the cavity of the third ventricle. It is covered by pia mater, which forms a delicate capsule for the organ and divides it into indefinite lobules. The gland contains clear rounded cells and neuroglial tissue. Small calcified bodies, called brain sand, are found throughout the substance of the gland.

It is difficult to demonstrate specific functions for this organ and there is some doubt whether or not the pineal is a true endocrine gland. There is evidence, however, that it may influence the processes of growth and ageing and tumours of the gland have been found occasionally in children showing precocious development of the gonads.

The thyroid gland (Figs 4.38, 5.38, 5.39)

This is the largest of the ductless or endocrine glands and is composed of many small vesicles lined by a cubical epithelium and filled with a colloidal secretion (Fig. 5.38). It lies in the antero-inferior part of the neck, well below the level of the hyoid bone (Fig. 5.39), and consists of two large lateral lobes, usually connected to one another across the middle line at the level of the upper part of the trachea (second to fourth rings) by a small mass of gland tissue, the *isthmus* (Fig. 4.8). From the isthmus a small pyramidal lobe usually projects upwards and is sometimes connected to the hyoid bone by a small slip of muscle (levator glandulae thyroideae). Each lateral lobe is about two inches in vertical height, extending from the lower part of the thyroid cartilage to about the level of the fifth or sixth tracheal ring. Each lobe can be described as having an anterior, lateral, posterior and a medial surface (Fig. 4.38).

The anterolateral surface is covered by the skin, subcutaneous tissue, the investing layer of the deep cervical and the pretracheal fascia, the sternohyoid, sternothyroid and sternomastoid muscles.

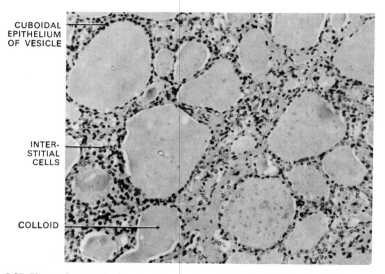

CUBOIDAL EPITHELIUM OF VESICLE

INTER-STITIAL CELLS

COLLOID

Fig. 5.38 Photomicrograph showing the structure of the thyroid gland × 135.

The posterior surface is related to the common carotid artery (within the carotid sheath) and the sympathetic nerve trunk (separated from the gland substance within its sheath by the prevertebral layer of deep fascia).

The medial surface is related from above downwards, to the thyroid cartilage, the lateral part of the cricothyroid membrane, the cricothyroid muscle and the side wall of the trachea. The posteromedial border of the gland on the left side usually comes into contact with the oesophagus, which lies somewhat to the left of the middle line in the neck region. On both sides this border of the gland is closely related to the recurrent laryngeal nerve and the inferior thyroid artery.

The thyroid gland is surrounded by a fibrous tissue capsule which sends septa into its substance and in which run the larger blood vessels. Outside the capsule the gland is surrounded by a sheath derived from the deep fascia of the neck (part of the pretracheal fascia). The sheath helps to hold the gland in position, for its deeper layer is firmly attached to the underlying cricoid cartilage and to the trachea.

The arteries of the gland are terminal branches of the superior thyroid artery, the inferior thyroid artery, and the thyroidea ima artery (either a branch of the brachiocephalic artery, or the aortic arch). The veins drain into the internal jugular veins (superior and middle thyroid veins) and into the brachiocephalic veins, chiefly the left brachiocephalic vein (inferior thyroid veins). Sympathetic fibres from all three cervical ganglia reach the gland along the superior and inferior thyroid arteries.

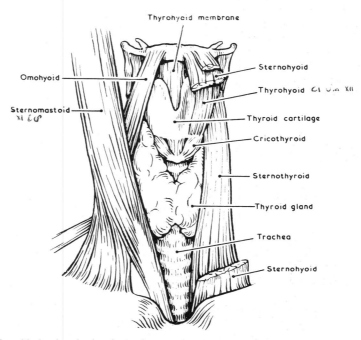

Fig. 5.39 The thyroid gland and related muscles.

The thyroid gland develops as a mid-line tubular downgrowth from the floor of the pharynx, and this *thyroglossal duct* is joined by derivatives from the endoderm of the fourth pharyngeal pouches. The greater part of the duct normally disappears but its upper end remains as the foramen caecum of the tongue. The lower part of the duct, however, frequently forms the pyramidal lobe of thyroid tissue above the isthmus and the strand of fibromuscular tissue forming the levator of the gland. Along the course of the thyroglossal duct small islands of thyroid tissue may develop in the mid-line. These are called accessory thyroid glands and can be found in the tongue or in the mid-line of the neck. Sometimes they will form fluid-filled sacs called thyroglossal cysts (page 342).

Functional anatomy
The thyroid gland regulates the rate of body metabolism. If the secretion (*thyroxin*) of the gland substance is deficient then cretinism results, characterized by physical and mental retardation. If overactive (hyperthyroidism) then Graves' disease is the result. This is characterized by increased metabolic activity and protrusion of the eyeballs, due to the increase in the amount of fatty tissue in the posterior part of the orbital cavity. These clinical features, together with an enlargement of the gland (goitre), gives the disease an alternative name— *exophthalmic goitre*. The activity of the thyroid gland is controlled by the thyroid stimulating hormone (TSH) of the adenohypophysis.

The gland receives its blood supply from the superior and inferior thyroid arteries and their relations are important surgically. The superior artery is accompanied only by its vein for the last 2·5 cm above the gland, but the inferior arteries are closely related to the middle cervical ganglion and the recurrent laryngeal nerve close to the posterior aspect of the gland. As much as the arterial supply has to be identified and interrupted during operations on the gland there is risk of damage to the recurrent laryngeal nerve, which will affect speech. The veins draining the thyroid gland do not have valves so haemorrhage is a problem in surgical approaches. It is important to know the position of the isthmus of the gland which lies in front of the trachea a short distance below the lower border of the cricoid cartilage, not only to avoid cutting it during tracheotomy to provide an emergency airway, but also to avoid the marked bleeding which can occur if the isthmus of the gland is damaged. Swelling of the thyroid gland may be the result of tumour formation or overactivity of the gland. If large, as in exophthalmic goitre, the swollen gland may press on major structures in the deep tissues of the neck including the internal carotid artery and jugular vein. More often the swelling causes a projection in the centre line at the lower part of the neck, just above the sternum and medial ends of the clavicles, because there is little resistance to subcutaneous enlargement.

The parathyroid glands (Fig. 5.40)
These small glands are usually two in number on each side of the neck and are sometimes embedded in the capsule of the thyroid gland. The upper parathyroid gland most commonly lies in relation to the posteromedial border of the middle part of the lateral lobe of the thyroid gland. The lower parathyroid gland is more variable in its position, it may be in relation to

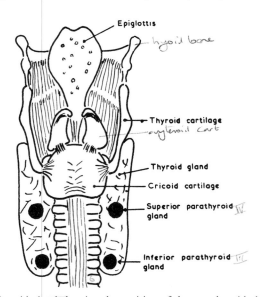

Fig. 5.40 Posterior view of the thyroid gland showing the position of the parathyroid glands.

the inferior pole of the thyroid gland, or at a lower level and separated from the thyroid tissue by a variable interval (Fig. 5.40). The blood supply to the parathyroid glands is from the ascending and descending branches of the inferior (and possibly the superior) thyroid arteries.

The glands are derived from the endoderm of the fourth and third pouches of the lateral pharyngeal wall, the fourth pouch giving rise to the superior parathyroid gland and the third pouch giving rise to the inferior parathyroid gland (page 340).

Functional anatomy

The parathyroid glands are essential to life and are concerned with the regulation of calcium and phosphorus metabolism. During surgical removal of the thyroid gland (*thyroidectomy*) great care is taken to ensure that at least one parathyroid gland is not removed, because of the vital role which they play in mineral metabolism. An excess of the parathyroid hormone increases the level of calcium in the circulating blood by extracting calcium from the bones and teeth. This results in osteoporosis, the development of bone cysts and calcium deposits in the kidneys. Low levels of parathormone have the reverse effect and removal of the glands is followed by a fall in blood calcium level with the onset of muscle spasms or tetany.

6. Anatomy of the Oral Region

THE ORAL CAVITY

The mouth cavity is bounded by the lips and cheeks at the sides and in front, the hard and soft palate above and the floor of the mouth below (Figs 6.9, 6.10). It contains the tongue and the teeth. Behind it opens into the oropharynx through the *fauces* or *oropharyngeal isthmus* bounded on either side by the palato-glossal folds which lie immediately in front of the tonsils. Into the mouth cavity open the ducts of the parotid, submandibular and sublingual salivary glands and those of numerous mucous glands. The teeth and their supporting alveolar processes divide the cavity into a *vestibular region* bounded by the lips and cheeks on the outer side of the teeth and gums, and the *oral cavity proper* within the dental arcades. When the teeth are in occlusion the vestibular region communicates with the inner oral cavity behind the dental arcades (behind the third molar teeth in the adult) and through any spaces resulting from the loss of teeth.

The mucous membrane of the oral cavity is tightly bound to the underlying bone over the alveolar processes and the hard palate, forming a mucoperiosteum. It is firmly united to the tongue musculature by the lamina propria and less firmly attached to the buccinator muscle, the lip musculature and the muscles of the soft palate. The attachment to the floor of the mouth and the vestibular region is much looser, which permits greater freedom of movement of the tongue, cheeks and lips. Throughout the oral cavity, the epithelium of the mucous membrane is of the stratified squamous variety (Fig. 6.1). The epithelium consists of the following layers:

1. *Stratum germinativum* or the basal cell layer rests on a basal lamina and continuously gives origin to more superficial layers of the epithelium by mitotic division of its cells (Fig. 6.2).

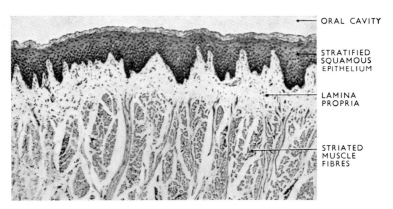

ORAL CAVITY

STRATIFIED SQUAMOUS EPITHELIUM

LAMINA PROPRIA

STRIATED MUSCLE FIBRES

Fig. 6.1 Photomicrograph showing the structure of oral mucous membrane taken from the under surface of the tongue of a 40 millimetre human embryo × 50.

2. *Stratum spinosum* in which the cells are closely related to one another but show 'intercellular' processes or bridges which seem to cross the intercellular spaces. Electron micrographs show that there is no protoplasmic continuity between adjacent cells, but contact at specialized regions of the cell membrane which form attachment plaques or desmosomes (Fig. 6.2).

3. *Stratum granulosum* where the cells are more flattened and contain keratohyaline granules, the precursor of keratin.

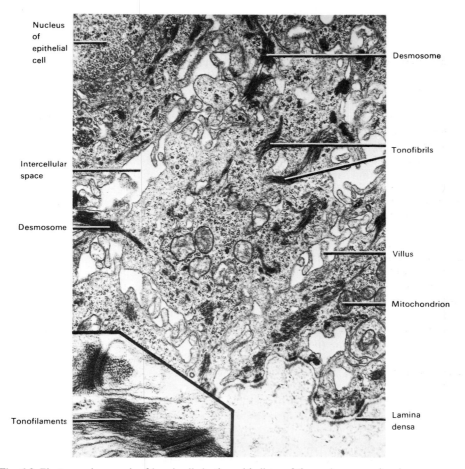

Nucleus of epithelial cell

Desmosome

Tonofibrils

Intercellular space

Desmosome

Villus

Mitochondrion

Tonofilaments

Lamina densa

Fig. 6.2 Electron micrograph of basal cells in the epithelium of the oral mucosa showing intercellular attachment plaques or desmosomes × 5,300. The insert shows a higher magnification of two desmosomes × 16,000.

4. *Stratum corneum* consisting of dead, flattened, cornified, structureless cells which are abundant over the gums, hard palate and dorsum of the tongue. The stratum corneum is well developed in these locations because they are subject to greater frictional and abrading forces compared with other parts of the oral mucous membrane. The surface cells are shed continually and replaced from the deeper layers of the epithelium.

The pink colour of the oral mucosa is due to the vascularity of the underlying lamina propria and the relative thinness of the epithelium. In regions where the stratum corneum is well developed the mucosa is paler in colour.

The deeper portion of the mucous membrane is called the dermis or lamina propria and the epithelial layers are sometimes referred to as the epidermis. The basal lamina serves to separate the deepest layer of the epithelium from the underlying dermis. In electron micrographs (Fig. 6.2) the electron dense line called the lamina densa is the most prominent feature of the basal lamina. The lamina propria contains collagen fibres, some elastic tissue, especially in regions such as the floor of the mouth, the soft palate and cheeks which are especially mobile. Small blood vessels and capillaries, sensory nerve endings, lymphatic vessels and mucous glands are abundant. The cells found in the lamina propria consists of fibroblasts, macrophages, mast cells, and cells derived from the blood stream, such as polymorphonuclear leucocytes. Over the gums and hard palate the deeper layers of the lamina propria are continuous with the periosteum of the bone, so that they form together a mucoperiosteum. In these regions the tissues are less vascular and less sensitive, with the exception of the region of the hard palate immediately behind the upper incisors which is richly endowed with nerve terminations.

The mucoperiosteum covering the hard palate shows a series of transverse ridges, variable in number and form, and these are the *palatal ridges* or *rugae*. These are much more highly developed in animals such as rodents, carnivores and ungulates. Behind the central incisors and overlying the palatal opening of the incisive canal (Fig. 4.39) the mucous membrane forms a low mid-line elevation, the *incisive papilla*. Within the incisive canal are found epithelial remnants of the nasopalatine ducts of the embryo and collections of cornified cells called *epithelial pearls*. These can be found also along the mid-line of the palate and are derived from the epithelium of the fused palatal folds (page 348). They become less common after birth but may form cysts in the palate and incisive canal. Up until the time of eruption of the deciduous incisors the superior labial frenulum is usually attached to the front of the papilla and in those children in whom the frenulum is abnormally large it often passes between the central incisors towards the papilla. This results in a gap between these teeth which are normally in contact.

The sensory nerve supply of the oral mucous membrane is derived from branches of the maxillary and mandibular divisions of the trigeminal (fifth cranial) nerve (see page 247): its blood supply from branches of the facial, lingual, and maxillary arteries (pages 253–259). Lymphatics from the oral mucous membrane drain into the submental, submandibular and upper deep cervical lymph nodes (page 288).

The most sensitive regions of the oral cavity are the lips, the tip of the tongue and the region of the incisive papilla. The two latter areas being in functional juxtaposition makes possible accurate identification of the nature of food particles in the mouth.

The movements of the lower jaw which produce the masticatory action of the teeth are brought about by the muscles of the mastication. The size of the mouth cavity is regulated by the buccinator and mylohyoid muscles; the shape and movements of the tongue by its intrinsic and extrinsic muscles, and the position of the soft palate is regulated by the palatal muscles (page 195). The bones forming the skeleton of the mouth region are the maxillary, palatine, mandibular and hyoid bones (page 221).

Regions of the oral cavity

The oral cavity can be divided into a number of regions which are of importance in systematic examinations of the mouth for such procedures as the diagnosis of disease, and the designing of artificial dentures. The most important regions are:

The vestibule of the mouth

This is bounded by the cheeks and lips on the outer side, and the gums and teeth on the inner side (Figs 6.3, 6.4). Its anterior part is related to the lips; its posterior part to the cheeks. The mucous membrane lining the vestibule is continuous above and below with the gums covering the alveolar bone (alveolar gingiva) and is closely related to the buccinator muscle and the musculature of the lips. Opening into the vestibule at the level of the upper second permanent molar in the adult is the parotid duct (page 153). One or more folds of mucous membrane may pass between the lateral (buccal) wall of the vestibule and the gum; these contain connective tissue and sometimes small slips of muscle tissue. The upper and lower lips are attached to the gums by superior and inferior labial frenulae. The former may pass between the central incisors to be attached to the incisive papilla, in which case the incisors are separated by a space, or diastema.

The tongue

The parts of the tongue seen in the mouth cavity are its upper or dorsal surface, particularly in its anterior two-thirds, and the lower or inferior surface (Figs 6.9, 6.10). In examining the tongue its colour, size, degree of mobility and surface texture should be noted. The dorsum is roughened by the presence of minute papillae, whilst the under surface is smooth. Two thin serrated folds, the fimbriated folds, are visible on the inferior surface, as well as the frenulum, which runs towards the top of the tongue and controls its range of movement. If the frenulum is very short then the condition 'tongue-tied' results. This is sometimes found in newborn infants. The palatoglossal folds merge with the sides of the tongue at the back of the oral cavity.

During foetal life and the first year after birth the tongue is relatively large within the mouth cavity and often extends between the gum pads, especially in front, to come in contact with the lips and cheeks (Fig. 16.6). With eruption of the teeth and an increase in vertical height of the oral cavity, the normal tongue becomes bounded by the lingual surfaces of the teeth and the adjacent gums. It adapts its shape when at rest to that of the dental arches and palatal vault with its tip in contact with the hard palate behind the upper incisors. If it is excessively large, as in Down's syndrome (or mongolism), it may produce an outward tilting, irregularity and spacing of the teeth.

The floor of the mouth

The mucous membrane lining the floor of the mouth is attached at its periphery to the inner surface of the body of the mandible where it is continuous with the mucoperiosteum of the gum on the lingual side of the teeth. Centrally it becomes continuous with the mucous membrane covering the anterior two-thirds of the tongue. In association with the great mobility of the tongue, the mucous membrane is itself freely movable, except at its mandibular attachment. It forms the roof of the slit-like sublingual spaces or compartments (page 447) between the body of the mandible and the tongue musculature. It contains the sublingual salivary glands, the deep portions of the submandibular glands and their ducts, the lingual arteries and nerves and the hypoglossal nerves (Figs 5.29, 6.3). The upper margins of the sublingual glands form ridges beneath the tongue, and the submandibular ducts open on either side of the frenulum of the tongue behind the lower incisor teeth (Fig. 6.11). At a deeper level than the mucous membrane and forming the floor of the sublingual compartments are the mylohyoid muscles forming the mylohyoid diaphragm (Fig. 6.3). This supports the tongue and the contents of the sublingual compartments, separating them from the submandibular compartment of the neck (page 447). At the side of the tongue the floor of the mouth extends backwards and becomes more shallow, as the mylohyoid attachment ascends towards the alveolar margin, to end in a partial cul-de-sac on the inner aspect of the molar tooth and in front of the palatoglossal fold. The extent and form of this region is of importance in the designing of lower dentures (p. 297).

The retromolar region

This important area extends from the back of the last lower molar below to the back of the last upper molar above. It is related to the retromolar triangle, the anterior border of the

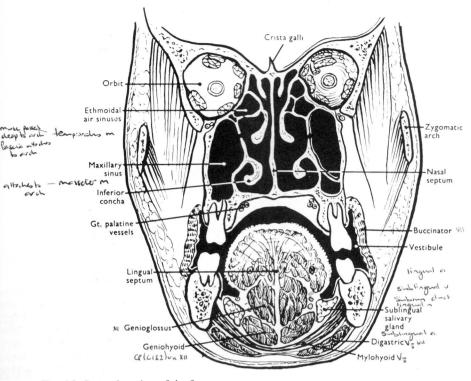

Fig. 6.3 Coronal section of the face.

ramus of the mandible, the buccinator and superior constrictor muscles, the palatoglossal fold, the anterior edge of the medial pterygoid muscle, the pterygoid hamulus and the tendon of the tensor tympani muscle and the maxillary tuberosity (Figs 5.15, 6.5). The mucous membrane is firmly attached to the underlying muscle and bone and contains some mucous glands. In the lower jaw the mucous membrane and its contained glands form a retromolar pad which overlies the bony retromolar triangle, whose borders become continuous anteriorly with the alveolar crests on the buccal and lingual aspects of the last molar tooth. This pad frequently becomes inflamed during eruption of the third molar tooth, particularly when this tooth is impacted (see page 295), and may become the site of a chronic bacterial infection. Also, the mandibular retromolar area is an important landmark in determining the correct site for the injection of solutions which will produce anaesthesia of the inferior dental and lingual nerves.

The roof of the mouth

This is made up of the hard and soft palate and is bounded in front and at the sides by the upper dental arch. In examining this region note the degree of arching and width of the hard palate, the palatal rugae, the line of attachment of the soft palate and the range of its movements. At the back of the mouth note the width of the oropharyngeal isthmus (fauces), and the size of the tonsils. The mucous membrane covering the hard palate is a mucoperiosteum firmly bound down to the underlying bone, except on the inner aspect of the alveolar processes of the cheek teeth. It is less firmly bound down at the sides where the greater palatine vessels and nerves run forwards from the greater palatine foramina (Fig. 6.3). The mucous membrane covering the under surface of the soft palate is attached to the underlying muscles except in the anterior central region where the palatal mucous glands intervene between the muscles

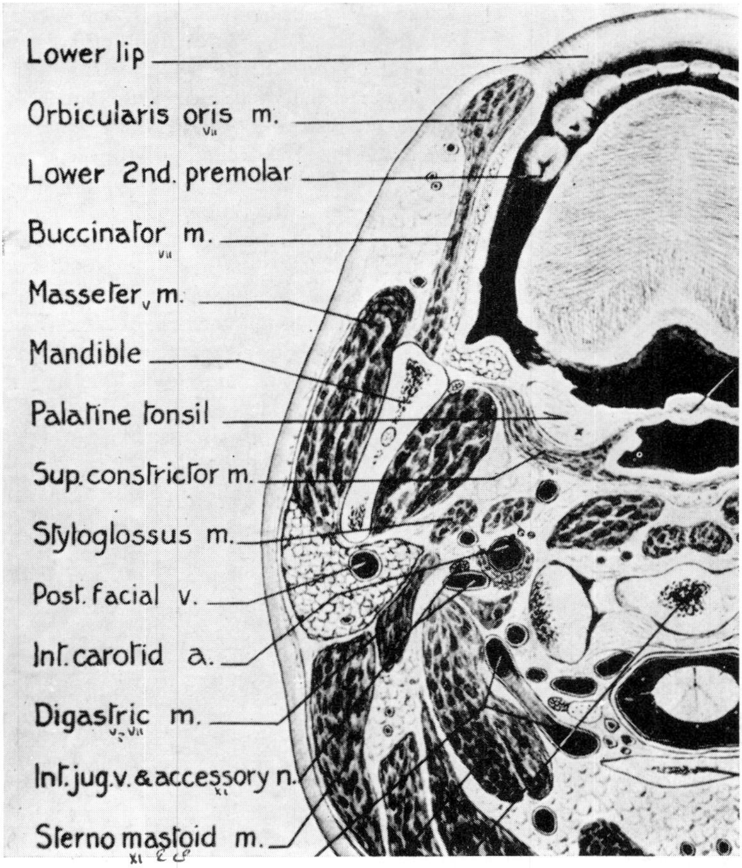
Lower lip
Orbicularis oris m.
VII
Lower 2nd. premolar
Buccinator m.
VII
Masseter, m.
V
Mandible
Palatine tonsil
Sup. constrictor m.
Styloglossus m.
Post. facial v.
Int. carotid a.
Digastric m.
V, VII
Int. jug. v. & accessory n.
XI
Sterno mastoid m.

Fig. 6.4 Diagram representing horizontal section through the mouth cavity. (By courtesy of The Anatomical Society of Great Britain and Ireland.)

and mucous membrane. The upper (nasal) surface of the soft palate is covered by a pseudostratified columnar ciliated epithelium. Nasal and oral type epithelia meet at its free posterior edge, with the oral type covering the edge, which abuts against the pharynx during swallowing. The soft palate is more vascular, more sensitive and contains more lymphatics than the hard palate. In its resting position the soft palate lies against the pharyngeal surface of the dorsum of the tongue forming a seal between the oral cavity and the oral part of the pharynx. During the swallowing of food, or in mouth breathing, the soft palate is elevated and the oral cavity is in communication with the pharynx.

The teeth and gums

The upper and lower teeth, supported by alveolar bone in which their sockets are situated, form arches or arcades which conform to a catenary curve. In man there are no natural spaces between the teeth. Each tooth is composed of calcified tissues, enamel, dentine, cement (cementum) and a central pulp cavity composed of connective tissue, vessels and nerves (Fig. 2.13). The *crowns* of the teeth are covered by enamel (Fig. 6.10) and their *roots* are socketed in the alveolar bone of the jaws. There are 20 deciduous, or milk, teeth and 32 permanent teeth. The permanent series can be grouped into incisors, canine (eye-teeth), premolars and molars.

The deciduous teeth are grouped into incisors, canines and molars. Premolars and molars are characterized by the presence of tubercles or cusps on their occlusal (biting) surfaces (Fig. 5.17). Incisors, canines, the lower premolars and the upper second premolars have single roots. The upper first premolar usually has two roots (palatal and buccal); the upper molars three roots (one palatal and two buccal); the lower molars two roots (mesial and distal). The mesial surface of the tooth is that nearest to the mid-line of the dental arch, the distal surface faces in the opposite direction.

The gums, or gingiva, form a specialized region of the oral mucous membrane and show at their free margins a union between oral epithelium and the epithelium covering that part of the enamel of the tooth not exposed in the mouth cavity. Through this epithelium the gum is attached to the necks of the teeth by the *epithelial attachment* (Fig. 6.21). The lamina propria of the gingiva is a mucoperiosteum, like that covering the hard palate. It contains bundles of collagenous fibres which unite the gum to the underlying alveolar bone and to the cementum covering the roots of the teeth immediately adjacent to the enamel margin. There are no glands in the gingiva. The gums of the incisor-canine area are more sensitive than in the region of the cheek teeth. Inflammation of the gums (gingivitis) is a very common disease both in civilized and primitive communities. (For further details, regarding the anatomy of the individual teeth, the dental arches and gingiva, the student should consult a textbook on Dental Anatomy.)

The oral musculature
The individual oral muscles are described in various sections throughout the text. Here some of them are grouped on a functional basis:
   1. The muscles of the lips and cheeks (see page 140).
   2. The tongue muscles (see page 244).
   3. The muscle of the floor of the mouth (mylohyoid and geniohyoid (see pages 243)).
   4. The muscles of the soft palate (see page 195).
   5. The muscles of mastication (see page 237).

All these muscle groups play a part in mastication, swallowing and speech (pages 315, 321, 323).

Normally the lips when at rest meet one another outside the teeth. If, however, the upper incisors protrude, the lower lip may come to lie on the inner side of the upper incisors between them and the lower teeth. In this position the 'trapped' lower lip tends to displace the lower incisor backwards and the upper incisors forwards and the malocclusion becomes progressively worse.

The lips and cheeks together make up a muscular sheet lying on the outer side of the dental arches. This is balanced by the muscular tongue lying within the dental arches. Abnormalities of muscle action on the part of the lip, cheek, or tongue musculature may play a part in the causation of deformed dental arches, by upsetting the normal balance of muscles around them.

At the back of the oral cavity the buccinator is continuous with the superior constrictor by virtue of their common attachment to the pterygomandibular raphe, so that the mouth and the pharynx are sometimes described as being surrounded by a horizontally disposed muscular sphincter made up of the superior constrictor, buccinator and lip muscles. It should be remembered, however, that the elements making up this 'sphincter' can and do act as separate elements and that in the buccinator the direction of the muscle fibres is not entirely horizontal (see page 141 and Fig. 4.48).

In the region of the soft palate there are two important sphincter mechanisms:

The *palatoglossal or post-oral sphincter*, which reduces the opening between the mouth and pharynx. This is closed when the soft palate is depressed to lie in contact with the posterior (pharyngeal) surface of the tongue and the palatoglossal folds are approximated.

The *palatopharyngeal sphincter* formed by the palatopharyngeal muscles and a horizontal ridge formed by some of the upper fibres of the superior costrictor (Passavant's ridge). This helps to reduce the communication between nasopharynx and oral pharynx which is closed when the soft palate is elevated.

The capacity of the mouth cavity can be increased when the mouth is closed and the lips are in contact

1. By the descent of the mandible
2. By the descent of the mylohyoid or oral diaphragm and the tongue
3. By the relaxation of the buccinator muscles. These muscles in cooperation with those of the lips and tongue also act together in such activities as suckling and in the playing of wind instruments.

The tongue can exert pressure against the hard and soft palate as in swallowing. It can also exert a considerable pressure against the teeth, especially the upper or lower incisors. If this becomes a habit, marked deformities of the anterior segments of the dental arches can be produced.

The muscles of mastication not only move the lower jaw at the mandibular joints and maintain the position of the mandible against gravity, they also play an important role in determining the position of the mandible relative to the upper facial skeleton. Prior to the eruption of the teeth there is no definitely established relationship and during this period the mandible can be said to grow as an independent unit of the facial skeleton. After the establishment of the deciduous, and later of the permanent, dentition a more definite relationship is established through the occlusion of the upper and lower teeth and the muscles learn to adapt the growing mandible to this position. If the occlusal relationship is abnormal the relationship between the jaws tends to be abnormal and the mandible may be too far back (known as Angle's Class II type of abnormality) or too far forward (Angle's Class III type of abnormality). In some cases, however, the abnormal dental relationship is secondary to an abnormal jaw relationship which may be the result either of abnormal growth of the facial skeleton or the establishment of an abnormal muscle posture, especially during the early years of life.

The normal position of the mandible, established during childhood through the occlusal pattern of the teeth and maintained by the habitual activity of the muscles of mastication both while at rest and during movement, is maintained in adult life by the muscles even after the occlusal surfaces of the teeth have been worn smooth by *attrition*. This occurs in dentitions which have been used in the continuous mastication of tough gritty foods. If artificial dentures are fitted which do not restore the previously established dental relationship there is often a considerable degree of muscular and joint discomfort before the muscles can adapt the jaws to the new relationship. As growth has ceased, or become considerably slowed down, a successful adaptation between the articular fossa, tooth occlusion and muscle action may not be possible.

THE FLOOR OF THE MOUTH, THE TONGUE, AND SUBMANDIBULAR REGION

This region contains the following important structures (Figs 6.3, 6.7):

1. The tongue with its intrinsic and extrinsic muscles.
2. The suprahyoid muscles, other than the tongue muscles; namely, the anterior bellies of the digastrics, the mylohyoid, and the geniohyoids.
3. The submandibular and sublingual salivary glands and their ducts.
4. Parts of the lingual and hypoglossal nerves; the submandibular parasympathetic ganglion.
5. Parts of the facial and lingual arteries with their corresponding veins.
6. The submental and submandibular lymph nodes with their afferent and efferent lymphatic vessels.

The region of the floor of the mouth is bounded above by the mucous membrane covering the tongue and the floor of the mouth cavity beneath the tongue; in front and at the sides by the inner surface of the body of the mandible; below by the suprahyoid part of the investing layer of the deep cervical fascia; by the platysma muscle and by the skin beneath the lower jaw. Behind, this region is related to the side wall of the pharynx, to the hyoid bone, to the great vessels of the neck and to the parotid gland.

The mylohyoid muscle

The two mylohyoid muscles together form a diaphragm across the floor of the mouth, the *oral diaphragm* (Fig. 6.5). They are attached in front and at the sides to the inner surface of the body of the mandible and behind to the hyoid bone. Between the back of the mylohyoid line of the mandible and the hyoid bone each mylohyoid muscle has a free posterior edge. Above the mylohyoid diaphragm (Figs 6.6, 6.7) lie the geniohyoid nerves, the tongue, its blood vessels, the hypoglossal and lingual nerves, the sublingual salivary glands, the upper (deep) portions of the submandibular salivary glands and their ducts. This region is known as the *sublingual compartment*.

Beneath the mylohyoid diaphragm on each side of the middle line are the anterior bellies of the digastric muscles, the submental and submandibular lymph nodes, the submental

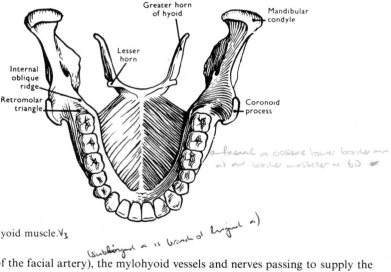

Fig. 6.5 The mylohyoid muscle. V_3

artery (a branch of the facial artery), the mylohyoid vessels and nerves passing to supply the mylohyoid and anterior belly of the digastric muscle, and the superficial (lower) parts of the submandibular salivary glands. This region is known as the *submandibular compartment*.

The superficial and deep parts of the submandibular salivary glands are continuous with one another around the posterior free margin of the mylohyoid muscle, between its attachment to the mandible above and laterally, and to the hyoid bone below and medially (Fig. 6.5). In this position also the chief vessels and nerves pass forwards from the side of the pharynx with the styloglossus muscle to enter the sublingual compartment above the mylohyoid diaphragm. The mylohyoid diaphragm is normally pierced only by lymphatic vessels draining the tongue and the floor of the mouth and a few small veins. In some cases however there are one or more hiatuses between the mandible and mylohyoid often containing herniations of sublingual gland tissue.

Although the mylohyoid muscles are supplied by the fifth cranial nerve through a branch of the inferior dental nerve, they are not primarily muscles of mastication. When they contract

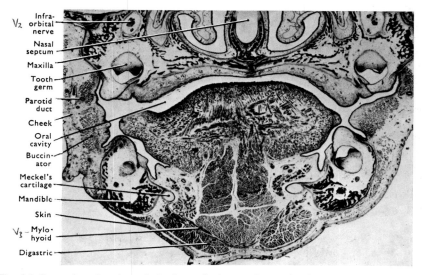

Fig. 6.6 Coronal section through the jaws of a human foetus showing structures related to the mouth.

they thrust the tongue against the palate and play an important part in suckling and swallowing (see page 321).

The submandibular salivary gland

The deep part of each submandibular gland lies above the mylohyoid muscle in the sublingual compartment with the hyoglossus muscle on its medial side, and below the mucous membrane covering the back part of the floor of the mouth (Fig. 6.7). The lingual nerve runs above this part of the gland in contact with the hyoglossus muscle while below it the hypoglossal nerve passes forwards to its distribution (Fig. 6.7). The lingual artery lies below the gland and is separated from it by the hyoglossus muscle. Where the deep part of the gland becomes continuous with the superficial part, behind the free edge of the mylohyoid, the medial surface of the gland rests against the side wall of the pharynx (here formed by the middle constrictor muscle and the stylopharyngeus).

The superficial part of the gland lies below the mylohyoid diaphragm in the submandibular compartment and comes into contact above with the inner surface of the mandible and the insertion of the medial pterygoid muscle. Below it is separated from the skin by the superficial layer of the deep fascia and the platysma muscle. The superficial part of the submandibular gland is separated from the lower lobe of the parotid gland by a thickening of the deep fascia, the stylomandibular band or 'ligament', which is part of the fascial sheath enclosing the parotid gland. The upper surface of the superficial part of the submandibular gland is grooved by the facial artery before this vessel leaves the neck to pass on to the face at the anterior border of the masseter muscle, around the lower border of the mandible. The anterior facial vein crosses superficial to the gland as it passes backwards to join the anterior division of the posterior facial vein to form the common facial vein. The latter enters the internal jugular vein deep to the anterior edge of the sternomastoid muscle (Fig. 6.8).

Parasympathetic nerve fibres supplying the gland come from the seventh cranial nerve, along the chorda tympani and the lingual nerve to the submandibular ganglion. Postganglionic branches are given off to enter the gland substance. Sympathetic fibres reach the submandibular gland from the upper cervical ganglion along the facial artery, the chief source of its blood supply. The submandibular lymphatic nodes lie in close relationship to the superficial part of the gland.

The duct of the submandibular gland
This is known also as Wharton's duct. It commences in the superficial part of the gland by the union of numerous ductules, and within the gland substance winds around the posterior edge of the mylohyoid diaphragm. After leaving the anterior end of the deep part of the gland, the duct runs forwards and upwards in relation to the upper (deep) surface of the mylohyoid muscle and lateral to the muscles forming the side of the tongue, namely the hyoglossus, genioglossus and styloglossus (Fig. 6.7). Its terminal part lies alongside the origin of the genioglossus and on the inner side of the sublingual salivary gland. The duct is crossed in this position by the lingual nerve before the nerve passes deeply into the substance of the tongue. The duct opens beneath the tongue on the floor of the mouth at the summit of a small papilla, a short distance behind and below the lower incisor teeth (Fig. 6.10). It is about 5 cm (2 inches) long and 3 mm in diameter.

The sublingual salivary glands
These are the smallest of the three major salivary glands (Figs 6.3, 6.7). They lie beneath the mucous membrane in the floor of the front part of the mouth in the sublingual compartment. Each gland is covered above by mucous membrane and rests below on the upper (deep) surface of the mylohyoid diaphragm. To the outer side of each gland lies the curved lingual surface of the body of the mandible and to the inner side of each gland lies the genioglossus muscle. Between each gland and the genioglossus runs Wharton's duct and the sublingual artery. In front the glands of each side may come into contact at the middle line while behind they may reach as far back as the anterior ends of the submandibular glands. Each sublingual gland has from 10 to 20 ducts which open into the mouth cavity by piercing the overlying mucous membrane. Some of the ducts, however, open into the duct of the submandibular gland, which develops in part as a closing over of the original floor of the mouth.

As in the case of the submandibular glands, parasympathetic nerve fibres reach the sublingual glands from the facial nerve via the chorda tympani, lingual nerve and the submandibular parasympathetic ganglion (Fig. 6.34). Sympathetic fibres reach the gland from the upper cervical ganglion along the lingual artery. The blood supply to each gland is from the sublingual branches of the lingual arteries.

The lingual artery
This is the chief artery supplying the tongue and the structures in the floor of the mouth (Figs 6.7, 4.27). It is a branch of the external carotid artery given off in the carotid triangle (page 132), close to the anterior border of the sternomastoid muscle. It sometimes shares a common stem of origin with the facial artery. At its origin it lies in relation to the lateral surface of the middle constrictor muscle of the pharynx, and then passes above the hyoid bone deep to the digastric, mylohyoid and hyoglossus muscles, lying on the stylopharyngeus and the lateral surface of the genioglossus muscle. At the anterior border of the hyoglossus it turns upwards and enters the substance of the tongue deep to the lingual nerve and Wharton's duct. Just before this it gives off a large sublingual branch which runs forwards on the inner side of the sublingual gland supplying the gland, the mucous membrane of the floor of the mouth and the gum covering the inner surface of the lower jaw. It also gives a branch to the frenulum of the tongue. At the posterior border of the hyoglossus muscle the lingual artery gives off one or more dorsal branches (dorsales linguae). These run upwards on the deep aspect of the muscle and supply the pharyngeal part of the tongue, the anterior pillars of the fauces, and the tonsils. The main vein draining the tongue passes superficial to the hyoglossus muscle running close to the hypoglossal nerve. It receives a sublingual branch and one or more dorsal

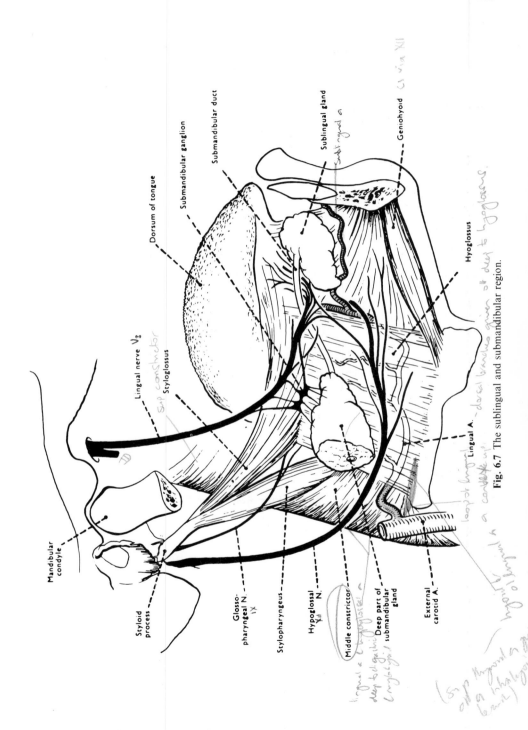

Fig. 6.7 The sublingual and submandibular region.

lingual veins at the posterior border of the hyoglossus muscle. It opens into the internal jugular vein (see also pages 166 and 263).

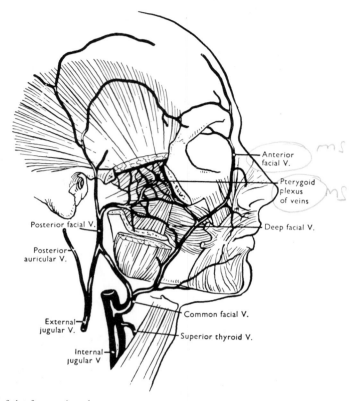

Fig. 6.8 The veins of the face and neck.

The tongue

The tongue is a movable muscular organ which plays an important part in mastication, swallowing, suckling and in speech. When at rest and when the mouth is closed it fills the mouth cavity, resting against the inner (lingual) surfaces of the teeth beneath the under surface of the hard and soft palate (Fig. 6.3). The tip of the tongue is usually in contact with the hard palate behind the upper incisor teeth. The back of the tongue faces towards the pharynx and forms part of the anterior wall of the oral pharynx (Fig. 4.8). The resting soft palate is in contact with its upper part.

The tongue is made up of intrinsic and extrinsic muscle fibres covered by a mucous membrane. The shape of the tongue can be altered by the activity of the intrinsic muscles which lie entirely within its substance. Their names—longitudinal, vertical and transverse—indicate the direction of their fibres.

In a coronal section the tongue is seen to be partly divided by a vertical fibrous tissue partition or septum united behind to the hyoid bone and to which some of the intrinsic muscle fibres are attached (Fig. 6.3). The muscle fibres are also attached to the deep surface of the covering mucous membrane. Only small blood vessels pass from one side of the tongue to the other through the mid-line septum. There is, however, a well-marked anastomosis between the lingual arteries at the tip of the tongue.

The position of the tongue relative to the oral cavity is altered by the action of its extrinsic muscles which have their origin outside the tongue and are inserted into it.

The upper surface, or dorsum of the tongue (Fig. 6.9), is covered by a mucous membrane continued on to its sides and on to its under surface in its free anterior part. Around the base of the tongue, the mucous membrane is continuous with that covering the floor of the mouth cavity. In the middle line a reflection of the mucous membrane passes from the under surface of the tongue to the floor of the oral cavity. This is the *frenulum* of the tongue (Fig. 6.10). It contains a terminal branch of the sublingual artery. When the frenulum is short a tongue-tied condition is produced.

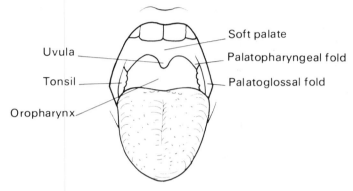

Fig. 6.9 Diagram of the oropharyngeal isthmus.

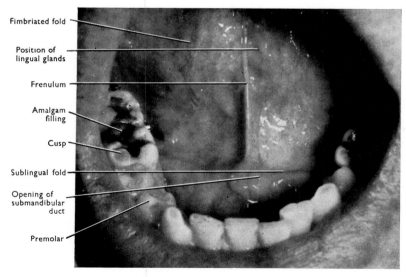

Fig. 6.10 The sublingual region of the oral cavity.

The mucous membrane covering the oral part of the dorsum of the tongue is not smooth as is the case with the membrane covering the under surface. It contains numerous fungiform and filiform papillae which give a rough texture to its surface (Fig. 6.11). At the junction between the oral and pharyngeal parts of the tongue there is a row of large circumvallate

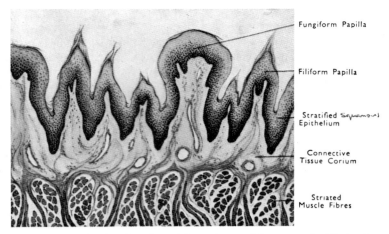

Fungiform Papilla

Filiform Papilla

Stratified Squamous
Epithelium

Connective
Tissue Corium

Striated
Muscle Fibres

Fig. 6.11 Vertical section through the dorsal surface of the oral part of the tongue (×40) showing fungiform and filiform papillae. (By courtesy of Miss Margaret Gillison.)

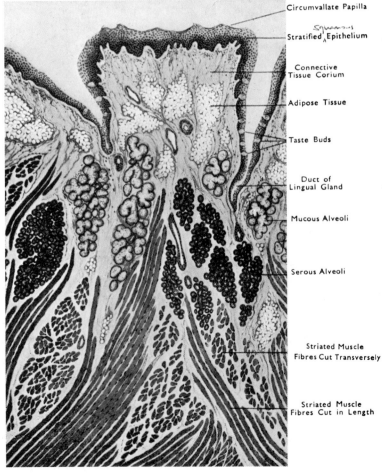

Circumvallate Papilla

Stratified Squamous Epithelium

Connective
Tissue Corium

Adipose Tissue

Taste Buds

Duct of
Lingual Gland

Mucous Alveoli

Serous Alveoli

Striated Muscle
Fibres Cut Transversely

Striated Muscle
Fibres Cut in Length

Fig. 6.12 Vertical section through the dorsal part of the tongue (×30) immediately anterior to the sulcus terminalis. Taste buds are visible in the epithelium of a circumvallate papilla and in the outer wall of the surrounding fossa. (By courtesy of Miss Margaret Gillison.)

papillae (Fig. 6.12) and immediately behind them there is a shallow V-shaped groove, the *sulcus terminalis*, marking the division between the oral and pharyngeal parts of the tongue, each of which has a different developmental origin (page 351). At the apex of the sulcus terminalis is the shallow foramen caecum which indicates the position of the downgrowth of the oral epithelium during foetal life which formed the thyroid gland (Fig. 6.13). This downgrowth, which is attached to the floor of the mouth by the thyroglossal duct, loses its connection with the oral cavity in late foetal life and its terminal part becomes one of the ductless, hormone-producing glands.

The pharyngeal part of the tongue is covered by a smooth mucous membrane overlying nodules of lymphoid tissue which, with the tonsils and lymphoid tissue of the nasopharynx, forms a field of lymphoid tissue encircling the pharynx.

Small mucus secreting glands lie in the mucous membrane of the pharyngeal surface and along the margins of the oral part of the tongue. Serous-type secreting glands open into the

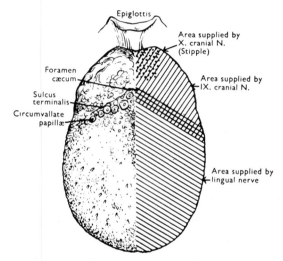

Fig. 6.13 The dorsum of the tongue.

'trenches' of the circumvallate papillae. Beneath the mucous membrane covering the under surface of the anterior part of the tongue, on either side of the frenulum towards the tip of the tongue, lie the larger mixed mucous and serous lingual glands of Blandin and Nuhn (Fig. 6.10).

The mucous membrane covering the tongue is sensitive to the ordinary forms of sensation (heat, cold, touch and pain) and to the special sense of taste. After passing through a nerve plexus in the lamina propria the nerve fibres from the endings subserving the ordinary forms of sensation in the tongue follow two pathways to the central nervous system. From the anterior two-thirds of the tongue (oral region) the sensory nerve fibres run with the lingual nerve (a branch of the fifth cranial nerve). From the posterior third of the tongue (pharyngeal region) the sensory nerve fibres run with the glossopharyngeal nerve. Taste fibres from the oral region of the tongue run at first with the lingual nerve but in the pterygoid region they leave the lingual nerve via a communicating loop, the chorda tympani, to join the facial nerve in the temporal bone (Fig. 4.26). Taste fibres from the pharyngeal region of the tongue and from the circumvallate papillae run with those of common sensation in the glossopharyngeal nerve. The ultimate destination of the taste impulses is described on page 426.

The musculature of the tongue except the palatoglossus receives its motor nerve supply through the twelfth cranial nerve (the hypoglossal nerve).

The tongue is attached to the hyoid bone by the hyoglossus muscle; to the styloid process by the styloglossus muscle; to the mandible by the genioglossus muscle, and to the palatal aponeurosis and back of the hard palate by the palatoglossus muscle (see page 244). Because the extrinsic muscles of the tongue are associated primarily with its lateral and inferior parts, large sections of the middle and upper parts of the tongue may be resected surgically without major impairment of tongue function. → Keyhole erosion for tongue reduction but bleeds !!!

The blood supply to the tongue is from the lingual artery and its branches, and the venous return is by the lingual vein, which drains into the internal jugular vein. Because the anterior two-thirds of the tongue develop from bilateral lingual swellings (page 351) there is a relatively poor blood supply in the midline of the tip of the tongue and surgical incisions or trauma result in relatively little bleeding. Generally tongue has a very good blood supply indeed

Beneath the mucous membrane of the tongue there is a rich plexus of lymphatic vessels in addition to blood vessels and nerves. From this plexus lymph vessels drain into the submental, submandibular and deep cervical lymph nodes (Fig. 6.14). The anterior draining, or afferent vessels, from the tip of the tongue and the under surface of the anterior part of the tongue, drain bilaterally pierce the mylohyoid diaphragm and most of them enter the submental nodes. A few cross the hyoid bone to end directly in the deep cervical chain (jugulo-omohyoid node) close to where the omohyoid muscle crosses the internal jugular vein. The middle draining, or afferent vessels, from the greater part of the dorsum and from the sides of the tongue, pierce the posterior part of the mylohyoid diaphragm to enter the submandibular nodes. Alternatively they pass backwards between the mylohyoid and hyoglossus muscle or deep to the latter muscle to enter the deep cervical nodes, which are related to the posterior belly of the digastric (jugulodigastric node) where this muscle passes across the internal jugular vein. The posterior

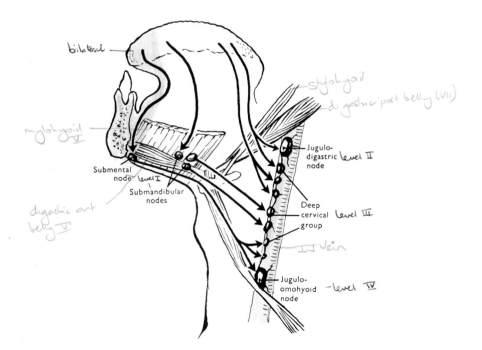

Fig. 6.14 Lymphatic drainage of the tongue.

See also p 51

drainage vessels pierce the side wall of the pharynx below the level of the tonsils and enter the upper group of deep cervical nodes.

All of the deep drainage of the jugular system passes to the jugulo-omohyoid nodes along with lymph draining from the submental region and the anterior part of the tongue. Efferent lymphatics from the jugulo-omohyoid nodes pass to the supraclavicular nodes and together they form the 'sentinel cervical nodes' used for the detection of cancerous or malignant lesions that have spread or metastasized from higher levels of the head and neck. Metastatic lesions from the tip of the tongue and the anterior part of the floor of the mouth usually spread along lymphatics associated with the anterior or external jugular venus plexuses, with likely spread through the carotid triangle into the internal jugular lymphatic system.

THE DENTITION

During foetal life and childhood before the deciduous teeth erupt, the gum pads of the upper and lower jaws are in contact with one another in the cheek region. They are usually—but not always—separated in front providing a space through which the tongue may be protruded. During this period the developing tooth germs (Fig. 6.15) lie beneath the surface of the gum

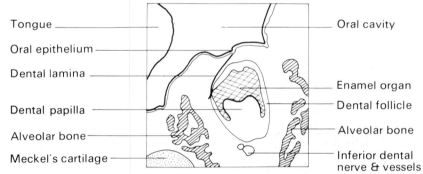

Fig. 6.15 Drawing of a developing tooth. Enamel organ at the bell stage. Coronal section of deciduous lower first molar from human foetus of 95 mm C.R. length.

(gum pads) within fibrous tissue sacs, the *dental follicles*, which are attached superficially to the lamina propria of the overlying mucous membrane. In the case of the deciduous teeth the tooth germs of the succeeding permanent teeth first occupy the same follicles. Later separate follicles develop, first as extensions of the deciduous follicle, and later as independent sacs. These are attached to the lamina propria of the gum through the *gubernacular cords*, which occupy canals in the alveolar bone. Each gubernacular canal of a successional permanent tooth passes from the roof of its crypt to an opening on the inner side of the deciduous tooth (Fig. 6.16). Sometimes the gubernacular canals open within the sockets of the deciduous teeth. The permanent molars, which have no deciduous predecessors, occupy crypts at the back of the mandible and maxilla (Figs 6.17, 6.18). The follicles surrounding the developing tooth germs in these crypts are attached to a molar gubernacular cord, which is united to the back of the oral mucous membrane in the region of the retromolar triangle in the lower jaw and the maxillary tuberosity in the upper jaw. The functional teeth, deciduous and permanent, are connected to one another by transseptal connective tissue fibres and each tooth is attached to the gum in the region of its neck by the gingival fibres of the periodontal ligament (Figs 2.13, 6.19). The follicles and gubernacular cords of the developing teeth and the transseptal ligaments and gingival attachments of the functional teeth unite the dental elements of each

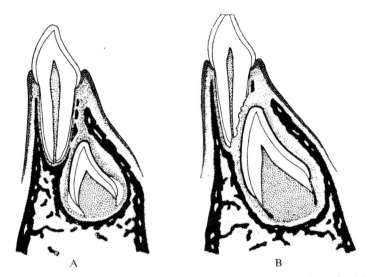

A   B

Fig. 6.16A Diagrammatic illustration of a sagittal section through the lower jaw of a child of two and a half years showing the relationship between a fully erupted deciduous incisor and its permanent successor. At this stage the deciduous tooth is contained in a bony socket. The crown of the permanent incisor is largely formed and is contained in a fully formed crypt.

Fig. 6.16B Diagrammatic illustration of a sagittal section through the lower jaw of a child of about seven years showing the relationship between a deciduous incisor and the replacing permanent incisor. With the beginning of the eruption of the permanent tooth the bone forming the roof of its crypt and part of the root of the deciduous tooth have been removed by resorption. Root formation has started in the permanent tooth.

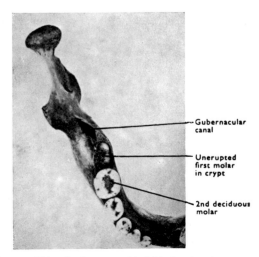

Gubernacular canal

Unerupted first molar in crypt

2nd deciduous molar

Fig. 6.17 The right half of the mandible of a five-year-old child showing the crypts of the first two permanent molars.

jaw into a developmental and functional unit. With growth of the jaws the whole dentition, including both the functional and the developing teeth, moves through the growing alveolar bone. The transseptal fibres help to maintain contact between adjacent teeth as interproximal wear occurs. Experimental studies in monkeys show that the forward drift of the cheek teeth (molars or premolars) that follows tooth extraction is dependent on the presence and further development of transseptal fibres across the healing socket. The fibre systems around the necks of the teeth are important for stabilizing the position of teeth and they may be involved in restoring the position of displaced teeth.

The *periodontal ligament or membrane* of each tooth is derived from its follicle (see also page 318). During the early stages of tooth development the follicles have no firm attachment to the developing alveolar bone. If the overlying gum is detached the follicles of the deciduous tooth germs are easily removed from their alveoli with the gum tissue. After tooth eruption commences the follicle becomes attached to the growing rims of the alveolar plates as these reach into the lamina propria of the overlying gum, with which the follicles are united. With growth of the root of the tooth the fibres of the follicle adjacent to the developing roots become reorientated and gain attachment to the alveolar bone as this changes from being an alveolus or crypt and becomes a tooth socket, and to the root of the tooth as cement is deposited in the deeper layers of the follicle.

The growth of tooth germs

The dental (enamel) organs and dentine papillae (the future pulp), which together make up the tooth germs (Fig. 6.15), and from which the dental tissues enamel, dentine and cement develop, grow rapidly within their fibrous follicles prior to the onset of calcification. The expanding follicles are surrounded by the bone of the alveolar compartments (deciduous teeth) or of the dental crypts (permanent teeth). The surface of each follicle contains numerous blood vessels forming an *external follicular plexus*. As the follicles expand, this vascular mesodermal tissue produces a resorption of the bone by the action of large multinucleated osteoclasts so that the bony compartments enlarge as the teeth grow in size.

With calcification of the crown portion of the tooth, expansion of the pulp no longer takes place in a centrifugal manner but only in the long axis of the developing tooth. The tooth and its follicle continue to grow in length but not in width. The parts of a developing tooth are shown in Figure 6.20.

Tooth eruption

Tooth eruption is essentially a movement of the developing tooth within its follicle as a result of differential growth of the pulp and follicle. The calcified crown of the tooth is lifted away from the base of the follicle and root formation follows from the continual growth of the *epithelial sheath of Hertwig* (the fused internal and external enamel (dental) epithelia at the root). The enamel of the tooth crown is covered by the reduced enamel epithelium derived from the cells of the dental organ after the completion of enamel formation. As the tooth moves within its follicle the epithelium covering the tooth crown approaches, and finally unites with the epithelium of the overlying gum (Fig. 6.21). Through the area of united epithelium the crown enters the mouth cavity and the epithelial attachment of the gum to the tooth is established. Tooth eruption and root formation continue as long as the apical foramen remains open. Tooth eruption does not depend solely upon root formation or on bone deposition in the tooth socket, but on the expansion of the pulp in relation to the base of the dental follicle. If as a result of an accident, such as an early jaw fracture or the establishment of a fibrous attachment following the healing of certain kinds of dental abscess, a tooth follicle is prevented from maintaining its correct relationship to the oral mucous membrane during the growth of the jaws and becomes buried in the developing face, it may erupt into such abnormal positions as the maxillary antrum, the nasal cavity, the pterygoid fossa or beneath the chin. Apart from these examples of gross displacements of teeth, certain teeth such as upper canines and third permanent molars sometimes fail to erupt although they develop and calcify. In such teeth the roots are often bent or distorted.

During their development within the jaws the incisor and canine teeth are crowded together in the anterior segments of the alveolar bone. With normal growth of the jaws they erupt into an arch of normal form. If bone growth is not adequate these teeth may consequently remain crowded and irregular after their eruption.

TIMES OF TOOTH ERUPTION

Deciduous dentition
    Lower central incisors—about 6 months
    Other incisors—7–9 months
    First deciduous molars—12–14 months
    Canines—16–18 months
    Second deciduous molars—24–30 months
Permanent dentition
    First permanent molars—about 6 years
    Central incisors—6–7 years
    Lateral incisors—8–9 years
    Premolars—10–12 years
    Canines—11–12 years
    Second permanent molars—12–13 years
    Third permanent molars—17–20 years

It is a useful guide to remember that in the average individual 12 teeth are present by the end of the first year. At the end of 3 years there are 20 teeth in the mouth and at the end of 6 years 24 teeth. The number increases to 28 teeth by the twelfth year and the full permanent dentition is in use by the twentieth year. Replacement of the deciduous teeth by the permanent teeth begins about the seventh year and is completed by about twelve years of age. The permanent molars have no deciduous predecessors.

These dates are however only approximate. There is a good deal of individual variation in regard to the dates of eruption, and in the permanent dentition in regard to the order of eruption especially among the premolars and canines. The teeth tend to erupt somewhat earlier in girls.

Tooth migration
The permanent molars develop in posteriorly directed extensions of the alveolar bone known as the *alveolar bulbs*. That of the lower jaw lies on the inner side of the anterior part of the ascending ramus of the mandible (Fig. 6.17); that of the upper jaw faces backwards into the lower part of the pterygopalatine fossa (Fig. 6.18). With forward migration of the dentition, the teeth which develop in the alveolar bulbs are drawn forwards by their molar gubernacular cords into the back of the mouth cavity. Forward migration of the teeth through the growing alveolar bone is a separate process from, and independent of, tooth eruption.

We have seen that functional and developing teeth are attached to the gum mucoperiosteum. As the bone-forming cells in the mucoperiosteum deposit new bone on the surfaces of the upper and lower alveolar processes the teeth will be carried forwards, downwards and outwards in the upper jaw and forwards, upwards and outwards in the lower jaw. Because of their attachment to one another through the transseptal fibre system the dentition in each jaw will move as a unit.

As the teeth migrate a continual reconstruction of their sockets takes place around the moving roots. This involves a continual formation of new periodontal fibres and their reattachment to the moving sockets. If there is an inadequate amount of forward movement of the teeth,

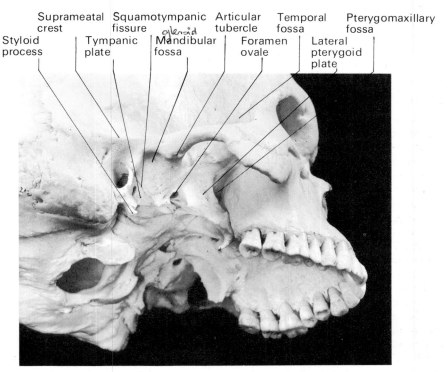

Suprameatal crest | Squamotympanic fissure | Articular tubercle | Temporal fossa | Pterygomaxillary fossa

Styloid process | Tympanic plate | Mandibular fossa | Foramen ovale | Lateral pterygoid plate

Fig. 6.18 The skull viewed from the lateral aspect showing the infratemporal region.

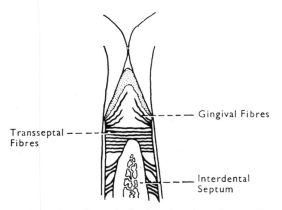

Transseptal Fibres

Gingival Fibres

Interdental Septum

Fig. 6.19 Diagrammatic illustration of a mesiodistal section through two adjacent teeth and the interalveolar septum to show the arrangement of the transseptal and gingival fibres of the periodontal ligament.

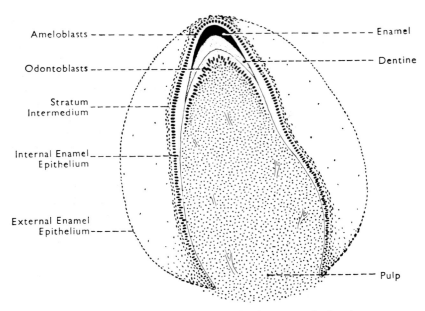

Ameloblasts

Odontoblasts

Stratum Intermedium

Internal Enamel Epithelium

External Enamel Epithelium

Enamel

Dentine

Pulp

Fig. 6.20 Diagrammatic illustration of a developing human tooth after the commencement of dentine and enamel formation. Note the lengthening of the cells of the internal enamel epithelium during their differentiation before odontoblast formation begins. Note also the subsequent shortening of these cells in becoming ameloblasts and starting the formation of enamel.

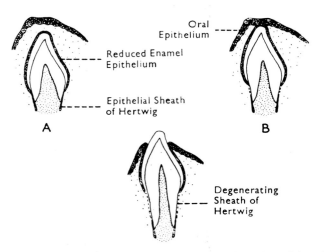

Oral Epithelium

Reduced Enamel Epithelium

Epithelial Sheath of Hertwig

A

B

Degenerating Sheath of Hertwig

Fig. 6.21 Diagram to illustrate the eruption of a tooth and the formation of the epithelial attachment. (A) Tooth just before the commencement of eruption. (B) Fusion of the reduced enamel epithelium with the oral epithelium as the tooth nears the mouth cavity. (C) The emergence of the tip of the tooth into the mouth cavity.

especially at the end of the period of facial growth, the lower third molars may become impacted behind the second molars. *Impaction* is especially common in the case of the lower molars because in development their crowns face forwards. If rotation does not occur during eruption the process of eruption will often bring them against the distal surfaces of the teeth in front. Impaction is less frequent in the upper jaw because the crowns of the developing upper permanent molars face backwards and downwards. But for the attachment of their follicles to the molar gubernacular cord which guides them forwards, they would tend to erupt in a backward direction and away from the mouth cavity.

Age changes in the dentition

After facial growth is complete and the permanent dentition is established tooth eruption and migration as described above ceases. The changes which occur in the adult dentition with age include:

Mesial (forward) movements of the teeth in each dental arch associated with the wear of the approximal surfaces (contact areas) of adjacent teeth. This is a process of local adjustment between adjacent teeth.

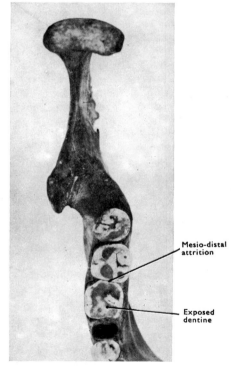

Mesio-distal attrition

Exposed dentine

Fig. 6.22 Right half of an adult mandible showing the effects of attrition (wear of the teeth).

Attrition of the enamel, and later of the exposed dentine, on the occlusal surfaces and incisive edges. This is correlated with reduction in size of the pulp chambers as a result of the continual deposition of additional or secondary dentine. In cases of severe attrition this secondary dentine may be exposed on the occlusal surfaces of the teeth (Fig. 6.22).

A *forward movement of the lower jaw* relative to the upper. This is a consequence of attrition of the cusps of the cheek teeth and tends to produce an edge to edge bite of the upper and lower incisors.

Recession of the gums. This is a common age change. The epithelial attachment between gums and teeth (page 291) retreats towards the root of the tooth and may reach the cement so that the enamel-cement junction is exposed in the mouth cavity As a result the area of attachment of the periodontal ligament is reduced and the margins of the sockets reabsorbed.

The *roots may be elongated* by the deposition of cement at the apical regions. This in part compensates for the recession of the gum in extending the area of periodontal attachment apically. The process of gum recession and elongation of the roots by the apical deposition of cement is known as passive eruption. The process, however, does not involve any movement on the part of the teeth. It is essentially an exposure of more tooth substance in the mouth cavity ('growing long in the tooth') associated with a compensatory deepening of the socket and extension of the roots so as to maintain the area of tooth attachment.

The edentulous mouth
The mouth cavity is edentulous (toothless):

1. Before the eruption of the deciduous teeth
2. After the loss of the permanent teeth.

During late foetal life and the first months of childhood the upper and lower gum pads, which overlie the developing deciduous teeth in their alveoli, are segmented to form a series of rounded swellings in each jaw. To the deep surfaces of these are attached the tooth sacs or follicles. In early foetal life the gum pads of the upper and lower jaws are separated by the relatively large tongue which extends between them so that its sides are in contact with the cheeks. With growth of the mouth cavity, however, the tongue comes to lie within the arch of the palate and the gum pads are in contact with one another in the cheek region when the mouth is closed. At the front of the mouth, however, the gum pads are usually separated even when the mouth is closed, leaving a space through which the tip of the tongue protrudes.

In early foetal life, the lower jaw often projects beyond the upper jaw but by the time of birth the interrelationship of the jaws is reversed in the great majority of individuals to assume the normal adult condition.

During suckling the gum pads grasp the breast so that the nipple lies within the oral cavity. The approximation of the lips to the breast forms a sealing mechanism and the milk is expressed by a combination of pressure exerted by the gum pads and suction produced by the lowering of the floor of the mouth and tongue and enlargement of the oral pharynx. Breathing is synchronized with the suckling mechanism so that the nasopharynx and larynx are closed off from the oral pharynx as the milk flows from the mouth to the oesophagus.

Changes in the mouth following loss of the permanent teeth
After the loss or removal of all the teeth the alveolar bone surrounding and supporting their sockets is gradually resorbed in both jaws. This leads to a number of important alterations in facial and intra-oral relationships.

1. Following loss of the teeth and resorption of alveolar bone, the jaws approximate closer to one another when the mouth is closed. This leads to a reduction of facial height. Also the cheeks and lips lacking the support of teeth and alveolar bone tend to fall inwards giving the face the characteristic features of old age.

2. Changes also take place in the mandibular joints. The ligaments, which limit closing movements of the mandible, become stretched and the condyle lies further back in the glenoid fossa in the rest position. Bone resorption may occur at the back of the head of the condyle, in the glenoid fossa and at the articular eminence.

3. In the mouth cavity the buccal and lingual surfaces of the gums become continuous over the reduced alveolar ridges. Normally the gum is firmly bound to the underlying bone, but

sometimes, if bone resorption is rapid, areas of unsupported pendulous gum tissue are found, especially in the incisor-canine regions.

4. Resorption of alveolar bone may be uneven leaving sharp ridges, especially in the lower incisor-canine region (Fig. 6.23). In the upper jaw alveolar resorption may proceed so far in the cheek region that the floor of the maxillary sinus is separated from the gum by only a very thin layer of bone. The degree of resorption in the region of the maxillary tuberosity is very variable and its extent is important in the design of upper dentures. In the lower jaw

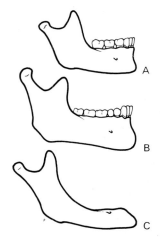

Fig. 6.23 Outlines of mandibles from A, a child; B, a young adult; C, an elderly person. Note the loss of alveolar bone in C following loss of the teeth.

alveolar resorption brings the attachments of the buccinator, mentalis, and mylohyoid muscles close to the upper margin of the jaw reducing the depth of the vestibule on the outer side and the orolingual sulcus on the inner side. In cases of extensive alveolar resorption, the mental foramina may open on, or close to, the upper margin of the jaw. Excessive bone resorption in the lower incisor region will bring the genial tubercles close to the jaw margin.

These changes are of considerable importance in the fitting of lower dentures for the denture bearing area is greatly reduced. Also there is greater tendency towards discomfort and pain for the margins of the denture may press upon the structures mentioned, e.g. the muscles or the mental nerve as it passes through the mental foramen. Furthermore, the discomfort is accentuated by the lack of stability of the denture due to the small area available to support it.

BLOOD VESSELS OF THE ORAL TISSUES

The blood supply of the teeth and their supporting structures

The maxillary artery (page 257) is a terminal branch of the external carotid divided into three parts by its relationship to the lateral pterygoid muscle (Fig. 4.20). The largest branch of the first part of the artery is the *inferior dental artery*. It descends between the inner surface of the ramus of the mandible and the outer surface of the medial pterygoid muscle with the inferior dental nerve. At the mandibular foramen it gives off the artery to the mylohyoid muscle and enters the mandibular canal. Within the canal it gives off ascending branches to the sockets and pulps of the teeth, branches to the medullary cavity of the mandible and branches to the cancellous bone of the mandibular ramus.

In the region of the premolar teeth the inferior dental artery terminates in a mental and an incisive branch. The former appears on the facial surface of the lower jaw through the mental foramen and supplies the adjacent structures, including the cheeks, lower lip and lower gum as far forward as the middle line of the face, anastomosing with branches of the facial artery. The incisive artery remains within the substance of the mandible and gives branches to the sockets and pulps of the incisor and canine teeth.

The veins drain from the sockets and tooth pulps into vessels which accompany the inferior dental artery. These drain backwards towards the mandibular foramen or forwards towards the mental foramen. At the mandibular foramen they join the pterygoid venous plexus, and at the mental foramen the anterior facial vein. The gum of the lower dental arch receives blood vessels from the mental arteries; the buccal artery (a branch of the second part of the maxillary artery) which reaches the gum after piercing the buccinator muscle on which it lies and also supplies; the facial artery and the lingual artery. These vessels are accompanied by veins draining into the anterior facial vein, the pterygoid venous plexus and the lingual vein.

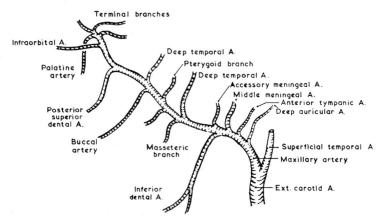

Fig. 6.24 Branches of the maxillary artery.

When a tooth is extracted the vessels to the tooth pulp, the periodontal membrane and the gum are ruptured and the socket becomes filled with mixed arterial and venous blood. This coagulates to form a clot and the blood clot is invaded by fibroblasts to form granulation tissue. This is overgrown by proliferating gum epithelium and eventually the fibrous tissue may be partly ossified. If a number of adjacent teeth are extracted a good deal of the alveolar bone which supported the teeth is resorbed and only the basal parts of the sockets become filled in with new bone.

The branches of the third part of the maxillary artery (Figs 6.24, 6.28) include the *posterior superior dental artery* and the *infra-orbital artery*. These vessels commence at the back of the maxilla and in the pterygopalatine fossa. From the infra-orbital artery, as it runs along the floor of the orbital cavity, there is given off the anterior superior dental artery and sometimes a middle superior dental artery. These reach the teeth in neurovascular canals situated in the bone forming the facial surface of the maxilla and in close relation to the lining mucous membrane of the maxillary antrum, to which they give branches.

The posterior superior dental artery is usually a single branch which soon divides into a number of smaller branches lying on the posterior surface of the maxilla. Some of these continue to descend on the surface of the bone to supply the gum of the upper permanent molar teeth and anastomose in the cheek with small branches of the buccal and facial arteries. Other branches enter one or more bony canals at the back of the maxilla and form a plexus above the apices of the molar teeth from which branches pass to the tooth sockets and the mucous membrane of the maxillary antrum. This plexus is joined in front by the anterior superior dental artery and, if present, by the middle superior dental artery. Veins run with the arteries and drain either forwards through the infra-orbital foramen to the anterior facial vein or backwards and upwards to the infraorbital veins, or directly into the pterygoid venous plexus (Fig. 6.8).

The maxillary artery in the pterygopalatine fossa also gives off a palatine artery which descends in the palatine canal to appear on the palate as a large branch emerging through the greater palatine foramen. A number of smaller branches pass through the lesser palatine foramina. The latter branches enter the soft palate where they anastomose with the ascending branches from the lingual artery (dorsales linguae), descending branches of the ascending pharyngeal and ascending palatine vessels, and with branches of the tonsillar arteries.

The *greater palatine arteries* pass forwards in a groove on the under surface of the hard palate through the mucoperiosteum towards the incisive foramen where they anastomose with the terminal branches of the nasal vessels (page 184), branches of the maxillary artery. The greater palatine artery supplies the mucoperiosteum and glandular tissue of the hard palate and the lingual gum of the upper teeth. Palatal veins accompany the arteries and drain into the pterygoid venous plexus and communicate through the incisive foramen with the veins of the nasal cavity. In operations for the repair of cleft palate it is important to know the details of the course, relations and distribution of the palatine vessels, for large flaps of mucoperiosteum have to be stripped from the underlying bone. The maintenance of a good blood supply is important for rapid healing of the soft tissues after an operation.

Other blood vessels supplying oral tissues

The *lingual artery* and its sublingual branch supplies the tongue and the structures at the floor of the mouth. It gives branches to the gum on the lingual side of the lower teeth. In extensive operations involving removal of parts of or the whole of the tongue, the lingual arteries may require to be ligatured in the neck (page 254). The lingual vein and its tributaries drain into the internal jugular vein.

The *facial artery* runs upwards and forwards through the tissues of the cheek passing close to the corner of the mouth. The labial branches run in the lips beneath the lining mucous membrane and may be ruptured when the lip is cut by a blow, especially when the incisor teeth are irregular (page 254).

The *buccal artery*, a branch of the maxillary artery, accompanies the buccal nerve from the pterygoid region and reaches the cheek from behind, lying on the buccinator muscle. It anastomoses with branches of the facial, mental and posterior superior dental arteries (Fig. 5.31).

The vessels draining the cheeks and lips drain either into the anterior facial vein or via the deep facial vein into the pterygoid venous plexus. The angular vein (terminal part of the anterior facial vein) communicates with the veins of the orbital cavity.

The various regions of the oral cavity therefore receive their blood supply as follows:

1. *Teeth and supporting structures.* These are supplied by the inferior dental artery in the lower jaw; the posterior, middle and anterior superior dental arteries in the upper jaw. These are all derived from the maxillary artery.

2. The *hard palate.* This is supplied by the greater palatine arteries and incisive vessels from the nasal arteries (especially the nasopalatine branch). These are all branches of the maxillary artery.

3. The *soft palate.* This receives its blood supply from
   a. The lesser palatine arteries, branches of the maxillary artery;
   b. A number of branches from the ascending pharyngeal and ascending palatine arteries, reaching the palate at the upper border of the superior constrictor muscle. The ascending

pharyngeal is the first branch of the external carotid artery and the ascending palatine is a branch of the cervical part of the facial artery,

   c. The dorsales linguae branches from the lingual artery.

The *cheeks*. These receive their blood supply from

   a. The buccal artery, a branch of the second part of the maxillary artery;

   b. Branches of the posterior superior dental arteries;

   c. The facial artery;

   d. The mental artery and the infra-orbital artery. The central part of the cheek has a relatively poor blood supply as all the vessels are small, and this area is sometimes the site of gangrene (cancrum oris) in children.

The *lips*. The lower lip receives its blood supply from the inferior labial branches of the facial artery and branches of the mental and submental arteries. The upper lip receives one or two superior labial arteries from the facial artery and the labial branch of the infra-orbital artery.

The *tongue* is supplied by the lingual artery and its branches.

The *floor of the mouth* is supplied by the sublingual branch of the lingual artery and, below the mylohyoid, by the submental branch of the facial artery and the mylohyoid branch of the inferior dental artery.

The facial artery at the front of the cheeks, the labial arteries in the lips, the greater palatine arteries on the under surface of the hard palate, and the sublingual arteries in the floor of the mouth are all large vessels which may be injured during operations in the mouth region.

The inferior dental arteries, the superior dental arteries, the infra-orbital arteries and the vessels to the nasal cavity (which pass through the sphenopalatine foramen), the palatine arteries in their canals, are all liable to be involved in fractures of the facial skeleton. These are all branches of the maxillary artery, itself a branch of the external carotid, and bleeding may be controlled if necessary by a ligature of this vessel in the neck (carotid triangle). The facial and lingual vessels can also be ligatured in the same position.

NERVE SUPPLY OF THE TEETH AND SUPPORTING STRUCTURES

The nerve supply of the *lower teeth* is derived from the inferior dental nerve, a branch of the posterior division of the mandibular trunk of the trigeminal (fifth cranial) nerve. Sensory nerve endings are present in the gums, periodontal ligaments and tooth pulps. The sensory somatic fibres, which commence in these nerve endings, make up the bulk of the inferior dental nerve which is, however, also joined by the fibres of the mental nerve coming from sensory nerve endings situated in the skin and mucous membrane of the chin and lower lip. The inferior dental nerve also contains a number of small, non-myelinated, sympathetic fibres from the superior cervical ganglion distributed to the smooth muscle of the blood vessels of the lower teeth, gums, lower lip and cheeks, and the mucous secretory glands of the lower lip and the front of the cheeks. Most of the sympathetic nerve fibres are distributed by perivascular nerve plexuses surrounding the main arterial vessels to the region. The plexuses contain post-ganglionic fibres which originate in the superior cervical ganglion. It also probably contains parasympathetic fibres from the otic ganglion distributed to the same blood vessels and glands.

The gum of the lingual side of the lower teeth is supplied with sensory fibres from the lingual nerve, and the gum on the buccal side of the lower cheek teeth is supplied in part by the buccal nerve which is the terminal (sensory) branch of the anterior division of the mandibular nerve. It runs on the outer surface of the buccinator muscle with its accompanying vessels (Fig. 4.20).

There are extensive *nerve plexuses* in the lamina propria of the oral epithelium which take the form of netting wire-like arrangements parallel to the surface of the oral epithelium (Figs

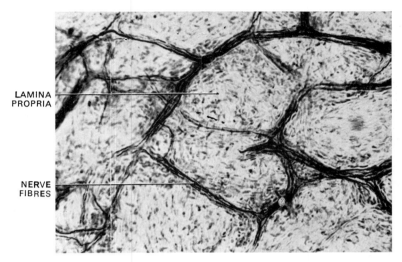

LAMINA
PROPRIA

NERVE
FIBRES

Fig. 6.25 Photomicrograph showing a small part of an oral nerve plexus prepared by a silver staining technique. The nerve fibres are represented by black lines in the photograph × 135. (By courtesy of *Archives of Oral Biology*.)

Fig. 6.26 Coiled nerve ending in the human palate. S. stem axon. E. epithelial cells. Silver impregnation method × 430. (By courtesy of *Archives of Oral Biology*.)

6.25, 6.26). They consist of bundles of myelinated and non-myelinated nerve fibres which send branches to the nerve endings (Figs 6.26, 6.27) and to the blood vessels. The myelinated fibres are branches of the maxillary and mandibular divisions of the trigeminal nerve.

Sensory fibres arising from nerve endings associated with the *upper teeth* and their supporting structures pass along the superior dental nerves which are branches of the maxillary division of the trigeminal (Fig. 6.28).

The *anterior superior dental nerve* is given off the infra-orbital nerve some distance behind the infra-orbital foramen. It runs in a canal which commences on the lateral side of the infra-orbital canal and passes close beneath the infra-orbital foramen in the facial surface of the maxilla. Close to the level of the nasal aperture the nerve descends towards the apices of the incisor teeth. In its course through the bone it is closely related to the mucous membrane lining the front of the maxillary antrum and gives off the following branches:

1. Posterior branches which communicate with the posterior superior dental nerve and the middle superior dental nerve if present, and supply the canine and premolar teeth.

2. Incisive branches to the incisor teeth.

3. A nasal branch to the nasal mucous membrane covering the atrial area of the nasal cavity (page 180).

4. Branches to the mucous membrane of the antrum.

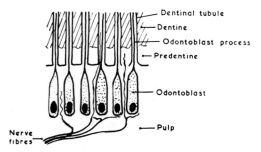

Fig. 6.27 The distribution of nerve fibres in the dental pulp.

The *middle superior dental nerve* is present in about 50 per cent of cases. It may be present on one side of the face and absent on the other. It is given off the infra-orbital nerve between the anterior superior dental branch and the inferior ophthalmic fissure. It usually descends in the side wall of the antrum deep to the zygomatic process of the maxilla and therefore at some distance from the surface of the bone. It joins the nerve plexus formed by the overlapping of the posterior superior dental nerve and the posterior branches of the anterior dental nerve. It supplies the premolar teeth and may help to supply the supporting tissues of the canine and first permanent molars. It gives branches to the mucous membrane of the antrum.

The *posterior superior dental nerve, or nerves*, are given off the maxillary division as this crosses the pterygopalatine fossa. They descend on the back of the maxilla with the corresponding blood vessels. Some branches continue on the surface of the bone to supply the gum related to the permanent molars and the cheek. Other branches enter one or more foramina at the back of the maxilla and run in horizontal bony canals below the zygomatic process of the maxilla towards the facial surface of the bone, forming there a plexus with the anterior superior dental nerve and the middle superior dental nerve. The posterior superior dental nerve supplies the permanent molar teeth and the mucous membrane of the antrum.

The plexus formed by the three superior dental nerves is closely related to the apices of the upper teeth and is formed by the overlapping of the distribution of the branches of the three nerves. The *incisors* are supplied from branches of the anterior nerve; the *canine* by the anterior nerve and by the middle nerve if present; the *premolars* by the anterior nerve and/or the middle nerve; the *first permanent molar* by the middle nerve if present and the posterior nerve, while the latter always supplies the *second and third molars.*

At birth the superior dental nerves have a short course to the developing teeth through the cancellous bone of the maxilla. With growth in height of the face as this bone becomes invaded by the expanding antrum and the nerves have to take a longer course to the teeth, running in canals in the anterior (facial) and temporal surfaces of the bone. The nerves are everywhere closely related to branches of the corresponding dental arteries (page 298). Sympathetic fibres to the upper teeth and supporting structures are derived from the superior cervical ganglion; parasympathetic fibres are derived from the sphenopalatine ganglion.

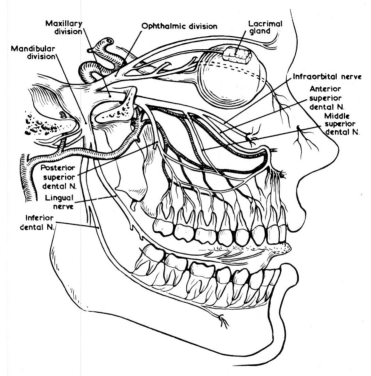

Fig. 6.28 The nerve supply and blood supply to the upper teeth. The main features of the ophthalmic and mandibular divisions of the trigeminal nerve are also shown.

In the pulp chambers the somatic nerve endings are naked nerve fibres with occasional small terminal swellings lying between the odontoblast and in the predentine (Fig. 6.27). It is uncertain whether nerve fibres enter the dentinal tubules and if they do so whether they penetrate for any great distance. The odontoblast cells of the pulp, which send protoplasmic processes into the dentinal tubules, are probably derived from the neural crest (ectomesodermal) tissue of the neck region (Fig. 7.6), and may have the function of pain conduction through dentine. Pain is the only sensation appreciated when stimulation of any kind (pressure, chemical, thermal) is applied to the pulp of a tooth.

The nerve endings in the periodontal ligament are in the form of either simple coils of nerve fibres lying between bundles of collagenous fibres, small knob-like expansions or naked nerve endings. The sensations appreciated in the periodontal ligament are of pressure and pain. Stimulation of nerve endings in the supporting tissues of the teeth also plays an important part in the proprioceptive reflex mechanisms which regulate masticatory movements (see page 316).

The pain associated with tooth decay (dental caries) or preparing a cavity in a tooth is produced by stimulating the dentine and the nerve endings in the pulp chamber of the tooth. The pain related to an abscess developing around the apex of a tooth, which is usually the result of bacterial infection of the pulp is produced by stimulation of nerve endings both in the pulp and periapical dental tissues. It is increased by pressure on the tooth as when it is driven into the socket during eating or by tapping (as in clinical diagnosis of the condition). Pain in an exposed pulp is sometimes relieved by cold as this reduces the pressure exerted by the inflamed pulp tissues within the rigid pulp cavity by causing constriction of the pulp vessels. The pain of a dental abscess is often lessened when it breaks through the bone into the mouth cavity or one of the fascial spaces, as the pressure exerted on nerve endings by the pus and inflamed tissue within the bone is reduced (see page 450).

Within the brain stem (mid-brain, pons and medulla oblongata (see page 395)) the incoming sensory fibres from all parts of the distribution of the fifth nerve are separated according to their sensory functions. Fibres from pain, surface pressure and thermal receptors pass to different parts of the long sensory nucleus of the nerve, which descends into the spinal cord. Fibres from endings in the gums, periodontal ligament and hard palate, and from the temporomandibular joint and muscles of mastication subserving the proprioceptive reflex mechanisms, pass to the separate mesencephalic (mid-brain) nucleus of the fifth nerve. These central connections of the trigeminal nerve are described more fully on page 419).

Other nerves to the oral tissues

The *lingual nerve*, like the inferior dental nerve, is a branch of the mandibular division of the trigeminal. It is given off the main trunk under cover of the lateral pterygoid muscle in front of the inferior dental nerve to which it is connected by a communicating branch. It also receives the chorda tympani which joins it from behind and through which it communicates with the facial nerve (page 249). The lingual nerve descends between the mandibular ramus and the medial pterygoid muscle. It passes to the floor of the mouth at the anterior edge of the medial pterygoid muscle where it comes to lie on the outer surface of the styloglossus above the mylohyoid and on the inner side of the body of the mandible below the level of the third permanent molar (Figs 4.25, 4.26). Here it is in close relationship with the mucous membrane of the back of the oral cavity.

Further forwards it lies at a somewhat deeper level, lateral to the hyoglossus muscle and after passing beneath the duct of the submandibular gland it passes upwards on the inner side of the duct to enter the tongue (Fig. 6.7). It gives off branches to the mucous membrane of the floor of the mouth and to the gum on the inner (lingual) side of the lower teeth. It carries sensory fibres of common sensation from the anterior two-thirds of the tongue, floor of the mouth and gums and also taste fibres from the anterior two-thirds of the tongue. The *taste fibres* leave the lingual nerve to join the facial nerve via the chorda tympani. The lingual nerve also contains secretomotor (parasympathetic) fibres which reach it from the facial nerve along the chorda tympani and pass to the submandibular parasympathetic ganglion. From the ganglion branches are distributed to the submandibular and sublingual salivary glands and mucous glands to the floor of the mouth and tongue (page 320).

The *buccal nerve* is the terminal branch of the anterior division of the mandibular nerve. It passes deep to the insertion of the temporal muscle onto the surface of the buccinator muscle and reaches the skin overlying it. It is the chief sensory nerve to the skin, the mucous membrane of the vestibule of the mouth and the gingiva on the buccal side of the teeth. Its distribution may extend as far forward as the angle of the mouth. Although branches of the buccal nerve intermingle with those of the facial nerve, the buccal nerve is sensory and local anaesthetic blocks to it will not interfere with muscle function. The buccal nerve contains some secretomotor fibres probably derived from the otic ganglion, which supply the buccal mucous glands.

The *palatine nerves* are branches given off the maxillary division of the trigeminal in the pterygopalatine fossa. They pass through the sphenopalatine ganglion and from it receive secretomotor (parasympathetic) fibres to the mucous glands of the palate. They also contain some taste fibres. Some branches of the palatine nerves emerge through the lesser palatine foramina and are the sensory nerves to the soft palate. A larger branch, the greater palatine nerve, emerges through the greater palatine foramen and runs through the mucoperiosteum covering the hard palate, with the greater palatine vessels, towards the incisive foramen. It supplies the mucous membrane of the hard palate and the gum tissue on the lingual side of the upper teeth.

A *terminal branch of the nasopalatine* (*long sphenopalatine*) branch of the nasal nerves also derived from the maxillary nerve in the pterygopalatine fossa usually passes through the incisive foramen and may supply an area of mucous membrane in the region of the incisive papillae. It may also help to supply the supporting tissues of the central and often the lateral incisor teeth.

The various regions of the oral cavity receive their sensory nerve supply as follows:

The teeth and supporting structures
The inferior dental nerve, a branch of the mandibular division; and the superior dental nerves, branches of the maxillary division of the trigeminal nerve.

The hard palate
The greater palatine nerves and the incisive branch of the nasopalatine nerve. Both are branches of the maxillary division of the fifth nerve.

The soft palate
The lesser palatine nerves, branches of the maxillary division of the fifth nerve, and the tonsillar branches of the glossopharyngeal nerve.

The cheeks
The buccal nerve and the mental nerve, branches of the mandibular division; and the posterior superior dental nerves, branches of the maxillary division of the fifth cranial nerve.

The upper lip
This is supplied by the labial branch of the infra-orbital nerve from the maxillary division, and the *lower lip* from the mental branch of the mandibular division. These nerves also send branches to the adjacent gum tissue.

The mucous membrane of the oral part of the tongue
The anterior two-thirds is supplied by the lingual nerve which also supplies the floor of the mouth and the gum on the inner side of the lower teeth.

The distribution of the secretomotor fibres to the salivary and mucous glands of the oral cavity is described on page 320.

The inferior dental nerve in its canal is in close relation to the lower third molar tooth and may pass on rare occasions through a foramen in one of the roots of this tooth. It is sometimes damaged in extraction of the tooth. The lingual nerve may also be damaged in operations on impacted lower third molars. The inferior dental, infra-orbital, superior dental and palatine nerves are liable to be damaged in fractures involving the facial skeleton.

LYMPHATIC DRAINAGE OF THE ORAL TISSUES

An understanding of the oral lymphatic drainage is important in the diagnosis of infective conditions and of cancer, which is most common in relation to the tongue and lips.

A subepithelial lymphatic plexus interconnects all regions of the oral cavity, and the lymphatic drainage of the mucous membrane of the oral cavity, the teeth and tongue, is through a system of lymphatic vessels draining into several groups of lymph nodes, including the submental, submandibular, parotid, retropharyngeal and upper deep cervical lymph nodes.

A circle of superficial nodes which extends from the occipital region to the chin is made up of a number of named groups. These are: the *occipital nodes*, one to three in number, placed on the back of the head close to the attachment of the trapezius muscle; the *retro-auricular* (*mastoid*) *nodes*, usually two in number situated on the insertion of the sternomastoid muscle, just behind the ear; the *anterior auricular nodes*, from one to three in number, immediately in front of the ear; the *parotid nodes* forming two groups in relation to the parotid salivary gland, one superficial group embedded in the gland and the other deep to the gland on the lateral wall of the pharynx. The *facial nodes* comprise three groups: the *infra-orbital* or *maxillary* nodes scattered over the infra-orbital region; the *buccal nodes* on the surface of the buccinator muscle opposite the angle of the mouth; and the *mandibular nodes*, on the outer surface of the mandible in front of the masseter muscle, close to the facial artery and the facial vein.

The regional lymphatic drainage of the oral tissues is as follows:

The teeth and supporting structures

The vessels of the incisor-canine region drain into lymphatics passing forwards towards the front of the face and join the anterior facial lymphatics which enter the submental and submandibular nodes. Lymphatics from the premolar and molar region drain backwards to pierce the buccinator and superior constrictor muscles to enter the deep parotid and upper deep cervical nodes. There are, however, numerous connections between the anterior and posterior lymphatic vessels.

The palate

Lymph from the anterior region of the hard palate drains forwards into vessels which join the anterior facial vessels and enter the submandibular nodes. Lymph from the back of the hard palate runs in vessels entering the upper deep cervical nodes, after piercing the superior constrictor muscle.

The rich lymph drainage of the soft palate passes to either side to join some of the lymphatics draining the tonsils and with them pierce the superior constrictor to enter the retropharyngeal, deep parotid and upper deep cervical nodes.

The cheeks

Lymphatic vessels draining the cheeks pierce the buccinator to enter the buccal lymph node or join the anterior facial lymphatics to terminate in the submandibular nodes.

The lips

They are drained by lymphatics which enter the submental and submandibular nodes. The central part of the *lower lip* is drained by vessels entering the submental nodes (with those of the lower incisor teeth and supporting structures), while vessels from their lateral parts run with the vessels of the upper lip to enter the submandibular nodes.

The floor of the mouth

The vessels of the anterior part of the floor of the mouth pass either directly to the lower nodes of the upper deep cervical group, or indirectly to this group via the submental nodes. Lymphatic vessels from the remainder of the floor of the mouth pass to the submandibular and upper deep cervical nodes. Many of the vessels pierce the mylohyoid muscular diaphragm on their way from the mucous membrane to their respective lymph nodes.

The tongue

The rich lymphatic system of the tongue drains into the submental, submandibular and deep cervical nodes (for details see page 288). All lymph from the tongue eventually reaches the jugulo-omohyoid node. The tip of the tongue drains bilaterally to the submental nodes and there is some overlap across the mid-line of the remainder of the tongue.

The glands of the oral cavity

The glands which open into the oral cavity include the three major bilateral salivary glands, whose ducts open into the vestibule (parotid) and the floor of the mouth (submandibular and sublingual). Of these the parotid is a serous secretory gland, the submandibular is mixed mucous and serous, while the sublingual is predominantly mucus secreting. Serous secretions are clear, watery fluids, in contrast to the thicker, slimy quality of the secretions of mucous glandular units.

The minor salivary glands can be grouped according to their location and the nature of their secretion as follows:

Labial and buccal glands, opening into the vestibule and predominantly mucus secreting.

Palatal glands, in both the hard and soft palates, which are entirely mucus secreting.

Anterior lingual glands (of Blandin, or Nuhn), at the side of the median line near the tip of the tongue.

Posterior lingual gustatory glands (of von Ebner), in the circumvallate papillae, which are entirely the serous type.

These glands are situated in the lamina propria of the mucous membrane or the adjacent submucosa. Between them they produce the saliva which is digestive, because of its content of the enzyme ptyalin (salivary amylase). Saliva also has cleansing and lubricating functions for the oral mucosa. Saliva provides in addition an adhesive quality necessary to hold the masticated food in the form of a bolus prior to swallowing (page 321). The role of salivary enzyme is not great since foodstuffs remain in the oral cavity for only a short period of time, and amylase activity is destroyed soon after the bolus of food enters the stomach.

THE MANDIBULAR JOINT

The features of the mandibular joint (see page 158) which require more detailed attention are:

1. The articular surface of the temporal bone.
2. The condyle of the mandible.
3. The joint capsule.
4. The articular disk.
5. The joint cavities.
6. Mandibular movements.
7. Structures related to the joint.

The articular surface of the temporal bone

In size, form and in the development of its various parts the articular surface of the squamous part of the temporal bone shows considerable variation. Average measurements are 20 mm in anteroposterior length, and 25 mm in mediolateral width. It consists of a concave posterior part (the *glenoid fossa*) and a convex anterior part (the *articular eminence*). The concave posterior region is bounded behind by the *squamotympanic fissure*. In some skulls there is a small downward projecting ridge or tubercle in front of the lateral part of the fissure; the *postglenoid tubercle*. Behind the squamotympanic fissure is the *tympanic plate* of the temporal bone forming the anterior wall and floor of the auditory canal (external auditory meatus) (Figs 4.41, 6.18, 6.29).

The *articular eminence* is an elongated transverse ridge of bone sometimes referred to as the anterior root of the zygomatic process. At its lateral end, at the junction of the anterior and posterior roots of the zygomatic process, is the *articular tubercle*. This gives attachment

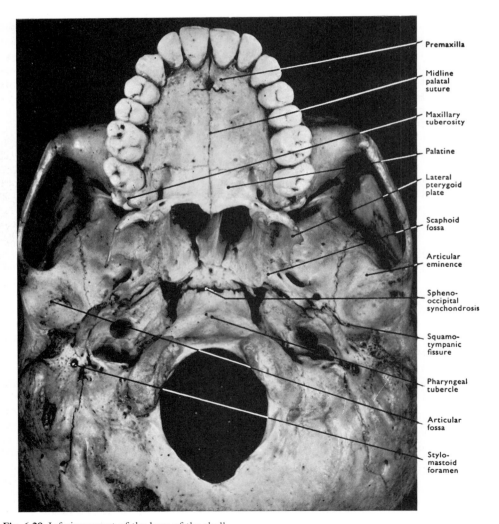

Fig. 6.29 Inferior aspect of the base of the skull.

on its outer surface to the temporomandibular ligament. The posterior root of the zygomatic process continues backwards. Its front part is an outward projecting shelf of bone, which forms the outer half of the glenoid fossa and above which lies the lower border of the temporal muscle (Fig. 5.22). Further back the posterior zygomatic root forms a ridge (suprameatal crest) above the external auditory meatus which is especially well developed in male skulls. The post-glenoid tubercle is sometimes referred to as the middle or intermediate root of the zygomatic process.

Medially the articular surface is bounded by the suture between the squamous temporal bone carrying the articular surface, and the great wing of the sphenoid. At the *spine of the sphenoid*, which projects from the posterior limit of the great wing, the suture turns inwards and runs forwards separating the petrous temporal from the great wing of the sphenoid. The great wing of the sphenoid therefore forms a triangular salient between the squamous and petrous parts of the temporal bone, while the spine of the sphenoid forms the apex of this salient (Fig. 6.29). The suture between the sphenoid and temporal bones on the base of the skull is part of the coronal suture system and reaches the mid-line cranial base on the lateral side of the spheno-occipital synchondrosis. Growth at this suture will displace the temporal bone (carrying the articular surface) outwards and backwards.

In the majority of adult human skulls the central area of the concave region of the articular surface consists of a thin plate of cortical bone separating the glenoid fossa from the floor of the middle cranial fossa which contains the temporal lobe of the cerebral hemisphere. This marks the position and approximate extent of the articular area at birth. At birth the glenoid fossa is a flat circular region measuring about 10 mm in diameter. The articular area increases in size mainly as a result of the growth of the two roots of the zygomatic process, the anterior root forming the greater part of the articular eminence, while the posterior root forms a large part of the concave region of the articular surface. This predominantly outward direction of growth on the part of the articular surface is correlated with the outward extension of the auditory canal. In foetal life the tympanic membrane lies on the under surface of the skull surrounded by a narrow ring-like tympanic bone. In the adult it is more deeply placed at the bottom of the auditory canal as the result of the lateral growth of the tympanic ring to form a bony tube.

The mandibular condyle

The articular surface of the head of the condyle measures about 20 to 25 mm in its mesiodistal diameter and about 10 mm in its greatest anteroposterior diameter. It is markedly convex from front to back and slightly convex from side to side. A lateral and medial pole delimit the larger mediolateral axis of the condyle (Fig. 6.22). The long axis of each condyle inclines somewhat backwards and medially so that a line drawn through the axis of one condyle would meet that of the other at an angle less than 180° facing forwards. The size of the condylar axis angle, however, varies considerably from one skull to another.

As seen from behind the condyle projects more inwards than outwards from the plane of the posterior border of the ramus. Below the line demarcating the articular surface at the back of the condyle, which is usually only faintly marked, is a triangular region for the attachment of the back part of the capsule and articular disk. The auriculotemporal nerve is closely related to this area of the bone (Fig. 6.32). In contact with this triangular area, separated from the bone by the capsular attachment, is the glenoid process of the parotid gland which passes inwards to a variable depth between the condyle and the tympanic plate. This region is also occupied by loose vascular connective tissue.

In front of and below the well-marked anterior articular margin of the condyle lies a triangular depression for the insertion of the lower fibres of the lateral pterygoid muscle (Fig. 5.15). The sharp bony margin in the sagittal plane connecting the coronoid and condylar processes, and forming the lower boundary of the mandibular notch, joins the anterior articular margin a short distance to the medial side of the lateral pole of the condyle. The two condylar poles usually lie at about the same horizontal level, but in some cases the medial pole is at a higher level. During foetal life and up to the time of birth, the lateral pole lies above the medial pole and the articular surface faces in a medial direction. This is associated with the oblique position of the glenoid fossa in the young skull.

By about 7 years of age the distance between the medial pole of one condyle and that of the other side has almost reached adult dimensions, but after this age each condyle continues to grow in an outwards direction in association with growth of the glenoid fossa.

Until about 20 years of age the articular surface of the condyle has a cap of modified hyaline cartilage covered by a fibrous perichondrium. During foetal life the condylar cartilage extends as a cone-shaped wedge of tissue into the ramus of the mandible. At birth it forms a large mass of cartilage limited to the head of the condyle (Fig. 4.23). It grows by proliferation of the deeper cells of the perichondrium and also by interstitial cell division. The deep surface of the cartilage is continually being replaced by bone. It is a *growth cartilage* corresponding to an epiphyseal plate in a long bone and is not a true articular cartilage as it is covered by a perichondrium. It is not constructed to bear heavy mechanical stresses. After about 20 to 25 years of age it is completely replaced by bone except for isolated islands of cartilage tissue. The perichondrium covering the articular surface becomes a periosteum overlying a layer of compact bone superficial to the cancellous bone which comprises the interior structure of the condylar head. During the period of full growth activity of the condylar cartilage the head of the condyle shows the following layers in section (Fig. 4.23).

1. A surface layer of collagenous tissue—the fibrous perichondrium.
2. A layer of reserve or quiescent cartilage cells—the cellular perichondrium.
3. A proliferative zone where cartilage cells are dividing rapidly and new cartilage matrix is being formed.
4. A zone of calcifying cartilage matrix in the process of being replaced by bone.
5. The trabecular bone of the condyle which has replaced the cartilage during growth.

At birth the head of the condyle is only a short distance above the level of the alveolar bone. In the adult, with growth of the mandibular ramus in vertical height it lies at a considerable but variable distance above the occlusal level of the teeth (Fig. 6.23).

The joint capsule
The attachment of the capsule to the articular area of the temporal bone is strong behind and at the sides, but weak in front. Laterally it is attached to the articular margin from the articular tubercle to the post-glenoid tubercle. Behind it is attached along the anterior margin of the squamotympanic fissure so that structures passing through the fissure lie outside the joint cavity. Medially it is attached to the temporal bone close to the suture between it and the great wing of the sphenoid, and in front to the anterior edge of the articular eminence.

The capsule is attached below around the articular margin at the head of the condyle. At the sides it is firmly attached to the lateral and medial poles, and at the back of the condyle to the triangular area below the articular margin. The capsule is strengthened by lateral and medial collateral ligaments. The *lateral ligament* is the strong fan-shaped temporomandibular band extending from the articular tubercle at the outer end of the articular eminence (anterior root of the zygomatic process) to the lateral pole of the condyle (Fig. 4.21). When the head of the condyle lies at the posterior concave part of the articular fossa (the rest position) the *lateral ligament* passes downwards and backwards and prevents any further backward displacement of the condyle. The ligament is also tightened when the condylar head reaches its extreme anterior position on the articular eminence, as when the mouth is wide open. The deeper fibres of the ligament are united with the outer fibres of the joint capsule.

The *medial capsular ligament* is less strong than the lateral. Above it is attached to the inner end of the articular eminence, which sometimes shows a small tubercle in this position, and below to the medial pole of the condyle.

These two capsular ligaments of each mandibular joint as well as limiting movements in the anteroposterior direction limit side to side movements of the jaw.

The *sphenomandibular ligament* (Fig. 4.20) is sometimes referred to as an accessory ligament of the mandibular joint. It extends from the spine of the sphenoid to the lingula of the mandible and develops from the perichondrium of Meckel's cartilage during foetal life (page 341). It is doubtful if it plays any important part in the mechanics of mandibular movement.

The *stylomandibular ligament* (Fig. 4.21) is likewise referred to as an accessory ligament of the joint. It is a condensation of deep fascia which extends from the styloid process of the temporal bone to the angle of the mandible between the attachments of the masseter and medial pterygoid muscles. It lies on the deep aspect of the parotid gland and is a thickened part of the gland capsule.

The articular disk (meniscus)

This important structure consists largely of dense connective fibres and fibroblasts with some elastic fibres in its posterior part. Although it may contain isolated groups of cells surrounded by a limited amount of cartilage-like matrix, it is not accurately described as fibrocartilaginous structure (Fig. 6.30). The number of chondroid cells increases with age.

Its central part is especially dense and usually shows a thickened anterior and posterior band and between these a thinner intermediate part (Fig. 6.31). At the sides the disk is firmly attached to the lateral and medial poles of the mandibular condyle along with the capsule

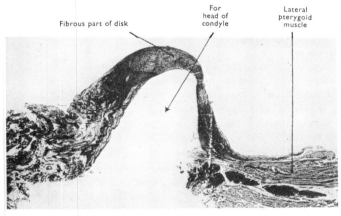

Fig. 6.30 Sagittal section through the intra-articular disk removed from the mandibular joint of an adult. The posterior or bilaminar aspect of the disk is to the left.

and the capsular collateral ligaments. Behind the disk becomes looser in texture and divides into an ascending temporal lamina attached with the joint capsule to the anterior margin of the squamotympanic fissure, and a descending mandibular lamina attached to the triangular area of the back of the condyle below the articular margin (Figs 4.22, 6.31). This bilaminar region of the disk contains a moderate amount of elastic tissue in its upper part which helps to draw the condyles and the disks backwards into their resting position in the articular fossa. It also contains blood vessels and sensory nerve endings of the freely ending variety in its lower part. These endings are one source of pain in abnormally functioning joints.

The bilaminar retrodiscal pad also contains Golgi tendon organs and encapsulated receptors which have important functions in maintaining normal joint positions through proprioceptive sensory pathways. Interference with this region during surgical procedures involving the joint hinders the patient's ability to sense positional changes and requires postoperative retraining of mandibular movements.

In front the disk also becomes looser in texture and divides into an ascending lamina attached to the anterior edge of the articular eminence and a descending lamina attached to the articular margin at the front of the head of the condyle. Between these ascending and descending lamina the upper fibres of the lateral pterygoid muscle gain attachment to the disk. The masseter and temporal muscles also gain a slight attachment to its outer anterior edge.

The joint cavity

The joint cavity consists of two compartments completely separated by the articular disk (Figs 4.23, 6.17, 6.31). The upper compartment, lying above the disk, is the larger as the attachment of the joint scale around the articular margin of the glenoid fossa is more extensive than its attachment around the articular surface of the head of the condyle.

In opening the mouth the disc moves forwards with the head of the condyle as this passes on to the articular eminence. The back of the capsule and the posterior superior lamina of the disk are stretched and with some of the retrocapsular fatty tissue are drawn forwards to occupy the back of the articular cavity as this is vacated by the head of the condyle (Fig. 6.31). In some people this is shown by dimpling of the overlying skin between the condylar head and the auricle.

The interior of each joint cavity is lined by a layer of endothelial tissue reflected from the capsule to the upper and lower surfaces of the disk, and over the fibrous tissue (perichondrium or periosteum depending on age) covering the articular surface of the condyle. The endothelium forms a synovial lining to the joint cavities which is most marked in the upper joint cavity, producing a synovial fringe at the posterior end of this compartment.

Mandibular movements

Two basic movements can take place at the temporomandibular joint:

a. A rotational or hinge movement when the head of the condyle pivots in the lower joint compartment relative to the articular disk.

b. A sliding or translational movement by which the articular disk glides along the undersurface of the temporal bone in the upper joint compartment. A combination of hinge and sliding movements is required when the mouth is opened widely. Hinge opening in the normal adult is in the range of 20 or 25 mm between the upper and lower anterior teeth. When combined with sliding movements the range of normal opening is increased to 35–45 mm. Defects in the capsular ligaments, which arise from either congenital or acquired defects, can result in hypermobility of the joint when frequent subluxations occur. When this happens the condyle of the mandible becomes displaced out of the mandibular fossa but can be replaced spontaneously. True dislocation of the temporomandibular joint results when the condyle of the mandible is displaced from the mandibular fossa in an anterior direction and cannot be repositioned to the normal resting state without the use of external forces. Inability to move the mandible within the normal range, or ankylosis, may result when fibrous or bony adhesions occur between the articular surfaces as a result of chronic inflammation, infection or trauma. Some conditions, such as inflammation of surrounding muscles or disorders of the nervous system, may also result in loss of joint movement or pseudo-ankylosis.

In the lower joint compartment the head of the condyle rotates beneath the under surface of the articular disk which is firmly attached to the poles of the condylar head. In the upper joint compartment the disk moves forwards with the head of the condyle on to the articular eminence to the extent permitted by the degree of stretching of the temporal attachment of the disk, the posterior part of the capsule, the collateral capsular ligaments and the posterior fibres of the temporal muscle. As the condyle rotates relative to the disk it lies successively in contact with the posterior thickened band, the thin intermediate zone, and finally the anterior band. As disk and condyle move on to the articular eminence the back of the upper compartment is occupied by soft tissues drawn in from the posterolateral aspect of the joint (Fig. 6.31).

When the mouth is fully opened the head of the condyle and the anterior band of the disk lie below the articular eminence. When the mouth is closed with the teeth in normal occlusion, the head of the condyle and the posterior band of the disc lie in the posterior concave portion of the glenoid cavity.

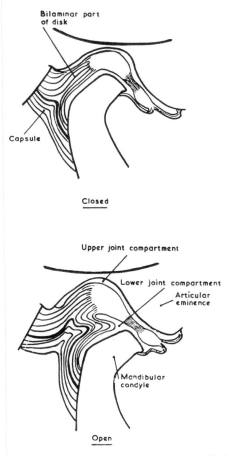

Fig. 6.31 Changes in the position of the mandibular condyle during mandibular movements. (After Rees.)

Structures related to the joint
These include:

Auriculotemporal nerve (Fig. 6.32). This passes on the outer side of the spine of the sphenoid lying close to the inner surface of the capsule of the joint. At the back of the joint it lies at a somewhat lower level in relation to the attachment of the capsule to the posterior surface of the condyle below the articular margin. As it passes outwards it usually enters the glenoid lobe of the parotid gland and here it gives off its secretomotor fibres to the gland (derived from the glossopharyngeal nerve via the otic ganglion). It then turns upwards over the posterior

lesser sup petrosal

root of the zygomatic process immediately behind and to the outer side of the post-glenoid tubercle. During its course it is closely related to the medial and posterior part of the joint capsule and gives off sensory branches to the capsule, the back of the articular disk and other joint structures.

The *chorda tympani nerve*. This emerges from the medial end of the squamotympanic fissure (Fig. 6.29). It passes obliquely downwards and forwards on the medial side of the spine of the sphenoid and the inner side of the joint capsule to join the lingual nerve deep to the lateral pterygoid muscle.

Auricular and tympanic branches of the first part of the maxillary artery enter the squamotympanic fissure behind the attachment of the joint capsule and give branches to the joint structures.

The *nerve and vessels to the masseter* lie in close relation to the anterior attachment of the

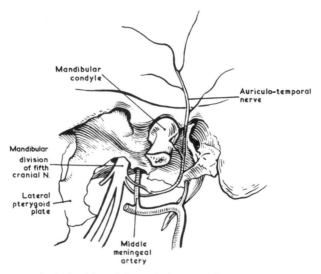

Fig. 6.32 The course and relationships of the auriculotemporal nerve.

articular disk to the front margin of the articular eminence above the fibres of the lateral pterygoid as these pass to their attachment to the front of the disk. The nerve and vessels reach the masseter muscle by passing below the zygomatic arch, through the mandibular notch and behind the posterior edge of the temporal muscle. They give branches to the front of the capsule and disk.

The *junction of the bony and cartilaginous parts of the auditory (Eustachian) tubes* lies about 5–8 mm to the inner side of, and somewhat anterior to, the joint capsule.

The *external auditory meatus* lies behind the glenoid cavity from which it is separated by the squamotympanic fissure and the tympanic plate (Fig. 4.41).

The *glenoid lobe* of the parotid gland is closely related to the back of the joint capsule between it and the tympanic plate of the temporal bone.

The *head of the condyle* is prevented from passing beyond the articular eminence into the temporal fossa by the collateral ligaments and the degree of shortening of the lateral pterygoid and relaxation of the posterior temporal fibres. Anterior dislocation sometimes occurs under anaesthesia due to a functional failure on the part of the proprioceptive reflexes which co-ordinate muscle activity. This is more common in older patients due to a flattening of the articular eminence.

The *roof of the glenoid fossa* is usually thin and the upper joint compartment is in close relationship to the middle cranial fossa and the temporal lobe of the brain.

The *maxillary artery* passes forwards between the neck of the mandible and the sphenomandibular ligament 10–15 mm below the medial pole of the condylar head. The middle meningeal branch enters the foramen spinosum some 5 mm in front of and medial to the upper attachment of the joint capsule. The foramen ovale through which the mandibular nerve emerges lies some 10–15 mm anteromedially to the articular surface. The pterygoid plexus of veins is in close relation to the joint capsule around the insertion of the lateral pterygoid muscle, and there is a rich venous plexus at the back of the joint capsule. The superficial temporal and transverse facial arteries are also closely related to the joint (Figs 4.14, 4.41).

Muscles. The upper head of the lateral pterygoid is continuous with the articular disk while the lower head is inserted into the pterygoid fossa of the neck of the condyle (Fig. 5.15). The posterior fibres of the deep portion of the masseter are closely related to the anterolateral aspect of the joint capsule. The posterior edge of the temporal muscle is closely related to the front of the capsule where it turns downwards at the anterior margin of the articular eminence deep to the masseter.

FUNCTIONAL ANATOMY

The oral cavity is the site of, or is closely associated with a number of important biological functions. These include the acceptance, preparation, storage, mastication, digestion and swallowing of food, respiration and speech. In many animals teeth are specialized for purposes other than mastication, such as offence and defence, prehension and grooming, excavating and even locomotion. The tongue is the site of the end organs for the special sense of taste. The lips and tongue have a predominant sensory and motor representation on the cerebral cortex and are among the earliest parts of the body to develop neural connections with the developing brain. Salivation is not only related to the mastication, digestion and swallowing of food but plays an important role in oral hygiene and the fluid balance of the body. The oropharyngeal lymphatic tissues play a part in the control of bacterial and virus infections.

These various functional activities involve many groups of muscles in complex correlated patterns of action, some requiring close conscious attention, others largely habitual or involuntary. The neural control system is situated mainly in the brain stem, involving the nuclei of the fifth, seventh and ninth cranial nerves which are related, moreover, not only to one another but also to closely associated centres concerned with the regulation of functions such as respiration, blood circulation control, body temperature, hearing, sight, taste and emotional behaviour.

Mastication

Mastication of the food is carried out between the incisive edges and occlusal surfaces of the upper and lower teeth and to a more limited extent by the action of the tongue against the hard palate. The movements of the mandible are a combination of protrusion and retrusion opening and closing and side to side movements. As the food is being masticated the secretions of the salivary glands are mixed with it and the masticated material is swallowed as a softened mass (bolus). The amount of mastication required by different foodstuffs varies to a considerable extent. Much of the food used by members of urban societies requires very little mastication and is swallowed almost as soon as it enters the mouth.

The incisor teeth are adapted for such purposes as tearing meat from bones, and cutting large or bulky objects into sections. In man the canines have no specialized function and act with the incisors. The cheek teeth (premolars and molars) are used for breaking hard objects and in disintegrating meat and vegetable material. The maximum crushing power is exerted by the teeth situated at the level of the key ridge (zygomatic process of the maxilla). In the adult these are the first and second permanent molars. The cusps of the premolar and molar teeth play an important role in the process of food disintegration.

The cusps of the upper and lower dental arches have a position of *normal occlusion* during which there is maximum contact between the various cusps and fossae. During mastication the teeth tend to return to this position and in doing so produce their maximum mechanical effect on the food. In primitive races who give their teeth hard wear owing to the nature of the diet, the cusps of the teeth become worn away. As a result the occlusal surfaces and incisive edges become flattened (*attrition*). The dentine is exposed until finally the occlusal surface of each tooth shows a central area of dentine surrounded by a marginal ring of enamel (Fig. 6.22). The enamel is first removed over the individual cusps and the degree of enamel wear is useful in estimating the age of skulls. As the permanent molars erupt at approximately 6, 12 and 18 years of age, the relative amount of cusp wear in these three teeth gives a fairly accurate estimation of age up to about the middle of adult life.

The tongue has a variety of functions in the masticatory process. With the co-operation of the muscles of the lips and cheeks it carries the food between the occlusal surfaces of the teeth and maintains the food in this position during mastication. It is used in the rejection of objects such as seeds, foreign bodies, fragments of bone and unpleasant tasting substances, and it moulds the masticated food mass (bolus) against the palate prior to swallowing. It also plays an important role in maintaining oral hygiene, to remove food debris from the gums, the vestibule and the floor of the mouth.

The muscles used in mastication can be divided into two groups:

Primary muscles of mastication. These are the masseters, the pterygoids, the temporal muscles and the digastric muscles. All these muscles except the posterior belly of the digastric (facial nerve) are supplied by the mandibular division of the fifth nerve.

The *accessory muscles of mastication.* These include the buccinators and lip muscles (supplied by the seventh cranial (facial) nerve), the mylohyoid muscle (supplied by the fifth nerve), the geniohyoid muscle (supplied by the cervical plexus), the stylohyoid muscles (supplied by the seventh nerve), the tongue muscles (supplied by the twelfth (hypoglossal) nerve), and the infrahyoid muscles (supplied by the cervical plexus).

Mandibular movements

The movements of the mandible in relation to the upper jaw can be classified as follows:

Protrusion. In the case of bilateral protrusion both mandibular condyles move forwards on the articular eminences and the teeth remain in close sliding contact. The prime movers are the lateral pterygoids, assisted by the medial pterygoid muscles. The posterior fibres of the temporalis muscles are the antagonists to the contraction of the lateral pterygoid muscles. The masseter muscles, medial pterygoids and the anterior fibres of the temporalis muscles maintain tonic contraction and prevent rotation movements of the mandible which would separate the teeth. The contraction of the lateral pterygoid muscles pulls the articular disks downwards and forwards onto the articular eminences. The posterior attachment of the disks to the back of the glenoid fossae limits the range of protrusive movement.

Retrusion. During the movement the heads of the condyles with their articular disks are carried back into the glenoid fossae by contraction of the posterior fibres of the temporalis muscles. The lateral pterygoids are the antagonist muscles and they relax. The remaining muscles of mastication maintain tonic contraction and keep the teeth in sliding contact. The elasticity of the posterior part of the articular disk and capsule of the temporomandibular joint keep the disk in its correct relationship to the head of the condyle as the condyle moves backwards.

Opening. The lateral pterygoids pull the condyles forward onto the articular eminences as in jaw protrusion. The posterior fibres of the temporalis muscles relax and this is followed quickly and smoothly by relaxation of the masseter muscles, the anterior fibres of the temporalis muscles, and the medial pterygoids. This permits the mandible to rotate around a horizontal axis, so that as the condyles move forward the mandibular angles move backwards. The chin is depressed and this is assisted by the digastric muscles which contract against a stable hyoid bone, held in place by the infrahyoid strap muscles. The axis around which the mandible rotates does not remain fixed during opening movements but moves downward and forward along a line passing from the mandibular condyle to the opening of the mandibular canal.

Closing. The prime movers are the masseters, temporalis, and medial pterygoid muscles. The jaws may close in a variety of positions, from closure in full protrusion to closure when the condyles are in their most posterior position in the glenoid fossae. Closure in protrusion requires contraction of the lateral pterygoids, assisted by the medial pterygoid muscles. The heads of the condyles remain in a forward position on the articular eminences. In retrusive closing the posterior fibres of the temporalis muscles act with the masseters to return the condyles to the glenoid fossae, so that the teeth meet in normal occlusion.

Side to side movements. In swinging the jaw from one side to the other to produce grinding movements between the occlusal surfaces of the premolars and molars, the condyle of the side towards which the mandible is moving is held in the rest position by the posterior fibres of the temporalis muscle and the tonic contraction of the other muscles of mastication on that side. On the other side the condyle and the articular disk are pulled forward onto the articular eminence by the contraction of the lateral pterygoid muscle, associated with the relaxation of the posterior fibres of the temporalis muscle. Swinging movements of the mandible from side to side are produced by alternating contraction and relaxation of the muscles of mastication which are involved in protrusive and retrusive movements.

In closure of the mouth the forces exerted by the muscles of mastication are transmitted chiefly through the teeth to the upper facial skeleton. The lateral pterygoid and the posterior fibres of the temporalis muscle tend to remove pressure from the head of the condyle when they contract, by depressing it slightly during hard clenching of the teeth. This is related to the fact that the axis of rotation of the mandible passes through the ramus somewhere in the vicinity of the opening into the mandibular canal. Controversy exists about whether or not the temporomandibular joint is a stress bearing joint. Recent work using photoelastic models studied by means of polarized light under various loading conditions suggests that the joint is directly involved in stress bearing.

As well as producing active movements the muscles of mastication have an important postural action in maintaining the position of the mandible against gravity. When the mandible is in the *resting position* the teeth are not in occlusion for there is a small gap between the upper and lower dental arches.

The primary muscles of mastication acting together in a pattern of movements determined by the nerve reflexes and adapted to the nature of the food undergoing mastication are responsible for the movement of the mandible relative to the fixed upper dental arch. The muscles of the cheeks (buccinators) and lips assist the tongue in keeping the food between the teeth. The mylohyoid diaphragm is the chief agent in elevating the tongue against the hard palate. The geniohyoid, stylohyoid and infrahyoid muscles fix the hyoid bone and enable the digastric and mylohyoid muscles to act to the greatest mechanical advantage.

The co-ordinated activity of all these muscles is regulated by a system of proprioceptive (muscle-joint) nerve reflexes. The sensory nerve endings are situated in the periodontal ligaments of the individual teeth; in the mucosa of the gums and palate; in the capsule and disk of the mandibular joint; and in the muscles themselves. The reflex arcs involve the sensory fibres of the mandibular and maxillary divisions of the fifth cranial nerve; fibre pathways in the brain stem; and a motor nerve outflow involving the upper cervical spinal nerves, the fifth, seventh and twelfth cranial nerves. Masticatory movements are controlled by mid-brain centres and can be produced in decerebrate animals by stimulation of the teeth or gums, but in intact animals mastication is also under the control of nerve tracts descending from higher centres in the cerebral cortex.

In relation to mastication there are certain important structural and functional adaptations of dentofacial anatomy.

Neuromuscular spindles
These have been demonstrated in many of the muscles of man, including those of the extremities and muscles which do not have typical antigravity functions, such as the diaphragm and levator ani. There is uncertainty about the presence of spindles in the muscles of mastication although their existence in the masseter, temporalis and medial pterygoid is established. Recent work has demonstrated a small number of spindles in the lateral pterygoid, where, as in the other masticatory muscles, they provide the nervous system with important information for normal masticatory function.

Opening and closing movements in chewing appear to depend on the interaction of proprioceptive nerve endings in the muscles which close the jaws with pressure receptors of the teeth and oral soft tissues. The stretch reflex mechanism in the law elevators initiates the sequence of nerve impulses which reverse the opening movement. Reflex opening is dependent on the influence of pressure receptors in the periodontal ligaments of the teeth and tactile receptors in the gums, palate and oral mucosa.

Tooth attachment
Each tooth is attached to its socket in the alveolar bone by its *periodontal ligament* or *membrane* (Figs 2.13, 6.33). This consists of bundles of collagenous fibres arranged so as to transmit the stresses of mastication from the tooth to the supporting bone. The greater part of the ligament is made up of oblique fibre bundles running from the compact bone lining the wall of the socket (lamina dura) to the cement covering the root of the tooth. The oblique fibres sling the tooth in its socket (or sockets in the case of a multirooted tooth). When pressure is exerted on the occlusal or incisive surfaces of the teeth these periodontal fibres are stretched and transmit the pressure to the lamina dura of the socket. The lamina dura is a thin layer

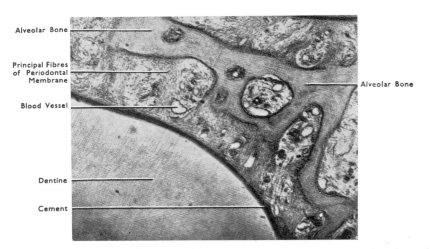

Fig. 6.33 Photomicrograph of the periodontal ligament seen in transverse section. Note the oval areas filled with loose tissue and containing blood vessels between the bundles of dense fibrous tissue. × 35.

of compact bone which lines the tooth socket. It is opaque to X-rays (see page 440) and there-fore shows as a continuous white line around the roots of healthy teeth. The lamina dura is interrupted if a dental abscess or dental cyst is present. With removal of the pressure there is a certain amount of recoil of the tooth brought about by a refilling of the blood vessels which run between the fibre bundles and by the return of the fibre bundles of the periodontal ligament to their resting position. The periodontal ligament also contains sensory nerve end-ings which are stimulated by the pressure acting on the teeth and which as well as regulating muscle action protect the teeth and their supporting structures from excessive strain. At the apex of each root where vessels and nerves enter the pulp cavity the periodontal ligament forms a protective buffer-like mass of more delicate connective tissues. As root formation continues for some time after tooth eruption the length of the roots of teeth may be determined to some extent by the functional activity of the dentition.

The gum is attached to the necks of the individual teeth. This attachment is of dual nature (Figs 6.19, 6.21).

The *epithelial attachment*. This comprises the union of the gum epithelium with the reduced enamel epithelium of the tooth crown. The latter is removed over the greater part of the crown surface with use of teeth after eruption but persists on that part of the crown which is embraced by the gum. The epithelial attachment plays an important part in the genesis of gingival disease.

The *subepithelial attachment*. Below the gum epithelium collagenous fibres pass from the cement of the tooth close to the enamel-cement junction, and from the alveolar bone, to the gum, forming a mucoperiosteum which is held firmly against the neck of the tooth.

Both the epithelial and fibrous attachments between the teeth and gums permit of local movements of the teeth in their sockets and between adjacent teeth.

In a complete dentition the individual teeth in each jaw are connected to the teeth in front and behind by strong transseptal fibres which traverse the interdental gum papillae (Fig. 6.19). Over the papillae the mesial and distal crown surfaces are in contact. The movements between adjacent teeth during mastication produces wear of the crown enamel at the contact areas which become larger with age. Their size indicates even more accurately than occlusal wear the degree of functional activity of the dentition. In dental arches where the teeth have been put to extensive use the wear of the approximal surfaces of the teeth (Fig. 6.22) may produce a considerable mesiodistal reduction in the length of the arch. The teeth, however, remain in contact as the transseptal fibrous ligaments are continually reconstructed so as to keep the individual units of the arch bound together.

Functional changes in the jaws

The supporting tissues of the teeth all respond to prolonged and adequate function: the perio-dontal fibrous bundles become thicker, the cortical bone of the lamina dura, to which the fibres are attached, becomes more dense, and cement deposition on the roots of the teeth increases. The surface epithelium of the gum keratinizes, the collagenous fibres in the gum mucoperiosteum increase in strength. The alveolar bone shows a better developed cancellous structure and increases in size; those parts of the facial skeleton (buttress system) which trans-mit the forces of mastication to the cranial base become better developed and stronger (Fig. 5.21).

The ability of the tissues to respond to function is, however, probably to some extent genetic-ally determined and varies from one individual to another so that the same amount of functional activity will not produce exactly the same response in any two individuals.

Salivation

There is a second reflex mechanism concerned in mastication which regulates the amount and quality of the salivary secretions. The sensory nerve endings are in the taste buds of the tongue and palate although the senses of smell and sight are also involved.

The secretomotor fibres to the various mucous and salivary glands come from different sources. These are (Fig. 6.34):

From the seventh nerve (superior salivary nucleus in the brain stem) via the great superficial petrosal nerve, and in the nerve of the pterygoid canal to the sphenopalatine ganglion (in the pterygopalatine fossa). From this parasympathetic ganglion post-ganglionic fibres are distributed along the palatine nerves to the mucous glands situated in the mucous membrane of the hard and soft palates, and to the lacrimal gland.

From the seventh cranial nerve (superior salivary nucleus) via the chorda tympani and lingual nerves to the submandibular parasympathetic ganglion. From this ganglion branches are distributed to the submandibular and sublingual salivary glands, the lingual glands (of Blandin and Nuhn) and the glands of the floor of the mouth.

From the ninth cranial nerve (inferior salivary nucleus) via the tympanic nerve, the tympanic

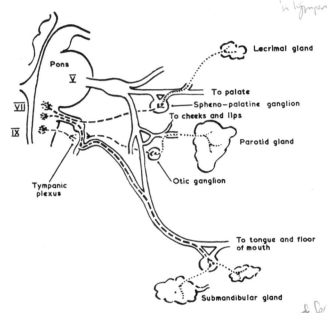

Fig. 6.34 Nerve supply to the salivary and lacrimal glands.

plexus of the middle ear, and the lesser superficial petrosal nerve to the otic ganglion. From this parasympathetic ganglion fibres are distributed along the auriculotemporal nerve to the parotid salivary gland and probably along the buccal and inferior dental nerves to the mucous glands of the cheeks and lips.

The secretion of the salivary glands depends upon and varies with the richness of the blood supply. This is regulated by the sympathetic nerves which reach the glands from the superior cervical sympathetic ganglion by branches in the form of nerve plexuses which run along the arteries including the external carotid, facial, lingual and branches of the maxillary artery.

The rate and type of secretion produced by the salivary glands is controlled by a combination of the influences of their sympathetic and parasympathetic nerve supplies.

In addition to the major salivary glands a series of minor glands is found in the submucosa of the lips, at the junction of the hard and soft palate, on the inner side of the cheek, in the retromolar triangle, the pillar of the tonsil and the undersurface of the tip of the tongue. These glands produce a thick secretion which is of limited significance except when their short ducts are injured and retention cysts may form. In contrast to this, the major salivary glands have long ducts which are near the surface and predispose them to obstructions and chronic inflammation of the glandular masses.

The mixed saliva derived from the three major salivary glands is a viscous colourless fluid with a slightly acid reaction. About $1\frac{1}{2}$ litres are secreted per day. Saliva moistens the food and facilitates swallowing. Through its enzyme ptyalin it aids in the digestion of starch. It subserves the sense of taste by acting as a solvent and assists in the rapid movements of the tongue, lips and cheeks by keeping their surfaces moist. Under certain conditions calcium carbonate and calcium phosphate are deposited on the teeth to form masses of salivary calculus.

The combined mechanical activity of mastication and the secretions of the oral glands play an important role in the maintenance of oral hygiene both in the cleansing of the teeth and the health of the gingival tissues. One of the results of the lack of mastication associated with the use of modern predigested foodstuffs is a greater tendency for food debris to adhere to the teeth and gums and this plays a part in the initiation of dental caries and inflammatory gum diseases (gingivitis).

Suckling

Between birth and the functional establishment of mastication with the eruption of the deciduous dentition the infant obtains its food by suckling either from the breast of its mother or from an artificial source. In breast feeding the lips, whose musculature is relatively well developed at birth, are applied to the breast around the nipple forming an efficient seal, the tongue lying below the nipple. Intermittent activity of lips and tongue draw the nipple into the oral cavity between the upper and lower gum pads at the same time exerting pressure on the adjacent breast. The capacity of the oral cavity is increased by movement of the mandible, depression of the tongue and dilation of the cheeks (buccinator muscles). As soon as the mouth is filled, the milk is swallowed and the cycle is continued. As in adult swallowing, movements of the soft palate and larynx prevent the milk from entering the nasal cavities or trachea.

Drinking from a spoon or cup is a modified form of suckling, the fluid placed in the mouth initiates the associated swallowing reflex. Food or fluid which is physically, thermally, tactilely or psychologically unacceptable is rejected from the mouth. This process can occur as the result of either reflex or voluntary action.

Swallowing

The swallowing of food is a complex co-ordinated activity of voluntary and involuntary movements in which a number of muscles are involved. It can be divided for descriptive purposes into three stages, although the act itself is continuous and rapid in its execution.

The first stage is the collection of the masticated food into a mass of cohering food particles (the bolus) on the dorsum of the tongue and the passage of the bolus from the mouth to the pharynx.

The second stage involves the passage of food through the pharynx to the beginning of the oesophagus and is associated with a temporary suspension of respiration, closure of the nasopharynx and the elevation of the larynx.

The third stage consists of the passage of the food along the oesophagus to the stomach.

At the end of mastication the food which has been broken up by the action of the teeth and moistened by the secretions of the salivary glands is gathered on the dorsum of the tongue, which becomes grooved to receive it, as a flattened mass of food material, the bolus. It is held against the under surface of the palate. The lips are closed, the teeth brought together in the position of occlusion and the act of swallowing commences. This is initiated by the contraction of the mylohyoid muscle diaphragm which elevates the tongue against the palate obliterating the bolus groove and drives the food backwards towards the pharynx.

Meanwhile the soft palate leaves its position of rest against the dorsum of the tongue and is elevated, thus opening the passage between the mouth and oropharynx and closing the passage between the nasopharynx and oropharynx. The capacity of the oral pharynx is increased by a drawing forwards of the hyoid bone and tongue and this produces a partial vacuum behind the oral cavity which assists in the rapid passage of food from the mouth through the pharynx. As the food descends through the pharynx it may pass on either side of, or above, the epiglottis acting as a 'lid' for the ascending larynx. As the larynx ascends towards the epiglottis which then helps to seal off the laryngeal orifice the muscles around its inlet contract and its opening is obliterated.

The oesophageal opening also ascends with the larynx, comes to lie at a higher level and opens to receive the descending bolus. The constrictor muscles of the pharynx contract from above downwards. The upper fibres of the superior constrictor and the palatopharyngeus muscles narrow the nasopharyngeal orifice and reduce the space required to be closed by the ascending soft palate and uvula.

A fuller analysis of swallowing can be considered in relation to the facial skeleton and the activity of individual muscles and muscle groups.

The mandible is elevated and the teeth brought into occlusion by the primary muscles of mastication. This enables the mandible to act as a fixed basis for the action of muscles such as the mylohyoid, superior constrictor, geniohyoid and tongue muscles. The hyoid bone is carried upwards and forwards. This movement is produced by the digastrics, the geniohyoids and the mylohyoid diaphragm, while the infrahyoid muscles relax. The tongue is drawn forwards by the genioglossus as the geniohyoids draw the hyoid bone forward. The thyroid and cricoid cartilages ascend so that the upper margin of the thyroid cartilage comes close to the ascending hyoid bone. This movement of laryngeal ascent is produced by the thyrohyoid, stylopharyngeus, palatopharyngeus and salpingopharyngeus muscles acting from the hyoid bone, the base of the skull, the back of the hard palate and the auditory tubes.

In swallowing the orifices of the auditory tubes are usually opened. During the ascent of the laryngeal cartilages the ligaments between them and between the rings of the upper part of the trachea are stretched by virtue of their elastic tissue. At the end of swallowing the recoil of this elastic tissue returns the larynx to its resting position. In the ascent of the soft palate the palatal aponeurosis is rendered tense by the tensor palati muscles and the elevation of the soft palate is carried out by the levator palati muscles.

The aryepiglottic and interarytenoid muscles close the opening to the larynx as this ascends. Contraction of the pharyngeal sphincters from above downwards appears to be a variable phenomenon. It is most marked if there is any obstruction or difficulty in swallowing. In some children the closing of the lips may be incomplete and the forward thrusting of the tongue which occurs in the early stages of swallowing may be exaggerated so as to exert abnormal pressure on the upper incisor teeth which may lead to malocclusion.

Neuromuscular co-ordination during swallowing involves the cricopharyngeus muscle. If this sphincter-like muscle (page 193) fails to relax in front of the wave of perstalsis which passes down the pharynx, the pressure exerted by muscular contraction around the bolus of food above the muscle affects the weakest part of the pharyngeal wall and an outpouching, or pharyngeal diverticulum can result. The weakest part of the pharyngeal wall is called the dehiscence of Killian, from the Latin word meaning 'gaping'.

The muscle groups involved in swallowing and their nerve supply may be summarized as follows:

The *muscles of mastication* fix the position of the mandible. Nerve supply: fifth cranial nerve.

The *mylohyoid muscle* drives the tongue against the hard palate and initiates the first phase. Nerve supply: fifth cranial nerve.

The *tongue muscles* form the bolus groove and draw the tongue forwards so as to increase the antero-posterior diameter of the pharynx. Nerve supply: twelfth cranial nerve (except the palatoglossus).

The *muscles of the lips* (close the mouth in front and probably help in maintaining the oropharyngeal vacuum. Nerve supply: seventh cranial nerve. Swallowing can occur with the lips apart but it is less smooth in its action and is often accompanied by an excessive forward thrusting of the tongue.

The *hyoid muscles*, geniohyoid, stylohyoid, and infrahyoid group. Nerve supply: the cervical plexus and seventh cranial nerve (stylohyoid).

The *elevating muscles of the larynx*, the stylopharyngeus, salpingopharyngeus and palatopharyngeus. Nerve supply: the ninth and eleventh cranial nerves.

The *elevating muscles of the soft palate*, the tensor and levator palati. Nerve supply: fifth and eleventh cranial nerves (via the vagus).

The *muscles closing the laryngeal inlet*. Nerve supply: the eleventh cranial nerve (fibres passing via the recurrent laryngeal branch of the vagus).

The *pharyngeal muscles*. In the initial stages these relax to open up the oropharynx. Later they may contract from above downwards. Nerve supply: eleventh cranial nerve (fibres passing via vagus to the pharyngeal plexus).

The mucous membrane of the mouth, pharynx and oesophagus, across which the food passes in swallowing, is supplied by sensory fibres of the fifth, ninth and tenth cranial nerves. The act of swallowing is the result of the co-ordinated activity of a number of reflex arcs involving the fifth, seventh, ninth, tenth, eleventh and twelfth cranial nerves, the upper spinal nerves, and mid-brain centres. The mid-brain centres regulate and co-ordinate the reflex mechanisms so that once swallowing commences it cannot readily be suspended by any voluntary action. Emotional disturbances, however, may interfere with the smooth working of the reflex mechanisms so that food may pass into the larynx, provoking a violent cough reflex, or into the nasopharynx, or an impression of pharyngeal obstruction may be experienced. This produces strong contraction of the pharyngeal sphincters. Difficulty in swallowing, or dysphagia, is a common complaint.

Food may be arrested in the upper part of the oesophagus and its removal may require the initiation of a number of supplementary swallowing acts. Once the food reaches the middle of the oesophagus, where smooth muscle takes over from striated voluntary muscle, swallowing is no longer a conscious process and the propulsion of food is by peristaltic waves. One is not aware of the entrance of the food into the stomach.

The anatomy of speech

The ability to speak depends on the development and proper function of certain areas in the cerebral cortex (page 403) and the use of certain muscle mechanisms in the larynx, pharynx and mouth. These mechanisms regulate the column of air which is expired from the lungs and air passages by the recoil of the thoracic cage and the ascent of the diaphragm.

Sounds are produced in the larynx chiefly by the alteration in tension and bulk of the vocal cords and to a less important degree by the vestibular folds and the sphincter mechanism of the laryngeal inlet. The determination of pitch is effected by the vibrations of the margins of the vocal cords in relation to variations in pressure of the escaping air current. The ability of the vocal cords to vibrate depends on their degree of tension and the thinness of their free margins. The tension and bulk of the cords are varied by the laryngeal muscles (page 200). The male larynx during puberty undergoes a considerable degree of enlargement resulting in a reduction in the pitch of the voice (breaking of the voice). The loudness of the voice depends on the amount of air passing through the larynx and on the resonator mechanism. The sounds produced in the larynx are feeble and poor in quality and their amplification depends on the resonator mechanism which consists of three parts:

1. The column of air in the air passages below the larynx.
2. The air in the pharynx.
3. The air in the mouth, nasal cavities and air sinuses.

During speech, the column of air is set into vibration by action of the vocal cords and the acoustic or resonant properties of the vocal tract modify the characteristics of the resultant speech sounds. Localized constriction of the vocal tract is used to produce hiss-like, or fricative sounds ('s' or 'sh') and momentary blocking of the air flow by the tongue against the teeth or palate, or closure of the lips, will result in plosive sounds, such as 'p' and 'g'. Some sounds can be produced with or without vocal cord vibration, hence 'voiced' and 'unvoiced' consonants. The tension and length of the vocal cords is continually altered during speech and results in alterations of frequency. The range of vocal cord frequencies used in normal speech is from 60 to 350 cycles per second, or more than two octaves.

Vowel sounds are produced by the larynx and amplified without obstruction in the oral or nasal cavities. Vowel sounds passing through the nasal cavities receive a nasal quality. Consonants are produced by the interruption of the air current above the larynx. This interruption may be produced by the soft palate, tongue, teeth or lips; producing various guttural, facial or dental sounds.

The muscle groups used in speech production are:

Diaphragm and intercostal muscles; nerve supply: cervical plexus (phrenic) nerve and thoracic spinal nerves.

Laryngeal muscles; nerve supply: eleventh cranial nerve through recurrent branch of vagus and external laryngeal nerve (to cricothyroid muscle).

Pharyngeal muscles; nerve supply: cranial nerve through pharyngeal plexus.

Muscles of the soft palate; nerve supply: eleventh and fifth cranial nerves.

Muscles of the tongue; nerve supply: twelfth cranial nerve.

Muscles of the cheeks and lips; nerve supply: seventh cranial nerve.

Muscles of mastication (to move the mandible and regulate the mylohyoid diaphragm); nerve supply; fifth cranial nerve.

Abnormal speech

In cleft palate speech is distorted by the passage of all sounds into the nasal cavity. If the larynx is removed, the upper end of the oesophagus can act in place of the vocal cords and the air column is provided by swallowing air and returning it through the oesophageal sphincter. Speech is also interfered with by enlarged tonsils, adenoids, engorgement of the nasal mucous membrane, gross irregularity of the teeth and after the loss or removal of the teeth, especially the incisors. A great deal can be done in cleft palate cases by speech training in early childhood.

Correct denture design is extremely important in maintaining the quality of a person's speech when extracted teeth are being replaced. The position of the teeth relative to the alveolar bone and the shape and thickness of the denture base can affect the person's ability to form certain sounds. The letter 's' can be especially troublesome and badly constructed dentures may cause the person to lisp or hiss when pronouncing certain words.

The balance of the head

In man the head is balanced on the upper end of the vertebral column in such a manner that the eyes look straight forwards and the hard palate is horizontal in position. The position of balance is maintained largely by the postural activity of the postvertebral muscles (page 176) through proprioceptive reflexes arising from skin of the neck and from the muscles themselves (tonic neck reflexes); from the inner ear apparatus (labyrinthine reflexes); and from the eyes (visual reflexes). It is against this postural background of head balance that the other movements of the head and neck such as eye movements, mastication, swallowing and speech take place.

Conditions such as spasticity, paralysis and abnormal habits involving muscle groups which maintain the postural balance of the head can result in abnormal actions of the facial, masticatory, pharyngeal and laryngeal muscles. These abnormal actions may give rise to a number of symptoms which have been termed collectively the cervicofacial syndrome.

The form of the dental arches

The dental arches lie between groups of muscles, the tongue on the inner side and the lips and cheeks on their outer side. It is commonly believed that the balance of these muscular forces plays a part in determining the form of the dental arches. This view holds that the position of the teeth and alveolar supporting bone is secondary to the morphology of the neighbouring soft tissues and that the width of the dental arches is largely dependent on the width of the tongue. However, studies of the developing jaws and teeth in foetal life show that the form of the dental arches is established in response to the form of the dental lamina, the jawbones and related parts of the cartilaginous facial skeleton.

In normal functional activity of the orofacial musculature the tongue has adequate space within the dental arches and readily adapts itself in the resting position to the structures which surround it. The balance between the forces created by the tongue on the one hand and the cheeks and lips on the other has potential effect on arch form during mastication, swallowing and speech. Abnormal patterns of swallowing can modify the shape of the dental arches and it should be remembered that swallowing occurs continuously in order to empty the mouth cavity of accumulating salivary gland secretions. During the act of swallowing the teeth come into contact, the lips are closed and the tip of the tongue exerts some pressure on the anterior part of the hard palate behind the upper incisors. The buccinator and mylohyoid muscles contract and the dorsum of the tongue is forced against the hard palate and the muscular soft palate, which has been tightened by the tensor palati muscle. The sides of the tongue may exert some pressure on the lingual surfaces of the premolar and molar teeth. Less is known about the forces applied to the teeth by the lips and cheeks during swallowing and the precise quantitative differences between these forces, but it appears that the tongue has a more pronounced effect than the lips and cheeks. Abnormal swallowing habits which involve thrusting of the tongue against the lingual aspect of the upper incisors, in combination with a relaxed upper lip and a lower lip placed between the upper and lower teeth, can play havoc with the normal relationship of the upper and lower incisors.

7. Developmental Anatomy

EARLY DEVELOPMENT OF THE EMBRYO AND FOETUS

At conception the fertilization of the *ovum* by a *spermatozoon* takes place in the outer part [*lateral*] (abdominal end) of the Fallopian (uterine) tube (Fig. 3.53). Almost immediately the ovum commences to divide until the original single cell becomes a multicellular spherical organism (the morula). In the midst of this cell mass a fluid filled cavity appears, the *blastocyst cavity*, and the early embryo has now entered the *blastocele stage* of its development (Fig. 7.1). A

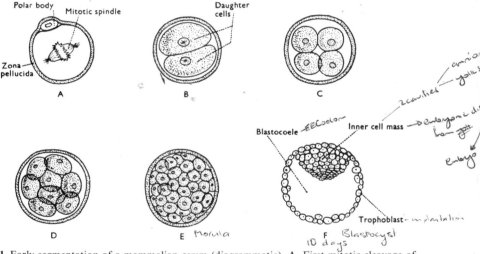

Fig. 7.1 Early segmentation of a mammalian ovum (diagrammatic). A. First mitotic cleavage of fertilized ovum. B, C, D. Stages of division into two, four and eight cells, the cells becoming progressively smaller. E. The morula stage. F. Blastocyst stage (in section). (By courtesy of Professor G. A. G. Mitchell and Dr E. L. Patterson.)

mass of cells projects into one aspect of the cavity forming the *inner cell mass*, while the peripheral cells bounding the growing spherical blastocyst form the *trophoblast layer*. At about this stage the blastocyst enters the uterine cavity and becomes implanted in its mucous membrane about eleven days after fertilization. Implantation is the primary function of the trophoblast cells. Later they form the outermost part of the foetal membrane, the chorion, which surrounds and protects the growing foetus, and also the placenta in which the maternal and foetal circulations come into intimate relationship but not actual continuity so as to permit the exchange of food, hormones, oxygen and waste products. The *embryo* develops from the inner cell mass. Among these cells two cavities appear:

1. A cavity proximal to the place of attachment of the inner cell mass to the trophoblastic cell layer, the *amniotic cavity*.

2. The second cavity is distal in position and forms the *yolk sac* (Fig. 7.2).

The blastocele cavity becomes the *extra-embryonic coelom*. Between the two embryonic cavities there is a disk of cells; those of the upper layer form the floor of the amniotic cavity, while the lower layer forms the roof of the yolk sac cavity. The bilaminar disk is the *embryonic disk* or *shield* from which the embryo itself develops. With further development the yolk sac becomes relatively reduced in size and the amniotic cavity enlarges so as to surround completely the developing embryo and foetus (Fig. 7.3). After the second month the embryo becomes known as the *foetus*. The amniotic cavity contains the amniotic fluid in which the

| Trophoblast | Chorionic villus | Yolk sac | Embryonic shield | Amniotic cavity | Extra-embryonic coelom |

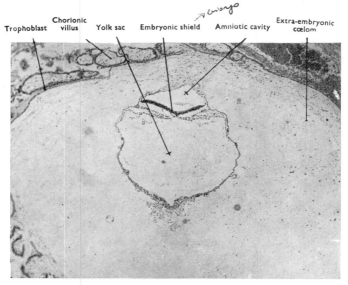

Fig. 7.2 Transverse section through an eleven-day human embryo (the Manchester embryo) and its membranes.

foetus floats in a protective fluid bath. When the mouth and oral orifices develop and the alimentary canal is established the foetus probably learns to swallow by drinking amniotic fluid.

The embryonic disk is at first a bilaminar structure (second week of development), the upper amniotic layer of cells being called the *ectoderm*, the lower, yolk sac layer, the *endoderm*. Along the mid-line axis of the disk, the ectoderm is thickened to form from behind forwards; the primitive streak, the notochord or head process and the prochordal plate (Fig. 7.4). From the sides of these mid-line structures a third layer of cellular tissue develops between the ectoderm and endoderm; this is the *mesoderm* (Fig. 7.3). The original two-layered structure now consists of three primary germ layers—the trilaminar stage (third-week).

From the ectoderm layer will develop the skin, the central nervous system and the peripheral nerves, the lining of the primitive mouth cavity and anal canal, and the enamel of the teeth. The endoderm layer will develop into the lining epithelium of the alimentary canal between mouth and anal canal, the liver and pancreas and the epithelium lining of the lungs. The heart and circulatory system, the bones and muscles, the dentine and cement of the teeth form from mesoderm.

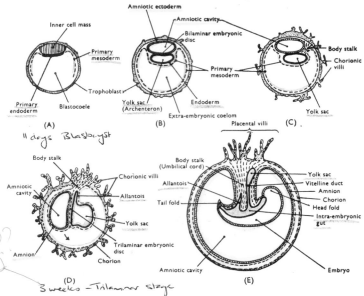

11 days Blastocyst (handwritten)

? 3d week ?? *? 3d week = 14th day* (handwritten)

3 weeks – Trilaminar stage (handwritten)

Fig. 7.3 Formation of the embryo and its membranes (diagrammatic). A. Blastocyst stage. B. Appearance of amniotic cavity and primary yolk sac and formation of bilaminar embryonic disk (disc): the body stalk and primary chorionic villi are also present. C and D. Enlargement of the amniotic cavity and rotation of the embryonic disk hinged at the body stalk; in D, intra-embryonic mesoderm has appeared forming a trilaminar disk. E. The amnion now lines the chorion and the embryo is suspended within it by the umbilical cord: the appearance of head, tail and lateral folds in the embryonic disk has divided the primary yolk sac into intra-embryonic and extra-embryonic parts. (By courtesy of Professor G. A. G. Mitchell and Dr E. L. Patterson.)

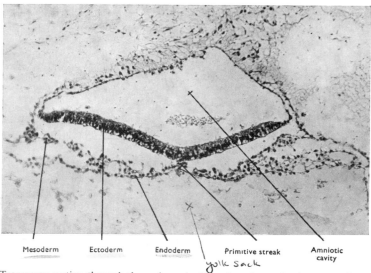

yolk Sack (handwritten)

Fig. 7.4 Transverse section through the embryonic disk of an eleven-day human embryo showing the formation of the three germ layers.

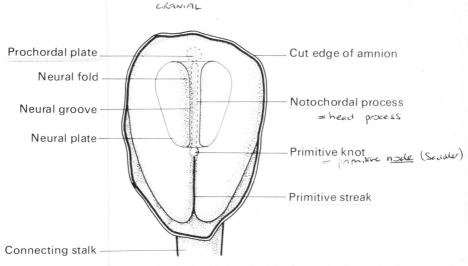

Fig. 7.5 Dorsal view of the embryonic disk in the third week of development. The amnion has been removed. Compare with Fig. 7.6.

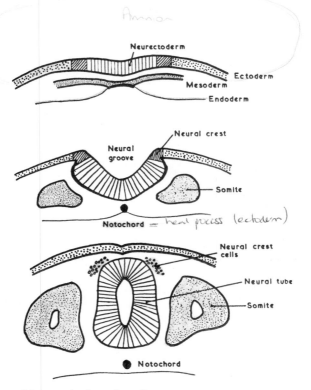

Fig. 7.6 The formation of the neural tube and somites.

After the formation of the primitive streak and notochord and the commencement of meso-derm formation, the ectoderm on either side of the mid-line forms an elongated anteroposterior fold, the neural groove (Fig. 7.6). This later becomes a tube, extending from the front end to the hind end of the disk. The neural tube differentiates into the brain and spinal cord and from its sides the cerebral and spinal nerves grow out into the adjacent tissues. The cavity of the tube becomes the ventricular system of the brain (page 407) and the central canal of the spinal cord.

The mesoderm immediately adjacent to the primitive streak and notochord first forms a

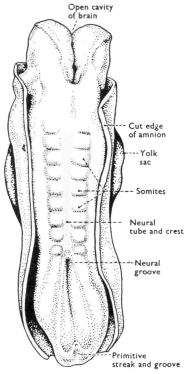

Fig. 7.7 Dorsal aspect of a human embryo of 2·1 mm (fourth week) with nine somites (after Eternod). The neural folds have united except at the cranial and caudal ends. (By courtesy of Professor G. A. G. Mitchell and Dr E. L. Patterson.)

compact elongated mass of tissue, the *paraxial* mesoderm (Fig. 7.6), which later becomes broken up into a series of segmental blocks, *somites*, extending from the head of the embryo to the hind end (Fig. 7.7). The formation of the paired cube-like somites begins about the twentieth day of development, the first pair appearing a short distance caudal to the cranial end of the notochord. Eventually, 42–44 pairs of somites develop and they show as distinct surface elevations on either side of the neural tube, which is closed opposite the somites but is open at either end of the embryo. These openings are the cranial (or rostral) and caudal neuropores. The ventromedial part of each somite forms the *sclerotome*. It lies immediately adjacent to the notochord and the neural tube and the elements of the vertebral column develop from it. Each sclerotome consists of a dense caudal part and a more loosely arranged cranial portion. The densely packed cells of one sclerotome fuse with the loosely packed cells of the one immediately caudal to it, to form the centrum of a vertebra. Thus each vertebra develops from two adjacent somites and is an intersegmental structure. The *intersegmental arteries* which arise from the developing aorta and are eventually destined to become the intercostal arteries thus come to lie opposite the middle of the vertebral bodies. By the same token, each *segmental nerve* which grows out from the neural tube into relation with a somite comes to lie between adjacent vertebrae in close relationship to the intervertebral disks. At the same time the sclerotome cells migrate dorsally to envelop the neural tube and form the neural arches of the vertebrae. Another stream of cells passes ventrally into the body wall to give rise to the costal processes and the ribs in the thoracic region.

After the sclerotome cells migrate medially from the somite the remainder of the somite is called the *dermomyotome*. The most lateral cells of this group, the *dermatome cells*, give rise to the dermis of the skin. From the medial part of each dermomyotome develop the muscles of the back and, by a process of migration, the muscles of the body wall (intercostal and abdominal muscles and certain muscles of the neck), of the limbs, of the orbital cavity, and the tongue. The motor roots of the growing segmental nerves made contact with the developing muscle cells, the *myoblasts*, at an early stage so that when the muscles migrate they carry their nerves with them. *Occipital myotomes carry XII ellers to tongue*

~sensory   *Dermatomes*, which are also segmental, are defined as areas of skin each supplied by a single spinal nerve and its dorsal root ganglion. Peripheral nerves grow into the limb buds during the fifth week of development and are distributed to segmental bands of skin on the dorsal and ventral surfaces of the developing extremities. As the limbs elongate, the cutaneous distribution of the nerves migrate along the limbs in an orderly sequence of distribution which carries through to the adult (page 391). The limb buds first appear as small elevations on the side of the body wall opposite the cervical, lumbar and upper sacral segments. Because the limbs, trunk and neck region undergo different rates of growth the limbs descend as they elongate and carry their nerve supply with them. This produces the oblique orientation of *?ventral* the brachial and lumbosacral plexuses in the adult.

Lateral to the somites the mesodermal layer forms an elongated mass of tissue, the *intermediate cell mass* (Fig. 7.8), from which develops in turn the primary (pronephros), secondary (mesonephros), and definitive (metanephros) kidneys and their ducts. The mesonephric duct persists in the male as the *vas deferens* and *ejaculatory duct*. The ureter and pelvis of the metanephric kidney develops as an outgrowth from the lower part of the mesonephric duct, close to its entrance into the cloaca (terminal part of the hind gut). This outgrowth, the *ureteric*

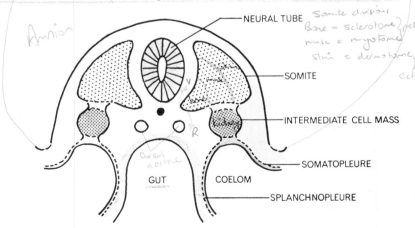

Fig. 7.8 Diagrammatic representation of a cross section through the body of a human embryo about the twenty-first day of development.

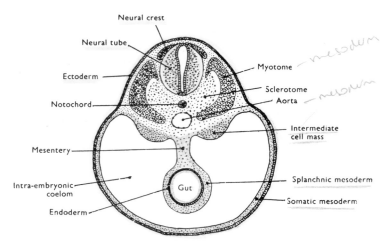

Fig. 7.9 Schematic cross section through the body of a human embryo of about four weeks, showing the arrangement of the mesoderm. (By courtesy of Professor G. A. G. Mitchell and Dr E. L. Patterson.)

bud, grows in a cranial and dorsal direction to penetrate the *metanephric blastema*, which is a solid mass of cells derived from the intermediate mesoderm. The blastema gives rise to the excretory units, or *nephrotomes* (the urinary tubules of the adult kidney) while the ureteric bud is concerned with the formation of the collecting tubules and excretory duct system of the kidney. The kidney, therefore, is in part a mesodermal structure and in part an endodermal structure.

Lateral to the intermediate cell mass the mesodermal tissue divides into a layer which remains in contact with the ectoderm (the *somatopleure*) and a layer which remains in contact with endoderm (the *splanchnopleure*) (Fig. 7.9). Between them a cavity appears extending from the future neck region to the future pelvic region. This is the *coelomic cavity* or intra-embryonic coelom from which develop the pleural, pericardial and peritoneal cavities of the adult. In the splanchnopleuric mesoderm, which covers the endoderm, there will develop the smooth muscle of the alimentary and respiratory systems.

The embryonic disk during its later development changes its form from that of a flattened disk to that of an inverted bowl. The growing edges, especially at the head and tail end, but also at the sides, commence to grow inwards towards one another on the attached aspect. As a result the upper part of the yolk sac becomes more and more enclosed within the developing embryo and the communicating stalk between the intra-embryonic and extra-embryonic parts of the sac becomes progressively narrowed. From the intra-embryonic part of the yolk sac develops the greater part of the gut or alimentary canal.

At the anterior end of the embryonic disk, in front of the termination of the notochord, the ectoderm and endoderm layers unite to form the *prochordal plate* (buccopharyngeal membrane) and in front of this develops the cardiac area where there is continuity of mesoderm across the middle line. With inturning of the anterior end of the embryo these regions come to lie beneath the developing brain (Fig. 7.14).

DEVELOPMENT OF THE HEART AND THE CIRCULATORY SYSTEM

Early development of blood vessels

The first indication of the developing vascular system is the appearance of cords of endothelial cells, which develop a lumen and run together to form a plexiform network. In man this process

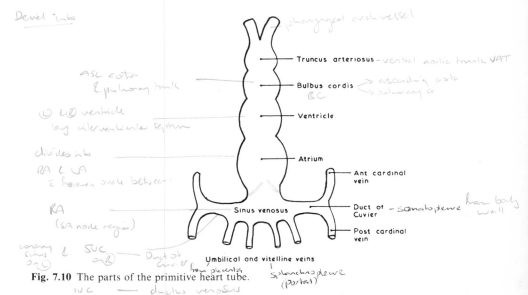

commences at about the same time in the chorionic membrane, yolk sac wall, and in the intra-embryonic mesoderm. From the union of these vascular networks develops the foetal circulation, partly within the foetus and partly in the umbilical cord and placenta.

The *vitelline veins* first develop in the wall of the yolk sac and become continuous with vessels which develop in the splanchnopleuric mesoderm surrounding the developing alimentary canal. They are joined in the septum transversum, within which the liver develops, by the veins bringing blood from the placenta via the umbilical cord (*umbilical veins*) to form the right and left *sinus venosus*. Each sinus venosus is also joined by the corresponding *duct of Cuvier* draining the veins of the body wall (somatopleure) (Fig. 7.10). The left and right

Fig. 7.10 The parts of the primitive heart tube.

(Figure labels: pharyngeal arch vessel; Truncus arteriosus; Bulbus cordis; Ventricle; Atrium; Ant cardinal vein; Duct of Cuvier; Post cardinal vein; Sinus venosus; Umbilical and vitelline veins)

(Handwritten annotations: Devel ints; asc aota & pulmonary trunk; LV ventricle by interventricular septum; divides into RA & LA; i foramen ovale between; RA (SA node region); coronary sinus & SVC; Duct of Cuv; Truncus arteriosus – ventral aortic trunk VAT; Bulbus cordis BC → ascending aorta → pulmonary a; Duct of Cuvier – somatopleure / body wall; IVC — ductus venosus; from placenta; splanchnopleure (portal); Fig 7.8)

sinus venosus undergo a partial fusion to form a common sinus venosus cavity, which communicates through a valvular opening with the next chamber of the developing heart, the atrium. The atrium in turn leads to the ventricle via the atrioventricular canal. The ventricle opens into the bulbus cordis leading to the ventral aortic trunk.

From the aortic trunk a series of pharyngeal arch vessels pass in the developing visceral arches of the neck to the dorsal aorta which runs towards the hind end of embryo beneath the notochord and above (behind) the developing alimentary canal. In the lower thoracic region the two dorsal aortae unite to form a mid-line vessel. The dorsal aorta in the abdomen divides into left and right iliac branches, each of which in turn divides into the external iliac artery (to the hind limb) and the internal iliac artery to the pelvis. The main branch of the internal iliac artery during foetal life is the umbilical artery, which returns the 'venous' blood of the foetus to the placenta, where it gives up its waste products, becomes oxygenated, and is returned to the foetus along with the umbilical vein (Fig. 7.11).

Development of the heart

The heart commences to develop in the mesodermal tissue at the anterior end of the embryonic disk in front of the prochordal plate and comes to lie beneath the foregut with the folding of the anterior end of the growing embryo. Within this mesoderm two endothelial tubes are formed which soon unite to produce a single heart tube, subdivided as described above into sinus venosus, atrial and ventricular compartments. The endothelial heart tube early becomes ensheathed by a mantle of mesodermal myoblasts from which the cardiac muscle develops

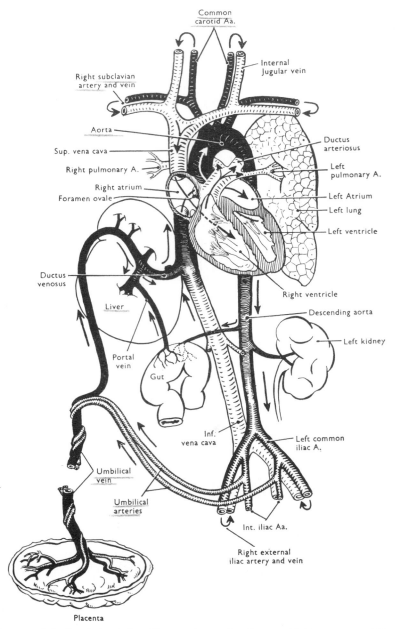

Fig. 7.11 Diagram of the foetal circulation. The vessels carrying the most highly oxygenated blood are shown in black. (By courtesy of Professor G. A. G. Mitchell and Dr E. L. Patterson.)

Cardiac musc doesn't beh as syncytium but a syncytial like struct.

DEVELOPMENTAL ANATOMY   335

(myocardium) and the first heart beats occur soon after these cells have formed a syncytium.

With later development the following changes occur:

1. Owing to alterations in the circulation of the venous blood (vitelline and umbilical veins) through the liver and the development within that organ of the *ductus venosus*, the right sinus venosus becomes larger than the left. The ductus venosus conducts a great deal of the blood from the left umbilical vein directly to the heart, thus bypassing the liver venous system (Fig. 7.11).

2. The single atrial cavity becomes divided into a right and left chamber by the appearance of two incomplete septa (*septum primum and secundum*), leaving an opening, the foramen ovale, between the one atrium and the other. The common sinus venosus chamber opens into the right atrium, the developing pulmonary veins into the left atrium (Fig. 7.12).

3. The sinus venosus becomes taken up into the right atrial cavity. The inferior vena cava of adult anatomy represents the terminal part of the ductus venosus; the superior vena cava, below the entrance of the vena azygos, represents the right duct of Cuvier, and the terminal part of the coronary sinus represents the left duct of Cuvier, which becomes cut off from the

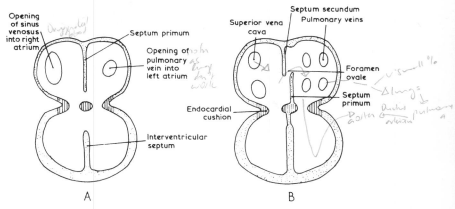

Fig. 7.12 Diagram illustrating the earlier (A) and later (B) stages of development of the heart chambers viewed from the anterior aspect.

body wall veins (left hemiazygous venous system) and limited to the coronary sinus. The left body wall veins make secondary connections with the vena azygos on the right side of the body to enter the heart via the superior vena cava (right duct of Cuvier). The left brachiocephalic vein crosses the middle line to join the superior vena cava.

The sinus venosus gives rise to the smooth posterior wall of the right atrium of the adult, in contrast to the ridged anterior wall of the chamber derived from the atrial portion of the primitive heart tube. In the same way the pulmonary veins are included in the posterior wall of the left atrium until normally four veins open into the cavity of the left atrium.

4. The ventricular chamber becomes divided into left and right ventricles by an interventricular septum which grows up from the apex of the heart and the bulbus cordis is divided by a spiral aortico pulmonary septum into an aortic trunk (ascending aorta) and a pulmonary trunk (pulmonary artery). The terminal ends of the two trunks communicate at the ductus arteriosus. The ductus arteriosus short circuits blood from the pulmonary trunk to the aortic arch because the lungs are not functional in foetal life (Fig. 7.11). The atrioventricular valves develop from endocardial cushions in the atrioventricular canal, which is divided into a right channel connecting the right atrium and right ventricle and a left channel connecting the left atrium and the left ventricle (Fig. 7.12).

Changes in the pharyngeal (aortic) arch arteries (Fig. 7.13)

In man a series of pharyngeal arch arteries arise from the front (cephalic) end of the ventral aortic trunk (aortic sac) and join the bilateral dorsal aortic vessels in the dorsal aspect of the neck. The following changes occur in the foetal arch system to give rise to the condition found in the adult:

1. The first and second arch arteries almost completely disappear. They may, however, contribute to the formation of the maxillary artery in the first arch region and the stapedial artery, a small vessel running through the footplate of the stapes, in the second arch region.

2. The segment of the dorsal aorta between the third and fourth arch arteries disappears on the left side. On the right side the whole of the dorsal aorta from the third arch artery to the place of union of the left and right dorsal aortic channel disappears.

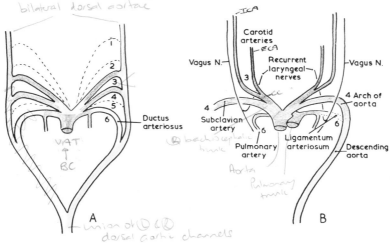

Fig. 7.13 The fate of the aortic arches.

3. From the third arch artery develops the common carotid and the commencement of the internal carotid artery. The dorsal aorta above (cephalic to) the third arch artery forms the remainder of the internal carotid artery.

4. The external carotid artery develops as a new structure.

5. The fourth arch artery on the left side forms the arch of the aorta. The ascending aorta develops from the bulbus cordis and ventral aortic trunk; the descending aorta from the left dorsal aorta beyond (caudal to) the fourth arch artery. The fourth arch artery on the right side forms a segment of the first part of the right subclavian artery. The second and third part of each subclavian artery develops from segmental branches which pass from the dorsal aorta to the growing limb buds.

6. The sixth arch artery sends branches to the lung buds developing from the foregut in the region of the future larynx and the artery itself help to form the left and right pulmonary arteries which join the terminal end of the developing pulmonary trunk.

The foetal circulation (Fig. 7.11)

From the placenta fresh 'arterial' blood comes to the foetus along the umbilical veins. In later foetal life only the left umbilical vein persists. In the liver most of the incoming blood is joined by the blood from the developing portal system of the foetus (vitelline veins) and passes via the ductus venosus to the right side of the heart (right atrium). In the right atrium there is a further mixing of the oxygenated blood from the placenta with venous blood from the body wall of the foetus (superior vena cava—right duct of Cuvier). Most of the blood entering the right atrium passes directly to the left atrium via the foramen ovale, and then via the left ventricle to the aortic trunk, hence bypassing the pulmonary circulation. A limited amount of blood, however, does pass into the right ventricle and to the pulmonary trunk. From the pulmonary trunk most of this blood reaches the aortic trunk directly via the ductus arteriosus, and only a limited amount passes through the lungs to return to the left atrium.

At birth the following changes occur:

1. With the cessation of blood flow along the umbilical vein the ductus venosus becomes obliterated to form a fibrous cord and the portal blood passes through the liver sinusoids. At the same time the umbilical arteries close and the foetal blood can no longer leave the body.

2. The ductus arteriosus is closed by a muscle sphincter action and all the blood in the pulmonary artery (from the right ventricle) must now pass through the pulmonary circulation and the left atrium to the left ventricle.

The degenerated ductus arteriosus forms the ligamentum arteriosum connecting the left pulmonary artery and the aortic arch in the adult.

3. The foramen ovale closes because the pressure of blood in the left atrium rises suddenly with the beginning of respiration and the commencement of a true pulmonary circulation. The septum primum is forced into contact with the septum secundum. At first the closure is purely mechanical but later the endothelium over the area of contact breaks down and fusion of the two septa takes place.

Congenital malformations of the heart and great vessels

Due to the complexity of the development of the heart and the great vessels associated with it, congenital abnormalities are quite common. Some of the more important defects of the heart, aorta, pulmonary trunk and aortic arches will be described.

One of the most common congenital malformations of the heart is the atrial septal defect, for in approximately 25 per cent of persons a probe can be passed from one atrium to the other through the upper part of the fossa ovalis. A patency of the foramen ovale which is sufficiently serious to produce clinical signs and symptoms is the result of abnormal resorption of the septum primum or incomplete fusion of the endocardial cushions with the lower margin of the septum primum. This defect allows blood returning to the right atrium to pass to the left side of the heart without going through the pulmonary circulation, so that the left ventricle passes poorly oxygenated blood into the systemic circulation. The skin of these infants is cyanotic, giving the appearance commonly described as 'blue baby.'

Septal defects in the ventricles are also common and may either be a defect of the membranous part of the septum or, less commonly, a deficiency in the muscular part of the septum. A membranous septal defect results from failure of the subendocardial tissue to fuse with the muscular part of the septum and the spiral septum of the bulbus cordis. Muscular septal defects appear in all parts of the septum and may be multiple. They appear to be due to excessive resorption of cardiac muscle tissue during formation of this part of the ventricular septum. If the spiral aorticopulmonary septum fails to develop, the truncus arteriosus, which develops from the bulbus cordis, is not divided into the aorta and pulmonary trunk. Sometimes the aorticopulmonary septum does not follow a spiral course, leading to a transposition of the great arteries.

If the pulmonary or aortic valve cusps fuse together during development this leads to stenosis of either valve and obstructs blood flow. Stenosis or atresia (complete obstruction of the mitral or tricuspid openings, or the pulmonary or aortic trunks) is frequently associated with other cardiac defects which are an attempt to compensate for the stenotic condition. For example, in the 'tetralogy of Fallot' there is pulmonary stenosis or atresia, ventricular septal defect, hypertrophy of the right ventricle and a consequent shunting of blood from the venous to the arterial side resulting in congenital cyanosis. Aortic overriding

Failure of the distal part of the left sixth aortic arch artery to close and form the ligamentum arteriosum after birth leads to the condition called patent ductus arteriosus. The malformation appears to be due to a failure in the muscle sphincter of the ductus and it is more common in females. Abnormalities of the aortic arch and its branches include double aorta, right aortic arch, abnormalities in the pattern of the subclavian and carotid arteries and abnormalities of the coronary arteries, which may arise from the pulmonary trunk.

Anomalies of cardiac development may produce little or no clinical manifestations or they may result in serious disabilities, as with large septal defects or stenoses of the pulmonary or aortic openings. Some cardiac abnormalities are quite well compensated for a number of years but they often cause clinical problems in early life due to insufficient compensation appropriate to the increased size of the individual, or superimposed infections.

DEVELOPMENT OF THE PHARYNGEAL REGION

The inturning head fold at the front end of the neural tube consists of two regions (Fig. 7.14).

1. An oval area immediately below the developing brain derived from the prochordal plate. This is the bucopharyngeal membrane where ectoderm and endoderm remain in contact without the interposition of mesoderm. With further inturning of the head end of the disk, the membrane comes to lie beneath the forebrain at the bottom of a slit-like depression, the *stomatodaeum*.

2. In front of (later below) the buccopharyngeal membrane is a region where the mesoderm separates the ectoderm and endoderm across the middle line. From this mesoderm develops the heart and this area is invaded by a forward extension of the coelomic cavity, forming part of the pleural and pericardial cavities. When the inturning process of the anterior head

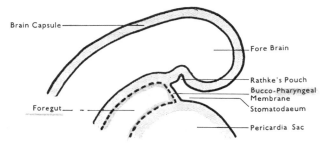

Fig. 7.14 Diagram of a sagittal section of the head of a human embryo of 2·5 mm.

end of the embryo is completed, the developing heart in the pericardial sac comes to lie beneath or ventral to the developing brain from which it is separated by the anterior end of the intraembryonic yolk sac (the foregut), the anterior end of the notochord, and the anterior somites. Behind the region of the developing heart the mesoderm forms an incomplete partition, the *septum transversum*, situated between the developing peritoneal and pleuropericardial cavities. In it the *liver* develops as an outgrowth from the foregut. From the septum there also develops the greater part of the *diaphragm*. The high position of diaphragm development and its subsequent migration accounts for its nerve supply in the adult descending from the third, fourth and fifth cervical segments of the spinal cord.

The development of the neck: At first the buccopharyngeal membrane lies immediately in front of and above the cardiac area after the infolding process is complete (Fig. 7.14), but very soon a series of visceral (pharyngeal) arches develop on each side to meet each other in the ventral mid-line separating the buccopharyngeal membrane from the cardiac region. In this way the neck comes into existence. In man six *pharyngeal or visceral arches* develop but of these the fifth is a transient structure (Figs 7.15, 7.16). The first arch is the *mandibular arch*, and from it develops the lower part of the face. Each arch contains a core of mesoderm. Between each arch lies an external (ectodermal) and an internal (endodermal) groove or furrow. The external grooves are called pharyngeal or *visceral clefts*, which soon become obli-

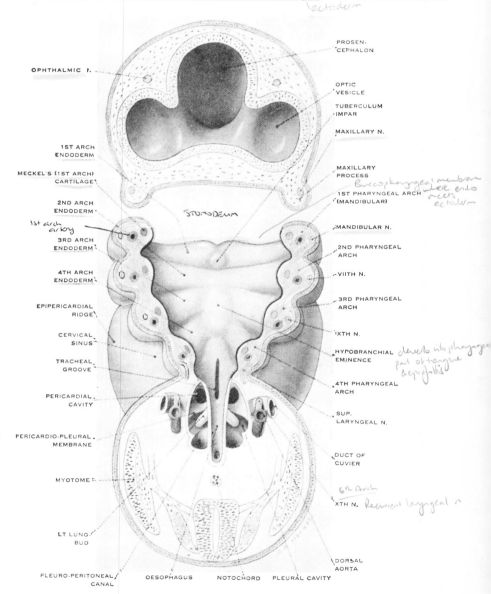

Fig. 7.15 A horizontal section through a reconstruction of a 5 mm human embryo (seen from above) to show the pharyngeal arches, the structures in the floor of the developing pharynx and the primitive pleural cavity. The rudimentary sixth pharyngeal arches bound the tracheal groove. (By courtesy of Professors Hamilton, Boyd and Mossman and W. Heffer & Sons Ltd., Cambridge.)

terated with further development; the internal grooves form the *pharyngeal pouches*, and lie in the side wall of the anterior end of the foregut (pharynx). Between each visceral arch at the floor of each visceral cleft and pharyngeal pouch, the ectoderm and endoderm are at first very close together with only a small amount of mesoderm between them.

From each pharyngeal pouch a number of important structures develop:

From the *first pharyngeal pouch* develops the pharyngotympanic tube, middle ear and tympanic antrum. Part of the first visceral cleft remains as the external auditory canal and the tympanic membrane persists as a region where ectoderm and endoderm remain in close relationship to one another.

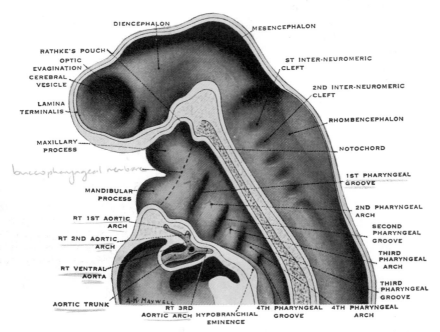

Fig. 7.16 A drawing of the right half of a sagitally sectioned reconstruction of the cephalic region of a 4·2 mm human embryo (modified from His). The interrupted line A represents the approximate site of the previous attachment of the buccopharyngeal membrane, i.e. the boundary between ectoderm and endoderm. (By courtesy of Professors Hamilton, Boyd and Mossman and W. Heffer & Sons Ltd., Cambridge.)

The *second pharyngeal pouch* gives origin to the tonsillar fossa. This later becomes filled with the lymphoid tissue of the tonsil.

Each *third pharyngeal pouch* forms a ventral and dorsal diverticulum. From the ventral diverticulum develops the thymus gland; from the dorsal diverticulum the inferior parathyroid gland (parathyroid III).

The *fourth pharyngeal pouch* also develops a ventral and dorsal diverticulum; the former may contribute to the thyroid gland; from the latter develops the superior parathyroid gland (parathyroid IV). The ventral diverticulum of the fourth pouch develops into the *ultimobranchial body*, which fuses with the developing thyroid gland and gives rise to the parafollicular or C cells of the thyroid gland. These are the cells which produce thyrocalcitonin, the hormone required for the maintenance of normal calcium level in the circulating blood and body fluids.

Neural crest cells → Bones → Skeleton of face
mesoderm → muscle → disappears / Artery / Cartilage / proximal + distal plate
?neurectoderm or neural crest → Nerve

The fate of the pharyngeal arches

Each pharyngeal arch contains a cartilaginous skeleton, a mass of muscle tissue, an artery and a nerve trunk. All of these structures, except the nerve trunk, are derived from the mesoderm of the pharyngeal arch. The nerve trunk is an outgrowth from the developing nervous system. The fate of these structures in each of the pharyngeal arches is as follows (Fig. 7.17):

→ mandible which is intramembranous & not endochondral ossification. & incus

First (mandibular) pharyngeal arch. The cartilage (Meckel's cartilage) forms the malleus of the middle ear, the malleolar-sphenoidal and sphenomandibular ligaments. Its proximal part becomes incorporated in and replaced by the bony mandible. From the muscle mass develop the muscles of mastication including the mylohyoid and anterior belly of the digastric, and also the tensor tympani and tensor palati muscles.

The nerve trunk is the mandibular division of the fifth (trigeminal) cranial nerve. It supplies the muscles which develop within the arch. The first arch artery, which carries blood from the ventral to the dorsal aorta of the embryo, does not persist (Fig. 7.13).

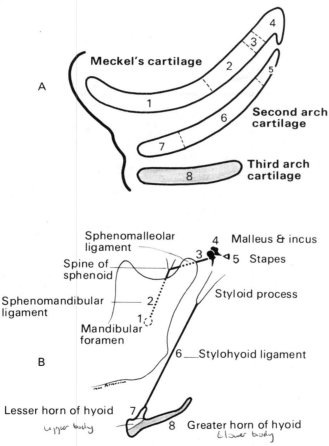

Fig. 7.17 Diagram to show the fate of the cartilages of the first three pharyngeal arches. A. The embryonic form of the cartilages. The front of the developing pharynx is to the left. B. Structures in the adult derived from the cartilages. The numerals correspond to those shown in A. The anterior part of Meckel's cartilage (1 in A) is absorbed into the developing mandible and thus does not appear as an adult structure.

Second (hyoid) pharyngeal arch. The cartilage forms the stapes of the middle ear, the styloid process, the stylohyoid ligament, the lesser cornu and the upper part of the body of the hyoid bone.

From the muscle mass develops the stapedius muscle, the posterior belly of the digastric and the stylohyoid, the platysma, the facial muscles of expression, and the muscles of the scalp. The facial and scalp muscles migrate from the neck and branches of the seventh nerve migrate with them.

The nerve trunk of the second arch is the seventh (facial) cranial nerve. The second arch artery does not persist.

Third pharyngeal arch. The cartilage of this arch forms the greater cornu and lower part of the body of the hyoid bone. The remainder of the cartilage does not persist.

The muscle tissue forms the stylopharyngeus muscle, and the nerve of the arch is the ninth (glossopharyngeal) cranial nerve.

The third pharyngeal arch artery forms the common carotid and part of the internal carotid artery.

Fourth pharyngeal arch. The cartilage probably forms the thyroid cartilage. The muscle mass develops into the pharyngeal muscles (except the stylopharyngeus) and the soft palate muscles (except the tensor palati). The nerve of this arch is the pharyngeal and superior laryngeal branches of the vagus, which belong, however, to the cerebral or cranial part of the accessory nerve.

On the left side the fourth arch artery forms the arch of the aorta; on the right side part of the right subclavian and brachiocephalic arteries.

The fifth pharyngeal arch does not develop as a separate structure in man.

Sixth pharyngeal arch. The cartilage probably forms the smaller laryngeal cartilages (cricoid and arytenoid). From the muscle mass develop the muscles of the larynx. The nerve of the arch is the recurrent laryngeal. The sixth arch artery contributes to the formation of the pulmonary artery, which migrates into the thorax with the development of the limb buds.

Congenital abnormalities of the neck region
These abnormalities arise during the sequence of developmental changes which transform the pharyngeal arches into the adult neck structures.

Branchial or *lateral cervical* sinuses open externally on the side of the neck and are the result of failure of the second pharyngeal cleft to obliterate. A blind channel remains which opens through the skin along the anterior border of the sternomastoid muscle in the lower part of the neck. *Branchial fistulae* are abnormal channels which communicate between the pharynx and the external surface of the neck. This is the result of persistence of the second pharyngeal cleft and pouch. The internal opening of the fistula is usually in the tonsillar fossa and the tract passes obliquely downwards between the internal and external carotid arteries to open externally in the lower part of the neck. Saliva and other fluids from the pharynx can drain to the surface by this pathway and it has to be removed surgically. Remnants of either the second pharyngeal cleft or pouch may remain to form cysts which may develop anywhere along the line of a branchial fistula, but most often they are found just below the angle of the mandible.

The thyroid gland develops as a mid-line diverticulum between the first and second pharyngeal arches which grows downwards in the floor of the pharynx as the thyroglossal duct. The lobes of the thyroid gland develop as expansions of the lower end of the duct. During the descent of the thyroid gland from the region of the developing tongue the thyroglossal duct atrophies and disappears, but pieces of it may persist and give rise to thyroglossal cysts anywhere along the line of the duct, but usually in the tongue or in the mid-line of the neck below the hyoid bone. Such ectopic thyroid tissue is quite functional.

DEVELOPMENT OF THE RESPIRATORY SYSTEM

The lower part of the respiratory system begins its development in the fourth week from a median *laryngotracheal groove* in the floor of the primitive pharynx. A diverticulum develops from the groove and soon becomes separated from the developing foregut to form the oesophagus and the laryngotracheal tube. The endoderm which lines the tube gives rise to the epithelium and superficial glands of the trachea and bronchi. The mesoderm surrounding the tube forms the cartilage, muscle and blood vessels of these structures. The largyngeal muscles and the cartilaginous skeleton of the larynx are derived from the more caudal pharyngeal arches.

As the laryngotracheal tube continues to grow downwards towards the thorax it divides into two lung buds. The one on the left forms two buds and that on the right forms three buds, to establish the lobes of the adult lung. The terminal alveoli develop their squamous epithelial lining after approximately twenty-six weeks of development, at which time the capillary network has also formed, so that the lungs are sufficiently well developed to allow survival if the foetus is born prematurely. The lungs become greatly expanded at birth as air is drawn into them; inflation is complete about three days after birth.

DEVELOPMENT OF THE ALIMENTARY CANAL

The alimentary canal from the pharynx to the rectum develops from that part of the yolk sac which becomes incorporated in the developing embryo (page 327). The part of the mouth in front of the buccopharyngeal membrane (page 338), and the lower part of the anal canal develop as invaginations of the surface ectoderm of the embryo. After the rupture of the buccopharyngeal and anal membranes, the alimentary canal becomes continuous from the mouth to the anus. Certain parts of the alimentary canal show a relatively greater development in size such as the stomach and caecum; other parts such as the appendix show a relative retardation in their growth.

The foregut gives rise to the pharynx and its derivatives (page 338), the lower respiratory tract, the oesophagus, the stomach, duodenum, the liver and pancreas.

The oesophagus develops with the formation of the tracheo-oesophageal septum and it elongates rapidly during the first six or seven weeks. The striated muscle in the upper part of the oesophagus comes from the caudal pharyngeal arches while the smooth muscle of the lower part of the oesophagus develops from the surrounding mesoderm, which is of splanchnic origin. The stomach first appears as a dilation of the caudal part of the foregut and its dorsal border grows much faster than the ventral border to produce the greater curvature. As the stomach continues to develop it rotates clockwise through ninety degrees, so that the dorsal border or greater curvature moves to the left and the ventral border or lesser curvature moves to the right. The stomach is suspended from the dorsal wall of the abdominal cavity by a mesentery.

The duodenum develops from the most caudal part of the foregut and the adjoining part of the midgut. It grows rapidly to form a C-shaped loop that projects ventrally with the apex of the loop representing the junction between the foregut and midgut. The liver and bile apparatus arise as an outgrowth from the caudal part of the foregut. This is the hepatic diverticulum which extends into the septum transversum where it enlarges and divides into two parts, one of which forms the liver and the other the gall bladder and cystic duct. The pancreas develops from two buds that arise from the caudal part of the foregut. The dorsal bud forms the greater part of the pancreas, while the ventral bud forms the uncinate process. The pancreatic ducts are formed from the stalks of the two buds and often communicate with one another.

The region of the gut between the duodenum and rectum shows a relatively great increase in length and comes away from the posterior abdominal wall to form a series of loops which occupy the abdominal cavity. For a time some of the loops project into the umbilical cord. This part of the gut remains attached to the posterior abdominal wall by a continuous mesentery through which it receives its blood supply.

Later, certain segments of the alimentary canal develop a secondary attachment to the posterior wall of the abdominal cavity and lose their original mesenteries. These segments are the duodenum and the ascending and descending parts of the colon. The wall of the canal, external to the endodermal lining, from the middle third of the oesophagus to the lower end of the rectum, develops from splanchnopleuric mesoderm (page 332). The mesoderm differentiates into circular and longitudinal layers of smooth muscle. It is penetrated by the endodermal outgrowths which give rise to the major glands that eventually lie external to the alimentary canal and pass their secretions through ducts into its lumen.

Congenital malformations of the digestive system have limited relevance but mention should be made of *pyloric stenosis*. This condition is the result of hypertrophy of the muscular sphincter which surrounds the lumen of the pylorus of the stomach. It is much more common in males and may depend on genetic factors. The infant is not able to retain milk when the stomach fills, for it cannot pass into the duodenum and lower parts of the intestine. A projectile type of vomiting is very characteristic and the condition is relieved surgically.

DEVELOPMENT OF THE CENTRAL NERVOUS SYSTEM

After the completion of the neural tube (page 330) this structure extends along the dorsal surface of the embryo in close relation to the adjacent notochord (Figs 7.7, 7.9). Its lateral walls thicken and become divided on each side into a dorsal and ventral part by a groove called the *sulcus limitans*. Nerve cells develop from primitive neuroblast cells and their processes commence to form tracts ascending or descending within the walls of the neural tube or passing out to associate themselves with the adjacent somites. Essentially, the cell bodies form the grey matter and their processes the white matter of the developing central nervous system.

Three layers can be recognized in the wall of the primitive neural tube, an inner *ependymal zone*, which is the germinal layer for the cells of the nervous system; an intermediate or *mantle* zone into which neuroblasts migrate from the ependymal zone; and an outer or *marginal* zone which contains the protoplasmic processes of the developing nerve cells. In the adult the ependymal layer persists as the lining epithelium of the nervous system, the mantle layer forms the grey matter and the marginal layer the white matter of the central nervous system. In parts of the system, particularly the cerebral hemispheres and cerebellum, mantle cells migrate into the marginal zone to form a superficial layer of grey matter. The cell bodies tend to become grouped in clumps in certain regions to form the nuclei of cranial and spinal nerves. Those developing in the dorsal half of the neural tube (alar laminae) are related to the sensory component of the somatic nervous system, while those developing in the ventral half (basal laminae) are related to the motor component (Fig. 7.18). From cells which develop in immediate relationship to the sulcus limitans are derived the sensory and motor components of the autonomic nervous system.

Neuroblasts send processes through the mantle layer which become grouped to form rootlets at the surface of the developing hindbrain and spinal cord. These are the motor or anterior roots of the cranial or spinal nerves. At the same time, neuroblasts in developing ganglia, which have originated from neural crest cells (Fig. 7.9) send out processes which divide almost at once into central and peripheral fibres. The central fibres enter the developing brain and constitute the sensory or afferent root of the cranial and spinal nerves, while the peripheral processes extend outwards to nerve terminations in the skin, oral mucous membrane and visceral structures. Afferent and efferent fibres can be distinguished as early as the fifth week of embryonic life.

The development of the trigeminal ganglion follows this sequence of events and establishes the main sensory system for the teeth, gums, oral mucous membrane and skin of the greater part of the face. The central processes of the ganglionic cells form synaptic connections with other parts of the trigeminal system in the developing brain stem and these are described in the adult on page 419.

While this process of differentiation is taking place the anterior end of the tube enlarges more rapidly than the remainder to form three primary vesicles; the forebrain (*prosencephalon*), midbrain (*mesencephalon*), and hindbrain (*rhombencephalon*) vesicles. The forebrain projects forwards beyond the anterior end of the notochord (Fig. 7.16). From its side walls develop

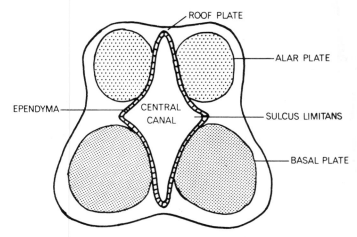

Fig. 7.18 Diagram to show a cross section through the developing spinal cord. The alar and the basal plates give rise to the sensory and motor elements of the grey matter respectively. Compare with Fig. 8.13.

the optic vesicles which, with the lens placodes (Fig. 7.15), form the eyes; and the cerebral vesicles which give origin to the *cerebral hemispheres*. Between the midbrain and hindbrain a deep flexure develops (*pontine flexure*), while the growing cerebral hemispheres grow backwards on each side of the central brain stem. From the hindbrain region develop the *pons*, the *cerebellum* and *medulla oblongata*. From the mid-line portion of the forebrain, lying between the cerebral hemispheres (the *diencephalon*) develop the *thalamus*, the *hypothalamic nuclei* and the neural portion of the pituitary gland. The *corpus striatum* develops in the floor of each cerebral vesicle and later becomes divided into two parts, the *caudate nucleus* and the *lentiform nucleus*, by the development of tracts passing to and from the cerebral cortex which has developed as an elaboration the side wall and roof of the vesicle. These tracts form the *internal capsule* (Figs 8.15, 8.20).

It is the great development of the cerebral hemispheres, and especially of the cortical areas, which is the characteristic feature of the human brain. The sensory tracts from the spinal cord pass to the developing cortex after relaying in the thalamus, and motor tracts from the cortex descend to end around the nuclei of the cranial and spinal nerves. The ventricles of the brain and the canal of the spinal cord develop as modifications of the central canal of the neural tube. Closely related in development to the beginning of function of the central nervous system but incompletely understood, is the process of myelinization of the nerve fibres (page 41).

From the neural crest tissue (Fig. 7.9) develop the posterior root spinal ganglia and the corresponding ganglia of the cranial nerves (fifth, seventh, ninth and tenth nerves). From the neural crest there also develops the ganglia of the autonomic system; the medullary cells of the suprarenal (adrenal) gland which secrete adrenaline and noradrenaline; and, some believe, the precursors of the odontoblast cells of the dental pulp.

DEVELOPMENT OF THE FACE

After the head end of the embryo becomes flexed around the anterior end of the notochord (page 338) and has reached a crown-rump length of 3 mm (about the twenty-fifth day after conception), the primitive mouth cavity (stomatodaeum) is a narrow slit-like space bounded by the brain capsule above, the pericardial sac below, and the mandibular and maxillary processes at the sides.

Very quickly the mandibular processes sweep medially to form the primitive lower jaw and separate the stomatodaeum from the pericardial sac. At the same time the brain capsule is separated from the primitive mouth cavity by the formation of the frontonasal process. Neural crest cells have migrated from their initial position at the sides of the neural tube and form a sheet of cells deep to the ectoderm which, when it reaches the region of the developing eye, splits into two parts. The anterior stream of cells enters into the formation of the mesoderm of the frontonasal process, while the posterior stream contributes to the mesoderm of the pharyngeal arches. Subsequent enlargement of the embryonic facial processes is the result of continuing proliferation of their mesoderm, in which the bones of the face will develop. The boundaries of the facial processes are demarcated by the grooves or furrows that lie between them. The grooves are obliterated by the differential growth of the processes, leading to recognizable features of the face as we know them in the infant.

Failure of the facial processes to grow properly and thus merge with one another, which involves a fusion or obliteration of their ectodermal coverings where they come into contact, results in developmental defects known as *facial clefts*. Clefts are the result of disruption of one or several stages in the fusion process, including abnormal induction by neural crest cells, some abnormality in the migration sequence or mesodermal merging. The most common congenital defect of the face is unilateral cleft lip and results when the maxillary and frontonasal processes fail to merge on one side. This defect apparently has a genetic basis because there is a strong tendency for cleft lip to recur in the same family. Bilateral clefting of the lip is less common and produces a wide defect of the upper lip in the mid-line. Median cleft lip is extremely rare and is due to incomplete formation of the two medial nasal processes and the proper fusion of the maxillary processes with them. If the maxillary and mandibular processes fail to fuse to the normal extent at the angle of the mouth, the mouth opening is much wider than normal, termed macrostomia. Conversely, the fusion process can extend beyond normal limits so that the opening of the mouth is much smaller than usual, or microstomia. If the furrow between the maxillary process and the lateral nasal process (Fig. 7.19) persists then an oblique facial cleft is produced which extends from the medial corner of the eye to the alar groove at the side of the nose. Sometimes the fusion process is almost complete so that the defect is a minor one limited to the formation of developmental cysts. These are ectodermal lined cavities which can be removed surgically. Those which occur along the line of a potential oblique facial cleft are called nasolabial cysts.

At the back of the primitive mouth cavity is the buccopharyngeal membrane forming a thin septum between the mouth cavity and the anterior end of the foregut. The latter extends forwards as a diverticulum of the yolk sac, between the neural tube and notochord dorsally and the developing heart ventrally, and from this region of the foregut there will develop the pharynx, larynx, trachea, lungs and oesophagus (page 343). The buccopharyngeal membrane slopes upwards and backwards so that the roof of the primitive mouth cavity is more extensive than the floor (Fig. 7.14). Immediately in front of the place of attachment of the buccopharyngeal membrane to the roof of the mouth at the anterior end of the notochord, there is a slit-like pouch lined by oral ectoderm. This extends towards the base of the developing brain and is known as *Rathke's pouch* and from it develops the anterior part of the hypophysis, or pituitary gland. The posterior part of the gland develops as a downgrowth of the forebrain vesicle from a region known as the diencephalon. Later the anterior part loses its connection with the mouth cavity, although remains of the epithelial tract may give rise to cysts of the cranial base region. Soon afterwards the buccopharyngeal membrane becomes perforated, sieve-like and rapidly disappears so that the primitive mouth cavity becomes directly continuous with the developing pharynx.

Meanwhile the epithelium covering the projecting frontal region of the brain capsule becomes thickened in two regions on each side forming the nasal and lens *placodes*. Soon afterwards the nasal placodes lie at the bottom of two *nasal pits* bounded by slight elevations, the lateral and medial *nasal folds or processes* (Fig. 7.19). The area between the two olfactory

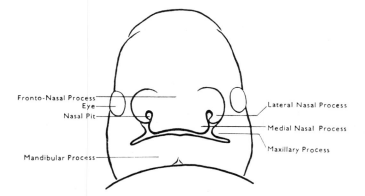

Fronto-Nasal Process
Eye
Nasal Pit

Lateral Nasal Process

Medial Nasal Process

Mandibular Process

Maxillary Process

Fig. 7.19 Diagrammatic illustration of the head of a human embryo of 6 mm (fifth week).

pits is the *frontonasal* process or primary nasal septum, which at this stage forms the greater part of the upper margin of the mouth opening.

The *mandibular processes* of each side soon meet one another in the mid-line, above the pericardial swelling, to form the lower boundary of the mouth opening. The *maxillary processes* develop as buds from the mandibular processes in the region of the angle of the mouth opening, and grow forwards on each side of the face beneath the developing eyes to make contact with the lateral nasal processes and soon afterwards, by further growth, with the lower ends of the medial nasal processes. The intervening epithelium in the areas of contact disintegrates and maxillary process mesoderm becomes continuous with frontonasal mesoderm. An alternative theory postulates the forward migration of maxillary process mesoderm towards the frontonasal process between existing layers of epithelium. According to this concept facial clefts are due to failure of mesodermal penetration and rupture of the unsupported epithelium.

The development of the upper lip

There is still some lingering uncertainty about the development of the human upper lip and there are two conflicting views about the part played by the maxillary mesoderm. One view is based on classic research by Frazer and holds that the lip is formed entirely from the maxillary processes. The maxillary mesoderm appears to invade the lower part of the frontonasal process until the two streams from each side meet in the middle line. In this way the mesoderm of the frontonasal process becomes cut off from taking any part in the further development of the upper lip and the anterior part of the palate; all of which are believed to develop from maxillary process mesoderm (Figs 7.19, 7.20). The end result is a continuous mass of maxillary mesoderm which extends without interruption from one side of the face to the other and from this the lip is separated off by the formation of the labiodental sulcus. The advantages of this explanation of lip development are:

1. The explanation is in agreement with what is known about the comparative anatomy and embryology of the lip.

2. It enables us to make satisfactory explanation of the development of congenital anomalies of the region, such as cleft lip and palate.

3. It helps to explain the nerve supply of the entire upper lip from the maxillary nerve through its infra-orbital branch.

The other more current view about the development of the human upper lip is based on the classical concept of His who, with many other embryologists at the turn of the century, believed that the central part of the lip including the depressed area called the philtrum is derived from the frontonasal process, and the lateral parts from the maxillary processes. Investigation by Warbrick (1963) provided evidence in favour of this earlier description. Reconstructions of the head region of human embryos confirm that the upper margin of the mouth opening is formed centrally by the frontonasal process and laterally by the maxillary processes. The boundary between the two embryonic processes is indicated by a vertical surface groove which extends in the sagittal plane from the lower end of the nasal pit to the border of the mouth. This groove is filled out by proliferating underlying mesoderm, with the result that the processes appear to merge with one another. As long as the groove persists it maintains its original position between the processes and there is no indication of the maxillary process growing medially over the frontonasal process. When the groove is flattened out at a later stage of development it is no longer possible to distinguish between the frontonasal and maxillary processes.

If this theory is considered to be correct, a cleft of the lip to one side of the mid-line, which is the commonest variety, can be explained as occurring between maxillary and frontonasal embryonic processes. The nerve supply to the upper lip has to be explained on the basis of the maxillary nerve sending branches onward towards the mid-line through the mesoderm of the frontonasal process. Migration or growth of nerve tissue from one pharyngeal arch derivative into another mesodermal mass is not without precedent in comparative embryology of this region of the developing face.

The middle part of the upper lip, or philtrum, develops a groove directly below the nasal septum. The philtrum is characterized by a heaping up of mesoderm to form two ridges, one on each side of the median groove. Later, the ridges become the sites of insertion for those fibres of the orbicularis muscle which have crossed the mid-line from the opposite side. Each philtral ridge also receives fibres from the levator labii superioris muscle of the same side (Fig. 4.12).

The development of the palate

At about this time the nasal pits extend backwards and gain a secondary posterior opening into the stomatodaeum. In this manner the primitive nasal cavities are formed. They are bounded below by the mesial extensions of the maxillary processes and elsewhere by frontonasal mesoderm (Fig. 7.20).

Meanwhile maxillary process tissue also extends beneath the developing brain in the roof of the primitive mouth cavity. This extension of mesoderm from each side meets in the middle line and then commences to extend downwards as the secondary nasal septum (Figs 7.21, 7.22), continuous in front with the primary nasal septum of the frontonasal process. At a

Fig. 7.20 Diagrammatic illustration of the head of a human embryo of 20 mm (seven weeks) seen from the primitive mouth cavity. The arrow shows the connection of the nasal cavity with the mouth cavity above the primary palate. The future position of the secondary part of the nasal septum is shown in broken line.

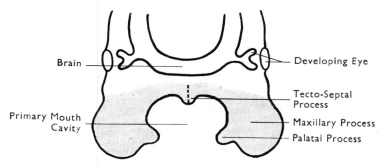

Fig. 7.21 Diagrammatic illustration of a coronal section through the head of a human embryo to show the tectoseptal, palatal and maxillary processes.

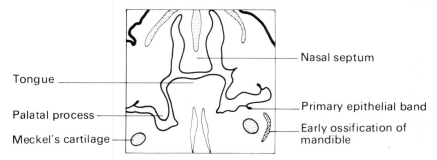

Fig. 7.22 Drawing of a coronal section of the head of a human embryo of 22 mm C.R. length (about seven weeks). The palatal processes lie vertically on each side of the tongue.

lower level two further extensions of the maxillary processes grow inwards and downwards as the *palatal processes* or shelves which come to lie on either side of the developing tongue (Fig. 7.22). With further growth in the size of the mouth cavity, especially in vertical height, the tongue sinks downwards and the palatal processes come to lie above the dorsum of the tongue, and are thus able to unite with one another and the lower end of the nasal septum. In this manner the primitive mouth cavity becomes divided into three parts: a right and left nasal cavity above the developing palate on each side of the nasal septum, and a definitive mouth cavity below the palate. As a consequence of this formation of the palate the posterior openings of the nasal cavities are carried backwards so that they no longer open into the mouth cavity but into the upper part of the pharynx (nasopharynx).

Later the united palatal processes are invaded by bone in front and by muscle behind. The bone grows from the premaxillary, maxillary and palatine centres of ossification to form the hard palate. The muscle tissue is derived from two sources: the tensor palati muscles from the mandibular arches, while the other palatal muscles are probably derived from the mesodermal tissue of the third and fourth misceral arches. The different origin of the muscle explains the difference in their nerve supply; the tensor palati by the mandibular division of the trigeminal nerve, the other muscles by the cerebral part of the accessory nerve through the pharyngeal branch of the vagus.

It will be seen therefore that both the nasal septum and palate develop in two stages.

1. The primary nasal septum is derived from the frontonasal process; the primary palate forms by mesial extensions of maxillary process mesoderm which merge with the lower end of the frontonasal process.

2. The nasal septum and the secondary palate are both derived from tissue of the maxillary process, behind the frontonasal process.

The parts of the face which develop from the frontonasal process have as their sensory nerve supply branches of the ophthalmic division of the trigeminal nerve (ethmoidal and external nasal nerves), while those parts derived from the maxillary process have as their sensory nerve supply branches of the maxillary division of the trigeminal nerve (the nasal, nasopalatine, palatine and infra-orbital nerves).

The cartilaginous facial skeleton

Before the formation of bone and also during the early stages of bone formation, the skeleton of the face is formed by cartilage. Meckel's cartilage develops within the mandibular arch

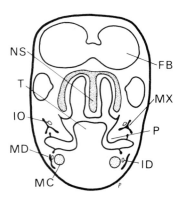

Fig. 7.23 Drawing of a coronal section through the facial region of a human embryo about the sixth week of development. FB, forebrain vesicle; ID, inferior dental nerve; IO, infraorbital nerve; MC, Meckel's cartilage; MD, mandible; MX, maxilla; N, nasal septum; P, palatal process; T, tongue.

and extends from the developing cranial base in the region of the otic capsule to the middle line of the future chin region where it is united with the cartilage of the opposite side. The cartilage of the nasal capsule develops in the maxillary process tissue and extends forwards into the frontonasal process. Behind it becomes continuous with the cartilage of the cranial base. Within the primary and secondary parts of the nasal cavity it forms the common primordial skeletal framework. The lateral part of the capsule on each side of the developing face forms the outer skeletal framework for the nasal cavities; its lower free end turns inwards as the developing inferior turbinate process (Fig. 7.23). Within the two parts of the nasal septum it forms the common septal cartilage.

At a later stage the maxillary and premaxillary bones develop on the outer side of the cartilage of the nasal capsule (page 369), and the mandible develops on the outer side of Meckel's cartilage (page 356), while the vomer develops in relation to the lower edge of the septal cartilage. The lateral (turbinate) masses of the ethmoid replace part of the lateral region of the nasal capsule and the perpendicular plate of the ethmoid extends downwards and forwards in the septal cartilage. The inferior turbinate bone (concha) replaces the cartilage in the inferior turbinate process. The remainder of the nasal cartilage atrophies except for the cartilage forming the front of the nasal septum and the alar cartilages of the nose. Remains of the nasal cartilage may, however, persist and become in rare cases the sites of origin of facial chondrosarcoma, which is a malignant tumour (cancer).

Tongue

The tongue develops in two parts (Fig. 7.24):

1. An anterior (oral) part, derived from three swellings of the mesoderm of the mandibular arches just within the mouth cavity. These are the *lateral lingual swellings* and a mid-line structure in the floor of the mouth, the *tuberculum impar*, which occupies the groove between the mandibular and hyoid arches.

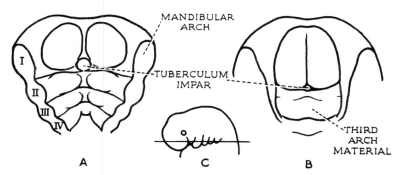

Fig. 7.24 Diagram to show the development of the tongue. (A) Floor of the mouth at 9 mm C.R. length. Anterior part of the tongue appearing as paired lateral swellings from the mandibular arch. (B) Floor of the mouth at later stage when third arch material has moved forward over the second arch to form the posterior (pharyngeal) part of the tongue. (C) Figure showing the plane of section of (A) and (B). (Modified from Hamilton, Boyd and Mossman.)

2. The posterior (pharyngeal) part of the tongue is derived mainly from the third pharyngeal arch and grows forwards over the second (hyoid) pharyngeal arch on the floor of the mouth to join the back of the anterior part of the tongue. This is the *copula* or *hypobranchial eminence*, from the back part of which develops the epiglottis.

The different origin of the parts of the tongue explains the sensory nerve supply of the mucous membrane on the dorsum of the tongue; the anterior two-thirds by the lingual branch of the trigeminal nerve, the posterior third by the glossopharyngeal nerve—the nerves of the first and third pharyngeal arches respectively. The nerve of the second arch (the facial nerve) contributes taste fibres to the anterior (oral) part of the tongue. The muscles of the tongue appear between the sixth and eighth week, and by this time the tongue fills the greater part of the mouth cavity. In the groove between the tongue and the mandibular process develop the sublingual and submandibular salivary glands as downgrowth from the covering epithelium. They are probably endodermal in origin.

NB Tongue musc devel from occipital myotomes i brings it in (XII) z it

Cheeks

The cheeks consist of tissue derived from both the mandibular and maxillary processes. On either side of the mouth in the cheek region a narrow outpouching of the mouth cavity passes outward for some distance between the maxillary process above and the mandibular process below. The outer limit of this pouch on either side is where the mouth epithelium passes from the maxillary to the mandibular process and lines the inner surface of the cheek (Fig. 7.22). Later, the vertical height of this part of the mouth cavity (*the vestibule*) is further increased. At the back of the vestibular region on each side of the mouth cavity there develops the parotid salivary gland as a bud-like projection of mouth epithelium into the adjacent mesoderm.

The salivary glands

The salivary glands all begin their formation as solid cords of cells from the epithelium of the primitive mouth during the sixth and seventh weeks. The parotid gland buds are the first to appear, arising from the ectodermal lining of the stomatodaeum on the inner side of the developing cheek near the angle of the mouth. The buds grow backwards towards the ear region, first as solid, branching cords of cells which later canalize to form the acini and duct system. The capsule of the gland is derived from the surrounding mesoderm.

The submandibular glands develop from the endoderm covering the floor of the primitive mouth and grow backwards on the lateral aspect of the developing tongue. A groove forms at the side of the tongue which later closes to form the submandibular duct. The further development of the acini and duct system is the same as that for the parotid gland. The sublingual glands appear at a slightly later stage, in the eighth week of development, and develop in a similar fashion to the other glands from the endoderm at the side of the tongue.

Early in their development the cheeks and lips are invaded by muscle tissue derived from the second visceral arch. This muscle tissue, part of the facial musculature, forms the buccinator muscle and the various elements which make up the complex muscular apparatus of the lips. Their nerve supply is the nerve of the second pharyngeal arch, the seventh cranial or facial nerve. As the buccinator muscle sheet grows backwards it surrounds the duct of the growing parotid gland, which then comes to pierce the muscle. At the back the mouth cavity the buccinator becomes united to the superior constrictor muscle at the *pterygomandibular raphe*.

pterygoid hamulus → medialedge retromola A

FACIAL DEFORMITIES

The forward growth of the maxillary processes to form the side of the face, is largely a growth of mesodermal tissue beneath the covering epithelium, but the union of the palatal processes (Fig. 7.25), involves the contact of epithelial-covered surfaces, and, if union is to occur, disintegration of the epithelium in the region of contact must take place to allow continuity of mesoderm. If epithelial disintegration does not take place there can be no fusion of adjacent processes, or if there is inadequate mesodermal penetration rupture of unsupported epithelium may result. In this way deformities such as hare-lip, cleft palate and macrostoma will be produced. Incomplete disintegration of the epithelium may produce one or more epithelial-lined cysts which may be situated in the lips, palate or cheeks.

Clefts of the lips and anterior alveolar region in front of the incisive foramen develop earlier than those of the palate behind the foramen. This is the basis of the classification of orofacial clefts into three main categories.

1. Those in front of the incisive foramen.

2. Those behind the foramen.

3. Those extending from the lip region to the uvula (complete palate clefts) with unilateral or bilateral hare-lip.

Cleft palate, in its various forms, is one of the most common developmental anomalies in children, occurring in one in every six hundred live births. Fortunately, although cleft palate causes great anxiety for the parents and severe management problems in feeding of the infant, it can be corrected surgically. Many advances in the treatment of this condition have been made in recent years and the combined efforts of the plastic surgeon, orthodontist and speech therapist at appropriate stages of the development and growth of the child can lead to excellent cosmetic and functional results. Anatomical knowledge of facial growth mechanisms is of great importance to all these specialists in their diagnosis and treatment plans.

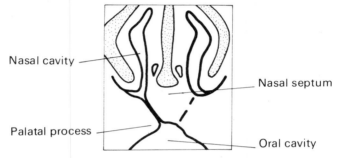

Fig. 7.25 Diagram of a coronal section of the head of a human embryo showing the region of fusion of the palatal processes with the nasal septum. On one side complete fusion has taken place, on the other side the mesodermal tissue of the nasal septum and the palatal process are still separated by epithelium.

DEVELOPMENT OF THE SKULL AND FACIAL SKELETON

The chondrocranium
During early foetal life the primary skeleton of the skull is laid down in cartilage as the chondrocranium (Fig. 7.26). This consists of a mid-line basal plate extending from the foramen magnum to the region of the foramen caecum. The posterior part of this plate of cartilage (the parachordal plate), as far forward as the pituitary fossa, is related in its early development to the anterior end of the notochord. On either side of this region of the basal plate develop the cartilaginous auditory (otic) capsules which contain the inner ear apparatus and become attached to the mid-line basal plate cartilage between the jugular foramen and the foramen lacerum.

Below the prepituitary, anterior, segment of the basal plate and attached to its under surface, there develops the nasal capsule consisting of a mid-line septum and bilateral side walls from which develop on their inner sides cartilaginous turbinate processes. Meckel's cartilage extends from the middle ear region behind, to the region of the future mandibular symphysis in front and forms the primary skeleton of the lower jaw. The primary jaw joint is situated between the back of Meckel's cartilage (the future malleus) and the developing incus. The stapes and styloid processes develop in cartilage within the second (hyoid) visceral or pharyngeal arch (see page 342). For further details on the chondrocranium and its derivatives see page 367.

Development of skull bones
Soon after the appearance of the cartilaginous elements which make up the chondrocranium the bones of the skull commence to appear as separate and isolated ossification centres. Some

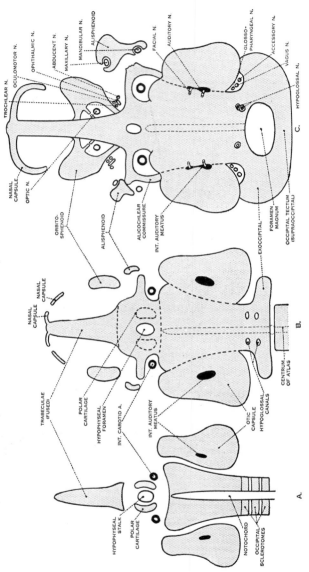

Fig. 7.26 Three stages in the development of the neural portions of the chondrocranium (based on de Beer, 1937). Right side in C is at a later stage than left side. (By courtesy of Professors Hamilton, Boyd and Mossman and W. Heffer & Sons Ltd, Cambridge.)

bones such as the parietal and palatine bones are developed from single centres of ossification, others such as the temporal and sphenoid develop from a number of centres which unite with one another at a later date. The bones of the skull may be classified as follows:

1. Those which appear as centres of *endochondral* ossification and develop within, and ultimately replace, portions of the chondrocranium. These include the basi-occipital, the petrous part of the temporal bones, the body and lesser wings of the sphenoid, the perpendicular plate of the ethmoid (mesethmoid), the lateral masses of the ethmoid (the facial ethmoid) and the inferior turbinates (conchae).

2. Those which appear within the *fibrous tissue capsule* which surrounds the developing brain and is attached at the skull base to the chondrocranium. These include the parietals, the temporal and occipital squamous elements, the great wings of the sphenoid beyond the foramina ovale and rotundum, and the frontal bones. These are sometimes referred to as *membrane bones* as they develop within connective tissue membranes and not within a cartilaginous matrix.

3. Bones which develop around, and in close relation to, the cartilage of the nasal capsule and Meckel's cartilage. These include the maxillary bones, the palatine, lacrimal, zygomatic, and nasal bones, the vomer and the mandible. See page 365 for further details.

Sutures

As the bony elements develop in a definite relationship to the growing chondrocranium, brain and eyeball, they come into a definite topographical relationship to one another at the various cranial and facial sutures. These are remarkably constant in position in the human species and among the primates and members of the mammalian order generally. Each suture is a meeting place, not only between two or more bones, but also between the fibrocellular periosteal capsules within which the bones develop. Therefore, during growth each suture shows

Uniting

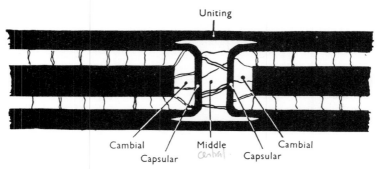

Cambial   Middle   Cambial
   Capsular   Capsular

Fig. 7.27 Diagram of a suture showing the various layers in its construction. (By courtesy of Professor J. J. Pritchard, Dr F. G. Girgis, and *Journal of Anatomy*.)

five zones of tissue between each bony elements. These are the cellular osteogenetic tissue at the edge of each of the growing bones, the fibrous capsular layers, and between the capsular layers an intermediate (middle) zone which is later the site of union between the sutural elements and contains numerous blood vessels (Figs 7.27, 7.28). At a suture, therefore, there are two growth centres, one for each bony element, and the rate and extent of growth may vary to a considerable extent at each centre. After active growth has ceased in a suture the division into five layers becomes less distinct and fibre bundles pass directly from one bone to the other. The sutures thus become predominantly sites of union between bones rather than growth centres.

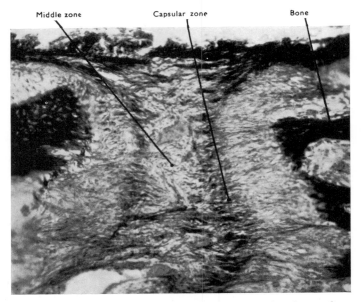

Fig. 7.28 Photomicrograph showing the structure of a cranial suture in a human foetus.

EARLY DEVELOPMENT OF SOME FACIAL BONES

Development of the mandible

The mandible is formed in the lower or deeper part of the first pharyngeal or visceral (mandibular) arch. It is preceded there by Meckel's cartilage, which represents the primitive vertebrate lower jaw. In the human embryo Meckel's cartilage attains its full form by the 15 mm C.R. stage (six weeks) and then stretches downward and forward as an unbroken rod from the cartilaginous otic capsule to the middle line. There its ventral end turns upward in contact with the cartilage of the opposite side, to which it is joined by mesenchyme. It is surrounded in its whole length by a thick investment of fibrocellular tissue. The dorsal end of the cartilage gives rise to the malleus of the middle ear; the remaining part of the cartilage is largely associated with the development of the membrane bone, which forms the replacing skeletal structure, the mandible.

Meckel's cartilage at this stage of development has a close relationship to the mandibular nerve, the nerve of the first pharyngeal arch, and its branches, acting as their skeletal support. The main nerve trunk issues from the skull medial and ventral to the dorsal end of the cartilage and comes into direct relationship with it about the junction of its dorsal and middle thirds (Fig. 7.29). Here, after giving off its other branches, it divides into the lingual and inferior dental nerves. The lingual nerve passes forward on the medial side of the cartilage, whereas the inferior dental nerve lies lateral to its upper margin, and running forward parallel to it terminates by dividing into mental and incisive branches; the incisive branch continues its course parallel to the cartilage.

The body of the mandible

The further history of Meckel's cartilage is bound up with the development of the bony mandible in which, however, it takes very little direct part. The mandible first appears as a band of dense mesodermal fibrocellular tissue on the lateral side of the inferior dental and incisive nerves. Ossification occurs in this tissue at the 17 to 18 mm C.R. stage (seventh week) in the angle formed by the incisive and mental nerves (Fig. 7.29); that is, the region of the future

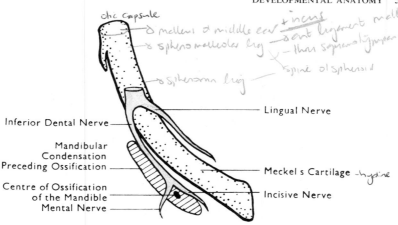

[handwritten annotations:] otic Capsule — & malleus & middle ear + incus — ant ligament malleus / & sphenomalleolar lig — & ant ligament — thus sphenomandibular fissure — spine of sphenoid / & sphenomandibular lig

Lingual Nerve

Inferior Dental Nerve

Mandibular Condensation Preceding Ossification

Meckel's Cartilage *[handwritten: hyaline]*

Centre of Ossification of the Mandible

Incisive Nerve

Mental Nerve

Fig. 7.29 Diagram to illustrate the early development of the right side of the mandible. Viewed from above.

mental foramen, and from this centre the formation of bone spreads rapidly backward below the mental nerve, which then lies in a notch in the bone, and on the lateral side of the inferior dental nerve. By the 19 mm C.R. stage the bone in the region of the notch for the mental nerve has grown medially below the incisive nerve and soon afterwards spreads upward between this nerve and Meckel's cartilage; in this way the incisive nerve is contained in a trough of bone formed by lateral and medial plates which are united beneath the nerve. At the same stage the notch containing the mental nerve is converted into the mental foramen by extension of bone over the nerve from the anterior to the posterior edge of the notch (Fig. 7.30). The bony trough grows rapidly forward towards the middle line where it comes into close relationship with the similar bone formation of the opposite side but from which it is separated by connective tissue. Union between the two halves of the bony mandible takes place before the end of the first year. The growth of bone over the incisive nerve from the lateral and medial plates converts the trough of bone into the incisive canal.

A similar spread of ossification in the backward direction produces first a plate of bone in relation to the whole of the lateral aspect of the inferior dental nerve, then a bony trough in which the nerve lies, and very much later the canal for it. Thus by these processes of growth the original primary centre of ossification produces the *body* of the mandible as far back as the mandibular foramen and as far forward as the symphysis; this is the part of the mandible which surrounds the inferior dental and incisive nerves—the *neural element*. At this stage the

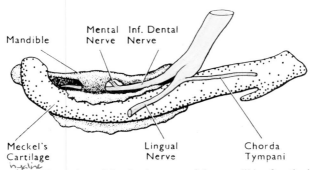

Mandible

Mental Nerve

Inf. Dental Nerve

Meckel's Cartilage *[handwritten: hyaline]*

Lingual Nerve

Chorda Tympani

Fig. 7.30 Diagrammatic illustration of the development of the mandible after the bridging over of the mental and incisive nerves has commenced. Viewed from the lingual aspect.

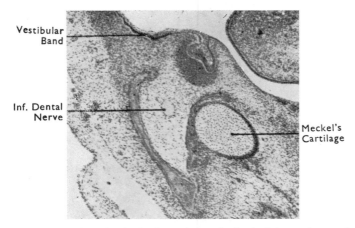

Vestibular Band

Inf. Dental Nerve

Meckel's Cartilage

Fig. 7.31 Coronal section of the developing lower jaw at the level of the tooth germ for the deciduous first molar. The mandibular bone forms a well-defined trough containing the inferior dental nerve. Note in this part of the mouth the shallow vestibular band and its relative remoteness from the dental lamina. Human foetus of 27 mm C.R. length × 45.

developing tooth germs lie some little distance superficial to the mandible and are not contained by it (Fig. 7.31).

As the dental organs of the deciduous tooth germs commence to differentiate the bone of the mandible begins to come into close relationship to them. This is brought about by the upward growth, on each side of the tooth germs, of the lateral and medial plates of the mandibular bone above the level where the roof of the canal for the incisive and inferior dental nerves is formed to form the lateral and medial *alveolar plates* (Fig. 7.33). By this growth the developing teeth come to lie in a trough of bone. This trough is later divided into separate small basins or alveoli for the teeth by the formation of bony septa between its medial and lateral walls (Fig. 7.34).

The fate of Meckel's cartilage
With the exception of the ventral terminal part of Meckel's cartilage at the middle line, the anterior part of the mandible in the incisor-canine region includes the cartilage in its substance. This part of the cartilage is first surrounded by an extension of bone from the medial plate and then is gradually resorbed and replaced by an extension of ossification from the membrane bone around it. During the later foetal period and at least until the time of birth one or two nodules of cartilage are seen in the fibrous tissue of the symphysis; these nodules are remnants of the ventral end of Meckel's cartilage. The rest of Meckel's cartilage disappears completely except for a part of its fibrous covering which persists as the sphenomandibular and sphenomalleolar ligaments. The most dorsal part of the cartilage ossifies to form the malleus, and this is attached to the spine of the sphenoid by the sphenomalleolar ligament which passes through the squamotympanic fissure of the temporal bone (Fig. 7.32). This becomes the anterior ligament of the malleus in the adult.

The ramus
The backward extension of the mandible to form the ramus is produced by a spread of ossification from the body, behind and above the mandibular foramen. From this region the mandible diverges laterally from the line of Meckel's cartilage. Just as the body of the mandible is first indicated by a fibrocellular condensation in which ossification occurs, here, too, the ramus and its processes are first mapped out by a backward extension of this condensation. The

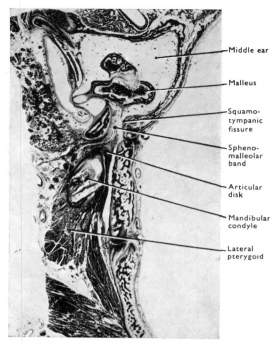

Fig. 7.32 Horizontal section through the mandibular joint region in a five-month human foetus showing continuity of the lateral pterygoid muscle, articular disk, sphenomalleolar ligament (band) and malleus through the squamotympanic fissure.

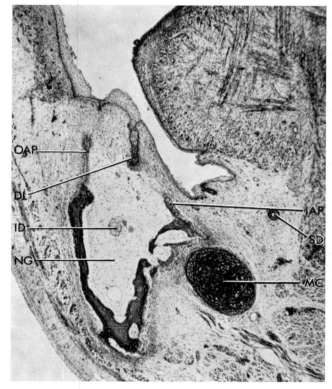

Fig. 7.33 Coronal section through the mandible of a 50 mm C.R. length human embryo. IAP, inner alveolar plate; DL, dental lamina; ID, inferior dental nerve; MC, Meckel's cartilage; NG, neural groove; OAP, outer alveolar plate; SD, duct of submandibular gland.

formation of bone in this tissue occurs rapidly so that the *coronoid* and *condylar* processes and the region of the *angle* are to a large extent ossified by the 40 mm C.R. stage (tenth week).

The further growth of the first and second of these processes is modified by the appearance of secondary cartilages. These cartilages, which occur at various sites in the region of membrane bone formation, are described as secondary or accessory, since they are not part of and have no connection with the primary cartilaginous skeleton (to which Meckel's cartilage belongs). They also differ in their behaviour and histological appearance from the typical hya-

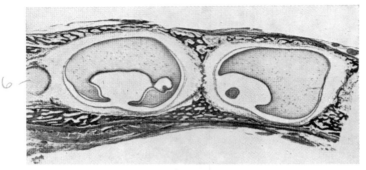

Fig. 7.34 Horizontal section through a human foetal mandible of six months showing the deciduous molars in their alveoli. Immediately behind the deciduous molars (left) is the tooth germ for the first permanent molar × 7.

line cartilage of which the primary cartilaginous skeleton is composed. The secondary cartilages increase in size by the proliferation and transformation of the cells of the thick layer of fibrocellular tissue covering them. These cartilages have larger cells and less intercellular matrix than hyaline cartilage and they may be associated with the development of cartilaginous tumours in adult life.

The formation of the coronoid and angular processes of the mandibular ramus is initiated by the development of major muscles of mastication. The temporalis muscle gains attachment to the future coronoid process, while the differentiating fibres of the masseter and medial pterygoid are associated with the region of developing bone matrix which becomes the angle of the jaw.

The condylar cartilage
In the mandible there are three main sites of secondary cartilage formation. The largest and first of these to appear is the *condylar cartilage*, and it is of great importance in the growth of the mandible. It first appears at about the 50 mm C.R. stage (twelfth week). At this stage it is seen as a fringe of cartilage on the superior and lateral aspects of the bone in the condylar process, which merges on one side into this bone and on the other into the fibrocellular layer which delimits the condylar region. Through additions from the cells of this covering layer of fibrocellular tissue, the cartilage soon forms a cone-shaped mass which not only occupies the whole of the condylar process but reaches forward and downward into the ramus as far as the level of the mandibular foramen.

By the fifth month of foetal life the original cone of cartilage is largely replaced by bony trabeculae, which runs through the ramus and contrasts strongly with the membrane bone of the ramus (Fig. 7.35). The zone of cartilage left beneath the articular surface of the condyle persists throughout not only the whole foetal period but until about 20 years of age (Fig. 4.23). During this period the thickness of the zone of cartilage gradually diminishes as the proliferative activity of the cells of its covering fibrocellular layer grows less, until eventually

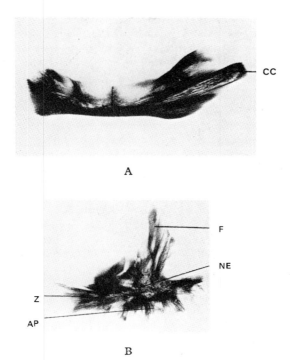

Fig. 7.35 Microradiographs of the mandible (A) and maxilla (B) from a mid-term human foetus. AP, alveolar process; CC, condylar cartilage (ossified); F, frontal process; NE, body of maxilla; Z, zygomatic process. (By courtesy of *The Dental Practitioner*.)

the cartilage disappears and the replacing bone forms the whole of the condyle. Since the cartilage remains during the whole of the normal growth period it increases the length of the mandible throughout that time.

By the fifth month of foetal life large vascular canals have appeared in the condylar cartilage; these are still present at birth. They are probably related to the nutritive requirements of the rapidly growing cartilage enabling blood vessels to reach the proliferative cells on the surface of the condylar head beneath the fibrous covering.

Other secondary cartilages
The coronoid cartilage forms a strip along the anterior border and summit of the coronoid process. It first appears about the 80 mm C.R. stage. The cartilage is covered superficially by a thick fibrocellular layer and rests on the membrane bone below. All trace of the cartilage has disappeared long before birth.

The third of the main secondary cartilages of the mandible appears after the 100 mm C.R. stage at the symphysial end of each half of the bony mandible. The two symphysial cartilages are separated from each other by the connective tissue of the symphysis, the cells of which add to the cartilages. The symphysial cartilage is entirely independent of Meckel's cartilage and its perichondrium. The union which takes place at the symphysis between the two halves of the mandible shortly after birth obliterates them so that they take no further part in the growth of the mandible.

The mandible at birth
At birth the mandible, though perfectly recognisable as such, differs in several respects from

the adult bone. The chief differences are the obtuse mandibular angle, the small size of the ramus compared with the body, and the absence of a true compact layer of bone on its surface. The body consists of a neural and alveolar element while the ramus consists of the coronoid, angular and muscular processes built around the central core of the ramus developed from the condylar secondary cartilage.

Development of the maxilla

The maxilla proper (excluding the premaxilla) is developed in the maxillary process of the mandibular arch. Like the mandible its first appearance is as a membranous ossification, but unlike the mandible its further development and growth are little affected by the appearance of secondary cartilage. Ossification in the maxilla commences slightly later than in the mandible, about 18 mm C.R. stage. The centre of ossification first appears in a band of fibrocellular tissue on the outer side of the cartilage of the nasal capsule, and immediately lateral to and slightly below the infra-orbital nerve, where it gives off its anterior superior dental branch. The ossification centre thus lies in the angle formed by the two nerves and is situated above that part of the dental lamina from which develops the enamel organ of the canine tooth germ.

From this centre ossification spreads backward towards the developing zygomatic bone below the orbit, and forward in front of the anterior superior dental nerve below the terminal part of the infra-orbital nerve towards the developing premaxilla. At this stage the forming bone takes the shape of a curved strip, arranged vertically with the convex side directed medially. From the anterior extension there develops the upward directed frontal process which, with a corresponding process of the premaxilla, forms the frontal process of the adult bone. The developing facial and frontal processes of the premaxilla and maxilla rapidly unite with one another so that from an early stage no suture appears between them on the face. Early in development the maxilla forms a bony trough for the infra-orbital nerve and, by downward growth, an outward alveolar plate in relation to the canine and deciduous molar tooth germs. The maxilla continues to grow mainly upward, downward, and backward and with the development of a palatal process also spreads towards the midline in the substance of the anterior part of the united palatal folds.

About the 27 mm C.R. stage a mass of secondary cartilage appears in the zygomatic (malar) process, and by its proliferation for a time adds considerably to the bulk of this part of the maxilla. This area of cartilage is still present at 40 mm (Fig. 7.36). During this period the palatal process extends backward; at the union of palatal process and the main part of the developing maxilla a large mass of bone is produced. From this region, on the inner side of the dental lamina and tooth germs, the medial alveolar plate develops somewhat later than the lateral alveolar plate. The trough of bone thus formed is still later divided by septa into alveoli, as happens in the mandible. Small areas of secondary cartilage may develop along the growing margins of the alveolar plates as in the mandible, and in the middle line of the developing hard palate, between the two palatal processes.

The maxilla at birth

The maxilla, which from mid-term shows all the adult elements (Fig. 7.35), differs however at birth in several respects from the adult bone. Chief differences are the small size of the alveolar process, the lack of depth (which is related to the small size of the maxillary air sinus), and the cancellous nature of the bone. The maxillary sinus at birth forms a small depression (about the size of a small pea) on the medial aspect of the bone. The maxilla can be described as being composed of neural, alveolar and palatal elements and the zygomatic processes.

Comparison of mandibular and maxillary development (Fig. 7.37, Tables 7.1 and 7.2)

Both bones commence as centres of ossification in close relation to a corresponding nerve at the place of bifurcation; both bones have neural and alveolar elements; both develop secon-

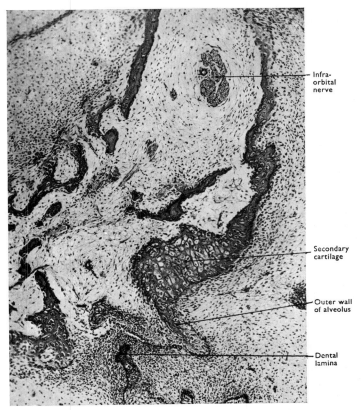

Infra-
orbital
nerve

Secondary
cartilage

Outer wall
of alveolus

Dental
lamina

Fig. 7.36 Photomicrograph of the developing maxilla in a 44 mm C.R. length human embryo showing secondary cartilage at the base of the zygomatic process.

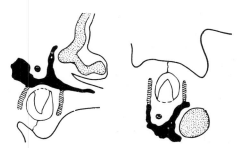

Fig. 7.37 Diagram comparing the developmental parts of the maxilla (on the left) and mandible. Solid black indicates the neural groove; hatching, the alveolar plates; stipple, cartilage. (By courtesy of *The Dental Practitioner*.)

dary cartilages in their backward extensions. Whereas the condylar cartilage, however, remains active as a growth centre for a long period in relation to the mandibular joint, the zygomatic (malar) cartilage of the maxilla is restricted in its appearance and activity to a limited period of foetal life and is related to the suture between it and the zygomatic bone. The maxilla has no muscular processes and the mandible no palatal process. In its growth the maxilla depends upon surface deposition and on growth at the sutures where it articulates with adjacent bones; the mandible depends for its growth after the first year on surface deposition and the replacement of cartilage by bone.

Table 7.1 Structures Related to the Developing Jaws

Mandible	Maxilla
Inferior dental nerve	Infra-orbital nerve
Meckel's cartilage	Nasal capsule
Tooth germs	Tooth germs

(By courtesy of *The Dental Practitioner*.)

Table 7.2 Developmental Elements

Mandible	Maxilla
Neural	Neural
Alveolar	Alveolar
Ramal	Zygomatic
Muscular	Palatal
Cartilaginous	Cartilaginous

(By courtesy of *The Dental Practitioner*.)

Development of the premaxilla

The premaxilla is formed in the region of the junction of the maxillary and frontonasal processes, though it is impossible to decide to what extent either of these processes contribute to the premaxilla since they have fused and so lost their identity before bone formation commences.

The premaxilla usually commences to ossify after the maxilla. It has two centres of ossification. The one responsible for the bulk of the bone is the first to appear. It starts close to the external surface of the nasal capsule in front of the anterior superior dental nerve and above the anlage (cellular condensation) for the second deciduous incisor (Fig. 7.38). From

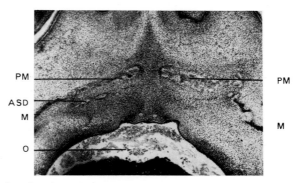

Fig. 7.38 Horizontal section through the premaxillary region of a 25 mm C.R. length human embryo. ASD, anterior superior dental nerve; M, maxilla; O, oral cavity; PM, premaxilla. (By courtesy of *The Dental Practitioner*.)

this centre bone formation spreads above the tooth germs of the incisors and then downward behind them, eventually forming the posterior wall of their alveoli and the palatal part of the premaxilla. In an upward direction the spread of ossification quickly produces the anterior half of the frontal process of the adult maxilla, and in a backward direction it rapidly unites with the maxilla.

The palatal part of the primary premaxillary ossification is joined by the infravomerine or paraseptal centre which appears about the end of the tenth week. It lies medial to the paraseptal cartilage which is a small plate of cartilage close to the middle line at the lower end of the nasal septum.

Though on the facial aspect all trace of union between the maxilla and premaxilla has disappeared before birth, on the palatal aspect the suture between the premaxilla and maxilla can still be seen until after birth extending from the region of the incisive foramen forward to the alveolar process between the canine and lateral incisor. Man is an exception in this respect for in other primates and lower animals the premaxillary-maxillary suture is visible on the face at birth and usually persists throughout life.

Development of the palatine bone

The palatine bone develops in membrane on the medial (inner) side of the cartilaginous nasal capsule. Ossification commences in the seventh to eighth week of foetal life in the region of the tuberosity in close relation to the descending palatine nerves. Ossification extends upward as the vertical plate and horizontally as the palatal process. By the end of the second month (30 mm C.R. length) all the processes of the bone are visible. At first the palatine bone is separated from the maxilla by the back part of the lateral wall of the nasal capsule. With atrophy of this part of the cartilage the vertical plate of the growing palatine bone comes to overlap the inner side of the nasal surface of the maxillary bone and helps to form the medial wall of the maxillary sinus. Eventually it forms the posterior boundary of the opening into the sinus.

The development of other facial bones

The *vomer* develops from two centres within the perichondrium covering the inferior margin of the septal cartilage during the ninth week of foetal life. These soon unite beneath the cartilage and extend backwards and forwards forming a trough of bone embracing the lower free margin of the cartilage. It reaches the inferior surface of the body of the sphenoid during the fourth month. With further development it changes its form in the coronal plane from a U-shaped structure embracing the septal cartilage to a Y-shaped structure with a stem descending towards the hard palate.

The *lacrimal* bone commences as a centre of ossification in membrane on the lateral side of the nasal capsule at about the end of the third month of foetal life.

Each *nasal* bone appears as a single centre of ossification in membrane on the surface of the anterior part of the roof of the nasal capsule towards the end of the second month of foetal life. The cartilage persists until after birth beneath the two nasal bones.

The *zygomatic* bone appears as a single centre of ossification in membrane just below and lateral to the eyeball at the end of the second month of foetal life. The growing bone comes quickly into contact with the temporal and maxillary bones but does not reach the frontal bone until somewhat later (Fig. 7.47).

A centre for each facial element of the *ethmoid* appears in the upper part of the cartilage forming the lateral wall of the nasal capsule during the fifth month of foetal life. Ossification has extended into all parts of the cartilage in the ethmoidal region by the time of birth. A separate centre appears in the upper part of the septal cartilage during the first year, extending upwards into the crista galli and downwards towards the vomer. The three parts of the bone, the bilateral facial elements and the mid-line cranial element (mesethmoid) unite across the cribriform plate region by the third year.

The *inferior concha* develops from a separate centre of ossification within the lower inturned end of the side wall of the nasal capsule about the fifth month of foetal life.

The cartilage of the capsule between the ethmoidal and inferior concha atrophies and in this region the developing maxillary antrum commences to invade the maxilla.

THE DEVELOPMENT OF THE MANDIBULAR JOINT AND MUSCLES OF MASTICATION

The mandibular joint is first indicated by the growth of the tissue condensation of the developing mandible, which everywhere precedes ossification, towards the corresponding condensation for the temporal region. The mandibular condensation maps out the shape of the condyle. At this stage the mandibular and temporal elements of the joint are still separated by a wide interval. The closer approximation of the mandible to the temporal region is brought about by the development of the secondary cartilage in the condylar process. After its appearance the cartilage produces a marked increase in the size of the condyle. By this rapid growth the previously wide interarticular interval is largely obliterated. The only intervening tissue left is a strip of dense tissue immediately above the upper surface of the condyle. This tissue appears at the same time as the condensation for the condyle and it is connected to the lateral pterygoid muscle from its first appearance. The strip of tissue becomes the *articular disk* (Fig. 7.39). The formation of the joint cavities above and below this strip of tissue occurs as the condyle becomes approximated to the temporal element of the joint. Joint cavity formation is virtually complete between the 65 mm and 70 mm C.R. stages. Small areas of secondary cartilage appear in the temporal region. They appear later than the condylar cartilage and disappear before birth.

At an early stage of foetal life the primitive jaw joint is situated between the posterior part

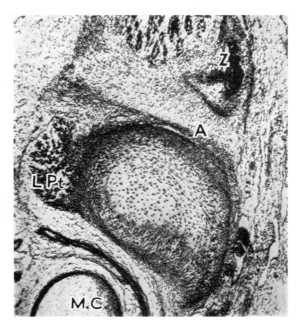

Fig. 7.39 Developing mandibular joint. The lower joint cavity has just started to appear above the cartilage of the mandibur condyle; (A) articular disk; (L Pt) lateral pterygoid muscle; (M.C.) Meckel's cartilage; (Z) Zygomatic arch. Coronal section. Human foetus of 57 mm C.R. length × 60.

of Meckel's cartilage (in which the malleus later develops) and the developing incus. Later with the development of the mandibular joint there is a brief phase when there are two joints in action on each side. With the disintegration of Meckel's cartilage the primitive jaw joint becomes the joint of the middle ear between the malleus and incus and mandibular movements are limited to those between the condyle and the temporal bone.

The major muscles of mastication develop as a single mass within the mandibular arch. During the second month of foetal life the individual muscles commence to differentiate, beginning with the medial pterygoid.

Early in foetal life the fibres of the upper head of the lateral pterygoid muscle can be seen to be continuous with the articular disk. At the back of the joint region many of the fibres composing the disk pass through the squamotympanic fissure to gain attachment to the malleus in the middle ear (Fig. 7.32). Later with closure of the fissure the majority of the fibres of the disk become attached to its bony margins, especially to its anterior (squamous) lip and the functional continuity of the lateral pterygoid muscle with the malleus through the articular disk is lost.

Before and after birth the temporal muscle migrates up the side of the skull increasing the area of its attachment. The lateral pterygoid and temporal muscles undergo continual re-attachment to their sites of insertion with growth changes in the neck of the condyle and coronoid process.

The dental arches lie between the tongue on the inner side and the lips and cheeks on their outer aspect. Study of the developing jaws and teeth from early embryonic life onwards has demonstrated that the form of the dental arches is first determined by the form of the dental lamina, the developing jaw bones, by the form of the cartilaginous facial skeleton, particularly Meckel's cartilage, and the lower rim of the nasal capsule. Except for the considerable and obvious difference in size, the form of the dental arches in the human foetus does not vary a great deal from that found in the adult.

DEVELOPMENT OF THE CHONDROCRANIUM

The cartilaginous or primordial skeleton of the human skull commences to develop about the second month of intra-uterine life with the appearance of centres of chondrification beneath the developing brain and in relation to the head end of the notochord. When fully developed at about the fourth month of foetal life, the chondrocranium consists of the following parts (Fig. 7.40).

A *central stem* made up of three regions from behind forward:

1. Parachordal region, which develops in close relation to the front end of the notochord;
2. Trabecular (prechordal) region, which develops in relation to the pituitary gland, and extends forwards as the:
3. Interorbitonasal septum, between the eyes and to which is attached the nasal capsule.

Cartilaginous sense capsules, the auditory (otic) and nasal capsules. These develop in close relation to the central stem and become united to it.

Lateral structures and various commissures uniting these to the central stem. The chief lateral structures on each side are:

1. The exoccipital process
2. The ala temporalis
3. The ala orbitalis.

All the cartilaginous sense capsules and the lateral elements of the chondrocranium become united to the mid-line central stem to make up an irregular cartilaginous plate lying below and supporting the developing brain. The lateral processes also unite with one another, leaving

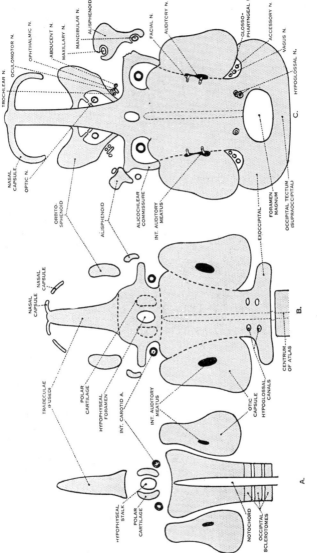

Fig. 7.40 Three stages in the development of the neural portions of the chondrocranium (based on de Beer, 1937). Right side in C is at a later stage than left side. (By courtesy of Professors Hamilton, Boyd and Mossman and W. Heffer & Sons Ltd, Cambridge.)

various foramina and fissures between themselves and the mid-line region for the passage of cranial nerves and blood vessels.

Two bilateral cartilaginous bars are attached to the lower aspect of the chondrocranium. These are Meckel's cartilages which form the primary skeleton of the mandibular arch, and the cartilage of the second pharyngeal or hyoid arch.

From about the middle of foetal life various centres of ossification appear within the cartilage of the chondrocranium, giving rise to a number of parts of the adult skull which are listed on page 370. At birth the remaining unossified parts of the chondrocranium are the nasal septum and the spheno-occipital synchondrosis.

The parachordal region (Fig. 7.40)

Two centres of cartilage appear on each side of the terminal part of the notochord and soon unite around it so that it comes to occupy a canal in the cartilage. The growing cartilage extends backwards into condensed connective tissue derived from the three uppermost (occipital) somites (see page 330). In this process the cartilage surrounds the roots of the hypoglossal nerve, which becomes included in the cranial series of nerves.

Chondrification of the auditory capsules commences first around the developing semicircular canals and then around the cochlear part of the inner ear. Later cartilaginous union takes place between the inner side of the auditory capsule and the parachordal plate in front of the emerging ninth, tenth and eleventh cranial nerves (jugular foramen). An extension of the parachordal plate, the exoccipital process, joints the auditory capsule behind the emerging nerves. The eighth nerve enters the auditory capsule to terminate in relation to the inner ear structures. The seventh nerve passes outwards over the upper surface of the auditory capsule, between the semicircular canal region behind and the cochlear region in front.

The trabecular (prechordal) region

On each side of the pituitary gland two centres of chondrification appear; the front pair are the trabecular cartilages; the back pair are the polar cartilages. In man these soon unite beneath the gland and become continuous with the parachordal (basal) plate behind and continue forwards as the nasal septum (interorbitonasal septum).

The ala orbitalis (orbitosphenoid) appears as a separate centre of cartilage formation on either side of the central stem. It becomes united to the stem in the region of union between the prechordal region and the interorbitonasal septum by two roots embracing the optic nerve, so forming the optic foramen.

The ala temporalis (alisphenoid) appears as a separate centre of chondrification which becomes united to the side of the central stem in the trabecular region. At first the union between the parts has the appearance of a developing diarthrodial joint. The alisphenoid is thought by some authorities to be derived from part of the primitive skeleton of the upper jaw—the pterygoquadrate cartilage. The developing ala temporalis embraces the emerging maxillary division of the fifth nerve, while the mandibular division passes through a notch in its posterior border. This later becomes the foramen ovale. The ala temporalis gives origin to some fibres of the developing lateral pterygoid muscle on its under surface.

Between the ala orbitalis and ala temporalis the third, fourth, sixth cranial nerves and the ophthalmic division of the fifth nerve leave the cranium. Between the ala temporalis and the front of the auditory capsule the internal carotid artery enters the cranial cavity.

The nasal capsule (Fig. 7.23)

The nasal septum is the forward continuation of the central stem (the interorbitonasal septum). The lateral parts of the nasal capsule forming the side walls of the primitive nasal cavities unite with the nasal septum around a large mid-line aperture through which pass the branches of the olfactory nerve in the region of the future cribriform plate. In man and the higher primates there is no connection between the lower border of the nasal septum and the lateral parts as in other animals (lamina transversalis anterior and posterior).

By the end of the second month the chondrocranium is fully differentiated and the cranial base, nasal capsule and Meckel's cartilage are growing rapidly as a result of cartilage cell proliferation. The nasal septum doubles the length attained at 10 weeks by 14 weeks, trebles it by 17 weeks, and is six times as great by 36 weeks. Other dimensions of the growing skull show a similar rate of growth but the parachordal region grows less rapidly, increasing by about four times by the 32nd week. The dominant feature of facial growth during foetal life is the nasal capsule. Upper facial growth progresses in a downward and forward direction from the anterior part of the cranial base and the nasal capsule appears to act as a pacemaker for these growth changes. At birth the maxilla is a relatively small bone and the upper tooth germs lie close to the floor of the orbital cavity. Increase in height of the maxilla is correlated with expansion of the maxillary sinus (Fig. 7.45).

Later, centres of ossification appear in the chondrocranium from which many of the bones of the skull and face, or parts of these bones, develop. The parts of the adult skull which develop in this way (endochondral ossification) are:

1. The nuchal region, condylar regional (exoccipital) and basal part (basioccipital) of the occipital bone.
2. The petromastoid part of the temporal bone (page 210).
3. The body, lesser wings and the roots of the greater wings of the sphenoid (page 215).
4. The perpendicular plate and crista galli of the ethmoid (mesethmoid).
5. The lateral masses (facial parts) of the ethmoid.
6. The inferior turbinate bones of the nasal cavity.

The malleus, incus, and stapes develop from Meckel's cartilage, the pterygoquadrate bar, and the hyoid arch respectively. These are cartilaginous structures developing in close relationship with the chondrocranium. The styloid process also develops from part of the hyoid arch cartilage, and the hyoid bone develops from parts of both the second (hyoid) and third pharyngeal arch cartilages.

Other bones develop in membranous condensations in close relation to the chondrocranium. These include:

1. The greater wing of the sphenoid to the outer side of the foramina rotundum and ovale, and the lateral pterygoid plate. This part of the sphenoid develops in maxillary arch mesoderm lying immediately beneath the brain coverings and unites with the ala temporalis.
2. The tympanic plate of the temporal bone develops in membrane on the outer side of the middle ear ossicles.
3. The maxillary, zygomatic, nasal, lacrimal and palatine bones and the medial pterygoid plates develop in relation to the outer wall of the nasal capsule.
4. The mandible develops in relation to Meckel's cartilage (see page 356).
5. The vomer develops in relation to the free lower edge of the cartilaginous nasal septum.

Certain bones of the skull vault develop in membrane. In man these bones are not closely related to cartilage. In other animals, however, cartilage contributes more extensively to the formation of the developing cranial vault. The membrane bones are the frontal, parietal, the squamous part of the temporal and the interparietal part of the occipital.

After birth the only regions of the skull remaining as cartilage derived from the chondrocranium are at the spheno-occipital synchondrosis and the nasal septum (Figs 7.46, 7.47). The foramen lacerum is also filled in during life by fibrocartilage derived from the same source.

GROWTH OF THE HUMAN BODY

The nature of growth

Between the time of fertilization and the attainment of adult life the human organism increases tremendously in size as well as in complexity. Increase in size is the result of cell multiplication and the accumulation of certain products of cell activity within the body such as body fluids, connective tissue, and the matrix of bone and cartilage. The marked difference in the size of various animal species is the result of a greater or lesser number of cells being produced from the single ovum. Changes in form are the consequence of the pattern of cell division in the various regions of the body. Both rate and pattern of cell division is under considerable genetic and hormonal control, but growth also depends significantly on adequate nutrition and a favourable intra-uterine and postnatal environment.

With increasing age the rate of cell division becomes progressively reduced and tends to be localized in well defined regions such as epiphyseal plates, bone surfaces, including the sutures, the deeper layers of the skin and mucous membranes. The total number of highly specialized cells such as neurons and muscle cells is already determined by the time of birth. Further growth is a matter of increase in the size of these cells or the length of their processes (e.g. nerve cells and odontoblasts). Concomitantly, highly specialized cells tend to lose their capabilities for repair after injury, being usually replaced by fibrous tissue, if damaged.

Measurement of growth

Growth can be estimated by linear measurements or from the weight of the body as a whole, its component parts or organs. There are two types of growth study, one based on changes

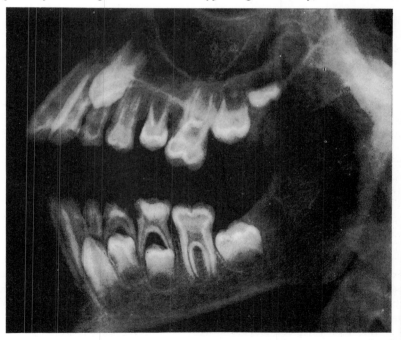

Fig. 7.41 Lateral radiograph of the jaws of a boy at six years. The first permanent molars have erupted behind the deciduous molars, while the second permanent molars are still buried. The permanent bicuspids (premolar) teeth are straddled by the roots of the deciduous molars, and farther forwards are the permanent canine and incisor teeth lying adjacent to the partly absorbed roots of their deciduous counterparts. (By courtesy of Professor G. A. G. Mitchell and Dr E. L. Patterson.)

in the same individual at different stages over a period of years (longitudinal growth studies), the other on the measurement of a number of individuals each of the same age or within the same age period (cross-sectional growth studies). The former is essential in determining the growth pattern of any given individual; the latter is useful in the study of population trends.

Linear growth measurements can be taken directly from the body, from photographs or radiographic plates, the latter being the basis for cephalometric studies of special importance in the study of craniofacial growth (page 443).

The precise amount, rate and pattern of growth can be demonstrated by the use of a variety of graphs. Those in most common use show the change from period to period by comparing the increase in a selected dimension with increasing age, or graphs in which the rate of gain between each interval is compared with increasing age (velocity curve). Another useful form of graph shows the percentage of adult size reached at various ages.

The stage of development of an individual can be measured not only in relation to chronological age but in relation to skeletal age, such as is indicated by the stage of development of the small bones of the wrist, or dental age as indicated by the developmental condition of the deciduous or permanent dentitions (Figs 6.23, 7.41). These do not always coincide with the findings based on chronological age and are useful adjuncts, especially in conditions of abnormal development involving the skeleton and dentition.

Patterns of growth
Before birth the maximum velocity for body length is reached about the middle of foetal life followed by a period of relative slowing of the rate of growth, until immediately after birth when the rate increases again.

After birth the growth pattern is somewhat different for different tissues:

1. Bone and muscle tissue, including the heart, grow rapidly, but at a decreasing rate until about the third year. This is followed by a period of steady, although less rapid growth until about the twelfth year, when there is a spurt in growth rate associated with puberty (the adolescent growth spurt). This begins and reaches its maximum about two years earlier in girls than in boys, being one explanation why adult men are usually larger than women.

2. The tissues of the brain and eyeballs and the parts of the skeleton associated with them (cranium and upper part of the face) grow much more rapidly up to about 7 years of age, before assuming a much slower rate until adulthood. They show little or no evidence of an adolescent spurt. The growth of the lower part of the face (below the Frankfort plane, see page 445) shows a growth curve which is intermediate in form between that demonstrated for the general skeleton and the orbitocranial part of the skull.

3. The reproductive organs, both internal and external, show a curve almost the reverse of that for general skeletal structures. For the first 12 years growth is slow. Following puberty, at about 12 years, growth is rapid for several years reaching adult dimensions at about 20 years of age.

4. The lymphoid tissues of the tonsils, adenoids, thymus, spleen and intestines grow rapidly until about 10 years of age, thereafter demonstrating an extensive reduction in amount (atrophy).

5. Subcutaneous fatty tissue increases in amount during the first year, then shows a decrease, especially in boys, until about 7 years when it again increases in quantity. During adolescence boys, but not girls, show a further decrease of fatty tissue until the attainment of adulthood. During adult life many people show a further increase in the amount of body fat, usually at the expense of muscle tissue, and accentuated by incorrect diet or insufficient exercise.

It will be seen that the growth pattern of children differs considerably in the first and second decades.

Nutrition and exercise
Adequate nutrition, both in regard to amount and a well balanced diet, is of course essential

for normal growth. Prolonged malnutrition or starvation leads to the slowing down, cessation and reversal of growth and ultimately to death. Children, however, have great powers of recuperation, provided the starvation state is not carried too far or is too prolonged. After a period of malnutrition the rate of growth is more rapid than normal until the normal rate for the child is again reached, demonstrating a compensatory 'rebound'. This phenomenon is better demonstrated in girls than in boys, the female of the species being more resistant than the male in this respect.

Muscular exercise can increase the size of voluntary muscle. This is produced by an enlargement of individual muscle fibres (hypertrophy). In association with this process there is an increase in the diameter, but not the length, of associated limb bones brought about by subperiosteal bone deposition. With cessation of exercise, however, both the muscles and bones tend to return to their previous dimensions.

There is also some evidence that psychological factors may affect the rate and amount of growth, but all aspects of growth, and especially the determination of the ultimate size and form of the body, are regulated by the genetic constitution of each individual. The genes work through the intermediary action of enzymes and hormones, especially those of the pituitary, adrenal and sex glands. The time of onset of the adolescent growth spurt, however, appears to be determined primarily by the hypothalamic centres of the brain, at a time when growth and maturation of the brain is to all intents complete.

Body build

This is a matter of the proportional size of the body as a whole and, as already stated, is predominantly a matter of heredity. Classifying people in accordance with their body form has a long history in medicine and anthropological studies. The most frequently used contemporary system is that of Sheldon in which individuals are classified as to their content of the three basic elements of *endomorphy*, *mesomorphy* and *ectomorphy*, each element being represented on a seven unit scale, so that an extreme endomorph would have a rating of 7–1–1, an extreme mesomorph of 1–7–1, and an extreme ectomorph of 1–1–7. Common combinations of these elements or somatotypes would include 4–3–4 and 3–5–3. It is claimed that with each body form there is associated a corresponding temperament and tendency to certain types of disease.

Classification is carried out mainly by inspection of standardized photographs taken to show front, side and back view of the unclothed body.

Endomorphs show a round head, smooth face, large abdomen and relatively weak limbs, tend to be good-natured, fond of food and company, and are liable to metabolic disorders.

Mesomorphs show predominance of skeletal and muscle tissues, have rugged facial features and strong limbs, tend towards aggressiveness of action and manner, are extrovert in temperament and liable to cardiovascular disorders.

Ectomorphs are lean, non-muscular with narrow faces and liability to crowding of the teeth. They tend towards introversion and psychosomatic disorders.

Some individuals show a significant variability in the classification of the various parts of their body. This condition, known as *dysplasia*, is estimated by noting the separate rating of each body component. The overall constitutional form of the body does not appear to alter much with age or changes in diet. An endomorph deprived of food becomes a starved endomorph, not an ectomorph.

GROWTH OF THE SKULL

Growth of the skull can be estimated either by various measurements on the living body taken by means of calipers or by measurements between certain landmarks on standardized lateral and frontal X-rays of the skull. The latter method (cephalometric analysis) is especially valuable in growth studies of the face during childhood and early adult life. Various points are selected between which lines and planes are drawn (Figs 9.7, 9.8), which in some cases meet one another at angles situated inside or outside the craniofacial skeleton. Growth of the skull as a whole or of various craniofacial regions (page 381) can be estimated by:

1. Linear increase of the various planes.

2. Changes in the size of the various angles.

3. Proportional studies in which the growth of various planes is compared with a selected standard plane, such as the distance between the midpituitary point (sella point) and nasion situated at the bridge of the nose: the S–N plane (Fig. 9.8).

In this manner the amount, rate and direction of growth can be noted throughout childhood. This is of special value in diagnosis and treatment planning in orthodontics.

There are three ways by which growth of bone in the skull occurs:

Growth by the conversion of cartilage into bone

Centres of ossification appear within the chondrocranium and grow by replacement of cartilage, as in typical long bones (page 25). Areas of growing cartilage between individual bony centres form synchondroses which are important sites of craniofacial growth. They are similar in structure and function to the epiphyseal plates of long bones, but differ from them in that growth occurs on both sides of the synchondrosis, so that they function as 'double' epiphyseal plates. The most important are:

1. The spheno-occipital synchondrosis in the mid-line cranial base between the occipital and sphenoid bones (page 381). It persists through puberty, until 16 years of age or later, closing somewhat earlier in girls than in boys (Figs 2.14, 5.30).

2. The sphenoidal synchondrosis in the mid-line cranial base between the anterior and posterior parts of the sphenoid. In man it closes just before birth, in contrast to the condition in many primates where it persists for some time after birth, continuing to function as an important growth centre.

3. The bilateral synchondrosis between the body and great wing of the sphenoid, i.e. between the membranous and cartilaginous elements. This synchondrosis also closes at about the time of birth in man.

Two important regions in the mandible must not be overlooked, where growing cartilage is converted into bone during the growing period. The condyle is an important growth centre until the beginning of adult life and the cartilage found at the mandibular symphysis contributes to mandibular growth until the first year after birth. It should be understood that these mandibular cartilages are not part of the primary cartilaginous skeleton or chondrocranium, but are 'secondary' cartilages which develop later and quite independent of the cartilage of the cranial base.

Growth by surface deposition

By this method new bone is laid down beneath the surface periosteum and quite independently of any cartilage formation. This is the mechanism which causes increase in the girth of long bones and all the bones of the skull, including the facial skeleton, depend on this method for part of their growth even though they may have begun their existence in cartilage, as endochondral bones. Surface deposition of bone matrix becomes more important during the later periods of skull growth and is often associated with a corresponding resorption of previously formed bony tissue. The nasal cavities and paranasal sinuses grow by this means, new bone being added on the outer surfaces of the face and removed on the inner aspect of these cavities. The result is basically an increase in size of the face and the contained air-filled cavities. The continuing proper balance between bone deposition and resorption is a very important factor in changing the shape of individual bones of the skull.

Suture growth

This mechanism is comparable to surface deposition, but takes place at the edges of adjacent bones which form the boundaries of a suture. The five layers of tissue within a growing suture are described on page 355. At each suture there are in effect two growth centres, one for each of the adjacent bones. As a result one bone may grow faster than the other at a given suture and this results in movement or drift of the suture during the growing period.

There are two views about the mechanism responsible for the separation of the bones which surround a suture. According to one view the proliferation of the soft tissues in the suture separates the bones bounding it. The other view holds that separation of the bones at sutures is produced by the growth of other organs, such as the brain or eyeball, or by the growth of cartilage at a synchrondrosis between two bony elements in close relationship to one or more suture systems. If one accepts this view, the bones grow into the spaces produced by separating mechanisms which may be located some distance from the growing sutures.

SUTURE SYSTEMS AND SKULL GROWTH

The various craniofacial sutures can be grouped together in a number of suture systems. At each system growth takes place to a predominant extent in a definite direction. The major suture systems are as follows (Figs 7.42, 7.43, 7.47):

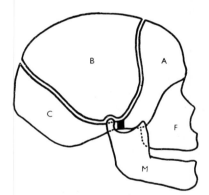

Fig. 7.42 Diagrammatic representation of the cranial segments and their relation to the upper and lower parts of the face. (A) the anterior cranial segment. (B) the middle cranial segment. (C) the posterior cranial segment. (F) the upper part of the face. (M) the lower jaw. The spheno-occipital synchondrosis is shown in solid black.

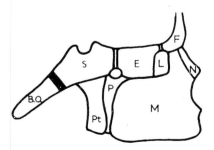

Fig. 7.43 The relation of the maxilla to the other bones of the face and the cranial base in man. (M) maxilla. (N) nasal bone. (F) frontal bone. (L) lacrimal bone. (E) ethmoid. (P) vertical plate of palatine bone. (S) body of the sphenoid. (Pt) pterygoid plate. (Bo) basi-occipital.

The lambdoidal suture system

This separates the occipital bone behind and the parietal and temporal bones in front. The occipital bone makes up the *posterior cranial segment*, and the parietal and temporal bones the *middle cranial segment*.

The coronal suture system

This separates the middle cranial segment from the *anterior cranial segment* made up of the frontal, sphenoid and mesethmoid elements of the cranium (Fig. 7.44).

Both the lambdoidal and coronal suture systems meet at the base of the skull at the foramen lacerum, on each side of the spheno-occipital synchondrosis. This important growth centre

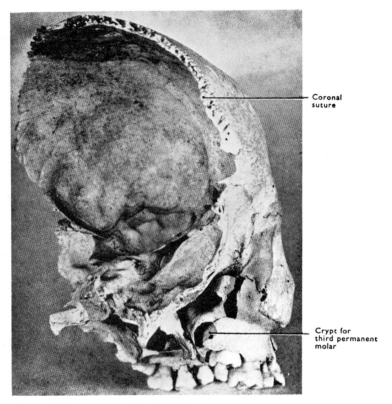

Coronal
suture

Crypt for
third permanent
molar

Fig. 7.44 The anterior cranial segment seen from behind.

between the basi-occipital and basisphenoid growth bones separates the posterior and anterior cranial segments (Figs 2.14, 7.46).

At the pterion an anterior limb of the coronal suture system passes in front of the great wing of the sphenoid and separates the sphenoid from the frontal, mesethmoid and zygomatic bones. At the base of the skull it separates the body and lesser wings of the sphenoid from the frontal (orbital plates) and ethmoid (cribriform plates) in the floor of the anterior cranial fossa.

The pterion is the small area on the lateral surface of the skull where the frontal, parietal and sphenoid bones meet one another. It is an important clinical landmark because it indicates the position inside the skull of the trunk of the middle meningeal artery, which is related to the lower end of the motor area of the cerebral cortex (page 404).

The main stem of the coronal suture system crosses the middle cranial fossa separating the squamous temporal bone from the great wing of the sphenoid. In this position the suture runs from before backwards to behind the foramen spinosum and then forward to the foramen lacerum.

Growth at all these suture systems will regulate the anteroposterior growth of the cranial cavity. Growth at the coronal suture as it crosses the middle cranial fossa at the base of the skull will also regulate the outward growth movement of the middle cranial segment (temporal bone) and produce separation of the glenoid fossae (see page 383).

The craniofacial suture system

This separates the anterior cranial segment (frontal, mesethmoid and sphenoid) from the circummaxillary facial bones; the lacrimal, facial ethmoid, vomer, palatine, and zygomatic. It consists of an anterior horizontal limb running within the orbital cavity (between the frontal and the lacrimal and ethmoid, and between the sphenoid and the palatine and zygomatic); and a posterior vertical limb in the pterygopalatine fossa region separating the medial pterygoid plate from the vertical plate of the palatine. In the nasal septum the body of the sphenoid is separated from the vomer. Growth at this suture system will thrust the facial bones downwards and forwards from the bones of the anterior cranial segment.

The maxillary suture system

This separates the maxillary bones from the circum-maxillary facial bones, the lacrimal, facial ethmoid, palatine vomer and zygomatic, and from the frontal bone (frontomaxillary suture) and nasal bones. Like the craniofacial suture system, it consists of an anterior horizontal limb within the orbital cavity and a posterior vertical limb in the pterygopalatine fossa. Growth at this suture system will thrust the maxilla downwards and forwards relative to the circummaxillary facial bones (Figs 5.13, 7.43). Studies of human skulls show that some of the sutures between the maxilla and the surrounding bones do not fuse in the third decade, as was once believed, but instead may not unite until old age and then only partially. This finding is of value in the treatment of adults who have protruding maxillae and other related facial deformities that require correction by orthognathic surgery.

The sagittal suture system (Fig. 7.47)

At birth this extends from behind forwards between the parietal bones, the frontal bones (metopic suture), the nasal bones, the maxillary bones (midpalatal suture) and the two halves of the mandible. At the base of the skull the sagittal suture system during foetal life is bilateral in relation to the mid-line cranial base. It is made up on each side of the fronto-ethmoidal suture, the synchondrosis between the great wing of the sphenoid and the body of the sphenoid, the suture between the petrous temporal and the basi-occipital and the exoccipital synchondrosis. By the end of the first year in the majority of skulls the frontal bones begin to unite, the mandible has become a single bone, and the greater wings have united with the body of the sphenoid. During foetal life and the early postnatal period growth at this suture system will regulate growth in width of the face and cranium.

MECHANISMS OF SKULL GROWTH

The presence of sutures permits a considerable degree of local autonomy and regional adjustment in the growth of the skull. Growth of the *cranial vault* is closely associated with that of the brain. The bony elements develop in a fibrous tissue or membranous capsule (hence they are defined as membrane bones). Ossification extends from a number of centres within the growing capsule, the sutures and fontanelles being areas where growth continues until the brain reaches adult size. Premature closure of sutures or suture systems in the infant leads to deformities of skull form but not necessarily to overall reduction of skull size, as the mechanism of surface resorption and deposition is still intact.

Premature closure of the sagittal suture results in elongation of the anteroposterior diameter of the head, so that it becomes long and narrow with a bulging forehead and occipital region. This developmental anomaly is called *scaphocephaly*. Premature closure of the coronal suture causes a relative shortening of the antero-posterior diameter of the head, with a compensatory increase in width. The skull is then described as *brachycephalic*.

Growth of the *cranial base* is less closely related to that of the brain. In this region ossification centres develop in preformed cartilage (the chondrocranium), and this tissue has its own independent pattern of development and growth. After birth, however, the only remaining cartilaginous growth centre in the human chondrocranium is the spheno-occipital synchrondrosis (page 382). Here again the mechanisms of bone resorption and deposition permit local adjustments throughout the growth period.

Growth of the *maxillary bones* (the midfacial region) is associated with that of the cartilaginous *nasal septum*, which is continuous with the cartilage of the chondrocranium. This midfacial structure thrusts the maxillary elements downwards and forwards relative to the cranial base, allowing growth to occur at the circum-maxillary suture system. There is, however, another mechanism involved. Both the orbital and posterior (infratemporal) surfaces of the maxillary bones are closely related to masses of compact fibro-fatty tissue. Surface deposition of bone in relation to this tissue will also result in a downward and forward translation of the maxillary bones.

Growth in width of the postnatal facial skeleton is largely a matter of surface deposition and internal resorption in the orbital, nasal, paranasal and oral cavities. There may be some maxillary-zygomatic lateral translation associated with growth of the temporal muscles, because the sutures related to these bones remain open long after the bilateral elements of the frontal and mandibular bones have united. The zygomaticotemporal sutures (in the zygomatic arches) are regions of adjustment between the cranial and mid-facial regions of the skull.

The *mandible* (lower facial skeleton) is, in its development and evolution, an element of the subcranial branchial arch system. It is the most mobile of the cranial elements and its position relative to the other cranial components depends not only on the growth of the condylar cartilages, but on the balance of muscle action (compare with the hyoid bone) and the nature of the occlusion between the upper and lower teeth.

Growth at the mandibular condyles in a posterosuperior direction increases the total anteroposterior dimension of the mandible and the height of the ramus. Bone is added by surface deposition along the posterior border of the ramus and resorbed from its anterior border. These changes progress at rates which are proportional to the amount of condylar growth and lead to an increase in the depth of the body of the mandible and the width of the ramus as growth progresses.

PERIODS OF SKULL GROWTH

Skull growth in man falls into three periods:

1. During foetal life and up to about the third year after birth.
2. From about 3 years to about the end of the first decade.
3. After 10 years of age (second decade).

During the first period (foetal life to about 3 years of age) growth is active at most of the suture systems and is related to the rapid growth of the chondrocranium, brain, eyeballs, ear structures and tongue. The various parts of the occipital, temporal and sphenoid bones unite to form the single bones of adult life. The fontanelle between the frontal and parietal bones is usually obliterated by the end of the second year. An ossification centre appears after birth in mesethmoid and commences to invade the back of the nasal septum. The mesethmoid (perpendicular plate) and the lateral masses of the ethmoid, which are ossified at birth, commence to unite across the cribriform plate between 2 and 3 years of age. This union between a cranial and facial element indicates a slowing down of active growth in the craniofacial suture system.

During the second period (3 to about 10 years of age) growth at the sutures is considerably reduced. By the end of the first decade brain growth has reached almost adult dimensions, the orbital cavities are close to adult size, the cribriform plate and petrous temporal bones have reached adult size and growth is much reduced in the craniofacial and maxillary suture systems. During this period the perpendicular plate of the ethmoid unites with the vomer and the regions of active cartilaginous growth become limited to the spheno-occipital synchrondrosis and the mandibular condyles. Growth at the former of these sites may continue until about 16 years of age and continues to control growth at the lambdoidal and coronary suture systems so that the upper facial skeleton, carried by the anterior segment (Fig. 7.44), continues to grow forward relative to the vertebral column. Meanwhile growth of the mandible in the forward and downward direction maintains the jaws in normal relationship.

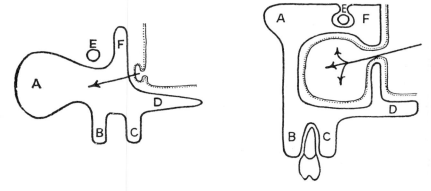

Fig. 7.45 Diagram illustrating how increase in height of the maxilla is correlated with the expansion of the maxillary sinus. *Left*, foetal maxilla; *right*, adult maxilla; A, zygomatic process; B, outer alveolar wall; C, inner alveolar wall; D, palatal process; E, neural groove; F, inner orbital plate. (By courtesy of *The Dental Practitioner*.)

During the third period (from about 10 years to adult life) growth at the facial suture systems is of much less significance although it usually shows a slight spurt during the adolescent period. Growth of the condylar cartilage is still active but less intensive. Otherwise growth of the facial skeleton is predominantly the result of surface deposition. This takes place on the facial surfaces of the mandible, maxillary, zygomatic and frontal bones and is especially marked in the alveolar region of the jaws as the dentition continues to migrate forwards and outwards until the permanent dentition is established with the eruption of the third molars. Associated with this process of surface deposition of bone, which begins shortly after birth and becomes predominant after about the seventh year, there is a co-ordinated process of internal absorption of bone in the facial skeleton so that the nasal cavities grow in height by resorption of bone from the upper surface of the hard palate as bone is deposited on its under (oral) surface and the air sinuses continue to invade and excavate the adjacent bones (Fig. 7.45). The maxillary antrum grows chiefly downwards and forwards and to some extent backwards from the position of its original out-pouching from the nasal cavity in the region of its opening. At 3 years of age it is related to the second deciduous molar (to the first if large) and the crypt of the developing first permanent molar. It continues to grow forwards and downwards with descent of the alveolar process so that in the adult its floor extends forwards to the second premolar which has replaced the second deciduous molar. The second and third permanent molars develop in the alveolar bulb and with their eruption pass downwards and forwards so as to lie in relation to the floor of the antrum.

At birth the tongue is relatively large in relation to the mouth cavity and may pass between the gum pads and come in contact with the lips and cheeks (this is the normal relationship in foetal life). After birth the alveolar processes and the supporting facial skeleton grow rapidly and by the time the deciduous dentition is complete the tongue should have adequate room within the dental arches.

The division of skull growth into three periods must not be accepted in too rigid a manner for there is considerable individual variation.

SKULL GROWTH AND MUSCLE FUNCTION

Certain regions of the skull are developed in relation to the attachments and functional activity of groups of muscles. These include:

1. The nuchal lines and fossae on the squamous portion of the occipital bone, above and behind the foramen magnum, for the attachment of the posterior neck muscles (Fig. 5.3).

2. The mastoid process and styloid process for the attachment of the sternocleidomastoid, digastric and styloid muscles.

3. The temporal lines on the side of the skull which delimit the area of attachment of the temporal muscle and its covering temporal fascia (Fig. 5.2).

4. The zygomatic processes and lateral pterygoid plates for the origin of the masseter and pterygoid muscles.

5. The ramus of the mandible, especially the region of the coronoid process, for insertion of the temporal muscle, and the angle, for the insertion of the medial pterygoid and masseter.

6. The mylohyoid ridge and genial tubercles on the inner surface of the body of the mandible.

The extent to which these regions of the skull are developed depends to a considerable extent on the functional activity of the various muscle groups concerned.

Muscle action also influences in a less direct manner the growth of the facial buttress systems and the alveolar processes, in that it determines the functional activity of the whole masticatory apparatus (Fig. 5.21).

Table 7.2 The joints of the skull

Main group	Type	Example
Synarthroses (fixed joints)	Synchondrosis (cartilage union)	Spheno-occipital (fuses 11–20 years)
	Syndesmosis (fibrous union)	Metopic suture (fuses 1–2 years) Coronal suture (begins to fuse 30 years)
	Gomphosis (fibrous union)	Tooth attachment (never fuses)
Diarthroses (movable joints)	Simple (single cavity)	Atlanto-occipital joint (condyloid joint)
	Compound (double cavity)	Temporomandibular joint Upper cavity—plane joint Lower cavity—hinge joint

REGIONAL GROWTH OF THE SKULL

Cranial vault

Variations in cranial form have long been used in the classification of racial groupings and subgroupings. Growth of the cranial cavity involving the lambdoidal, coronal and saggital suture systems, is very rapid during late foetal life and the first year of life. By the seventh year it is almost complete. A certain amount of growth appears to take place especially at the coronal suture system after this time, and is probably related to a thickening of the bones of the vault (by internal and external deposition) and the development of the diploë (cancellous tissue) between the inner and outer cortical layers or tables of the skull.

The bones making up the cranial vault are the frontal and parietal, the squamous parts of the occipital and temporal, and the great wings of the sphenoid (Fig. 5.2). This region is closely related in its growth to that of the brain. In microcephalic idiocy it is small; in hydrocephalus it is large. In anthropoid apes, but not in man, the cranial vault may have superimposed upon its basic form various crests and ridges which extend the attachment of the neck and temporal muscles. In man it is invaded to a limited extent by the frontal air sinuses. In other animals such as the pig and elephant air sinuses invade the cranial vault to a considerably greater extent. The thickness of the bones of the cranial vault is probably in some degree related to the development of the muscles of mastication. It is a useful generalization to say that the vault consists of an inner (endocranial) plate related closely to the brain and its covering membranes, and an outer (ectocranial) plate related to the degree of development and functional activity of the neck and jaw muscles. The sinuses and cancellous diploic tissue develop between these plates.

The form of the cranial vault, unlike its size, is probably not entirely determined by brain growth but depends on independent genetic factors, which determine dolichocephaly (long headedness), mesocephaly and brachycephaly (short headedness). These variations in endocranial form are found in anthropoid apes as well as in man. Apart from the genetic factors determining the form of the skull vault, its shape can be considerably modified if the head is bandaged in infancy or if there is a premature closure (synostosis) of the cranial sutures. In these circumstances the vault is often markedly deformed.

The cranial base

The cranial base as seen from within the skull (Fig. 5.6) is made up of a mid-line portion and the three cranial fossae. The bones forming the floor of the anterior cranial fossa are the orbital plates of the frontal, the cribriform plates of the ethmoid, and the lesser wings of the sphenoid. The foramina in this region are the foramen caecum, the optic foramina and the foramina of the cribriform plates (Fig. 5.11).

The bones forming the floor of the middle cranial fossa are the body and great wings of the sphenoid, and the upper surface of the petrous temporal. The foramina in this region are the foramen rotundum, foramen ovale, foramen spinosum, all in the great wing of the sphenoid; the foramen lacerum and the anterior (cranial) opening of the carotid canal (Fig. 5.11). The bones forming the posterior cranial fossa are the occipital in its floor and the postero-medial surfaces of the petrous temporal in front. The chief foramina are the foramen magnum, the jugular foramina, the anterior condylar canals and the internal auditory meatuses.

The *mid-line cranial base* extends from the foramen magnum (basion) to the foramen caecum and consists of the basioccipital, sphenoid body, and cranial ethmoid (mesethmoid). If the cranial base is taken to extend forward to the nasion, as is often done, it will contain at its front end part of the frontal bone (Fig. 5.19). Except for the frontal part, its various bony elements develop in the cartilage of the chondrocranium in front of and behind the pituitary fossa. In man it contains the *spheno-occipital synchondrosis*, an important site of growth during childhood. In many young animals there is a second cartilaginous growth centre between the postsphenoid and presphenoid elements. In man these unite just before birth. Growth of the cranial base is independent of that of the brain; it can be of normal size in microcephalic skulls. Its growth is seriously affected, however, in such conditions as cretinism, Down's syndrome, and achondroplasia. It does not grow equally in all parts. The middle segment (from pituitary fossa to foramen caecum) has completed its growth by about seven years of age; its anterior (frontal) segment from foramen caecum to nasion continues to grow until adult life and is related to the growth of the frontal air sinus; its posterior segment (from foramen magnum to pituitary fossa) also grows until adult life is reached. In anthropoid apes, but to a much lesser extent in man, growth changes also occur at the foramen magnum as this moves backwards.

The mid-line cranial base is sometimes described as consisting of two parts: from foramen magnum to pituitary fossa and from pituitary fossa to nasion.

Growth at the cranial base involves growth of the cranial, facial, and pharyngeal regions. Continual growth at the spheno-occipital synchondrosis enlarges the cranial cavity by separating the bones of the *coronal and lambdoidal suture systems* (Fig. 7.42) beyond the requirements of brain growth. This allows of bone deposition on the inner surfaces of the bones of the cranial vault. As a result the vault bones thicken and diploic tissue develops between the inner and outer cortical layers.

The auditory capsule

In man the auditory capsule ossifies to form a single bony element, the *petrous temporal bone* (Fig. 5.5). In the adult this is united with the squamous temporal, tympanic plate, and styloid process to make up the temporal bone of topographical anatomy. The petrous temporal of each side forms part of the cranial base. It is, however, a separate structure from the mid-line cranial base (occipital and sphenoid bones). Each petrous temporal bone replaces the cartilage of the foetal auditory capsule and comes to contain the complex inner ear apparatus. This reaches adult dimensions early in life and even in foetal life the inner ear structures are relatively fully developed. The middle ear cavity develops between the squamous, tympanic and petrous parts of the temporal bone (page 211).

The nasal region (Figs 4.39, 5.30, 5.31)

The *nasal region* can be divided into two parts: an upper interorbital or ethmoidal portion, and a lower, maxillary part. The upper portion reaches its full growth early in childhood and is related to the olfactory mucous membrane. The lower part continues to grow until the end of childhood and is related to the requirements of respiration. Growth in height of the ethmoidal region is by suture growth between the ethmoid, maxilla, and palatine bones, and is regulated by the growth of the cartilage of the nasal septum. Growth of the ethmoid is complete by about the end of the first decade. Growth in height of the maxillary region is brought about by suture growth in early childhood and later by a descent of the hard palate produced by bone deposition on the oral side, and bone resorption on the nasal side. This is seen in the increasing depth of the inferior meatus which forms the chief air passage through the nasal cavity.

A characteristic feature of young children is the depressed or flattened appearance of the bridge of the nose. Later the nasal bones are carried further forward by growth of the supporting septal cartilage. Failure of growth in this region is often associated with growth failure in other parts of the cranionasal chondrocranium. The nasal air sinuses, although they invade the frontal, ethmoid, sphenoid, and maxillary bones, and sometimes involve parts of the palatine, occipital, and zygomatic bones, have little influence on the form of the bones which they invade. In spite of statements to the contrary, the growth of the maxillary air sinus plays no active part in the eruption of the upper teeth.

The alveolopalatal region

The alveolar processes of the mandible and maxillae develop around the developing tooth germs during foetal life. If tooth formation is deficient and teeth fail to erupt (as in anodontia) the alveolar processes of the jaws do not develop. With eruption of the teeth the alveoli and crypts within which they develop in the alveolar processes become replaced by sockets to which the roots of the teeth are attached by the periodontal ligament. The changes in the vertical height of the mandible and maxillae, and of the face as a whole, are largely the result of alveolar bone growth after the third year and almost entirely the result of this process after the first decade.

During facial growth, as a result of growth at the mandibular condyles, the mandible is thrust downward and forward from the glenoid fossae (which are carried by the middle cranial segment). The space between the body of the mandible and the upper facial skeleton (below the level of the orbital cavities) is filled in by the growth of the alveolar bone which carries the teeth with it and maintains the two dental arches in normal occlusion. To some extent the alveolar processes are independent of the bone which supports them (body of the mandible and the palatal vault) as can be seen in the variation in the degree of alveolar prognathism in different races and especially in anthropoid apes.

During foetal life and the first year after birth when the interfrontal midpalatal, and mandibular (symphysial) sutures make up a *sagittal suture system*, growth in width of the oral cavity is partly the result of separation of the bones bounding this suture system. By the end of the first year the frontal and mandibular sutures are uniting, and although the midpalatal suture remains, it is unlikely to be an important site of growth after this time.

Further growth of the oral cavity is the result of bone deposition along the alveolar margins, on the under-surface of the palate and on the facial surface of the maxillary and mandibular bones. There may be a certain amount of resorption on the lingual surfaces of the alveolar processes, but increase in the width of the palate is largely the result of the downward and outward direction of growth of the alveolar process.

The palatal vault, as well as bounding the oral cavity, is part of the facial buttress system whereby the alveolar processes carrying the teeth are supported, and the stresses of mastication distributed to the cranium and the cranial base.

Glenoid fossae

At birth the glenoid (articular) fossae are small circular areas situated at the base and side of the cranium in front of the tympanic membranes. Each fossa grows forwards and outwards as a result of surface deposition at the anterior and posterior roots of the zygomatic process. At birth there is no articular eminence; its position is marked by the poorly developed anterior rim of the fossa but it is well established by the seventh year (Fig. 7.46) and continues to grow in vertical height until the establishment of the full permanent dentition. The glenoid fossae on each side of the skull are separated from one another chiefly by growth at the suture between the great wing of the sphenoid and the squamous part of the temporal bone (which carries the fossa). This suture belongs to the coronal suture system and in this position runs from before backwards in the middle cranial fossa and on the base of the skull. Growth of the mandible must be correlated with this process of separation of the glenoid fossae. During foetal life and the first year after birth this is produced mainly by growth at the mandibular symphysis but when this unites (during the first year) the separation of the two condyles must be the result of an outward as well as a backward direction of growth. Although the distance between the glenoid fossae (and the mandibular condyles) approaches the adult dimensions by the end of the first decade, the distance continues to increase at a slower rate until the end of facial growth. This later growth period is associated with the continued growth at the cranial base and the condylar cartilages of the mandible.

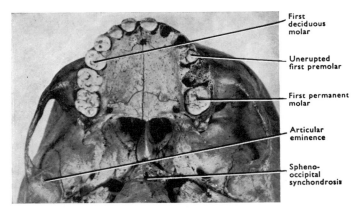

First
deciduous
molar

Unerupted
first premolar

First permanent
molar

Articular
eminence

Spheno-
occipital
synchondrosis

Fig. 7.46 The anterior part of the base of the skull in a child about six years old.

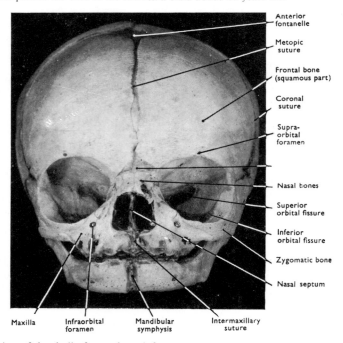

Anterior
fontanelle

Metopic
suture

Frontal bone
(squamous part)

Coronal
suture

Supra-
orbital
foramen

Nasal bones

Superior
orbital fissure

Inferior
orbital fissure

Zygomatic bone

Nasal septum

Maxilla   Infraorbital   Mandibular   Intermaxillary
           foramen   symphysis   suture

Fig. 7.47 Anterior view of the skull of a newborn infant.

The orbital cavities (Figs 4.44, 7.49)

Growth in size of the orbital cavities is determined largely by the growth of the eyeball and other orbital contents. Growth of the eyeball is similar to that of the brain, and is almost complete by about seven years of age. The shape of the orbital cavities is much less under the influence of its contents, and like that of the cranium is probably determined by independent genetic factors. Growth in height and width of the orbital cavities is partly the result of growth at the sutures of the bones bounding them. The bones are separated from one another by the growth of the orbital contents and of the septal cartilage of the nose. The growth of the sphenoid bone in late foetal life also contributes to the form and size of the orbital cavities. Suture growth probably ceases to be of importance by the end of the first decade, if not earlier.

Growth in anteroposterior depth of the orbital cavities is largely the result of the surface deposition of bone on the facial aspect of the orbital orifice. This is especially well seen in anthropoid apes, in which the orbital cavities move forward relative to the front of the cranial cavity. As this growth is taking place at the base of a cone, the diameters of the orbital orifices will continue to increase Interorbital width (across the body of the nose) is another of the facial dimensions which reach adult dimensions early and is relatively advanced in newborn infants.

Facial growth potentials

The part of the facial skeleton which lies below the Frankfort plane is about one-eighth of the size of the skull at birth. In the adult it has increased to about one-third of the size of the skull. Putting it another way, the infra-orbital part, or that part of the facial skeleton which is involved in mastication, grows much more after birth than the cranium, the olfactory and orbital regions of the face. Growth of the brain and eyeballs and, coincidentally, those parts of the skull which are associated with these structures and their growth, have reached more than 90 per cent of their adult size by the tenth year. Some parts of the skull, including the cribriform plate and the petrous part of the temporal bone have attained their adult size even earlier.

The width of the face at birth is about two-thirds of the adult size; the height of the face about one-half and facial depth about one-third of adult size. It should be recognized that these are average figures and that there is great variability among individual skulls in a given population. Genetic factors which appear to influence the cartilaginous parts of the skull in the embryo and foetus, namely the chondrocranium and nasal capsule, determine the morphological features of the skull in later life. The growth and function of muscles attached to appropriate parts of the skull play a great part in the full development of the skull and the muscles of mastication are especially important. Because of their attachments, these muscles influence the full development of the lateral pterygoid plates, the angle and coronoid processes of the mandible, the width of the ramus of the lower jaw, the size and form of the zygomatic arches, the facial buttress systems and the alveolar processes.

Between the time of birth and adult life the human face must grow enough to contain and protect the growing eyeballs; give attachment to the muscles of mastication; provide space for the tongue and the paranasal air sinuses; as well as provide attachment for the full number of permanent teeth in normal occlusion.

Cephalometric studies of serial lateral radiographs of the skull of the same individual superimposed at different ages by the use of various planes and points (page 443) demonstrates that the upper part of the face which is situated above the Frankfort plane tends to grow forward and upwards, the middle region of the face grows downwards and forwards, the chin moves forwards and downwards more steeply than the middle upper face and the gonial region grows downwards, or downwards and backwards. As a very general statement, it can be said that the human face grows along a series of radiating lines which have their common centre at the pituitary fossa.

The pharyngeal region (Figs 4.8, 4.37)

This is bounded by the cervical vertebrae behind, the posterior segment of the cranial base above, and the back of the facial skeleton (nasal and oral regions) in front. Growth in height of the pharynx is regulated to a large extent by the epiphyseal cartilages of the cervical vertebrae. The rapid growth of these structures relative to the growth of the larynx and the trachea in postnatal life carries the epiglottis downwards in relation to the soft palate. This downward descent of the larynx, a primate and especially a human characteristic, is related to the downward as well as forward growth of the mandible relative to the cranial base.

Growth in anteroposterior depth of the nasopharynx is related to growth of the cranial base at the spheno-occipital synchondrosis. It is this dimension of the pharynx which is mostly concerned in the blockage due to the growth of adenoid tissue. Another factor contributing to the relative restriction of the anteroposterior depth of the human nasopharynx is the bending of the cranial base which carries the facial skeleton backwards towards the vertebral column. Less is known about the growth of the pharyngeal region, however, than any other part of the face.

Table 7.3 Bones contributing to various craniofacial regions

Bone	Region
Parietal	Cranial vault
Frontal	Cranial vault, cranial base, orbital cavities, nasal cavities, buttress system
Occipital	Cranial vault, cranial base, pharyngeal region
Temporal	Cranial vault, auditory capsule, buttress system, muscular processes, pharyngeal region, 2nd pharyngeal arch (styloid process)
Sphenoid	Cranial vault, cranial base, orbital cavities, nasal cavities, pharyngeal region, buttress system, muscular processes
Ethmoid	Cranial base, orbital cavities, nasal cavities
Zygomatic	Orbital cavity, buttress system, muscular processes
Lacrimal	Orbital cavity, nasal cavity
Vomer	Nasal cavity
Inferior turbinate	Nasal cavity
Palatine	Orbital cavity, nasal cavity, buttress system, oral cavity (alveolopalatine region)
Maxilla	Orbital cavity, nasal cavity, buttress system, oral cavity (alveolopalatine region)
Mandible	Buttress system (basal element), muscular process, oral cavity (alveolopalatine region)

Table 7.4 Growth of the craniofacial regions

Cranial vault	1. Endocranium: Rapid growth to 3rd year. Growth almost complete by 7th year.
	2. Ectocranium: Growth slow to about 7th year. Rapid during adolescence.
Auditory capsule	Growth rapid during late foetal life.
Cranial base	1. Anterior (frontal) segment: Growth continuous until early adult life.
	2. Middle segment: Little growth after 7th year.
	3. Posterior segment: Growth continuous until early adult life.
Nasal region	1. Ethmoidal region: Complete by about 7th year.
	2. Maxillary region: Growth continuous until early adult life.
Orbital cavities	1. Height and width: Growth rapid up to 3rd year. Almost complete by 10th year.
	2. Depth: Continuous until early adult life.
Alveolopalatal region	Continual growth up to adult life with growth spurts related to the eruption of (1) permanent incisors, (2) permanent molars.
Facial buttress system	Continual growth throughout childhood. Period of rapid growth during adolescence (eruption 2nd and 3rd molars).
Muscular process	Continual growth throughout childhood. Period of rapid growth during adolescence (eruption 2nd and 3rd molars).
Pharyngeal regions	Continual growth to adult life

8. The Nervous System

The *central nervous system* is composed of the brain and spinal cord, contained within the skull and vertebral canal, acting as a central controlling system for every part of the body. The connections between the system and the periphery are by means of *afferent* or *sensory nerve fibres* which bring information to the central system where it is integrated or correlated, often with previous experience, before new nervous impulses are initiated which pass out along *efferent*, or *motor nerve fibres* to muscles and organs. The motor impulses are such that they produce the best or most suitable response in the effector tissue in keeping with the nature of the sensory stimulus whether it be, for example, the relaxation or contraction of the muscles of mastication to alter the force of chewing, increased activity of the salivary glands, constriction or the dilatation of the branches of arteries. The afferent and efferent nerve fibres comprise the *peripheral nervous system* and form the major part of the functional unit of the nervous system—the *reflex arc* (Fig. 8.1). In the simplest, rather automatic forms of nervous

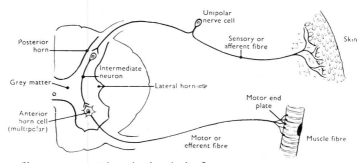

Fig. 8.1 Nerve fibre arrangement in a simple spinal reflex arc.

system in lower animals the association of sensory and motor components is a direct one but in the more highly evolved systems such as the human nervous system additional intermediate nerve fibres are included in the sequence. These are known as *intercalated nerve fibres* and are the processes of nerve cells or neurons which, by making connections with other neurons, are essential for and enhance integration between different levels of the nervous system.

The nervous system as a whole can be divided into two major functional components:

1. The *somatic nervous system* concerned with the reception of stimuli of an exteroceptive or proprioceptive nature and the initiation of stimuli to voluntary striated muscle, which result in co-ordinated voluntary motor movements.

2. The *autonomic nervous system* concerned with interoceptive impulses and the regulation of the involuntary activities of viscera and vessels. Sympathetic and parasympathetic fibres are recognized within this system (see page 416).

The basic arrangement of nerve cells and their processes is essentially similar in the two systems and they are closely associated anatomically. The main difference lies in the position of the cell bodies of the effector cells.

Essentially, the cell body of a typical somatic motor cell is found in the anterior horn of grey matter (Fig. 8.1), while the corresponding cell body in the autonomic system is found in an autonomic ganglion situated outside the central nervous system. Those associated with the sympathetic component form ganglia close to the vertebral column, the sympathetic chain of ganglia, while parasympathetic effector cell bodies form small ganglia, close to the structure which their processes innervate, e.g. salivary glands (Figs 8.3, 8.27).

The nervous impulses transmitted to the central nervous system by the afferent nerve fibres begin at sensory *nerve endings* which are structural specializations of the most peripheral part of termination of the nerve fibre (Figs. 8.2, 8.3). Impulses resulting from touch, temperature, pain, and pressure are termed *exteroceptive* for they are produced from outside sources. Also

Vibration (tuning fork)

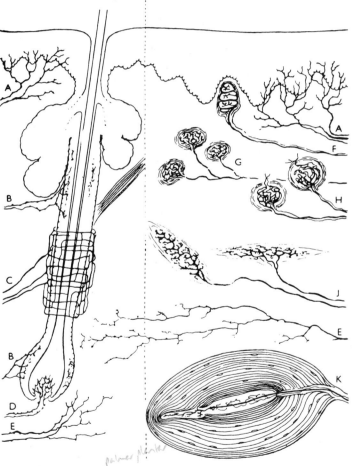

palmar plantar

Fig. 8.2 Diagram of the sensory nerve endings in the skin. The left part of the diagram represents hairy skin and the right part hairless skin. (A) free branching nerve endings in upper dermis and lower layers of epidermis. (B) free nerve endings in epidermis of hair follicle. (C) pallisade endings round follicle. (D) nerve endings in papilla of hair follicle. (E) free branching nerve endings in lower dermis. (F) Meissner's corpuscle. (G) end bulbs of Krause. (H) genital end bulbs. (J) end organs of Ruffini. (K) pacinian corpuscle. The relative positions and sizes of the different nerve endings are indicated but, for the sake of clarity, they are not drawn strictly to scale. (By courtesy of Miss Margaret Gillison.)

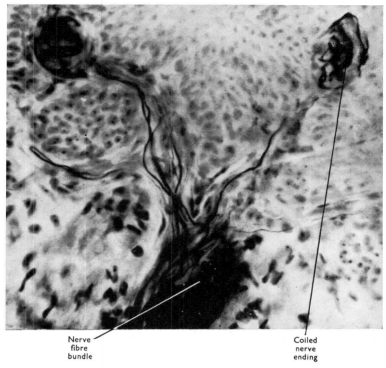

Nerve   Coiled
fibre   nerve
bundle   ending

Fig. 8.3 Coiled nerve endings in the lamina propria of the mucoperiosteum of the hard palate ×
540.

included in this group are impulses concerned with the special sense of sight, hearing, taste
and smell. Impulses arising from viscera and blood vessels are termed *interoceptive* whilst those
originating in joints, muscles, tendons and ligaments are classed as *proprioceptive*. The proprioceptive impulses are particularly important in the co-ordination of movement and posture
and play a vital part in the jaws by the regulation of masticatory forces through reflex activity.

Pain
The nerve endings subserving pain are naked, nonmyelinated-freely ending twigs situated in
the deeper layers of the epithelium of the skin and mucous membranes (superficial pain), in
the lamina propria of the skin and mucous membranes, the pulps of teeth, periodontal ligaments, periosteal coverings of bone, articular structures, deep fascia, and in muscle tissues
(deep pain). Parietal pleura is sensitive, ? pulmonary pleura isn't? lect 25/10/90

Touch Tm
The nerve endings subserving touch are considered to be Meissner's corpuscles, and the
basket work of nerve fibres around the hair follicles. These are found in the skin of the face,
the lips, oral mucous membrane and in the deeper tissues. Similar endings in muscle tissue,
peri-articular structures, and in the periodontal ligaments are concerned in the proprioceptive
regulation of muscle action.

Heat and cold
Warmth is supposed to stimulate Ruffini's corpuscles, and cold, the end bulbs of Krause.
It is probable that definite areas of skin and mucous membrane are supplied by a single

sensory neuron. Large myelinated fibres of neurons branch to form a network of smaller fibres which end in the 'organized' end organs. Branches of other neurons subserving pain sensation terminate in the same area and may also send fine myelinated fibres to the specialized end organs. Each neuron unit retains its functional individuality, although there may be much overlapping in the distribution of adjacent neurons. The composite of impulses reaching the central nervous system from the various neuron units is sorted out in the spinal cord and brain and the nature, extent, and distribution of the various sensations recognized. The specificity of nerve endings, except those for pain, is to be regarded with some doubt. Until we know more about their function they are best classified in some simple manner, such as *free* and *coiled nerve endings*, of which the latter can be subdivided further into simple, complex and compound subtypes.

THE PERIPHERAL NERVOUS SYSTEM

The peripheral nervous system consists of the spinal and cranial nerves which arise from the spinal cord and the brain and of the ganglionated trunks, plexuses and nerves (e.g. the sympathetic trunk, its ganglia and branches) which form the peripheral parts of the autonomic nervous system.

Spinal nerves (Fig. 8.4)

There are 31 pairs of spinal nerves arising from each side of the spinal cord and they are subdivided regionally into 8 *cervical*, 12 *thoracic*, 5 *lumbar*, 5 *sacral* and 1 *coccygeal* nerves.

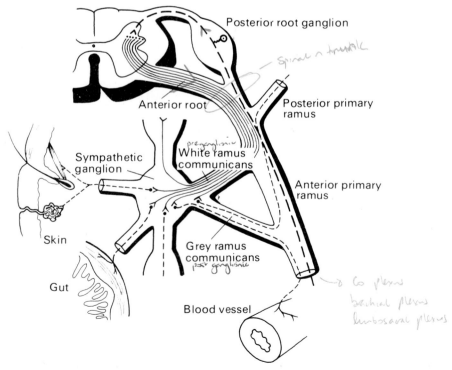

Fig. 8.4 The formation of a spinal nerve and the distribution of the sympathetic nerve fibres which run in it. Preganglionic fibres are represented by solid lines.

They are attached in series along the length of the side of the cord by anterior (ventral) and posterior (dorsal) *roots*. The anterior roots transmit the motor fibres and the posterior roots the sensory variety. Within or close to the intervertebral foramina the roots join to form the *spinal nerve trunk* and close to the region of fusion the posterior roots possess a ganglion composed of the cell bodies of the sensory nerve fibres. In the upper part of the vertebral canal the nerve roots pass almost horizontally outwards to the corresponding intervertebral foramina but in the lower lumbar and sacral regions of the canal the nerve roots have to descend abruptly to reach the appropriate foramina. This angulation of the roots is due to the difference in the length of the vertebral canal compared with the length of the contained spinal cord, which is much shorter.

The *spinal nerves* divide almost immediately they have formed from the fusion of the anterior and posterior roots into anterior and posterior primary divisions or *rami* through which nerve fibres pass to and from the skin and voluntary muscles of the trunk and limbs (Fig. 8.4). Each spinal nerve in its cutaneous distribution supplies an area of skin from the mid-line of the body posteriorly to the mid-line anteriorly in a segmented fashion, after the manner of the segmentation of the spinal cord. These areas, called *dermatomes*, are regularly arranged in the trunk but with embryological modification those in the region of the limbs are distorted in outline (Fig. 8.5). In addition, it is important to appreciate that the dermatomes overlap one another to some extent so that loss of sensory function in a spinal nerve does not result

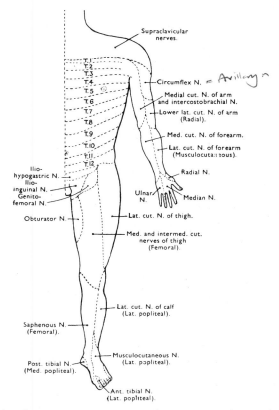

Fig. 8.5 Diagram showing the cutaneous sensory areas dermatomes on the anterior surfaces of the trunk and limbs supplied by the various spinal cord segments and by individual nerves. (By courtesy of Professor G. A. G. Mitchell and Dr E. L. Patterson.)

in denervation over the entire area of the appropriate dermatome. In addition to sensory and motor somatic fibres spinal nerve rami contain some sympathetic fibres for the supply of vessels and glands and some of the sacral nerves (S. 2. 3. 4.) also contain a parasympathetic component. The anterior primary rami are larger than the posterior rami and frequently form *plexuses* such as the cervical, brachial and lumbosacral plexuses described on pages 431 and 434. The posterior rami may interconnect by means of nerve loops but do not form plexuses. These rami supply the muscles and skin on the back of the head, neck and trunk.

Cranial nerves

The functional analysis of cranial nerves is more complicated than for spinal nerves which contain somatic and visceral afferent and efferent fibres. In addition to these four categories cranial nerves also contain special visceral efferent, special somatic afferent and special visceral afferent fibres. The special visceral efferent fibres supply muscles derived from the pharyngeal arches. The special somatic afferent fibres innervate special sense organs of the internal ear and are distributed through the eighth cranial nerve. The special visceral afferent fibres are associated with the special senses of taste and smell, via the seventh, ninth, and tenth cranial nerves. Other features of the cranial nerves and their distribution are described on pages 418–425.

THE CENTRAL NERVOUS SYSTEM

The central nervous system composed of the *brain and spinal cord* is a complex system of nerve cells and nerve fibres derived from the embryonic neural tube (Fig. 7.6).

The spinal cord

It is convenient to begin with a study of the spinal cord, this being the simplest part of the central nervous system in the structural sense for it shows evidence of division into segments. Each segment is comparatively similar to neighbouring segments in its internal fibre and cell arrangement.

The spinal cord averages 18 inches (45 cm) in length in the adult extending from the foramen magnum, where it is continuous with the medulla oblongata, to the level of the intervertebral disk between the first and second lumbar vertebrae. The exact level of the lower end which is tapered (*conus medullaris*) depends on the degree of flexion or extension of the vertebral column (Fig. 8.6). curl up for epidural

The spinal cord is not perfectly cylindrical but is flattened anteroposteriorly especially in the cervical region and this is reflected in the shape of the vertebral foramina. Neither is it of uniform diameter throughout its length being thicker in the cervical and lumbar regions where the nerves to the limbs are attached (*cervical and lumbar enlargements*).

The outline of a transverse section through the cord is seen in Figure 8.7. Several longitudinal grooves are to be observed on the surface of which the antero-median and postero-median fissures are the most obvious. The posterior fissure is shallow but the anterior fissure extends into the cord for about one-third of its thickness. The spinal nerves are attached by their roots into shallow grooves on its lateral aspect.

The spinal cord in section

When cut transversely the cord shows two conspicuous zones (Fig. 8.7) called the *grey* and the *white matter*.

The *grey matter* is darker in the fresh state for it contains fewer nerve fibres with myelin sheaths and a great many more nerve cells than the white matter. The grey matter, H-shaped on section, surrounds the central or *spinal canal* which is continuous with a complicated and more extensive cavity in the brain called the ventricular system (see page 407).

The anterior part of the grey matter called the *anterior horn* (or anterior column when considered in three dimensions) contains many large multipolar cells from which nerve fibres pass out in the anterior roots of the spinal nerves. The *posterior horn* of grey matter is thinner and into it pass the central processes of the unipolar sensory nerve cells situated in the posterior

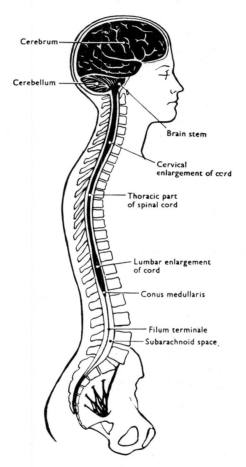

Cerebrum

Cerebellum

Brain stem

Cervical enlargement of cord

Thoracic part of spinal cord

Lumbar enlargement of cord

Conus medullaris

Filum terminale
Subarachnoid space

Fig. 8.6 The brain and spinal cord *in situ*. (By courtesy of Professor G. A. G. Mitchell and Dr E. L. Patterson.)

root ganglion. In the thoracic part of the cord a third or *lateral horn* of grey matter projects from the lateral aspect of the H-shaped area. This region contains the cell bodies of the neurons whose processes pass out of the spinal cord in the anterior roots of the thoracic spinal nerves to form the sympathetic trunks of the autonomic nervous system (Fig. 8.4).

The *white matter* is divided into anterior, lateral and posterior columns containing accumulations of nerve fibres ascending or descending along the length of the cord for a variable distance. These tracts or nerve fibre pathways contain fibres having a similar function and occupy quite definite portions within the white matter. Some fibres of the tracts enter or leave the cord through spinal nerves on the opposite side of the cord and therefore have to cross over the mid-line at certain levels. They do so through the *white* or *grey commissures* (Fig. 8.7).

The tracts within the cord are bilateral and are of two kinds:

Intersegmental or association tracts which connect various levels of the cord.

Projection or itinerant tracts which connect the spinal cord with the brain or *vice versa*.

The projection tracts are extremely important for they bring information to high centres in the brain and exercise control over the functions of the body by impulses passing downwards in the descending tracts.

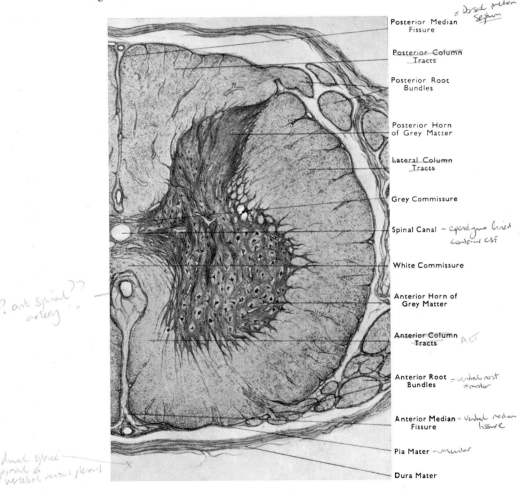

Posterior Median Fissure

Posterior Column Tracts

Posterior Root Bundles

Posterior Horn of Grey Matter

Lateral Column Tracts

Grey Commissure

Spinal Canal

White Commissure

Anterior Horn of Grey Matter

Anterior Column Tracts

Anterior Root Bundles

Anterior Median Fissure

Pia Mater

Dura Mater

Fig. 8.7 Transverse section through upper lumbar region of spinal cord (× 13). Note the typical arrangement of the grey and white matter. The faint strands visible here and there between the dura and pia mater represent the delicate arachnoid sheath. The subarachnoid space between the arachnoid and pia mater contains the cerebrospinal fluid. (By courtesy of Miss Margaret Gillison.)

THE BRAIN

Compared with the spinal cord the brain is a greatly enlarged and much more complex part of the central nervous system. The brain occupies virtually all the available space in the cranial cavity to such an extent that certain surface features form ridges or depressions on the internal aspect of the skull bones. The brain becomes continuous with the spinal cord (medulla spinalis) at the foramen magnum and consists of five major parts which are in ascending order: the *medulla oblongata*, the *pons*, the *mid-brain*, the *cerebellum* and the *cerebrum* (cerebral hemispheres) (Fig. 8.8). The first three parts mentioned are collectively known as the *brain stem* in which is found many of the central connections of the cranial nerves and their cell stations or nuclei. The cranial nerves emerge from or enter the brain substance at certain well defined sites and these can be seen when the brain is viewed from the inferior aspect (Fig. 8.9).

except IV

The medulla oblongata (Fig. 8.8)

This is the direct continuation of the spinal cord. It is about 1 inch in length, conical and begins at the level of the foramen magnum and ends at the lower border of the pons. Anteriorly it is in close relationship to the basilar part of the occipital bone. Its lateral surface shows an oval swelling (the olive) corresponding to the grey matter of the inferior olivary nucleus which lies just below the surface (Fig. 8.10). In front of the olive, between it and the anterior *descend* median fissure, is a longitudinal ridge through which pass the corticospinal nerve fibres. This ridge which is bilateral is called the *pyramid* and suggests the name pyramidal tract. Where the pyramidal fibres decussate the antero-median fissure is interrupted as the bundles of nerve fibres cross the mid-line. On the posterior aspect of the medulla the fasciculi gracilis and cuneatus produce two low ridges on either side of the posterior median fissure. At the upper end of these ridges are two tubercles produced by the underlying nuclei of the same name. Immediately above these features the central canal of the medulla, which is continuous with the central canal of the spinal cord, opens out into a large diamond-shaped space known as the *fourth ventricle* (page 408).

Certain of the cranial nerves emerge from the surface of the medulla in relation to the olive:

1. The *hypoglossal nerve rootlets* (twelfth cranial nerve) emerge in linear fashion between the olive and the pyramid. *preolivary sulcus*

2. The *glossopharyngeal* (ninth), *vagus* (tenth) and *accessory* (eleventh) *nerve rootlets* emerge in that order from above downwards in a groove behind the olivary swelling. *post olivary sulcus*

The internal structure of the medulla is complex but some of the more important features are shown in semidiagrammatic form in Figure 8.10.

The *decussation of the pyramidal tracts* is an obvious feature of sections through the lower part of the medulla (Fig. 8.10a). The tracts are crossing *descending* the mid-line in a dorsolateral direction to reach the lateral white column at the spinal cord. The fasciculus *ascending* gracilis and fasciculus cuneatus occupy a dorsal position and carry touch and proprioceptive sensation to higher levels of the brain. On their lateral aspect are the spinal tract and nucleus of the trigeminal nerve. *ascending posterior*

The *sensory decussation* is an important feature of sections through the middle of the medulla (Fig. 8.10b), otherwise known as the decussation of the medial lemniscus. The crossing fibres come from the gracile and cuneate nuclei and they ascend to the thalamus in a compact bundle, the medial lemniscus, which is a larger structure in sections through the upper part of the medulla (Fig. 8.10c). The spinal tract of the trigeminal nerve lies superficial to its nucleus at all levels of the medulla and these structures are associated with pain and temperature impulses reaching the brain stem through the root of the trigeminal nerve.

The olivary nucleus is a concave strip of grey matter immediately behind the pyramids and from it fibres cross the mid-line to reach the cerebellar hemisphere of the opposite side. These fibres are associated apparently with the maintenance of muscle tone.

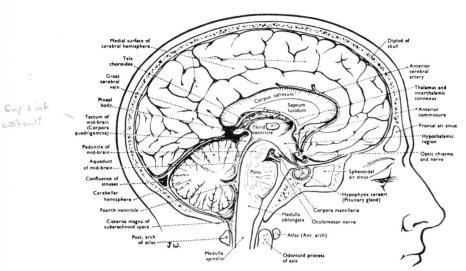

Fig. 8.8 Median sagittal section through skull and brain. The *falx cerebri* which lies in the mid-line in the longitudinal fissure of the brain has been removed to show the medial surface of the left cerebral hemisphere. The *tentorium cerebelli* overlies the cerebellum and is shown as a double line between the cerebral and cerebellar hemispheres. (By courtesy of Professor G. A. G. Mitchell and Dr E. L. Patterson.)

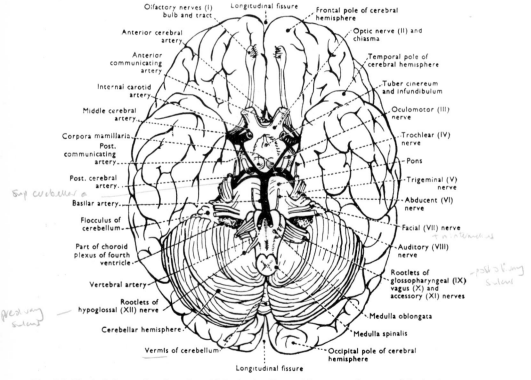

Fig. 8.9 The inferior or basal surface of the brain. From this aspect all parts of the brain are apparent. The superficial origins of the cranial nerves and the initial parts of the main cerebral arteries are visible. Each internal carotid artery gives off anterior and middle cerebral arteries and the posterior cerebrals arise from the basilar artery. The two anterior cerebrals are united by a short anterior communicating artery and the two posterior cerebrals are joined to the ends of the internal carotid arteries by the two posterior communicating arteries, so completing the arterial circle of Willis. (By courtesy of Professor G. A. G. Mitchell and Dr E. L. Patterson.)

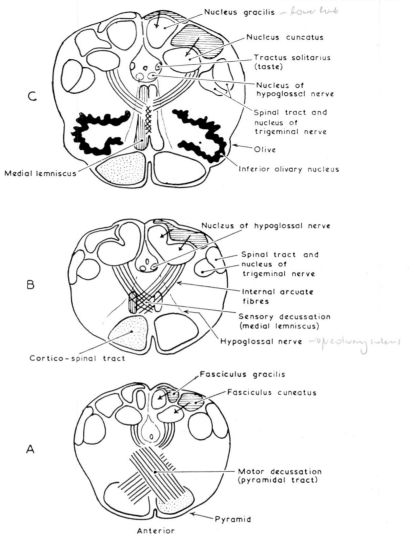

Nucleus gracilis ‒ *lower limb*

Nucleus cuneatus

Tractus solitarius
(taste)

Nucleus of
hypoglossal nerve

Spinal tract and
nucleus of
trigeminal nerve

Olive

Inferior olivary nucleus

C

Medial lemniscus

Nucleus of hypoglossal nerve

Spinal tract and
nucleus of
trigeminal nerve

Internal arcuate
fibres

Sensory decussation
(medial lemniscus)

Hypoglossal nerve ‒ *of evolvony nucleus*

B

Cortico-spinal tract

Fasciculus gracilis

Fasciculus cuneatus

A

Motor decussation
(pyramidal tract)

Pyramid

Anterior

Fig. 8.10 Diagrams made from transverse sections through the medulla oblongata: A, at its lower end through the motor decussation; B, at a slightly higher level through the sensory decussation; C, at the level of the olive. The main motor and sensory pathways are similarly shaded or stippled throughout these diagrams and those in Figures 8.12 and 8.14 so that they may be traced through various levels.

The pons (Figs 8.8, 8.9)

The pons lies between the medulla and the mid-brain anterior to the cerebellum. Many of its fibres pass horizontally across the mid-line connecting the two cerebellar hemispheres. Where the pons becomes continuous with the cerebellar substance the horizontal fibres form a well-defined bundle, the *middle cerebellar peduncle*, represented in transverse section (Fig. 8.11) as the largest of three circles, opposite to the nucleus of the facial or seventh cranial nerve and the salivary nuclei. The smaller circles lying to the medial side of the middle peduncle represent the *superior* and *inferior cerebellar peduncles*, which also transmit nerve fibres connecting the brain stem with the cerebellum.

On the anterior surface of the pons is a mid-line furrow occupied by the *basilar artery* (Figs 8.9, 8.23). Both the pons and the basilar artery are closely related in front to the upper sloping surface or *clivus* of the occipital bone.

The posterior surface forms the upper part of the floor of the fourth ventricle and is immediately in front of the cerebellum.

Certain cranial nerves are to be observed in relation to the pons:

The *abducent nerve* (sixth cranial nerve) emerges at the inferior border of the pons just above the pyramid of the medulla. *medullopontine jn*

The *facial and vestibulocochlear nerves* (seventh and eighth cranial nerves) emerge together, the facial occupying the more medial position, from the inferior border of the pons just above the olive. *medullopontine jn*

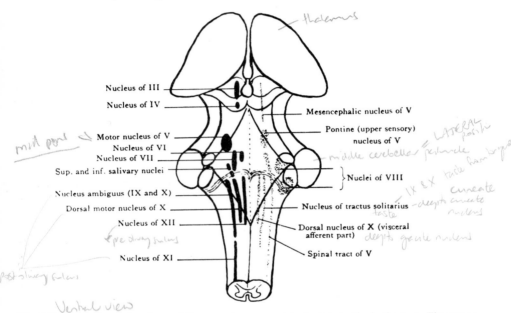

Fig. 8.11 Approximate locations of the cranial nerve nuclei of origin in the brain stem. The motor nuclei are indicated in solid black on the left side and the sensory nuclei are stippled in on the right. (By courtesy of Professor G. A. G. Mitchell and Dr E. L. Patterson.)

The *trigeminal nerve* (fifth cranial nerve), the largest of the cranial nerves, is attached to the lateral part of the anterior surface of the pons midway between its upper and lower borders by a large sensory root and a small motor root (which carries fibres for the innervation of the muscles of mastication).

The internal structure of the pons is of particular importance as it contains many of the nerve cells belonging to the main sensory nerve to the face, the trigeminal. These sensory and motor nuclei are shown in the diagram (Fig. 8.12) in addition to some of the other important features. Notice the close relationship between the motor nucleus of the trigeminal nerve, the nucleus of the facial nerve, and the sensory nucleus of the trigeminal nerve. The association illustrates a trend in the evolution of the nervous system—that nuclei giving rise to nerve fibres which innervate peripheral structure of associated function migrate in the direction of the chief sensory nuclei from which they receive impulses (a process known as *neurobiotaxis*). In this case the motor nuclei are concerned with the innervation of the muscles of mastication and expression respectively.

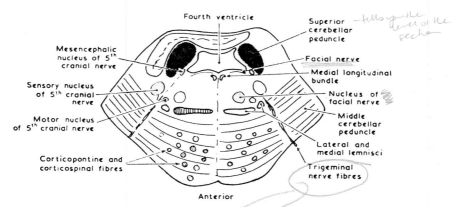

Fourth ventricle

Superior cerebellar peduncle

— tells up the level of the section

Mesencephalic nucleus of 5ᵗʰ cranial nerve

Facial nerve

Medial longitudinal bundle

Sensory nucleus of 5ᵗʰ cranial nerve

Nucleus of facial nerve

Motor nucleus of 5ᵗʰ cranial nerve

Middle cerebellar peduncle

Lateral and medial lemnisci

Corticopontine and corticospinal fibres

Trigeminal nerve fibres

Anterior

Fig. 8.12 Diagram of a transverse section through the middle of the pons. Note particularly the position of the nuclei of the trigeminal nerve. Only a small proportion of the corticospinal fibres are indicated.

Relationship of sensory and motor nuclei (Fig. 8.12)

It will be noticed also that motor nuclei are situated medial to the sensory nuclei throughout the brain stem. This relationship can be explained by the development of this part of the central nervous system. As in the spinal cord (page 344) the brain stem develops a distinct basal and alar plate on each side of the mid-line. The sulcus limitans, which in the spinal cord forms a boundary between the alar and basal plates, is present in the developing brain stem also, where it forms a division between motor and sensory areas.

The basal plate contains the motor nuclei, which can be divided into a medial somatic efferent (e.g. the hypoglossal nerve cells) and a lateral visceral or branchial efferent group (e.g. the glossopharyngeal, the vagus, and part of the accessory nerve).

The alar plate contains the sensory nuclei which, as in the basal plate, are divided into a lateral and medial group. The lateral somatic afferent group receives impulses from the surface of the head by way of the trigeminal nerve. The medial visceral or branchial afferent groups receives impulses from the taste buds of the tongue, by way of the glossopharyngeal and vagus nerves. These cells form the nucleus of the tractus solitarius (page 426). The position of these nuclei is shown diagrammatically in Figure 8.13.

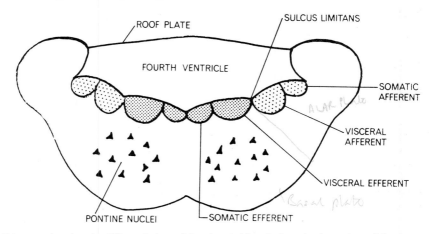

ROOF PLATE

SULCUS LIMITANS

FOURTH VENTRICLE

SOMATIC AFFERENT

ALAR PLate

VISCERAL AFFERENT

VISCERAL EFFERENT

Basal plate

PONTINE NUCLEI

SOMATIC EFFERENT

Fig. 8.13 Diagram showing the differentiation of the alar and basal plates in the region of the developing medulla oblongata. Compare with Fig. 7.18.

The mid-brain (Fig. 8.8)

The mid-brain forms the upper part of the brain stem, is rather less than one inch in length, and is perforated lengthwise by the central canal or aqueduct. This narrow canal connects the cavity of the fourth ventricle with that of the *third ventricle*, which lies between the cerebral hemispheres. The part of the mid-brain substance behind the aqueduct is known as the *tectum* or roof and presents four elevations, the *corpora quadrigemina*, which contain central cores of grey matter. The corpora are arranged in pairs. The upper pair, the superior corpora, are concerned with reflex activity associated with sight or vision. The lower inferior corpora are concerned with hearing and balance reflexes.

The part of the mid-brain substance in front of the aqueduct is formed by the *cerebral peduncles*, two large nerve fibre columns which leave the upper border of the pons, diverge from one another and enter the inferior aspect of the cerebral hemispheres. Each peduncle consists of three parts (Fig. 8.14): the basis pedunculi, the substantia nigra and the tegmentum.

The *basis pedunculi* is chiefly composed of nerve fibres passing from the cerebral cortex to the pons or spinal cord in the corticospinal or corticopontine tracts. The corticopontine tract fibres end in the pons around aggregations of cells forming the nuclei pontis. From these

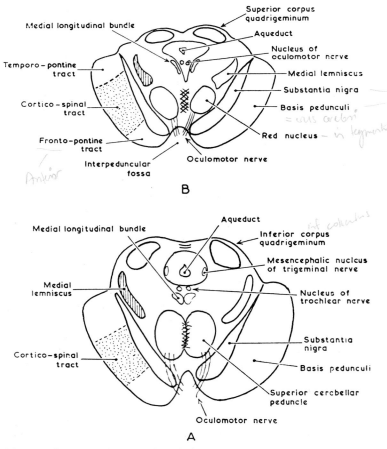

Fig. 8.14 Diagram of transverse sections through the mid-brain: A, at the level of the inferior corpus quadrigeminum; B, at the level of the superior corpus quadrigeminum.

cells new fibres pass into the cerebellar hemisphere of the opposite side through the middle cerebellar peduncle.

The *substantia nigra* is so called for it has a dark grey appearance in the freshly cut brain and is in the form of a slightly curved band. The functional significance of this pigmented area is not yet fully understood. It may be concerned in muscle tone, the slight degree of contraction evident in muscles in the resting state. *? Reduce dopa → caudate nucleus — Parkinsons*

The *tegmentum* which blends posteriorly with the tectum of the mid-brain contains nerve fibres which are predominantly ascending to higher levels, thus many of the ascending tracts observed in the posterior part of the pons are represented. These are shown in the diagram (Fig. 8.14). The nuclei of the nerves to the eye muscles—the oculomotor and trochlear nerves— have their cell stations in the grey matter around the central aqueduct, and here also is found the mid-brain or mesencephalic nucleus of the trigeminal nerve which is associated particularly with proprioceptive impulses from the muscles of mastication, the periodontal ligaments of the teeth and the capsule and disk of the mandibular joint.

Two oval masses of grey matter, which are a pinkish colour in the fresh brain, occupy a central position in the tegmentum. These are the *red nuclei* and have important connections with the cerebellum via the superior cerebellar peduncle.

Certain cranial nerves leave the surface of the mid-brain:

The *trochlear nerve* (fourth cranial nerve) emerges from the posterior aspect of the mid-brain just below the inferior corpora quadrigemina. It is the only cranial nerve to leave this aspect of the brain stem.

The *oculomotor nerve* (third cranial nerve) arises from the mid-brain in the interpeduncular fossa just above the pons.

From the foregoing description of the parts of the brain stem it will be appreciated that it transmits all the sensory and motor pathways passing between the brain and the spinal cord. Many of the nuclei within the brain stem control the vital activities of the body, e.g. respiration and the circulation. These significant areas are often called the *vital centres* and injury to them is frequently fatal.

The cerebellum (Figs 8.8, 8.9)

This occupies the greater part of the posterior cranial fossa and is separated from the posterior part of the cerebral hemispheres by a fold of dura mater, the *tentorium cerebelli*. It is related anteriorly to the posterior aspect of the pons and medulla, the fourth ventricle intervening. Like the cerebrum it consists of two large lobes or *hemispheres*, one on either side of the mid-brain connected by a constricted part called the *vermis*. The surface or cortex of the cerebellum is characteristically folded and consists of cellular grey matter, so that the internal arrangement of white and grey matter is reversed when compared with that in the spinal cord and medulla. The ridges of cerebellar cortex are narrow and lie parallel to one another in the horizontal plane.

The cerebellum is connected to the remainder of the brain by the three cerebellar peduncles or stalks through which efferent and afferent tracts pass to higher and lower centres of nervous activity. The main function of the cerebellum is the co-ordination of muscular activity both in movements and posture so that injuries of the cerebellum are characterized by loss of co-ordinating power, evidenced by difficulty in speech, staggering gait, twitching of muscles and loss of muscle tonus.

Internal structure of the cerebellum

The grey matter is mainly located at the surface of the cerebellum which permits a vast increase in the numbers of nerve cells but some islands of grey matter are also found in the centrally placed white matter, e.g. the dentate nucleus which closely resembles the inferior olivary nucleus in appearance.

The dentate nucleus is connected to the mid-brain and thalamus by the dentatorubral fibres which pass through the superior cerebellar peduncle. The cerebellum is connected to the pons and cerebrum via the middle cerebellar peduncle. The inferior peduncle connects it to the olivary nucleus of the opposite (contra-lateral) side, the nuclei of the auditory nerve (in particular the vestibular nucleus) and the nuclei gracilis and cuneatus of the same or homolateral side.

The cerebrum (Figs 8.8, 8.9)

This forms more than three-quarters of the whole brain and is divided into two equal-sized bilateral hemispheres. As in the cerebellum each hemisphere consists of an outer layer of grey matter forming the *cortex* and an inner core of *white matter* composed of myelinated nerve fibres. Embedded in the white matter are large accumulations of nerve cells, including the

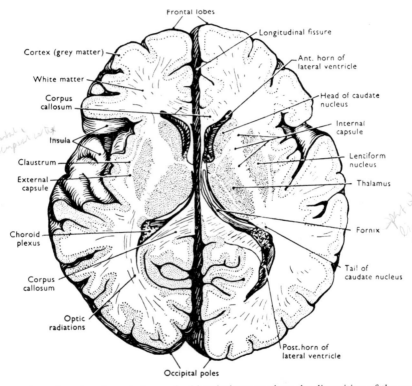

Fig. 8.15 Horizontal section through the cerebral hemispheres to show the disposition of the grey and white matter and various internal features such as parts of the lateral ventricles, the thalami, the caudate and lentiform nuclei, the internal capsule. The section is at a slightly different level on the two sides. (By courtesy of Professor G. A. G. Mitchell and Dr E. L. Patterson.)

caudate and lentiform nuclei, these together comprising the *corpus striatum*. These are best seen in horizontal sections through the hemispheres (Fig. 8.15) and are members of the group of *basal nuclei*. In addition the *thalamus*, a great sensory cell station, lies in the wall of the third ventricle in close relation to the basal nuclei. Between the thalamus and the caudate and lentiform nuclei the white matter, known here as the *internal capsule*, consists of fibres which include the pyramidal tract, frontopontine and thalamocortical fibres (Fig. 8.20).

The cerebral cortex

The cortex of the cerebral hemisphere is thrown into folds or *gyri* which increase its surface area and between these rather irregular convolutions are clefts or *sulci* of variable depth. Each hemisphere has three surfaces; superolateral, medial and inferior. The position of the main sulci and gyri on these surfaces is shown in Figures 8.6, 8.7. For descriptive purposes the cerebral surface is divided into lobes by some of the major sulci. These are the frontal, parietal, occipital and temporal lobes.

The *frontal lobe* lies in front of the central sulcus and above the lateral sulcus. Its anterior extremity is rounded and forms the *frontal pole*.

The *parietal lobe* is situated behind the central sulcus and above the lateral sulcus extending as far back as the parieto-occipital sulcus.

The *occipital lobe* lies behind the parieto-occipital sulcus and its posterior extremity is pointed to form the *occipital pole*. The anterior limit of the occipital lobe on the infero-lateral border is a broad notch, the preoccipital notch.

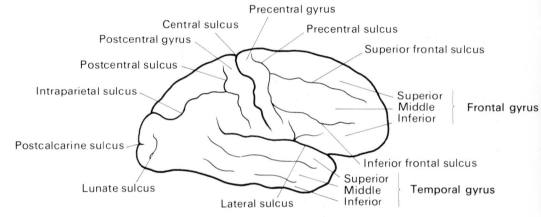

Fig. 8.16 Diagram of lateral surface of cerebral hemisphere showing the main sulci and gyri.

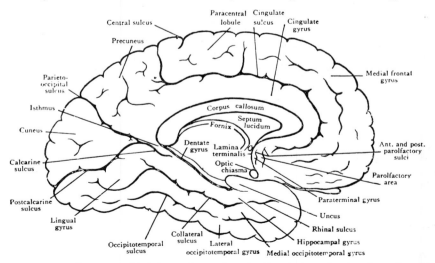

Fig. 8.17 Diagram of medial and inferior surfaces of cerebral hemisphere showing the main sulci and gyri. (By courtesy of Professor G. A. G. Mitchell and Dr E. L. Patterson.)

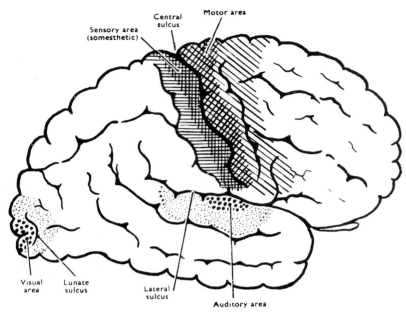

Fig. 8.18 The approximate cortical areas of motor, somesthetic, visual and auditory localization on the lateral surface of the cerebral hemisphere. The heavily shaded or stippled areas are the main afferent or efferent regions and the lighter adjacent zones are associated 'psychic' areas concerned with comparison, correlation and integration. (By courtesy of Professor G. A. G. Mitchell and Dr E. L. Patterson.)

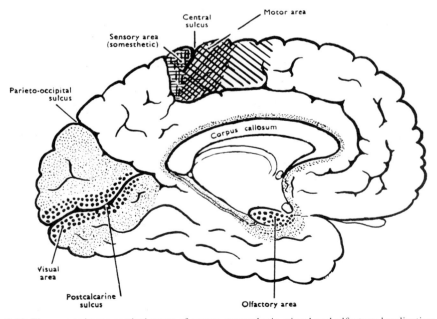

Fig. 8.19 The approximate cortical areas of motor, somesthetic, visual and olfactory localization on the medial and inferior surfaces of the cerebral hemisphere. (By courtesy of Professor G. A. G. Mitchell and Dr E. L. Patterson.)

The *temporal lobe* is situated in front of the preoccipital notch and below the lateral sulcus. Its anterior limit is the *temporal pole*.

The frontal and temporal lobes occupy the anterior and middle cranial fossae respectively. The occipital pole lies above the tentorium cerebelli and the parietal lobe is deep to the parietal region of the skull.

Cortical structure and connections

The cerebral cortex consists of a very large number of nerve cells arranged in layers of different thicknesses. The total thickness of the cortex is also different in different regions being thinnest at the occipital pole (about 1·5 mm) and thickest in the precentral gyrus (about 4 mm). The nerve fibres originating in and having connections with these cells are of three types:

Association fibres connecting gyri in the same hemisphere. Some of these bundles only extend from one gyrus to the next while others extend the full length of the hemisphere from the frontal to the occipital poles.

Commissural fibres join one hemisphere with the other and the most prominent commissure is the *corpus callosum* which lies at the bottom of the median sulcus between the two hemispheres. The two hemispheres are functionally integrated by these fibre connections.

Itinerant or projection fibres connect the cortex with other parts of the nervous system, such as the main motor pathway from the precentral gyrus via the internal capsule and cerebral peduncle to the anterior column of grey matter in the spinal cord. The main sensory pathway is a further example, passing upwards from the posterior white columns via the medulla and thalamus, the fibres ending in the sensory area of the cortex, which is the postcentral gyrus.

Localization of function in the cerebral cortex

The fact that the cortex does not have a uniform histological structure throughout all parts of the hemisphere is associated with differences in function. The various areas of cortex with modifications of structure have been mapped out into more than one hundred areas. Little is known about the function of many of these areas but those concerned with motor, sensory, visual and auditory impulses have been localized accurately (Figs 8.18, 8.19). In the precentral gyrus various regions of the body are represented in an inverted manner. Thus the head area in this strip of cortex lies near its lower end and the lower parts of the body, trunk, leg and foot are represented at its upper limit. The *motor area for speech* lies at the lower end of the precentral gyrus just above the stem of the lateral sulcus and is developed on the left side of the brain in right handed people. Within the head area the various parts of the face are not inverted, however, so that the area for the eyelid and eyeball lies at a higher level than for lips and jaws. The total area of cortex associated with a particular region of the body depends not on its size but on its functional importance. Thus, for example, the areas of cortex devoted to nerve cells associated with the mouth and thumb are considerably greater than those associated with movements of the neck and toes. The localization of function in the motor cortex is paralleled by a similar arrangement of the sensory cortex in the postcentral gyrus.

The frontal lobes of the cerebral hemisphere contain the higher centres for voluntary motor functions. They are interconnected with the hypothalamus and thalamus which allows association between somatic and visceral activities. Lesions of this part of the brain produce motor and autonomic disturbances, alterations in behaviour and personality, with a loss of the ability to perceive certain types of pain.

The parietal lobe is also an association area and in it a variety of somatic sensations are compared with those from the adjacent visual and auditory centres. Lesions in this area of the brain produce disturbances of the ability to detect different shapes, size and texture. Speech and writing difficulties are the special problems which result if the left inferior parietal region is involved.

The occipital lobe receives visual impulses and these are interpreted in adjacent areas of the occipital cortex. Injuries to this region of the brain result in disturbances of vision.

The temporal lobes contain the centres for hearing and smell and little is known about some of their parts. The sensations associated with alterations of balance are appreciated here.

The thalamus and basal ganglia

The thalamus is a large oval mass of grey matter which lies in the lateral wall of the third ventricle and is covered by ependymal epithelium (Figs 8.8, 8.15). It is a relay station for sensory impulses coming from lower levels of the central nervous system on their way to higher functional levels served by the cerebral cortex. The upper surface of the thalamus is exposed in the floor of the lateral ventricle (page 407); laterally, it is related to the posterior limb of the internal capsule (vide infra); the hypothalamus lies beneath it. It can be divided into a number of nuclei including:

1. The ventral nucleus, which is the terminal station for the medial lemniscus, the spinothalamic tracts and the trigeminal lemniscus.
2. The lateral nucleus which is situated in the dorsal part of the thalamus.
3. The lateral geniculate nucleus in which terminates the optic tract.
4. The medial geniculate nucleus which receives auditory impulses.
5. Medial, anterior and central nuclei.

The fibres which project to the cerebral cortex from the thalamus constitute the thalamic radiations. While the connections of some of the thalamic nuclei are known quite precisely, the nature of the impulses and the connections of other nuclei are completely unknown. Important connections exist between the thalamus and nuclei in the brain stem.

Other large masses of grey matter are found in the central part of the cerebral hemisphere (Figs 8.15, 8.20). These are the basal ganglia or *corpus striatum*, made up of the caudate and lentiform nuclei, the claustrum and the amygdaloid body which is embedded in the temporal lobe. The corpus striatum is a motor mechanism which functions in conjunction with other motor systems in the brain and plays a part in the integration of involuntary movements,

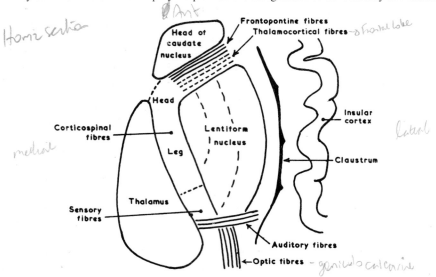

Fig. 8.20 The position of the main nerve tracts in the internal capsule of the cerebral hemisphere seen in transverse section. The internal capsule lies between the caudate nucleus, the lentiform nucleus and the thalamus.

particularly those associated with body posture, and with emotional reactions. Little is known about the functions of some of the elements of the basal ganglia and we have no knowledge of the purpose of the claustrum. The connections of the corpus striatum are diffuse and therefore difficult to analyse experimentally. The principal connections are to the cerebral cortex, hypothalamus, and nuclei of the brain stem including the reticular formation. The corpus striatum has been shown to exert control over voluntary muscle tone and fibres from the motor area of the cerebral cortex play an important part in regulating these changes. Consequently, pathological lesions of the corpus striatum manifest themselves in disturbances of muscle tone, with the onset of rigidity, tremors or spasmodic muscle movements. Parkinson's Disease is a good example and recent research suggests that altered cell structure in the caudate nucleus is the basis of schizophrenia.

The internal capsule (Fig. 8.20)
This is the layer of white matter between the head of the caudate nucleus and the thalamus medially and the lentiform nucleus laterally. It is bent to conform to the shape of the medial surface of the lentiform nucleus and is described as consisting of an anterior and posterior limb which meet at the genu (knee). The fibres within the capsule are arranged as follows:

1. The anterior limb contains frontopontine and thalamocortical fibres. The latter pass to the frontal lobe.

2. The genu and anterior two-thirds of the posterior limb contain the corticospinal fibres which pass from the prefrontal gyrus to the anterior horn cells of the spinal cord. Those in the genu carry fibres for the head end of the body.

3. The posterior third of the posterior limb is occupied in the main by sensory fibres, belonging to the great somatic sensory pathway, but includes optic and auditory fibres passing to the occipital and temporal region of the cortex respectively.

The *insula* is an area of cortex hidden from the surface by the lips of the lateral sulcus. Little is known about its function.

The ventricles of the brain
These are cavities within the brain tissue derived as enlargements of the central canal of the embryonic neural tube (Fig. 8.21). They are the lateral, third and fourth ventricles.

The *lateral ventricle* lies within the substance of the cerebral hemisphere and consists of a central part and three horns: anterior posterior and inferior. The anterior horn extends into the frontal lobe, the posterior into the occipital lobe and the inferior into the temporal lobe. In its entirety it takes the form of a letter C when viewed from the lateral aspect, with the concavity of the C facing forwards. The posterior horn projects backwards from the convex aspect. The lateral ventricle communicates with the third ventricle through the interventricular foramen (of Munro) situated in the anterior part of the lateral wall of the third ventricle below the anterior end of the fornix (Fig. 8.8).

The *third ventricle* (Figs 8.8, 8.21) is a rectangular slit-like cavity situated between the thalami. Its boundaries and related structures can be seen in the diagrams, particularly the optic chiasma (where optic or second cranial nerve fibres cross the mid-line on their way to the occipital cortex), and the *pituitary gland* suspended by the *stalk* or *infundibulum* from the floor. The infundibulum is thickened at its attachment to the floor of the ventricle into the *tuber cinereum*. Behind the tuber cinereum are two rounded bodies, one on either side of the midline, the *corpora mamillaria*, which lie immediately anterior to the bifurcation of the basilar artery to form the posterior cerebral arteries. Between the corpora mamillaria and the peduncle of the mid-brain (Fig. 8.8) the floor of the ventricle is thickened and is perforated by central branches of the posterior cerebral arteries—the *posterior perforated substance*. The anterior perforated substance lies to the lateral side of the optic chiasma, behind the posterior end of the olfactory tract (Fig. 8.27) and is hidden from the inferior aspect by the middle cerebral

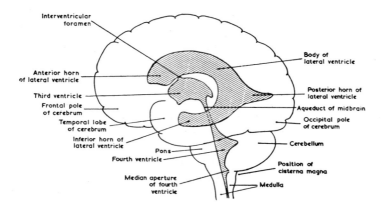

Fig. 8.21 The ventricular system of the brain. Cerebrospinal fluid produced in the ventricles passes into the subarachnoid space through the apertures of the fourth ventricle.

artery. Many of the small arteries through this region pass to the basal nuclei and thalamus. The third ventricle is continuous with the fourth ventricle via the aqueduct of the mid-brain.

The *fourth ventricle* (Figs 8.8, 8.21) is a tent-like cavity with a diamond shaped floor. Many of the nuclei of the cranial nerves lie in this region. It communicates with the subarachnoid space in the meninges through a *median* and two *lateral openings* (foramina of Luschka). The median aperture (foramen of Magendie) opens into an enlarged part of the subarachnoid space called the *cisterna magna* (between the cerebellum and the medulla).

The formation and circulation of cerebrospinal fluid
Extremely vascular fringes of the innermost layer of the meninges (the pia mater) project into the ventricles just described and produce a watery fluid, the cerebrospinal fluid, which bathes the cavities and escapes into the subarachnoid space through the apertures in the fourth ventricle. One of these fringes, from the *tela choroidea*, is shown in Figure 8.8. Eventually the fluid finds its way to the dural venous sinuses (see page 263) where it is reabsorbed into the circulation. Samples of cerebrospinal fluid may be drawn off from the cisterna magna via the suboccipital region or from the vertebral canal below the lower end of the spinal cord, usually via the interval between the second and third lumbar vertebrae.

The circulation of cerebrospinal fluid is continuous and it serves as a protective fluid buffer over the central nervous system which will prevent or lessen the effects of pressure, as in cases of intracranial haemorrhage. Blockage of any of the foramina in the ventricular system has disastrous results, such as can occur in *meningitis* with obstruction to the median and lateral openings of the fourth ventricle. The choroid plexuses continue to filter off cerebrospinal fluid from the blood and pressure builds up inside the ventricular system. Because the brain is contained within the rigid skull, the brain substance is compressed to a degree which may be sufficient to interfere with its blood supply. Increased pressure forces the hind brain into the foramen magnum causing disturbances of its function. The clinical signs and symptoms include headache, vomiting, disturbances of vision and coma. Similar results can arise from tumours that block the outlets from one part of the ventricular system to another.

Intracranial obstruction to the flow of cerebrospinal fluid in infants may be the result of deficient absorption of the fluid into the dural venous sinuses, brain tumours, or a fluid which has a high protein content. The rise in intracranial pressure causes considerable expansion of the brain, which can do so because the cranial bones are not yet fully ossified nor have they fused together. The condition is called *hydrocephalus* and is characterized by the enormous increase in the size of the cranial part of the skull while the facial skeleton and cranial base retain relatively normal proportions. If the condition is allowed to persist brain damage with mental retardation and eventually death are the results. Surgical treatment involves removal of the tumour, short circuiting the obstruction or diverting the fluid outside the cranial cavity for its absorption.

The meninges

The brain and spinal cord are surrounded by three layers of fibrous tissue: the *dura mater*, the *arachnoid* and *pia mater* (Fig. 8.22). These layers have a protective function, support parts of the brain and provide a pathway for the cerebrospinal fluid.

The *dura mater* is a strong fibrous sheet lining the skull bones. It can be further subdivided into an outer layer which acts as an endosteum for the skull bones and an inner layer, smooth and shining, which helps to protect the delicate brain surface. In certain well defined situations the inner and outer layers are separated to form the *dural sinuses*. These are pathways for

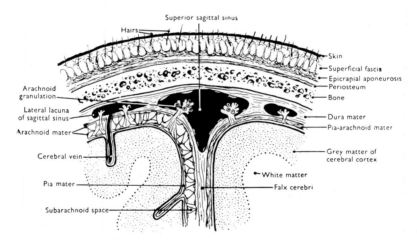

Fig. 8.22 A coronal section through the superior sagittal dural venous sinus showing the structure of the scalp and the arrangement of the cranial meninges. The subarachnoid space is probably not as extensive during life as that represented on the left of the diagram, being more correctly represented as the pia-arachnoid on the right half of the illustration.

venous blood and are described on page 263. The inner layer is prolonged into a double fold between the cerebral hemispheres (the falx cerebri) and between the occipital poles of the cerebral hemispheres and the cerebellum (the tentorium cerebelli). These septa support the brain tissue and render it less subject to damage. The dura mater is innervated by meningeal branches of the trigeminal and vagus nerves.

The *falx cerebri* is a sickle-shaped fold which is narrow at its anterior attachment to the crista galli of the ethmoid bone, becoming deeper as it is traced posteriorly. Its upper border contains the superior sagittal sinus and is attached in the mid-line from the foramen caecum, in front of the crista galli, to the internal occipital protuberance. Its free lower border is curved in conformity with the corpus callosum of the cerebral hemispheres (Fig. 8.8) and contains the inferior sagittal sinus which is of small size. The anterior end of the falx cerebri is often perforated (cribiform) and may contain fragments of bony deposits. The straight sinus lies along the line of attachment of the posterior end of the falx cerebri to the tentorium cerebelli.

The *tentorium cerebelli* is crescentic in shape and forms a tent-like roof for the posterior cranial fossa. The attached border of the tentorium encloses the transverse sinuses, from the internal occipital protuberance to the postero-inferior angle of the parietal bone, and the superior petrosal sinus along the upper border of the petrous temporal bone. The anterior or free border is sharply curved and with the dorsum sellae forms the *tentorial notch* within which lies the mid-brain. Antero-medial to the apex of the petrous part of the temporal bone the free and attached borders cross one another, the trochlear nerve leaving the cranial cavity at this point. The free margin is attached to the anterior clinoid process whilst the attached border passes more medially to gain attachment to the posterior clinoid process of the sphenoid bone.

The *arachnoid* is a delicate trabeculated layer. Between it and the pia mater is a potential space through which the cerebrospinal fluid can circulate—the *subarachnoid space*.

The *pia mater* is the only layer of the meninges which follows the convolutions and sulci of the nervous system. It is extremely vascular and many of the vessels within it send central branches into the brain tissue.

The pia mater which invests the lower end of the spinal cord is prolonged as the *filum terminale* as far as the tip of the coccyx. The arachnoid membrane lines the bony vertebral canal and thus below the lower end of the cord there is an extensive subarachnoid space. The subarachnoid space is also enlarged in three places in relation to the brain to form subarachnoid cisterns—the *cerebellomedullary cistern* (cisterna magna), the *pontine cistern* in front of the pons and the *interpeduncular cistern* at the base of the brain.

The blood supply of the brain and spinal cord
The arteries which supply brain tissue (Figs 4.28, 8.23) branch freely on the surface and anastomose with one another. The arteries entering the brain substance, however, are small and have little communication with neighbouring vessels so that damage to one of them will result in degeneration of the nervous tissue which it supplies. The main arteries of supply are derived from the terminal parts of the internal carotid and vertebral arteries whose branches unite to form an arterial *circle (of Willis)* at the base of the brain. From this arterial circle anterior, middle and posterior cerebral arteries are given off to the cerebral cortex:

The *anterior cerebral artery* supplies the medial surface of the cerebral hemisphere so far back as the parieto-occipital sulcus and a narrow strip of cortex along the supero-lateral surface at its superior border.

The *middle cerebral artery* emerges on the lateral surface of the brain through the lateral sulcus and supplies the greater part of that surface except the occipital pole.

The *posterior cerebral artery* supplies the inferior surface and the occipital lobe of the cerebral cortex.

The cerebral arteries give rise to *central* and *cortical* branches which do not anastomose with one another. The cerebral arteries supply the deep structures of the brain, including the thalamus and internal capsule. The cortical branches ramify in the pia mater and supply the cerebral cortex and underlying areas of the brain tissue. The central vessels are all branches from the arterial circle of Willis and are 'end' arteries, because they terminate without forming any anastomoses with neighbouring vessels, except capillaries, and blockage of one of them causes degeneration of an area of the brain.

The cortical vessels are the terminal branches of the anterior, middle and posterior cerebral arteries and their branches penetrate the brain tissue in a perpendicular manner, devoid of communications with neighbouring arteries, so that they also are 'end' arteries.

The veins of the brain have extremely thin walls and no valves and they end in the venous sinuses of the dura mater (page 263). There are three main groups on each side of the brain—

the superior, middle and inferior cerebral veins. Branches from the anterior and middle cerebral veins unite to form the *basal vein* which ends in the great cerebral vein.

The deeper parts of the cerebral hemispheres are drained by the internal cerebral veins which eventually join together to form the *great cerebral vein* (of Galen) which drains into the beginning of the straight sinus (page 263).

The veins which drain the cerebral cortex open into the dural venous sinuses which are described in the text on pages 263–264.

The brain stem and cerebellum are supplied by branches of the *vertebral* and *basilar* arteries.

The spinal cord receives its blood supply from *anterior* and *posterior spinal arteries* descending from the vertebral arteries. These are reinforced by vessels entering the vertebral canal from segmental arteries especially in the lower thoracic and upper lumbar region.

The veins of the spinal cord and medulla are arranged in a longitudinal manner and drain into the vertebral veins and the *vertebral venous plexus* which is a network of veins lining the inside of the vertebral canal outside the dura mater.

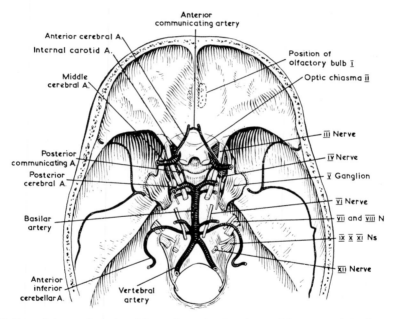

Fig. 8.23 Base of the cranium viewed from above showing the cranial nerves and the formation of the arterial circle of Willis.

Clinical anatomy

The arteries that contribute to the arterial circle of Willis are subject to anatomical variations and prone to aneurysms at their various bifurcations and junctions. The muscle coats of the vessels fail and dilatations occur as a response to rises in blood pressure. *Aneurysms* may lead to haemorrhage and signs of pressure on nearby structures such as the oculomotor nerve or disturbances of brain function. The central branches of the middle cerebral artery pass through the anterior perforated substance and supply the region of the internal capsule and the corpus striatum. These arteries are frequently ruptured in cerebral haemorrhage and the damage to the soft brain tissue leads to the condition known as a *stroke* with varying degrees of sensory and motor paralysis on the opposite side of the body (hemiplegia). This is because the sensory and motor pathways which pass through the internal capsule cross over to the opposite side in the brain stem and spinal cord. Depending on the amount of haemorrhage and thus the extent of damage to nerve tissue, persons with this condition show varying degrees of recovery and the same or associated vessels are likely to rupture on subsequent occasions leading to further clinical episodes.

Fractures of skull bones cause haemorrhage in the outer layer of the dura mater which forms the equivalent of the periosteum on the inside of the skull. Extradural blood clots may reach a sufficient size to cause death from increase in intracranial pressure. This can be avoided in many instances by surgical removal of the blood clot and coagulating or tying off the ruptured blood vessels.

THE SPINAL CORD

TRACTS IN THE SPINAL CORD (Fig. 8.24)

Only the major pathways in each of the three white columns will be described, for to do otherwise without special study would lead to confusion. A knowledge of a few major tracts will explain adequately the principles of the internal organization of the central nervous system. Some tracts have been named after the investigators who first described them but the better and easier method is to employ the terminology which indicates their place of origin, destination and direction in which the impulses pass.

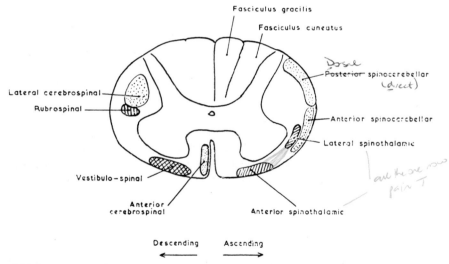

Fig. 8.24 Important nerve tracts in the spinal cord.

Ascending tracts

The *anterior spinothalamic tract* transmits tactile impressions from the cord to the thalamus (page 420). The fibres originate in the posterior column of grey matter on the opposite side and cross over in the anterior white commissure before ascending in the anterior white column.

The *lateral spinothalamic tract* conveys pain and temperature impulses to the thalamus, the fibres merging in the medulla with those from the anterior spinothalamic tract to form the *spinal lemniscus*. The fibres begin from cells situated in the posterior grey horn of the opposite side and principally ascend in the lateral column.

The *anterior and posterior spinocerebellar tracts* transmit to the cerebellum information which helps in the co-ordination of muscle groups (synergic activity) and muscle tone. The nerve fibres begin in the posterior horn of the same side and pass into the lateral white column.

The posterior columns of white matter are almost entirely occupied by nerve pathways concerned with the localization of touch and proprioception. The *fasciculus gracilis* lies close to the posterior median fissure and contains fibres predominantly associated with sensory fibres from the lower limb. The *fasciculus cuneatus* makes its appearance in the upper half of the spinal cord for its fibres are derived essentially in association with the nervous inflow from the upper limb and thorax. It lies to the lateral side of the fasciculus gracilis. The nerve fibres end in the medulla around collections of cells (called nuclei) of the same name. The further connections of these tracts is described on page 415.

Descending tracts

The *anterior corticospinal or cerebrospinal tract* (alternatively known as the *direct pyramidal tract*) consists of fibres which descend through the brain from the cerebral cortex. In the cord it lies close to the antero-median fissure and is concerned with the voluntary control of skeletal muscles. The tract becomes smaller as it descends and eventually almost disappears in the midthoracic region for by this level the majority of its constituent fibres have crossed to the opposite side to end by synapsing around anterior horn cells. The few remaining fibres terminate in the anterior grey column on the same side of the cord.

The *vestibulospinal tract* also descends in the anterior white column. It is concerned with balance and posture. The nerve fibres begin in the medulla on the same side from a collection of cells called the vestibular nucleus.

The *lateral corticospinal tract* (alternatively known as the *crossed pyramidal tract*) contains the majority of the fibres which control voluntary muscle movements. The fibres arise in the motor cortex (see page 404) and cross over or decussate with the corresponding tract of the opposite side in the medulla. Only one-tenth of the motor fibres do not cross over and these constitute the anterior cortico-spinal tract described above.

The *rubrospinal tract* lies immediately in front of the lateral corticospinal pathway, the fibres beginning in the mid-brain and descending to end around anterior horn cells. It is concerned in the control of muscle group action.

There are no important descending tracts in the posterior white columns. In all the white columns, but to a lesser extent in the posterior column, are *intersegmental fibres* which associate one cord segment with another. At the upper end of the spinal cord they are collected together to become continuous with the medial longitudinal bundle of the brain, which performs a similar function in relation to the nuclei of the cranial nerves by co-ordinating their activities.

The motor and sensory pathways

The most important tracts within the brain and spinal cord are undoubtedly those concerned with voluntary motor movements; general sensations of pain, temperature, and touch from sense organs in the skin and proprioceptive impulses from muscles and joints. From the foregoing description of the major features of parts of the nervous system it is now convenient to summarize the extent of the pathways.

The corticospinal or pyramidal tract or motor pathway (Fig. 8.25)

The tracts originate from the precentral motor cortex and fibres descend through the internal capsule in the genu and the anterior two-thirds of the posterior limb and constitute the most important motor tracts in the spinal cord because it is through them that voluntary movements are made possible. The convergence of fibres from the cortex to the internal capsule is known as the *corona radiata*. From the internal capsule the fibres continue downwards first in the central part of the basis pedunculi, then in the anterior part of the pons and the pyramid of the medulla oblongata. In the medulla the majority of the fibres decussate to take up a position in the lateral white column of the spinal cord—the lateral corticospinal tract. The remaining fibres which do not cross in the medulla continue for some distance on the same side as the direct pyramidal tract, but eventually by the mid-thoracic level these also have crossed to the opposite side. Only a few fibres remain on the same side and they terminate in the anterior column of grey matter. The crossed pyramidal fibres all end by synapsing with anterior horn cells, either directly or through intercalated neurons. The axons which follow

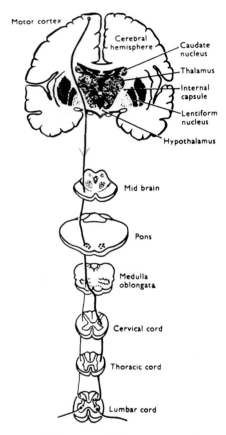

Fig. 8.25 The corticospinal or pyramidal tracts—the great motor pathways from the cortex (mainly the precentral gyrus) to the anterior horn cells in the cord. (By courtesy of Professor G. A. G. Mitchell and Dr E. L. Patterson.)

the path just described together with their cell bodies constitute the *upper motor neurons*. The axons from the anterior horn cells convey the impulses to the motor endplates in the skeletal muscles and constitute the *lower motor neurons*.

Voluntary muscle activities are not entirely controlled by the pyramidal tract but also by other pathways from the *corpus striatum* in the cerebral hemisphere. The corpus striatum is associated with certain brain stem nuclei, notably the red nucleus and the rubrospinal tract which passes from it into the spinal cord. These alternative tracts form the *extrapyramidal system* which has a functional relationship with motor neurons in the anterior column of grey matter of the spinal cord, perhaps to inhibit excessive muscular movements or to maintain a proper degree of muscle tone. In any event, the two systems complete their course through a final common pathway.

Lesions of the pyramidal tracts in the spinal cord cause a loss of voluntary movements below the level of injury, even though the muscles themselves are not paralysed. Lesions of upper motor neurons lead to an increase in the tone of the affected muscles with eventual rigidity or spasticity, possibly due to affects on the extrapyramidal system.

The sensory pathway (Fig. 8.26)

There are three neurons in the sensory pathway:

    1. The cells of the posterior root ganglia and their processes
    2. The neurons connecting these with the thalamus
    3. The neurons which connect the thalamus with the postcentral gyrus of the cerebral cortex.

They are often termed the *lowest, intermediate* and *highest sensory neurons*.

The fibres transmitting the finer variety of sensation (epicritic sensation) and proprioceptive impulses enter via the posterior nerve roots of the spinal nerves and ascend in the posterior white columns to the region of the nuclei gracilis and cuneatus. Here new fibres arise which pass inwards around the central canal of the medulla as the *internal arcuate fibres*. The internal

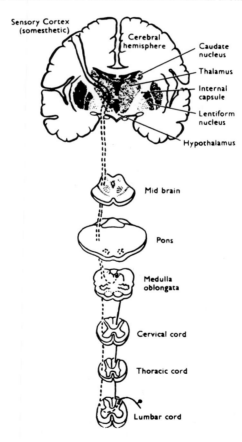

Fig. 8.26 The chief somesthetic sensory pathways transmitting impulses upwards to the cortex (mainly the postcentral gyrus). (By courtesy of Professor G. A. G. Mitchell and Dr E. L. Patterson.)

arcuate fibres cross the mid-line and interdigitate with those from the opposite side as the sensory decussation. Immediately the fibres ascend once again, as the *medial lemniscus* (ribbon), through the brain stem to the lateral part of the thalamus. New cells take over whose axons finally enter the sensory cortex.

Fibres carrying the coarser types of sensation of pain, heat, touch and pressure have a different course. The incoming sensory fibres relay in the posterior horn of grey matter in the spinal cord with nerve cells whose axons cross to the opposite side, either at the same level (pain and temperature) or a few segments higher (touch and pressure) to form the spinothalamic tracts. These unite in the medulla to form the *spinal lemniscus* which associates at a higher level with the medial lemniscus and the *trigeminal lemniscus* (consisting of sensory fibres from the trigeminal nerve area of distribution). Eventually the lemnisci relay in the thalamus from which the upper sensory neurons pass to the sensory cortex behind the central sulcus.

Proprioceptive impulses which result from the stimulation of receptors in muscles, joints and tendons and are associated with reflex coordination at the level of consciousness, pass to the cerebellum in the *spinocerebellar tracts* (Fig. 8.24). Through a combination of the proprioceptive pathways of the posterior white columns and the spinocerebellar tracts, the individual is able to recognize movements and appreciate the position in space of different parts of the body. A patient with lesions of the posterior columns is unable to describe the position of the feet unless they are actually seen. If asked to stand with feet together and then to close the eyes, the patient staggers and falls because the proprioceptive pathway has been interrupted. These neurological signs are evident in patients with advanced syphilis.

The functional significance of the sensory pathways can be summarized by considering the changes which occur following hemi-section of the thoracic part of the spinal cord. Below the level of the lesion on the opposite side of the body, the sensations of pain and temperature are essentially lost. On the same side, also below the level of the lesion, there will be loss of the sense of touch and the ability to judge position in space or to detect movement.

THE AUTONOMIC NERVOUS SYSTEM

GENERAL ARRANGEMENT

The autonomic nervous system regulates the body functions which are not under the control of the will, e.g. the regulation of circulatory, respiratory, alimentary and reproductive functions. It consists of central and peripheral parts.

The *central elements* are situated in the cerebral cortex, *hypothalamus* (the region below and in front of the thalamus), cerebellum, brain stem and spinal cord. These regions are connected by tracts comparable to those in the somatic or voluntary nervous system, with which system their activities are accurately co-ordinated.

The *peripheral parts* consist of ganglionated nerve trunks extending along the length of the vertebral column and various nerve plexuses such as the cardiac and coeliac plexuses.

The autonomic nervous system contains the same components as the somatic system in that afferent, efferent and intermediate neurons are linked to form reflex arcs. The essential difference is that in the autonomic system the axon of the intermediate or intercalated neuron passes outside the central nervous system to synapse with the efferent neuron in a peripheral *ganglion* (Fig. 8.4). This arrangement is the result of the migration of the efferent nerve cells from the neural tube, and consequently the intercalated axons during embryonic development.

The axons of the intercalated neurons are alternatively known as the *preganglionic nerve fibres*. They have well defined myelin sheaths which in bulk appear white and constitute the *white rami communicantes*. The axons of the efferent neurons are known as the postganglionic or *grey rami communicantes* for they lack myelin sheaths and have a greyish colour in the fresh state.

The autonomic nervous system is further subdivided into *sympathetic* and *parasympathetic* parts and most organs are innervated by nerve fibres from both sources which have opposing physiological effects. The transmission of nerve impulses across synaptic junctions in the ganglia and from the postganglionic fibre to the structure which it supplies is thought to depend on chemical activity (adrenaline and acetylcholine compounds) so that certain drugs which will affect sympathetic nerve endings will not affect parasympathetic endings and vice versa. Sympathetic fibres in the facial region are primarily concerned with the control of the diameter of blood vessels thus influencing the blood flow (vasomotor activity). Parasympathetic fibres form the secretomotor fibres controlling the secretions of glands, including the salivary glands. It is generally agreed that there are no specialized nerve endings in the autonomic nervous system but the terminal fibres break up into fine branches which form a delicate network or terminal plexus in relation to the innervated structure.

AUTONOMIC OUTFLOWS

The centres in the *brain stem* (Fig. 8.27) are related to the nuclei of certain of the cranial nerves (the oculomotor, facial, glossopharyngeal and vagus nerves) and the preganglionic fibres which leave the brain stem with them constitute the *cranial parasympathetic outflow*. The preganglionic axons end in a small ganglion related to the submandibular salivary gland; and the otic ganglion which is close to the parotid salivery gland. Thus the postganglionic fibres are characteristically very short.

The cells of the autonomic system in the *spinal cord* are situated in the lateral grey columns and their axons emerge through the anterior roots of the spinal nerves as preganglionic sympathetic fibres (*thoracolumbar outflow*), which pass to the ganglia of the paravertebral sympathetic trunks (see pages 85 and 174). Many of the preganglionic fibres or white rami communicantes synapse in adjacent ganglia but others ascend or descend for some distance before synapsing at a higher or lower level. The cervical sympathetic ganglia and trunk are formed in this manner.

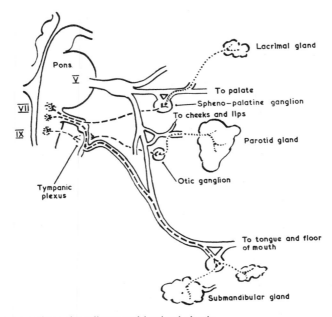

Fig. 8.27 Nerve supply to the salivary and lacrimal glands.

All the sympathetic preganglionic nerve fibres leave the cord from the thoracic or upper lumbar segments and the postganglionic fibres are distributed to all parts of the body. Therefore the majority of sympathetic ganglia lie at some distance from the structures which they innervate. Some of the postganglionic fibres pass directly to the viscera and vessels but others return to the spinal nerves in the grey rami communicantes and are distributed peripherally with their branches to vessels, sweat glands and smooth muscle in relation to hair follicles (arrectores pilorum muscles).

Many of the postganglionic sympathetic nerve fibres in the neck region pass through grey rami to the upper three or four cervical nerves and are distributed with them. Others form perivascular nerve networks around the internal and external carotid arteries and their branches. In this way the mouth and teeth receive their sympathetic supply via the branches of the maxillary artery and the soft tissues of the face derive their supply with the facial artery and its branches.

Nerve cells in the grey matter of the second, third and fourth sacral segments of the spinal cord give origin to preganglionic nerve fibres which constitute the *sacral outflow*. They are parasympathetic in type and pass to ganglia in the walls of the pelvic viscera as the *pelvic splanchnic nerves*. The postganglionic fibres are distributed to the tissues of these organs.

Afferent fibres in the autonomic nervous system. Sensory fibres exist in the autonomic system which have their cell bodies in the posterior root ganglia of the spinal nerves. They are not concerned with the ordinary modalities of sensation (touch, pain, etc.) but with alterations of pressure, crushing or a lowered concentration of oxygen in the circulating blood. The central processes of the sensory cells take part in autonomic reflex arcs or pass to the higher autonomic centres, in close relationship to the main somatic sensory pathways.

THE CRANIAL NERVES

The twelve cranial nerves are:
1. The *olfactory nerve*—special nerve for the sense of smell.
2. The *optic nerve*—special nerve for the sense of sight.
3. The *oculomotor nerve*—motor nerve to the muscles of the eyeball (except for the superior oblique and lateral rectus) and to the levator palpebrae superioris of the upper eyelid.
4. The *trochlear nerve*—motor nerve to the superior oblique muscle of the eyeball.
5. The *trigeminal nerve*—motor nerve to the muscles of mastication, tensor palati and tensor tympani muscles, sensory to the face, orbital cavities, nasal cavities, paranasal air sinuses, and oral cavity.
6. The *abducent nerve*—motor nerve to the lateral rectus muscle.
7. The *facial nerve*—motor nerve to the stapedius muscle, the platysma, stylohyoid, posterior belly of digastric and muscles of expression—sensory nerve for the special sense of taste.
8. The *vestibulocochlear nerve*—special nerve for the senses of hearing and balance.
9. The *glossopharyngeal nerve*—motor nerve to stylopharyngeus muscle—sensory to back of tongue, tonsil, upper pharynx, auditory tube, middle ear, mastoid air cells—special sense of taste.
10. The *vagus nerve*—motor nerve to smooth muscle of respiratory system, alimentary canal as far as transverse colon and to cardiac muscle. Sensory to tympanic membrane, lower pharynx and larynx.
11. The *accessory nerve*—motor nerve to sternocleidomastoid, trapezius, muscles of soft palate (except tensor palati), pharynx and larynx.
12. The *hypoglossal nerve*—motor nerve to muscles of the tongue. *except palato glossus (x)*
Parasympathetic fibres to the ciliary, submandibular, sphenopalatine and otic ganglia run with the third, seventh and ninth cranial nerves (see pages 191 and 320).

Central connections of the trigeminal nerve

The trigeminal nerve has deep connections within the brain stem which extend from the upper part of the mid-brain caudally to the upper part of the spinal cord. It has three nuclei or collections of nerve cell bodies, viz., the *sensory* nucleus, the *motor nucleus* and the *mesence-phalic*, or mid-brain nucleus (Fig. 8.12). 3 sensory nuclei —chief spinal mesencephalic
1 motor nucleus

The chief sensory nucleus (Fig. 8.28)

This is an oval mass of grey matter containing many nerve cell bodies which occupies a lateral position in the pons midway between its upper and lower borders. The incoming nerve fibres, whose cell bodies form the trigeminal ganglion (corresponding to the posterior or dorsal root Meckel's cave ganglion of a spinal nerve), enter the pons via the sensory root and arborize around the cells in the chief sensory nucleus. Within this nucleus the ophthalmic division fibres terminate ventrally, while the maxillary and mandibular components end successively in more dorsal areas of the nucleus. Some of the fibres descend into the medulla and upper part of the spinal cord, as the spinal tract of the trigeminal nerve, to end around cells of the spinal nucleus, which is in effect a downward extension of the chief sensory nucleus in the pons. The axons of the spinal nucleus are the equivalent of the intermediate sensory neurons in the spinal cord and have connections with the cerebral cortex via the thalamus and with the motor nucleus of the trigeminal nerve (for reflex activity). The three divisions of the trigeminal nerve continue to be represented in the nucleus of spinal tract in an inverted order and the most recent experimental studies show that all three divisions end together at the level of the upper part of the second cervical segment of the spinal cord.

The central processes of the nerve cell bodies which occupy the semilunar or trigeminal ganglion form the portio major of the trigeminal nerve, otherwise known as the root of the trigeminal nerve. This root is composed of myelinated and unmyelinated nerve fibres which are not aggregated together in any particular fashion but are scattered throughout the substance of the nerve root. The incoming fibres which terminate in the main or chief sensory nucleus are concerned only with tactile sensibility. The spinal tract and nucleus are chiefly concerned with pain and temperature impulses and it is generally agreed that the pain fibres are of the unmyelinated variety. The spinal tract and nucleus are situated in a superficial position on the lateral aspect of the medulla and this is of importance in the relief of pain by surgical interference with this pathway, known as the medullary tractotomy. By this procedure the neurosurgeon can produce differential relief of pain in the face by section of the appropriate parts of the spinal tract of the trigeminal nerve, with very little disturbance of other types of sensation.

An important concept of the arrangement of nerve fibres in the spinal tract of the fifth cranial nerve was described by Dejerine in 1914. All three divisions of the trigeminal nerve send fibres as far caudally as the upper segments of the cervical part of the spinal cord and they terminate in a pattern which matches facial areas arranged concentrically around the mouth opening. Five zones were described: a central or circumoral zone, three intermediate zones and a peripheral zone of the face, with all three divisions of the trigeminal nerve ending at all levels of the brain stem and spinal cord at which the spinal tract terminates. The nerve fibres supplying the most peripheral zone on the face descend to the lowest level in the spinal cord and those related to the area immediately around the nose and mouth end at the highest level in the brain stem. This arrangement of the descending fibres refers only to those that carry pain and gross temperature sensations. The functional pattern of concentric zones around the nose and mouth area has been referred to for many years as the *onion-skin pattern*.

The nerve fibres which come into the brain stem as central processes of trigeminal ganglion cells end either in the chief sensory nucleus or in the spinal nucleus of the trigeminal nerve. The spinal nucleus consists of second order neurons whose processes pass to higher levels of the central nervous system. Because the spinal nucleus can be subdivided into several parts it is often referred to as the spinal trigeminal nuclear complex. It can be subdivided into three fairly distinct regions which are, from above downwards beginning just below the chief sensory nucleus, the *subnucleus rostralis*, the *subnucleus interpolaris* and the *subnucleus caudalis*. The subnucleus caudalis lies at the level of the second cervical segment of the spinal cord (Fig. 8.28).

The axons of the cells which make up the spinal nucleus of the trigeminal nerve leave the nucleus on its medial aspect, many of them crossing the lower brain stem to the opposite side to ascend as the bulbothalamic tract. This secondary afferent pathway of the trigeminal is often designated as the *trigeminal lemniscus*. It can be divided into two parts, a ventral and a dorsal lemniscus. The ventral lemniscus is chiefly composed of crossed fibres which have

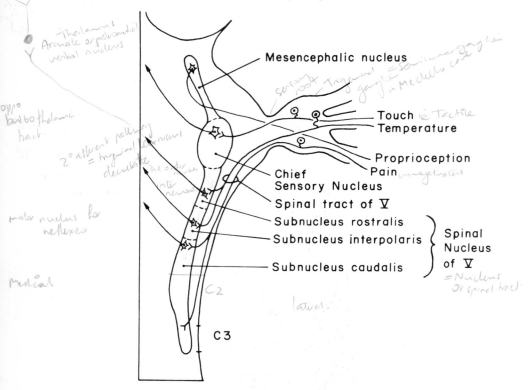

Mesencephalic nucleus

Touch

Temperature

Proprioception

Pain

Chief Sensory Nucleus

Spinal tract of V

Subnucleus rostralis

Subnucleus interpolaris

Subnucleus caudalis

} Spinal Nucleus of V

C2

C3

Fig. 8.28 Central connections of the trigeminal nerve.

arisen from the opposite spinal and main sensory nuclei and, in the medulla, it is closely associated with the lateral spinothalamic tract, which conveys pain and temperature impulses to the thalamus from other parts of the body (see page 412). These ascending fibres terminate in the arcuate or posteromedial ventral nucleus of the thalamus, medial to the termination of the pain fibres in the lower parts of the body, which have travelled upwards in the spinothalamic tracts. As far as is known at present impulses produced by painful stimulation pass from the thalamus in two divergent paths. Some fibres pass directly to the sensory cortex of the cerebral hemisphere of the same side, including the lower part of the postcentral gyrus. Other fibres pass to the hypothalamus and indirectly to the cortex of the frontal lobe. Some of these fibres are of special importance for they are involved in the emotional reactions to pain, by virtue of which different degrees of severity of pain can be recognized, e.g. of an unpleasant, unbearable or agonizing nature.

The trigeminal fibres associated with painful responses are probably of the unmyelinated variety. However, the occurrence of a significant number of unmyelinated fibres in the trigeminal ganglion suggests that still other routes may exist for an autonomic supply to the blood vessels and glands of the oral tissues. Some of the pathways which have been suggested are still hypothetical but are summarized by the illustration (Fig. 8.29).

Myelinated nerve fibres in peripheral branches of the trigeminal nerve pass centrally from nerve terminations in the oral mucosa, have their cell bodies (1 in Fig. 8.29) in the substance of the ganglion, and terminate within the brain stem as described above. Likewise, unmyelinated afferent fibres may pursue a similar course with cell stations (2 in Fig. 8.29) within the ganglion. These fibres are those associated with painful stimuli and are comparable to the visceral afferent fibres of spinal ganglia.

Unmyelinated fibres (3 in Fig. 8.29) which pass towards the periphery may arise as an additional outflow from the brain stem, comparable to the parasympathetic visual dilator

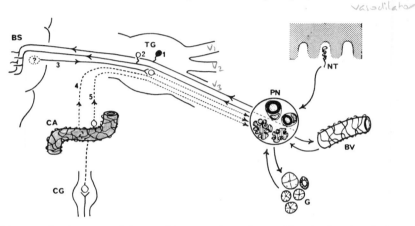

Fig. 8.29 Diagram showing some of the possible connections of nerve fibres in the trigeminal ganglion. BS. brain stem. BV, blood vessel. CA, cavernous sinus nerve plexus. CG, superior cervical ganglion. G, gland. NT, sensory nerve termination; PN, peripheral nerve; TG, trigeminal ganglion. (By courtesy of the Royal Society of Medicine.)

fibres, which are thought to form synapses in dorsal root ganglia. Postganglionic sympathetic fibres (4 and 5 in Fig. 8.29) pass through the trigeminal ganglion and have their cell bodies in the superior cervical ganglion, or in groups of ganglion cells in the sympathetic nerve plexus which surrounds the internal carotid artery in the cavernous sinus. These fibres are inconstant, but it seems probable that some of the unmyelinated fibres in the trigeminal ganglion are derived from these sources.

Sensation in the dental pulp. The primary afferent fibres from the dental pulp have their cell bodies in the trigeminal ganglion and the central processes of these cells terminate in the trigeminal sensory nucleus in the brain stem. Stimulation of nerve terminations within the tooth results in only a sensation of pain, provided no supplementary information about the nature of the stimulus reaches the subject from other receptors. The quality of the pain may be described differently by different subjects based on their previous experience, so that clinical studies of pain are extremely difficult to interpret. Since the caudal part of the trigeminal sensory nucleus appears to be involved in pain sensation and touch discrimination has been more often associated with the chief sensory nucleus, the primary or first order afferent fibres from the dental pulp might be expected to terminate only in the subnucleus caudalis. However, recent investigations suggest that the afferent fibres terminate around cells which form longi-

tudinal columns running the entire length of the trigeminal sensory nucleus. Electrophysiologi-
cal studies of the central connections of afferent fibres from the dental pulp show that these
fibres do indeed project to the main sensory nucleus, as well as the subnucleus caudalis. There
is general agreement that afferent fibres from the dental pulp are segregated to the medial
and ventral quadrant of the trigeminal nucleus.

The motor nucleus (Figs 8.11, 8.12)

This also is found in the pons, just to the medial side of the sensory nucleus. This relationship
of the motor to the sensory nucleus is typical of the arrangement of cranial nerve nuclei in
the brain stem and can be explained by its developmental history (see page 344). The motor
nucleus of the trigeminal nerve is derived in the basal plate while the sensory nucleus is a
derivative of the alar plate of the pontine region. The fibres from the nucleus pass out of
the pons in the motor root and are distributed almost entirely to the muscles of mastication,
through the mandibular division of the trigeminal nerve. These motor fibres are classified
as special visceral efferent because they innervate striated muscles derived from the mesoderm
of the pharyngeal arches (Fig. 8.13).

The mesencephalic nucleus (Figs 8.11, 8.14)

This is the third sensory nucleus associated with the trigeminal nerve, the others being the
chief and spinal sensory nuclei. The mesencephalic nucleus is concerned with proprioceptive
impulses from the muscles of mastication, the temporomandibular joint, the periodontal ligaments of the teeth, the oral cavity in general, and probably the extrinsic muscles of the eyeball.
The nucleus lies in the lateral part of the grey matter around the central aqueduct of the mid-
brain and in the ventrolateral angle of the upper part of the fourth ventricle.

Unlike the cells in the main sensory and spinal nuclei of the trigeminal nerve, which are
multipolar in type, the cells of the mesencephalic nucleus are unipolar, with axons which divide
almost at once into peripheral and central processes. The peripheral processes pass to proprioceptive terminations in the muscles of mastication and elsewhere, by way of the mesencephalic
root and the motor root, or portio minor, of the trigeminal nerve. Because the cells in the
nucleus are unipolar in type has led to the belief that this nucleus is developed from the
upper limits of the neural crest and that they fail to migrate from the region of the neural
tube, in contrast to the migration of cells which produce the true sensory ganglia of the spinal
nerves.

The trigeminal nerve is associated with a host of painful conditions arising from pathology
in any of the structures innervated by the nerve. A serious form of intractable facial pain
is *trigeminal neuralgia* which is characterized by its acute, paroxysmal nature. The onset of
this pain, whose cause is uncertain, is frequently initiated by mild stimulation of the skin of
the face or the oral mucosa. Various forms of treatment are employed to correct it, including
avulsion of the terminal nerve branches to the affected zone; injections of alcohol around
major divisions of the nerve or into the trigeminal ganglion; and surgical division of the sensory
root between the ganglion and the brain stem. Recent developments have shown that surgery
is not always necessary and relief can be obtained in many patients by the use of the newer
anticonvulsant drugs. Tegretol = carbamazepine

Complete division of the trigeminal nerve on one side results in extensive anaesthesia of
the face and scalp, the conjunctiva and the anterior two-thirds of the tongue on the same
side. There is also loss of sensation in the gums around the teeth in both the upper and lower
jaw, the mucous membrane lining the cheek and lips, hard and soft palates, as well as the
mucous membrane of the nose. The area of sensory loss extends to the mid-line and even
a short distance beyond it. The muscles of mastication are paralysed on one side and it is
important to remember that the buccinator escapes paralysis because it is supplied by the
seventh cranial nerve.

If divide V intracranially can still taste (VII & IX)
& salivate (VII & IX) alternate VII

Central connections of the facial nerve

The nucleus of the facial nerve lies on the medial aspect of the nuclei of the trigeminal nerve (Fig. 8.12). The fibres which originate in the nucleus course towards the floor of the fourth ventricle in a medial direction, through the tegmentum of the pons and around the caudal limit of the abducens nucleus. Then they turn sharply forward (cranially) beneath the floor of the fourth ventricle for a short distance when they again change course abruptly and pass laterally and ventrally through the tegmentum to emerge between the pons and the medulla. The part of the facial nerve which winds around the abducens nucleus is called its internal genu (meaning knee).

After the facial nerve leaves the brain stem it passes through the internal auditory meatus to enter the facial canal.

The facial nerve is liable to damage in operations around the parotid gland and the upper part of the neck. Surgical incisions behind the angle of the mandible must be only skin deep, otherwise the main trunk of the facial nerve is likely to be divided. Incisions on the face are made parallel to the branches of distribution (Fig. 4.16) and therefore radiate from the middle of the tragus. In young children the nerve can be divided by incisions below the ear which are used to gain access to the mastoid region of the skull, when the mastoid air cells are infected. Division of the cervical branch of the facial nerve sometimes occurs in operations to remove the upper deep cervical lymph nodes when, for example, they have become involved secondarily in the spread of cancer. This nerve injury causes paralysis of the platysma and the depressor labii inferioris muscles. The facial nerve may also be involved in disease (Bell's Palsy) which is believed to be the outcome of a viral infection of the nerve and results in facial paralysis on the affected side. When the facial nerve is permanently damaged by any of these circumstances nerve transplants may be attempted in the facial canal of the petrous temporal bone. Paralysis of the facial muscles may also follow cortical lesions of the brain, and when they affect the facial nerve they are distinguished from peripheral injuries of the nerve trunk in that the paralysis is incomplete on the upper part of the face, for this region has representation on the cortex of both cerebral hemispheres.

Surgical procedures in the region of the parotid gland are dictated by the possibility of facial nerve injury. The facial nerve divides the gland into superficial and deep parts, both of which can be removed without nerve damage. The surgeon exposes the trunk of the facial nerve as it emerges from the stylomastoid foramen and follows it forwards into the plane of artificial separation between the superficial and deep parts of the parotid gland. The part of the gland involved in a parotid tumour can then be dissected away without fear of injury to the branches of the facial nerve.

Central connections of the glossopharyngeal, vagus and accessory nerves

The nuclei of the glossopharyngeal, vagus and accessory nerves should be considered together because the fibres of all three nerves originate from the same columns of cells. They all contribute to the nerve supply to the pharyngeal arch musculature and the motor nucleus which is common to them is called the *nucleus ambiguus* (Fig. 8.11).

The nucleus ambiguus is a long column of multipolar cells extending from the level of the middle of the fourth ventricle to the level of the sensory decussation formed by the internal arcuate fibres (Fig. 8.10b). It lies deep in the lateral part of the medulla oblongata, posterior to the olivary nucleus (Fig. 8.10c). Fibres from it pass to the striated muscle of the pharynx, larynx, and upper part of the oesophagus. The part devoted to the glossopharyngeal nerve is quite small and is represented by the upper end of the nucleus ambiguus. General visceral motor fibres of the glossopharyngeal nerve, along with corresponding fibres of the vagus, arise in the *dorsal vagal nucleus* which lies lateral to the hypoglossal nucleus in the floor of the fourth ventricle (Fig. 8.11). These fibres provide the nerve supply to the heart, the muscles and glands of the oesophagus and the smooth muscle of the branchial system, as well as the upper part of the alimentary tract. Glossopharyngeal nerve fibres which innervate the parotid gland arise in the floor of the fourth ventricle from a group of cells which appears to be an upward extension of the dorsal vagal nucleus, called the inferior salivary nucleus. Special visceral sensory fibres of the glossopharyngeal and vagus nerves enter the medulla and turn downwards to form the *tractus solitarius*. This structure lies lateral to the dorsal vagal nucleus and extends throughout the length of the medulla. Alongside it is a thin column of cells, the nucleus of the tractus solitarius, in which fibres from the tract end at successive levels. It is concerned with gustation, recieving taste impulses from the tongue (Fig. 8.11).

The vagal nuclei play an important part in cardiovascular, respiratory and visceral motor functions. The vasomotor and respiratory centres, including reflex centres for swallowing and vomiting, are located beneath the lower part of the floor of the fourth ventricle at the level of the vagal nuclei.

Motor nerve fibres of the accessory nerve arise from the lowest part of the nucleus ambiguus in series with those belonging to the glossopharyngeal and vagus nerves (page 250). They form the *cranial root* of the accessory nerve and after it leaves the surface of the medulla it is joined by rootlets originating from the upper part of the spinal cord as the *spinal root*. The cell bodies of the spinal component are found in the lateral part of the anterior column of grey matter in the spinal cord, as far down as the fifth cervical segment.

If the vagus nerve is divided on one side there is unilateral paralysis of the muscles of the palate and larynx, as well as anaesthesia of the larynx on the same side. If both vagal nerves are paralysed, the heart rate quickens due to paralysis of the cardio-inhibitory nerves, as well as slowing and irregularities in breathing. The recurrent laryngeal branch of the vagus nerve may be damaged in aneurysm of the common carotid artery or tumours of the thyroid gland. The vocal cord on the same side is immobilized and the voice becomes hoarse. If both recurrent laryngeal nerves are divided speech is impossible.

Central connections of the hypoglossal nerve

The nucleus of the twelfth cranial nerve forms an elongated column of cells which lies on the floor of the lower part of the fourth ventricle close to the mid-line (Figs 8.10, 8.11). The nucleus is sufficiently distinct and compact to cause ridge in the floor of the ventricle, called the *trigonum hypoglossi*. The axons form distinct bundles which pass anteriorly through the medulla lateral to the medial lemniscus (Fig. 8.10) and they emerge on the surface of the medulla at the lateral border of the pyramid (Fig. 8.10a). *preolivary sulcus*

If the hypoglossal nerve is divided on one side the corresponding half of the tongue undergoes atrophy and wrinkling. When the patient is asked to protrude the tongue the contractions of the lingual musculature on the normal side cause the tip of the tongue to be deflected towards the paralysed side.

The olfactory nerve (Figs 8.9, 8.17, 8.19)

The olfactory nerves are short and delicate and extend from the olfactory area of the nasal epithelium (page 180) through the cribriform plate of the ethmoid bone (page 218) to end in the *olfactory bulb* situated on the under surface of the frontal lobe of the cerebral hemisphere (Fig. 8.9). From this ovoid structure the *olfactory tract* passes directly backwards towards the central region of the base of the brain, to the lateral side of the optic chiasma where it divides into a medial and lateral root. The *lateral root* passes towards the lateral sulcus through a region of cortex which is penetrated by central branches of the middle and anterior cerebral arteries (the anterior perforated substance) and ends in the *uncus* (Fig. 8.17).

The *medial root* ascends in front of the *lamina terminalis*, which is the thin anterior wall of the third ventricle, and ends below the anterior end of the corpus callosum. The olfactory fibres have further connections with some of the structures on the medial surface of the cerebral hemisphere; the dentate and hippocampal gyri, the fornix and the corpora mamillaria. Collectively these parts comprise the *rhinencephalon*, a part of the brain which is much more highly developed in lower animals.

The optic nerve (Figs 8.9, 8.18, 8.19)

Optic nerve fibres originate in the *retina* of the eyeball and travel backwards through the optic foramen towards the *optic chiasma*. In the chiasma the medial fibres in the nerve decussate and join the *optic tract* on the opposite side. The laterally placed fibres in the optic nerve, which come from the lateral side of the retina continue backwards in the optic tract of the same side. The optic tracts curve posteriorly round the sides of the cerebral peduncles, with which they fuse, and divide into medial and lateral roots. Many of the *medial root* fibres terminate in the superior corpus quadrigeminum. The *lateral root* fibres relay at the side of the cerebral peduncle and the majority of the succeeding fibres pass further backwards into the visual area of the cerebral cortex situated on its medial surface around the *postcalcarine sulcus*.

The remaining fibres enter the superior corpus quadrigeminum and relay with a further set of fibres concerned with visual reflexes via the nuclei of the nerves to the orbital muscles.

THE SPECIAL SENSES

The special senses are those of taste, smell, sight, hearing and the sense of balance. In each case the essential features are the specialized cells, which are modified in their structure for the purpose of responding to the specialized stimulus, and closely related to these, various forms of supporting or sustentacular cells. When stimulated the special sense cells initiate the passage of nerve impulses which reach the cortex of the brain by a variety of nervous pathways and are there co-ordinated, interpreted and associated with the neural mechanisms responsible for memory and thought.

Taste

Taste buds (Figs 6.12, 8.30) are found in the mucous membrane of the tongue (particularly in the fossae of the circumvallate papillae), palate, fauces, and pharynx. They are oval bodies

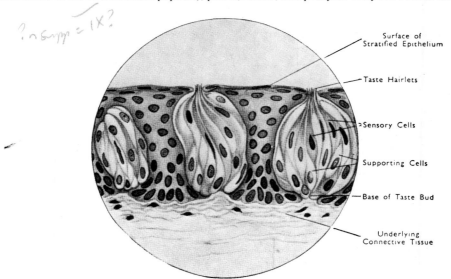

Surface of
Stratified Epithelium

Taste Hairlets

Sensory Cells

Supporting Cells

Base of Taste Bud

Underlying
Connective Tissue

Fig. 8.30 Diagram of taste buds × 390. (By courtesy of Miss Margaret Gillison.)

made up of groups of cells consisting of the neuro-epithelial cells and supporting cells. The former, from 4 to 20 in each bud, are rod-shaped elements with a peripheral hair-like process projecting into the taste pores at the surface of the mucous membrane overlying the taste bud. The terminal branches of the nerve fibres subserving taste, end in close relationship to these special cells. Recent evidence suggests that the cells of the taste buds undergo continual renewal and that the supporting cells are precursors of the neuro-epithelial cells.

The majority of the cells of mammalian taste buds have a life span of about 10 days. There is no scientific evidence for a second population of cells which have a different rate of turnover. The cells of the taste bud are innervated by about 50 nerves which enter at the base of the bud and branch frequently to give rise to some 200 terminal filaments. All but a few of the cells have anatomical contact with these nerve filaments.

At least three cell types can be identified in a taste bud. These are:

1. A rather dense cell or type I cell with characteristic granules at its apex, slender microvilli which project into the taste pore, and many processes which surround both nerve fibres and other cells

2. A light cell or type II cell, relatively empty, with short microvilli and numerous contacts with nerves

3. A type III cell with dark cored and synaptic vesicles. Of these, the type I cells are most numerous and probably have a supporting function, while the type III cells are almost certainly the sensory elements. The precise function of the type II cell is not known, but may represent a different form of another cell type or a different variety of sensory cell.

Much of the evidence for the role of taste cells as neural transducers comes from electron micrographs which reveal the presence of synapses between the nerve fibres and neuro-epithelial cells. The vast majority of nerve endings are in the lower part of the taste bud, at some distance from the taste pore which permits entrance of chemical substances from the saliva. The chemical stimulus initiates the changes which result in nerve impulses passing centrally along the taste pathway. Some believe that taste bud cells are not true receptor cells, but influence the transit of chemicals to the nerve endings as 'chemical filters'.

The nerve pathways for taste sensation are along the lingual, glossopharyngeal, vagus and palatine nerves, but, in the case of both the lingual and palatine nerves, the taste fibres pass into the brain stem along the chorda tympani and greater (superficial) petrosal branches of the facial nerve (Fig. 8.31).

The function of taste is mediated primarily by the chorda tympani branch of the seventh cranial nerve for the anterior two-thirds of the tongue. The glossopharyngeal (IX) nerve for the posterior one-third of the tongue and the superior laryngeal branch of the vagus (X) for the epiglottis and glottic taste receptors. The chorda tympani nerve carries visceral efferent fibres to salivary glands, as well as the afferent fibres for taste, and originates in the petrous temporal bone. It passes through the petrotympanic fissure, passes medial to the trigeminal nerve to exit from the skull into the infratemporal fossa where it lies deep to the temporomandibular joint and joins the lingual nerve. Surgical injuries to the lingual nerve usually involve the chorda tympani fibres and nerve regeneration after such injuries may not be ideal as cross-innervations develop between the general somatic fibres of the lingual nerve and the special visceral fibres of the chorda tympani. Surgical lesions of the chorda tympani nerve are commonly associated with damage to the lingual nerve and produce a clinical picture of reduced salivary flow and loss of taste from the anterior two-thirds of the tongue on the same side.

The central processes of the chorda tympani nerve bringing taste impulses from the anterior two-thirds of the tongue pass inwards through the facial and internal auditory canals and into the brain stem at the junction of the pons with the medulla. Within the medulla the fibres enter the *solitary tract*, through which the taste impulses are carried to the *nucleus solitarius*. Taste fibres from the posterior third of the tongue are from the glossopharyngeal nerve and terminate also in the nucleus solitarius. The taste buds of the epiglottis are innervated by the vagal nerve fibres whose cell bodies are situated in a nodose ganglion (see page 251) and whose central processes terminate once again in the nucleus solitarius. These various fibres passing to the nucleus solitarius form the solitary tract (tractus solitarius) and it will be appreciated that the chorda tympani, glossopharyngeal and vagus nerves are represented in the nucleus in succession from its cranial to its caudal extremity.

The secondary pathway for taste (from the nucleus solitarius to the thalamus) is included in the medial lemniscus of the opposite side and, on reaching the level of the thalamus, these secondary fibres end along with other secondary fibres from the head region. Tertiary, or third order, taste fibres project to the inferior part of the postcentral gyrus and the adjacent cortex

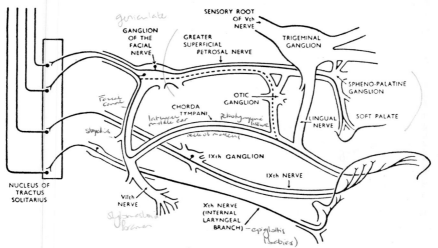

Fig. 8.31 Pathways of taste nerve fibres are shown in bold lines. The interrupted line is an alternative route from the anterior two-thirds of the tongue which appears to be present in some persons. (By courtesy of Professors Bell, Davidson and Scarborough.)

of the insula. The four fundamental varieties of taste sensation are sweet, bitter, sour and salty, but there appear to be no structural differences in the cells of the taste buds to account for these varieties of taste.

Studies of taste thresholds in human subjects commonly use sucrose for the sweet taste, urea for bitter responses, hydrochloric acid to produce sour taste sensations and sodium chloride to represent the taste of salt. Detection and recognition thresholds can be measured by applying the selected solution to precise regions of the oral mucosa. The tongue is most sensitive for salt and sweet tastes. Sour and bitter tastes can be recognized on the tongue but not as well as by the palate mucosa. Salt and sweet tastes can be appreciated on the palate also, but higher solution concentrations are required.

Smell

The special cells responsible for the initiation of the nerve impulses are scattered through the olfactory mucous membrane in the roof of the nasal cavities. They are bipolar in form with a central cell containing one or more nuclei. The peripheral process passes to the surface of the mucous membrane as a hair-like structure. The central process passes into the underlying connective tissue where it forms small bundles of nerve fibres with the processes derived from other olfactory cells. These pass through the openings of the ethmoidal cribriform plate to end in the olfactory bulb of the brain.

The cells which support the special olfactory cells contain a certain amount of yellow pigment and are said to show a deeper pigmentation in individuals with an acute sense of smell. Small mucous glands of the adjacent lamina propria keep the epithelial surface moist. See page 424 for a description of the central connections of olfactory nerve fibres.

Sight

The wall of the eyeball (Fig. 4.45) consists of three coats, the outer fibrous tunic or *sclera*, the intermediate vascular (nutritive) tunic, the *choroid*; and the inner sensory layer, the *retina*. The sclera is a complete layer but its anterior part, the cornea, is transparent so as to permit the passage of light. The front of the eyeball is covered by the conjunctiva, a modified mucous membrane, which is reflected on to it from the deep surface of the eyelids. The anterior part of the choroid and retinal layers are incomplete showing a circular opening, the pupil, the size of which is varied by the action of smooth muscle fibres in the iris diaphragm. Immediately behind this opening is the lens and its supporting structures making up the ciliary apparatus regulating the thickness of the lens and focusing the light on the retina within the eyeball. Between the lens and cornea is the anterior chamber through which circulates the fluid aqueous humor, while behind the lens, helping to support it and filling the eyeball, is the jelly-like vitrous body.

The choroid layer supports the retinal layer. It is deeply pigmented and contains blood vessels and capillaries supplying nutrition to the retina. At the front of the eyeball the choroidal layer is thickened to form the ciliary body containing smooth muscle arranged in circular and radial layers. The function of this muscle is to vary the tension of the suspensory ligament of the lens and thus regulate the thickness of the lens.

In front of the ciliary apparatus the choroid tunic forms the iris diaphragm containing smooth muscle in circular (sphincteric) and radial (dilatating) layers. This muscle regulates the size of the pupil and determines the amount of light falling on the retina. The iris also contains pigment which determines the colour of the eyes (blue, brown, black, etc.). In albinos this pigment is absent.

The smooth muscle of the ciliary apparatus and the iris diaphragm is controlled by sympathetic fibres from the superior cervical ganglion which reach the eyeball along the internal carotid and ophthalmic arteries and by parasympathetic fibres reaching the eyeball via the third cranial nerve (oculomotor) and the ciliary parasympathetic ganglion. The blood vessels to the choroidal tunic are branches of the ophthalmic artery.

The light-sensitive cells or photoreceptors of the retina (the rods and cones) are adjacent to the choroid, from which they derive their nutrition. They form the outer layer of the *retina* (Fig. 4.45). The intermediate layer is made up largely of bipolar neurons situated between the rods and cones. The central processes of the bipolar cells synapse with the processes of ganglion cells which form a narrow ganglion cell layer. The axons of the ganglion cells pass over the inner surface of the retina towards the optic disk as a fibre layer through which visual impulses are transmitted to the brain. Thus light must penetrate the inner and intermediate layers of the retina to reach the light sensitive cells of the outer layer.

At the *optic disk* where the optic nerve leaves the eyeball, the retina consists only of nerve fibres and this nonsensitive region is the 'blind spot'. Somewhat lateral to the blind spot is the macula, which is in the visual axis. Cone cells predominate here and this is the most light sensitive part of the retina. There are many more rods than cones and their distribution is related to different functions. The rods are very sensitive at low levels of illumination and they are able to register slight movements of objects in the field of vision under dim lighting conditions. The cones are most sensitive in light of high intensity and they are concerned with colour differences and fine visual detail.

The axons leaving the eyeball form the *optic nerve* which passes backwards to the optic chiasma. The optic nerve is penetrated a short distance behind the eyeball by the *central artery of the retina*, a branch of the ophthalmic artery, which reaches the retina through the centre of the optic disk and sends branches to all parts of the retina. It does not anastomose with any other arteries and if it is occluded by disease or otherwise damaged, the blood supply to the retina is completely cut off and blindness follows.

The *optic chiasma* (Fig. 8.9) is formed by the union of the two optic nerves and gives rise on either side to the optic tracts which pass backwards around the cerebral peduncle to the lateral geniculate nucleus. This is a relay station where retinal impulses are sorted out for projection to the visual area in the occipital lobe of the cerebral cortex. The optic nerve fibres partially decussate at the chiasma. Nerve fibres from the nasal half of the retina cross over the mid-line and pass into the optic tract of the opposite side. The fibres from the lateral, or temporal half of the retina are uncrossed and continue into the optic tract of the same side. Thus the visual cortex on one side of the brain receives impulses from the corresponding halves of both eyes.

The optic chiasma has anatomical relations which are of clinical importance. Above it lies the floor of the third ventricle; below it is close to the body of the sphenoid, immediately in front of the pituitary fossa. Anteriorly, it is related to the anterior cerebral arteries and posteriorly it lies close to the pituitary gland. Laterally, the internal carotid artery passes close to it on its way to the bases of the brain. A variety of pathological conditions will produce pressure effects on the optic chiasma and disturbances of vision, for example, aneurysm of the internal carotid artery, distention of the third ventricle or tumours of the pituitary gland.

Hearing and balance

The middle ear is described on page 211. The internal ear consists of a complex system of canals situated in the petrous part of the temporal bone (Fig. 8.32). The cochlear duct is concerned with hearing; the semicircular canals and the vestibular region are involved with the sense of position and balance. Both systems of canals contain a continuous membranous tube, the membranous labyrinth, containing a fluid called endolymph. The membranous labyrinth is partially surrounded by perilymph, which is similar in composition to the cerebral spinal fluid. Vibrations set up in the perilymph by movements of the stapes are transmitted to the endolymph of the cochlear duct.

Throughout the length of the cochlear duct is found the receptor organ for hearing, the

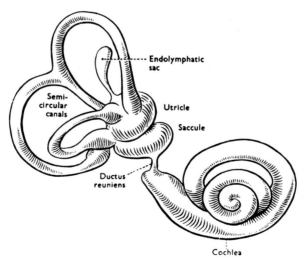

Fig. 8.32 The membranous labyrinth of the internal ear. Compare with Fig. 5.7. (By courtesy of Professor G. A. G. Mitchell and Dr E. L. Patterson.)

organ of Corti. It is composed of highly specialized epithelial cells, many of which have hair-like processes. These cells and the fibres attached to them, which belong to the cochlear part of the vestibulocochlear (eighth cranial) nerve, are stimulated by fluctuations of fluid in the cochlear canal. These movements are the result of vibrations of the tympanic membrane which actuate the foot piece of the stapes at the vestibular foramen. Thus, a sound causes air vibrations in the external auditory meatus that move the tympanic membrane and the ossicles of the middle ear, which in turn cause fluid movements in the inner ear apparatus, leading to nerve impulses in the cochlear part of the vestibulocochlear nerve. Finally, connections to the auditory cortex in the superior temporal gyrus result in the individual's perception of the particular sound.

Balance is controlled initially by the semicircular canal system and two compartments of the vestibular region, the utricle and the saccule (Fig. 8.32). Each of the three semicircular canals surrounds a corresponding semi-circular duct and they are arranged in planes at right angles to each other so that two are in the vertical plane and one lies horizontally. In certain regions specialized epithelial cells support small calcareous bodies, called otoliths. Changes in position and the movement of the fluid in the semicircular ducts produces changes in the position of the otoliths or stimulates the hair-like processes of specialized epithelial cells which have a sensory function. The evoked impulses pass into the central nervous system along nerve fibres of the vestibular part of the eighth cranial nerve. This pathway transmits information to the brain regarding the position of the head, and therefore the body, in space and plays an important part in the reflex mechanisms which maintain equilibrium or balance. Pathological changes in the delicate semicircular canal system, such as alterations in the composition of the contained fluids, are likely to manifest themselves as forms of dizziness or vertigo. Injury to the eighth cranial nerve in the vicinity of the internal ear results in disturbances of both equilibrium and hearing.

Table 8.1 Summary of the cranial nerves

Cranial Nerve	Functional Fibre Type	Functions
I. Olfactory	Special visceral afferent	Smell
II. Optic	No classification (in reality a brain tract)	Sight
III. Oculomotor	General somatic efferent	Eye movements
	General visceral efferent (parasympathetic)	Ciliary and pupillary movements
	General somatic afferent (in reality V nerve fibres)	Proprioceptive for extrinsic eye muscles
	General visceral efferent (sympathetic)	Blood vessel control in eyeball
IV. Trochlear	General somatic efferent	Eye movements (superior oblique muscle)
V. Trigeminal	General somatic afferent	Oral facial sensation; proprioception from muscles of mastication, TM joints and teeth
	General somatic efferent	Mastication
VI. Abducent	General somatic efferent	Eye movements (lateral rectus muscle)
VII. Facial	General visceral efferent	Salivary and lacrimal gland secretion
	General somatic afferent	Sensation from external auditory meatus tympanic membrane
	Special visceral efferent	Facial expression
	Special visceral afferent	Taste

Table. 8.1 (cont.)

Cranial Nerve	Functional Fibre Type *Nucleus*	Functions *Ganglion*
VIII. Vestibulocochlear	Special somatic afferent	Hearing and balance *Spiral +*
IX. Glossopharyngeal	General visceral efferent *inf. salivary* (parasympathetic) *Itto*	Parotid gland secretions
	General visceral afferent *Ambiguous*	Sensation from tongue *post 1/3* and pharynx
	Special visceral efferent *Ambiguus upper third*	Contraction of stylopharyngeus muscle
	Special visceral afferent *tractus solitarius*	Taste — *post 1/3 tongue*
X. Vagus	General visceral efferent *Dorsal vagal* (parasympathetic)	Autonomic control of heart, smooth muscle of lungs, alimentary tract *smooth musc. & of lower ds of oesophagus*
	Special visceral efferent *Ambiguus*	Pharyngeal muscle movements
	General visceral afferent	Sensation from heart, lungs, alimentary tract
	Special visceral afferent *Tractus sol. larius*	Taste — *Nodose*
	General somatic afferent *Ess*	Visceral reflexes; pain in ear, pharynx larynx
XI. Accessory	Special visceral efferent *Ambiguus* (cranial root) *lower end*	Movements of soft palate, pharynx, larynx
	General somatic efferent *lateral part* (spinal root) *ventral grey horn C1-5*	Contraction of sternomastoid, trapezius muscles
XII. Hypoglossal	General somatic efferent *hypoglossal*	Tongue movements.
	tongue	

SUMMARY OF THE SPINAL NERVES

These are attached by anterior motor and posterior sensory *roots* to each side of the spinal cord. There are 8 cervical, 12 thoracic, 5 lumbar, 5 sacral and 1 or 2 minute coccygeal nerves. The roots unite to form nerve *trunks* which then divide to form anterior and posterior primary *rami* (divisions). The posterior rami pass as individual branches to supply the postvertebral musculature and overlying skin (Fig. 3.7).

The larger anterior rami supply the musculature and skin of the anterolateral aspects of the neck, the body wall and of the limbs. In the thoracic region each nerve runs its separate course in the intercostal spaces (page 58); the lower nerves pass between the layers of the antero-lateral abdominal musculature.

In the cervical, lumbar and sacral segments, the anterior rami contribute to cervical, brachial, lumbar and sacral nerve *plexuses* from which many branches pass into the upper or lower limbs.

The *cervical plexus* (Fig. 8.33) is formed by the anterior primary rami of the first four cervical nerves, just beyond the points where they communicate with the superior cervical ganglion of the sympathetic trunk. Each nerve divides into an ascending and a descending branch. These branches form a series of loops which constitutes the plexus. The anterior primary ramus of the first cervical nerve communicates with the hypoglossal nerve, as well as the anterior primary ramus of the second cervical nerve. It supplies the prevertebral muscles and the fibres which join the hypoglossal nerve form the descendens cervicalis of the ansa hypoglossi. The branches of the cervical plexus are cutaneous and muscular. The cutaneous branches (Fig. 4.11) are the lesser occipital, great auricular, transverse nerve of the neck, and the supraclavicular nerves. The muscular branches include those to the infrahyoid muscles, except the thyrohyoid and the geniohyoid. The most important muscular branch is the *phrenic* nerve (C.3,4,5).

Each phrenic nerve provides the sole motor supply to its own half of the diaphragm. It also contains afferent fibres from the pleura, the pericardium, the peritoneal surface of the diaphragm, and the liver and gall bladder. Stimulation of these fibres may produce the sensation of pain referred to the area of skin supplied by the same segment of the spinal cord. For example, pericardial inflammation can manifest as pain on the side of the neck which radiates to the point of the shoulder.

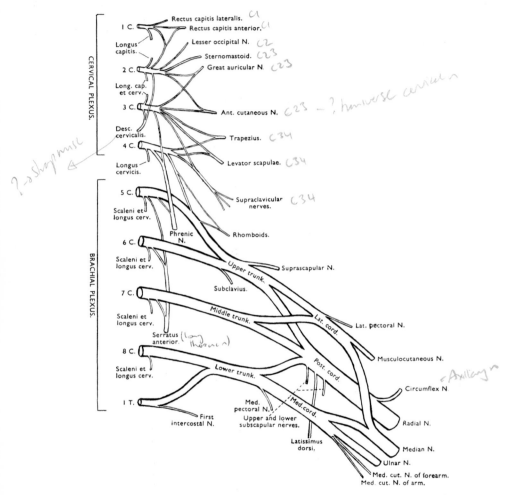

Fig. 8.33 Diagram showing formation of the cervical and brachial plexuses. (By courtesy of Professor G. A. G. Mitchell and Dr E. L. Patterson.)

The *brachial plexus* (Fig. 8.33) is formed in the neck from the anterior primary rami of C.5,6,7,8 and the anterior primary ramus of the first thoracic nerve. The pattern of union and redistribution of the nerve fibre bundles in the brachial plexus is quite complex. The roots unite to form three trunks which divide into ventral and dorsal divisions. The divisions unite to form three cords and from these the main branches of the plexus arise, namely, the musculocutaneous, median, ulnar, radial and the axillary nerves. The roots of the plexus have important connections with the sympathetic trunk and sympathetic nerve fibres are distributed through it to blood vessels, sweat glands and muscles of hair follicles in the upper extremity. The sensory innervation of the hand is from the median, ulnar and radial nerves. The motor innervation is shared by the ulnar and median nerves.

The brachial plexus is liable to be damaged as it curves across the upper surface of the first rib and it is subject to traction injuries. Injuries at the wrist frequently involve one or more of the three terminal branches of the brachial plexus. Injury to the radial nerve results in a loss of sensation in the skin on the back of the wrist and hand, the dorsum of the thumb and first two or three fingers. Trauma to the ulnar nerve leads to paralysis and wasting of the small muscles on the medial side of the hand, while the main effect of division of the median nerve at the wrist is loss of the ability to oppose the thumb to the fingers. These nerve

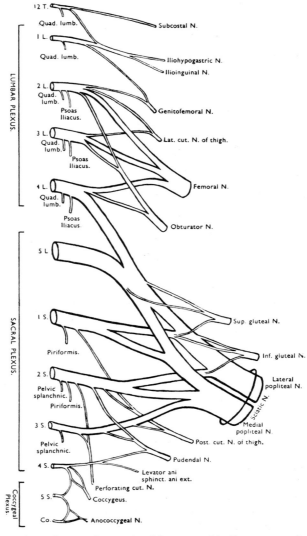

Fig. 8.34 The lumbar and sacral nerve plexuses (semidiagrammatic). (By courtesy of Professor G. A. G. Mitchell and Dr E. L. Patterson.)

distributions are obviously of great importance to the dentist, who is incapacitated if the fine movements of the hand should be lost.

The *lumbar plexus* (Fig. 8.34) is formed in the substance of the psoas major muscle by contributions from the upper four lumbar nerves (anterior rami). Its branches are the *iliohypogastric* and *ilioinguinal nerves* (L.1) which appear at the lateral border of psoas major, pass in front of quadratus lumborum to the crest of the ilium and then enter the lateral abdominal wall between transversus abdominis and internal oblique; the *genitofemoral nerve* (L.1,2) descends on the surfaces of psoas towards the inguinofemoral region; the *lateral cutaneous* nerve of the thigh (L.2,3) emerges from the psoas muscle and runs superficial to iliacus to enter the thigh beneath the lateral end of the inguinal ligament; the *femoral nerve* (L.2,3,4) enters the thigh near the medial end of the inguinal ligament, lateral to the femoral artery. The femoral nerve is the largest branch of the lumbar plexus. The last branch is the *obturator nerve* (L.2,3,4) which leaves the pelvic cavity through the obturator foramen with the obturator artery to supply the muscles of the medial side of the thigh.

The *sacral plexus* is formed on the posterior wall of the sacrum from contributions from the anterior rami of the fourth and fifth lumbar nerves and the upper three sacral nerves. Its branches are distributed to the flexor and extensor muscle compartments of the lower limb. Its main branch is the large *sciatic nerve* (L.4,5 S.1,2,3) which leaves the pelvis to reach the gluteal region and the back of the thigh. Other branches are the *gluteal nerves* to the hip musculature, the *pudendal nerve* to the perineal region (page 118), and the *posterior cutaneous nerve* of the thigh.

9. Clinical Anatomy

SURFACE ANATOMY OF THE FACE

Skeletal structures (Figs 9.1, 9.5)

The parts of the facial skeleton which are directly subcutaneous and can be felt readily beneath the skin are:

1. The posterior border of the ramus and the lower border of the mandible from the angle to the chin.

2. The zygomatic bone forming the prominence of the cheek. From it the zygomatic arch can be traced backwards to the external auditory meatus.

3. The orbital margin made up of the maxilla below and medially, the zygomatic bone laterally, and the frontal bone above.

4. The nasal bones forming the roof and bridge of the nose.

More deeply placed and covered by the facial musculature the alveolar processes of the jaws carrying the teeth, the facial surface of the maxilla and the outline of the nasal aperture can be palpated. The head of the condyle can usually be felt immediately in front of the external auditory meatus, especially if the mouth is slowly opened and closed.

The ramus of the mandible is more deeply placed and is covered by the massive masseter muscle and in its back part by the parotid gland. The anterior border of the ramus in the

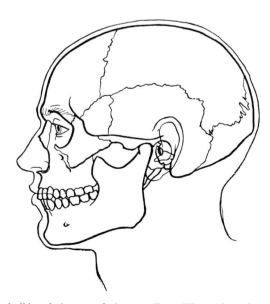

Fig. 9.1 Bones of the skull in relation to soft tissue outlines. (The teeth are in post-normal occlusion.)

adult usually lies to the outer side of the distal parts of the upper and lower third molar teeth. Deep to the ramus are also situated the pterygoid muscles, the inferior dental vessels and nerve, the lingual nerve, the maxillary artery and the pterygopalatine fossa. The angle of the mandible is superficial to the position of the tonsil from which it is separated by the medial pterygoid muscle and the superior constrictor. The maxillary antrum (sinus) occupies the body of the maxilla, while the frontal sinus extends to a varying degree into the frontal bone above the bridge of the nose and the inner third of the upper orbital margin.

The supra-orbital notch (or foramen) is situated at the junction of the inner and middle thirds of the upper margin of the orbit and can sometimes be felt below the skin of the eyebrow. The infra-orbital foramen lies about $\frac{1}{4}$ in. to $\frac{1}{2}$ in. (10 mm) below the middle of the lower margin of the orbital cavity. The mental foramen is situated in the body of the mandible about half way between its lower border and alveolar margin. Its relationship to the teeth varies from the level of the apex of the first premolar to that of the posterior root of the first molar. In about 50 per cent of cases it lies below the apex of the second premolar.

Vessels and nerves

The facial artery reaches the face at the anterior border of the masseter muscle where its pulsations can usually be felt. It takes a winding course through the tissues of the cheek passing close to the angle of the mouth to reach the groove between the nose and facial surface of the maxilla on the medial side of the infra-orbital foramen. By holding the front of the cheek between the finger and thumb its pulsations can usually be felt in the cheek and also those of its labial branches in the lips.

The anterior facial vein runs through the facial tissues behind the artery. It is, however, much less tortuous in its course and cannot be felt.

The superficial temporal artery becomes superficial at the root of the zygomatic arch between the external auditory meatus behind and the mandibular condyle in front. It runs upwards towards the temporal region of the scalp in front of the ear. The transverse facial artery, a branch of the superficial temporal, runs towards the prominence of the cheek a short distance below the zygomatic arch. The pulse can be recorded from the superficial temporal artery as it crosses the root of the zygomatic arch.

The position of the parotid duct occupies the middle third of a line drawn from the tragus of the ear (the projection in front of the external meatus) to half way between the ala of the nose and the red margin of the upper lip. At the anterior border of the masseter muscle it turns inwards to pierce the structures of the cheek obliquely and open into the vestibule of the mouth (page 153).

The *terminal branches of the facial nerve* appear at the upper and lower ends and at the anterior border of the parotid gland. They are:

1. The temporal branch crossing the zygomatic arch and passing towards the lateral orbital margin.

2. The zygomatic branch passing below the zygomatic arch towards the cheek bone.

3. The buccal branches running in the cheek below the level of the parotid duct.

4. The mandibular branch running towards the chin above or just below the lower border of the mandible.

5. The cervical branches pass downwards into the neck near the angle of the mandible (Fig. 4.16).

Branches of the trigeminal nerve reach the skin of the face:

1. Around the orbital margin (supra-orbital, supratrochlear, infratrochlear and lacrimal branches of the ophthalmic division).

2. Through the infra-orbital and zygomatic foramina (maxillary division).

3. Through the mental foramen; from beneath the anterior border of the ramus (buccal

nerve); and at the root of the zygomatic arch in front of the external auditory meatus (auriculotemporal nerve). These are all branches of the mandibular division.

The mental, infra-orbital, buccal and zygomatic branches of the trigeminal nerve form a nerve plexus in the facial tissues with the zygomatic buccal and mandibular branches of the facial nerve.

The skin over the angle of the mandible is supplied by the great auricular nerve, a branch of the cervical plexus (C.2.3) which also supplies the skin of the lower half of the auricle.

Facial regions

The face can be divided into the following regions:

1. The mandibular region related to the body and ramus of the mandible and including the lower lip and lower part of the cheek.

2. The auriculotemporal region above the zygomatic arch and behind the lateral orbital margin.

3. The midfacial region; including the area between the upper lip and lower orbital margin, the upper part of the cheek, the alar region of the nose and the cheek bone.

4. The orbitocranial region; including the orbital cavities, the bridge of the nose and the forehead.

The nerve supply to the skin of the mandibular and auriculotemporal regions is from branches of the mandibular division of the fifth nerve (the mental, buccal and auriculotemporal nerves) (Fig. 9.15). Its blood supply is from the facial and mental, buccal, transverse facial and superficial temporal arteries (Fig. 4.14). The lymph drainage is into the submental nodes (lower lip and chin), the submandibular nodes (cheek), and parotid nodes (the parotid, auricular and temporal regions) (Fig. 4.18).

The nerve supply of the middle facial region is from branches of the maxillary division of the trigeminal nerve (infra-orbital and zygomatic nerves). The blood supply is from the facial, infra-orbital and buccal arteries. The lymphatics drain into the submandibular nodes.

The nerve supply of the orbitocranial region is from branches of the ophthalmic division of the trigeminal nerve (supra-orbital, supratrochlear, infratrochlear, external nasal and lacrimal branches). Its blood supply is from branches of the ophthalmic artery corresponding to the nerves, and from the terminal part of the facial artery (the angular artery). Its lymph drainage is downwards and backwards to join the middle facial lymphatics to end in the submandibular nodes or more directly backwards to the parotid nodes.

INTERPRETATION OF RADIOGRAPHS OF THE SKULL

Study of radiographs of the skull should be accompanied by a study of the dried skull.

LATERAL VIEW

Cranial region (Fig. 9.2)

Note the form of the cranial vault, its length (glabella, the smooth area between the eyebrows, to the most prominent part of the occipital region), its height (external auditory meatus to either the highest point in the vertical plane or to the bregma, where the coronal and sagittal sutures meet), the slope of the forehead region, the prominence and form of the occipital region, the coronal and lambdoidal sutures, the thickness of the cranial bones, the grooves for the middle meningeal vessels, the diploic venous channels, the size, extent and degree of development of the frontal air sinuses.

Note in the cranial base region:

1. The position, size and relations of the pituitary fossa.

2. The petrous temporal bone forming a dense wedge-shaped shadow which sometimes shows the position of the internal auditory canal in its substance. The shadow partly obscures the external auditory meatus and the mandibular joint. Note the form of the glenoid fossa, the articular eminence, the position of the head of the mandibular condyle.

The mid-line cranial base is seen passing downwards and backwards behind the pituitary fossa. Its posterior part and the front of the foramen magnum (basion) are usually obscured by the petrous shadow. The notch at the back of the articular condyles of the occipital bone (the Bolton point) is an alternative landmark to basion in measurements of the cranial base. Measure the distance from the Bolton point to the middle of the pituitary fossa (posterior half of the cranial base). ~6cm

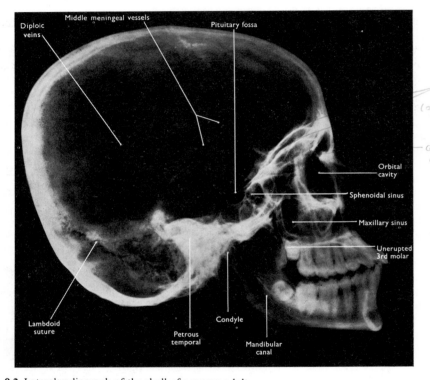

Fig. 9.2 Lateral radiograph of the skull of a young adult.

3. The anterior half of the cranial base can be measured by a line from the middle of the pituitary fossa to nasion (the point on the facial profile where the nasal bone articulates with the frontal). In front of the pituitary fossa two lines mark the extent and form of the anterior cranial fossa. The upper line indicates the roof of the orbital cavities (orbital plate of the frontal); the lower less distinct line, the cribriform plate and roof of the nasal cavity. Note the thickness of the frontal bone between the front of the cribriform plate and nasion, and again observe the size and extent of the frontal sinus. Note also the size and extent of the sphenoid air sinuses and their relation to the pituitary fossa.

Measure the angle between the anterior half of the cranial base (nasion to midpituitary point) and the posterior half (mid-pituitary point to Bolton point). This angle is an indication of the degree of flexion of the skull base.

Facial region

Measure the distance from nasion to the lower border of the mandible at the symphysis (menton). This gives the total facial height (the teeth should be in occlusion). Draw a line along the upper surface of the hard palate. This divides the facial skeleton into an upper orbitonasal area and a lower oral area.

Orbitonasal area

1. Trace the outline of the maxillary antrum (or sinus). The posterior and lower borders are usually easy to identify. Note the relationship of the antrum to the upper teeth.

2. Immediately behind the maxillary antrum note the form and position of the pterygopalatine fossa. Behind the fossa the shadow of the pterygoid plates are partly overlapped by the coronoid process of the mandible.

3. Trace the outline of the roof of the nasal cavity from nasion backwards. Note the relationships of the ethmoid and sphenoid regions.

4. Superimposed on the nasal cavity and maxillary sinus there can usually be identified parts of the external orbital rim, the zygomatic arch, and the zygomatic process of the maxilla.

Oral area

1. Note the number, occlusion and state of eruption of the teeth, their relationship to the line of the hard palate, the maxillary antrum and the lower border of the mandible. Note whether any teeth are missing or whether there are any extra teeth.

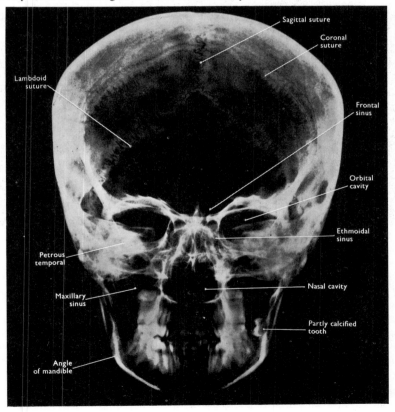

Fig. 9.3 Antero-posterior radiograph of the same skull as in Fig. 9.2

2. Trace the lower border of the mandible and the posterior border of the ramus to the condyle. Trace the outline of the anterior border of the ramus, the coronoid process and the mandibular notch.

3. Extend a line drawn along the lower border of the mandible backwards to meet:
   a. The backwards extension of the palatal line
   b. The backward extension of a line drawn through the lower border of the orbit and the upper border of the external auditory meatus (the Frankfort plane).

The lines meet at:
   a. The mandibular-palatal angle
   b. The mandibular-Frankfort angle.

These angles are used to analyse variations in facial form.

4. Note the outline of the symphysial region of the mandible (shown in section in lateral radiographs). Note the inclination of the lower incisor teeth to the mandibular line or plane and of the upper incisors to the palatal and/or the Frankfort plane.

5. Note the position of the hyoid bone and the shadow of the soft palate.

6. Draw a perpendicular line from the mid-pituitary point at right angles to the line of the anterior half of the cranial base (nasion to pituitary point). Note the relationship of the pterygopalatine fossa, anterior and posterior borders of the mandibular ramus, the articular eminence and external auditory meatus, and the most distal teeth (in the adult—3rd molars) to this posterior facial base line.

ANTERIOR VIEW
(Figs 9.3, 9.4)

Cranial region

Note the form and size of the cranial vault, the thickness of the bones, the coronal suture, the markings of the meningeal vessels, the mastoid processes and air cells, the roof of the orbital cavities, the size and extent of the two frontal air sinuses, the ethmoidal air cells, the degree of symmetry of the cranium and face.

Facial region

1. Note the outline of the nasal cavities, the position and deviation of the nasal septum, and the nasal conchae (turbinate processes).

2. Note the outline, size and form of the orbital cavities; their relationship to the upper part of the nasal cavities. Usually the superior orbital (ophthalmic) fissures and optic foramina can be seen.

3. Trace the outline of the maxillary air sinuses and of the lower border of the zygomatic process of the maxilla and the zygomatic arch. Note the angle between the process and arch; measure the distance between the external alveolar surfaces immediately below the zygomatic processes (external palatal width).

4. Trace the outline of the lower border of the body of the mandible and the posterior border of the ramus to the condyles. Measure the distance between the condyles of the two sides. Trace the outline of the mandibular notch and coronoid process. Note the form of the angle of the mandible.

5. Note the position, occlusion and state of eruption of the teeth. Note the relationship between the mid-line of the lower and upper dentitions (between central incisors), the nasal septum and the middle point between the orbital cavities.

Dental radiology

In the practice of dentistry radiographs of the skull and face are used to determine the rates and amounts of facial growth; to diagnose the position of unerupted teeth, including impacted molars; to determine the position, size and relations of dental and oral cysts; to find evidence

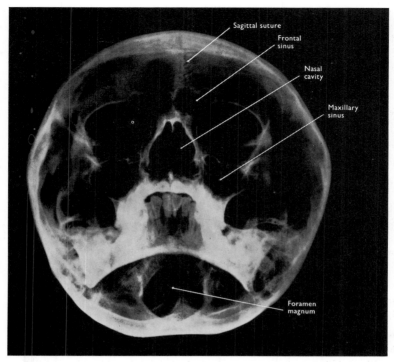

Sagittal suture
Frontal sinus
Nasal cavity
Maxillary sinus
Foramen magnum

Fig. 9.4 Radiograph taken obliquely from the front of the skull shown in the previous two radiographs. The frontal and maxillary air sinuses are shown more clearly.

of sinus infections. Dental conditions are evaluated in more detail by means of intra-oral radiographs. Reference to them is an important part of diagnostic techniques for diseases of the periodontal ligaments, early caries, fractured or misplaced teeth, cysts and dental tumours. A complete survey of the mouth involves taking a number of single intra-oral films. These are designed to examine the periapical region as well as the crowns of the teeth. For the latter purposes, bite wing films are used to give the most accurate interpretation of interproximal caries.

In addition to routine radiographs, specialized extra-oral methods are used including panoramic radiography and xeroradiography. In panoramic radiography the X-ray source and film are made to rotate around the patient so that the resulting radiograph shows a continuous view of the dental arches and their supporting bones from the region of one temporomandibular joint to the other. These panoramic views are extremely useful for general survey procedures and can be carried out quickly. However, the detailed information available from them does not compare with standard intra-oral or extra-oral films. Xeroradiography is a relatively new method which uses the principle of modern document copying machines. The conventional X-ray film is replaced by selenium plates which are reusable. The advantages of the technique include the greater detail of hard and soft tissues and the smaller dosage of radiation required to produce the image. A xeroradiograph of the facial region of a child is shown in Fig. 9.5. Although it is not in wide use in private dental practice, this technique is being developed rapidly and may soon become commonplace.

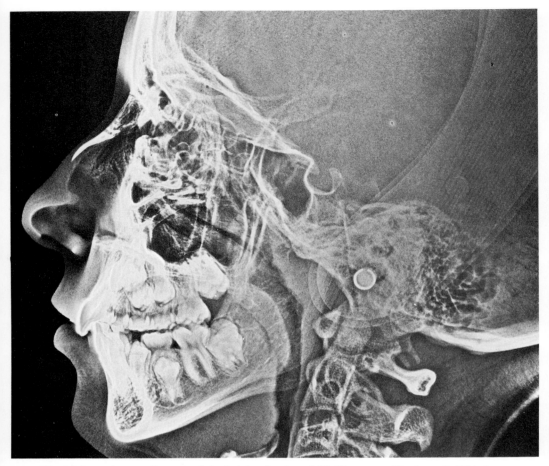

Fig. 9.5 Lateral xeroradiograph of the facial region in a child. (Courtesy of Professor S. White)

CEPHALOMETRIC ANALYSIS

A large number of points, planes and angles are used in cephalometric studies of facial growth and the number of methods used in these analyses is continually increasing. It is important in any cephalometric study of facial growth that there is a full understanding of what is being measured, the relative stability of the point landmarks which are being used, and the range of measurements at various stages of growth which can be accepted as normal.

The main points used in measurements of the craniofacial skeleton fall into two categories, *profile landmarks*, which fall on the outline of the skull as viewed in a lateral radiograph, and *internal landmarks*, which fall within the skull outline viewed from the same aspect. The most commonly used profile landmarks are shown in Figure 9.7.

Nasion. The termination of the frontonasal suture in the middle line on the surface of the skull.

Anterior nasal spine. The most anterior point on the bony projection of the floor of the nasal cavity.

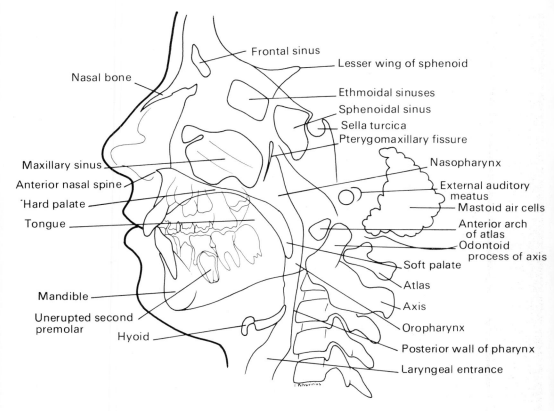

Fig. 9.6 Outline drawing of Fig. 9.5 to show some of the main features of the radiograph.

Point A (supraspinale). The deepest point on the concavity between the anterior nasal spine and prosthion.

Prosthion. The most anterior point between the upper central incisors at the mid-line of the upper alveolar process.

Point B (supramentale). The deepest point on the concavity between the anterior point of the lower alveolar process between the central incisors and pogonion.

Pogonion. The most anterior point on the bony chin.

Gnathion. The point where the anterior border of mandible joins the lower border of the lower jaw.

Menton. The lowest point of the mandible at the mandibular symphysis.

Gonion. The point on the angle of the mandible located by bisection of the angle formed by the intersection of the mandibular base line and the line of the posterior border of the ramus.

The chief internal landmarks of the craniofacial skeleton are:

Basion. The lowermost point on the anterior margin of the foramen magnum, in effect the posterior end of the mid-line cranial base.

Point R or *registration point*. The mid-point of a line from the sella turcica point perpendicular to the Bolton plane.

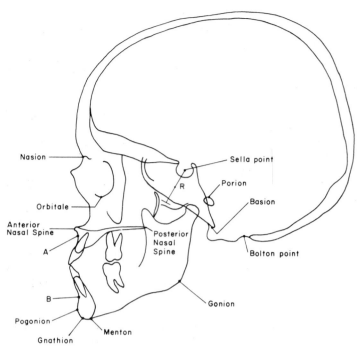

Fig. 9.7 Important points used in cephalometric analysis.

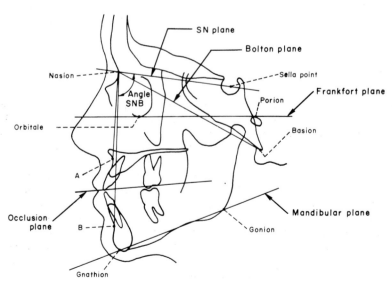

Fig. 9.8 Important planes used in cephalometric analysis.

Sella point. The mid-point of the sella turcica (pituitary fossa).

Orbitale. The lowermost point of the lower border of the orbital cavity.

Posterior nasal spine. The most posterior mid-line point of the hard palate.

Porion. The mid-point of the upper margin of the bony external auditory meatus.

Bolton point. The highest point on the curve between the occipital condyle and the lower border of the occipital bone.

Several planes can be drawn in cephalometric studies and these are important for the superimposition of a series of radiographs and for analytical measurements of growth changes. The chief planes are shown in Figure 9.8.

Frankfort plane. A line, customarily oriented in the horizontal plane, passing through the points porion and orbitale.

Sella nasion plane. A line passing through sella point and nasion.

Mandibular plane. A line which is tangential to the lower border of the mandible and passes through gnathion.

Occlusion plane. A line joining at the fissure between the mesial and distal cusps of the upper first permanent molar (or the second deciduous molar) and the incisal edge of the upper central incisor tooth.

Lines which join some of the points which are defined above, or are made by the intersection of planes described above, are frequently used as measures of the relationship of the jaws and teeth to one another. Three commonly studied angles are S-N-A, S-N-B, and S-N-Pg.

Growth changes involving various planes

The sella nasion plane can increase in length due to growth at the spheno-ethmoidal and fronto-ethmoidal sutures, a dimension which is stabilized by the end of the tenth year. Surface deposition of bone in the region of the nasion and reconstruction of the pituitary fossa, which carries it backwards and upwards, can also influence the length of the sella nasion plane.

The mid-line cranial base extending from nasion to basion, known as the Bolton plane, is influenced by growth at the spheno-occipital synchondrosis, the spheno-ethmoidal and fronto-ethmoidal sutures. The Bolton plane is increased in length by surface deposition at nasion and basion.

THE SPREAD OF INFECTIONS OF THE FACE AND NECK

It is important to understand that spaces do not exist between the fascial layers in the healthy person. There is loose connective tissue between them which is easily distended by fluid or an infective process. An infection in one space may spread to another or to several adjacent spaces by disruption of intervening fascial planes or along the course of blood vessels and nerves.

Infections arising in oral and pharyngeal structures can spread to areas remote from the original infected site and an understanding of the anatomical regions involved forms the basis for correct therapeutic measures to combat them. Infections may spread by three principal routes; by the blood stream, through the lymphatic vessels or by direct spread through adjacent tissues. The direct spread of infections is governed by the existence of the layers of cervical fascia and the potential spaces between these layers. Whether or not the infection is restricted from further spread by the fascial boundaries of a given space is determined by the nature of the fascia, whether it be of a loose connective tissue type or a dense connective tissue sheath.

The cervical fascia

Four layers of deep fascia may be described. The *superficial layer* of deep fascia, also called the general investing layer, surrounds all the structures of the neck with the exception of those contained in the subcutaneous tissues. The *pretracheal layer*, or middle cervical fascia, is divided into outer and inner parts which invest the infrahyoid group of muscles and fuse laterally with the *carotid sheath*, enclosing the carotid artery, internal jugular vein and the vagus nerve. The deepest layer of the cervical fascia, also known as the *prevertebral fascia*, lies immediately in front of the vertebral bodies and the deep muscles attached to them.

The circumoral fascial spaces are of considerable importance in determining the spread of dental infections. They are usually bounded by fascial layers, muscles, or bone, skin or mucous membrane. They contain blood vessels, nerves, lymphatics, lymph nodes and salivary glands. Between these structures the 'spaces' are filled in by loose connective tissue. The most important fascial spaces or compartments are (Figs 9.10, 9.11):

The superficial facial space (Fig. 9.9)
This is bounded superficially by the skin of the face and deeply by the buccinator muscle, the facial surfaces of the upper and lower jaws, and the surface of the masseter muscle. It is limited behind by the parotid space, above by the orbital margin and zygomatic arch, below by the lower border of the mandible, and communicates deep to the mandibular ramus with the pterygoid space. It contains the buccal pad of fat, the duct of the parotid gland, the facial artery and vein, the buccal lymph nodes and the openings of the mental and infraorbital foramina, branches of the facial and trigeminal nerves and the muscles of expression.

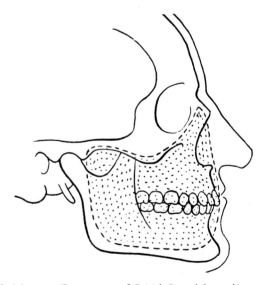

Fig. 9.9 Superficial facial space. (By courtesy of *British Dental Journal.*)

The masticator space
The masticator space, also called the masseteric space, adjoins the mandibular space and is bounded by extensions of the investing layer of the cervical fascia which cover the masseter muscle laterally and the medial pterygoid muscle on the inner aspect of the mandibular ramus. It is bounded behind by the posterior border of the ramus and extends forwards as far as the anterior borders of the masseter and medial pterygoid muscles. Above, it is limited by the attachment of the temporalis muscle to the temporal bone. The masticator space contains all the muscles of mastication, the ascending ramus of mandible and the zygomatic arch. Infections within it may spread to the temporal region, the parotid space, the parapharyngeal space

and the submandibular space. The swelling which usually follows the extraction of lower third molars is the result of inflammatory involvement of the masticator space.

The sublingual space

This is bounded by the lingual surface of the body of the mandible, the mucous membrane of the floor of the mouth and the upper surface of the mylohyoid muscle. It contains the submandibular salivary gland (deep portion), its duct, the sublingual salivary gland, the lingual and hypoglossal nerves and the lingual blood vessels (Figs 5.29, 9.10). The sublingual space of each side is continuous between the geniohyoid and genioglossus muscles (Figs 6.3, 6.21).

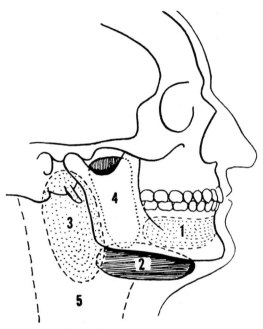

Fig. 9.10 Deep facial spaces. 1, sublingual space; 2, submandibular space; 3, parotid space; 4, pterygoid space; 5, parapharyngeal space. (By courtesy of *British Dental Journal.*)

The submandibular space (Fig. 9.10)

This forms a closed compartment on either side of the upper part of the neck. It is formed by a splitting of the superficial layer of the cervical fascia and lies on the medial side of the mandible below the posterior part of the mylohyoid muscle. It is bounded laterally by the superficial layer of the cervical fascia and the body of the mandible, behind by the stylohyoid muscle and the posterior belly of the digastric muscle, and in front by the anterior belly of the digastric muscle. It is bounded by the body of the mandible and the lower surface of the mylohyoid muscle above and the superficial layer of the deep cervical fascia below. The fascia is attached to the hyoid bone below, and to the mandible above. Each space contains the superficial part of the submandibular salivary gland, the submandibular lymph nodes, a part of the facial artery, its submental branch, and the common facial vein, the anterior bellies of the digastric muscles, and the submental lymph nodes (see page 151).

The submandibular and sublingual spaces communicate with one another and infections can spread rapidly across the mid-line, as well as to the parapharyngeal space. Infections of the spaces not only produce swellings on the side of the neck below the lower border of the mandible but also may produce serious swelling in the floor of the mouth which may interfere with normal breathing. Ludwig's angina is a serious bacterial infection of the submandibular region which involves the adjacent fascial spaces and spreads bilaterally. The mortality rate in this disease is quite high.

The parotid compartment

This contains the parotid salivary gland and lymph nodes. It is bounded by the posterior border of the ramus of the mandible, the styloid process and its muscles, the sternomastoid and the digastric muscles (Fig. 9.10). Through the parotid gland runs the posterior facial vein, external carotid artery and the facial nerve (see page 153). The covering fascia merges with the stylomandibular ligament which forms a boundary between the parotid space and the submandibular space. The parotid space can become involved by infections which extend backwards along the parotid duct as well as infections from the middle ear. Infections can spread from the parotid space into the pterygoid space and downwards into the submandibular space. The fascia covering the lateral surface of the parotid gland is extremely dense, thus infections of the parotid space show a tendency to spread medially.

Both the submandibular and parotid compartments are bounded by dense layers of deep fascia which separate their contents from the surrounding bony and muscular structures. Infective processes arising within these spaces are often localized there for considerable periods.

The pterygoid space

The pterygoid or infratemporal space is bounded anteriorly by the tuberosity of the maxilla and extends beyond it into the cheek on the lateral side of the buccinator muscle. Posteriorly, the space is limited by the condyle of the mandible, the temporalis muscle, the lateral pterygoid muscle and the parotid gland. The tendinous part of the temporalis muscle and the coronoid process of the mandible form its lateral boundary; medially the space is limited by the lateral pterygoid plate, the lower head of the lateral pterygoid muscle and the lateral wall of the pharynx. The infratemporal surface of the greater wing of the sphenoid bone forms the upper boundary. The inferior part of the space, often called the pterygomandibular space, lies between the pterygoid muscles and the ramus of the mandible. The space contains the pterygoid plexus of veins, the first two parts of the maxillary artery, the mandibular, mylohyoid, lingual, buccal and chorda tympani nerves, as well as the pterygoid muscles. Infections of this space can spread to the masticator space, the submandibular space, the parotid space or the parapharyngeal space.

Part of the pterygoid space lies posterior to the maxillary sinus, lateral to the lateral pterygoid plate and on the deep aspect of the temporomandibular joint. It contains the maxillary nerve, the termination of the maxillary artery and the sphenopalatine ganglion. Infections within it are usually the result of infections originating from the upper molar teeth.

The parapharyngeal space

This lies between the lateral pharyngeal wall and the vertebral column (retropharyngeal compartment) medially, and the deep cervical fascia and sternomastoid muscle at the side of the neck. It contains the deep cervical lymph nodes and through it, surrounded by the carotid sheath, run the carotid arteries, the internal jugular vein and parts of the ninth to twelfth cranial nerves. The inner portion of the pterygoid space is an upward continuation of the parapharyngeal space which also extends downwards into the thoracic cavity. It is the most dangerous region into which infective conditions can spread.

The retropharyngeal space

The space lies between the posterior wall of the pharynx and the prevertebral fascia. It is limited on the lateral aspect by the connective tissues of the carotid sheath and extends from the base of the skull above to the mediastinum below. Infections of this space are regarded by surgeons as especially dangerous, due to the loose nature of the connective tissue and thus the ease with which infection can spread into the upper part of the thorax.

The paratonsillar space
This lies between the wall of the pharynx and the mucous membrane of the fauces. It extends upwards into, and includes, the tissues of the soft palate. The tensor palati muscle passes from the deep pterygoid space into the soft palate around the hamulus of the medial pterygoid plate (Fig. 5.8).

Intercommunications between fascial spaces
   1. The *sublingual space* communicates with:
      a. The superficial pterygoid space along the course of the lingual nerve
      b. The deep pterygoid space (upper part of the parapharyngeal space) along the styloglossus muscle
      c. The paratonsillar space along the palatoglossus muscle (anterior pillar of fauces)
      d. The submandibular space around the posterior edge of the mylohyoid muscle.
Occasionally the sublingual and submandibular spaces communicate through a hiatus in the mylohyoid diaphragm.

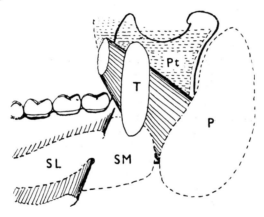

Fig. 9.11 The critical area. P, parotid space; Pt, pterygoid space; SL, sublingual space; SM, submandibular space; T, paratonsillar space. (By courtesy of *British Dental Journal*.)

   2. The *submandibular space* communicates with:
      a. The sublingual space
      b. The parotid space along the anterior branch of the posterior facial vein which passes from one space to the other
      c. The superficial facial space along the facial artery and vein
      d. The pterygoid space (upper part of parapharyngeal space) along the hypoglossal nerve and stylohyoid muscle.
   3. The *pterygoid space* communicates with:
      a. The parapharyngeal space below
      b. The facial space along the buccal vessels and nerve
      c. The sublingual space
      d. The cranial cavity via emissary veins draining into the pterygoid plexus
      e. The pterygopalatine fossa, and orbital cavities along the maxillary artery and through the inferior ophthalmic fissure

f. The parotid compartment along the maxillary artery and vein.

The posterior free border of the mylohyoid muscle is a critical area for the spread of dental infections. At this region, which is closely related to the lower third molar, the sublingual, submandibular, pterygoid, parapharyngeal and paratonsillar spaces come into close relationship with another (Fig. 9.11).

Lymphatic communications

1. Lymphatic vessels from the face drain into the submandibular spaces and parotid spaces and from these to the parapharyngeal spaces (deep cervical glands).

2. Lymphatic vessels from the sublingual spaces and tongue pierce the mylohyoid to enter the submandibular spaces (submandibular lymph glands) and also pass backwards to enter the retropharyngeal spaces (upper deep cervical glands).

3. Lymphatic vessels from the paratonsillar spaces pierce the superior constrictor muscle to enter the parapharyngeal spaces (upper deep cervical glands).

Lymphatics from the lower teeth and gums either pass outwards to join the facial lymphatics or inwards to join the sublingual and lingual lymphatics. Lymphatics from the upper teeth, gum, mucous membrane of the hard palate, soft palate, nasal cavities and maxillary sinuses, either pass outwards and forwards to join the lymphatics of the face or inwards and backwards to the upper deep cervical glands in the parapharyngeal spaces. They reach these by piercing the superior constrictor with the tonsillar vessels or by passing around the posterior edge of the medial pterygoid plate.

THE SPREAD OF DENTAL INFECTIONS

Infections around the roots of teeth are very common and whether these infections remain at the apex of a tooth or progress through the surrounding tissues depends a great deal on the virulence of the infecting organisms and the ability of the tissues to combat the infection.

The alveolar bone is the first barrier and a progressive periapical infection tends to spread concentrically through the bone until it perforates one of the cortical plates. The site of perforation can be predicted with some accuracy, if one knows the relations of the root apices to the alveolar bone, since perforation will naturally occur at the closest bony surface.

Once the infection has travelled through the bone and its covering periosteum, the anatomical arrangement of the adjacent soft tissues, including muscles and fasciae, will determine the subsequent route of spread and the possible location on the face or in the oral cavity. The mandible is surrounded by a periosteum of moderate thickness and density beneath which ineffective processes may spread for considerable distances before breaking through to reach the skin surface, vestibule or mouth cavity, or adjacent fascial spaces. The periosteum covering the maxilla is very much thinner except over the posterior surface where it is of considerable thickness. A dense mucoperiosteum binds the oral mucous membrane to the alveolar processes and to the hard palate, and in these regions there is but a limited possibility of subperiosteal spread once the bone surface has been broken through. The mucoperiosteum lining the nasal cavities and air sinuses is less dense than that of the mouth. Infective processes spread most easily through loose connective tissue and cancellous bone, and with greatest difficulty through dense connective tissue such a mucoperiosteum and fascial capsules such as those surrounding the parotid gland.

The fascial spaces in the head region which have been described on page 446 are not always the same sites in which one sees clinically the localization of dental infections. For example, some of the spaces are seldom involved completely in the inflammatory process. This is due to the fact that the fascial spaces are often subdivided by connective tissue septa which make spread from one part to another quite difficult. In other areas of the face where localization occurs, there are no strong sheets of fascia, but muscles and bone surfaces define the limits of the infection. Infections that are limited to single fascial spaces are usually controlled either by antibiotics or surgical drainage. Some strains of bacteria have developed resistance to antibiotics and appear to be able to spread throughout the neck and related structures despite the existence of the connective tissue fascial sheets.

The teeth fall into the following groups according to their relationship to the fascial compartments and skeletal cavities:

1. Lower incisors and canines.
2. Lower cheek teeth.
3. Lower third molars.
4. Upper incisors.
5. Upper canines.
6. Upper premolars and molars.

Lower incisors and canines (Fig. 9.12). The mylohyoid muscle attachment (mylohyoid line) begins as a forward continuation of the inner boundary of the retromolar triangle and passes towards the lower border of the mandible so that at the front of the jaw the genial tubercles are above the mylohyoid attachment. Here the apices of the incisors and canines lie above the attachment of the mylohyoid muscle, while at the back of the mouth the apices of the

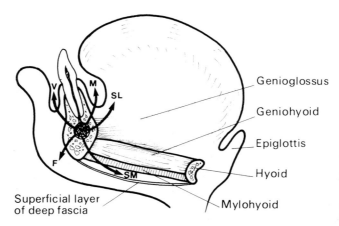

Fig. 9.12 Paths for the spread of infection from the apical region of a lower incisor. F, superficial facial space; M, floor of mouth; SL, sublingual space; SM, submandibular space; V, oral vestibule.

third lower molars are usually below it. Infection arising from the apical region of the lower incisors or canines may spread:

1. Downwards to the submental region (submental lymph nodes).
2. Outwards to the chin region or to the lower incisor vestibular region.
3. Inwards to the mouth cavity (subgingival) or if at a lower level, into the sublingual space.
4. Into the medullary cavity of the mandible.

The relationship of the mentalis muscles to the apices of the lower incisors determines the path taken by the infection. If the infection breaks through the bone above the attachment of the mentalis muscle, it will involve the oral vestibule. If the infection tracks below the mentalis, it will involve extraoral tissues. It may remain localized to the area of the chin or spread to the submental region. Here the infection will be contained by the anterior bellies of the digastric muscles laterally, by the mylohyoid muscle above and by the skin below. Clinically, the resulting swelling will be confined to the point of the chin and the region immediately below it.

Lower cheek teeth. Infection arising in relation to the lower cheek teeth may invade either the sublingual space or the submandibular space, the medullary bone of the mandible, or the superficial compartment. If the abscess points beneath the gingival mucous membrane, it may discharge either into the oral vestibule or mouth cavity proper (Fig. 9.13)

If the infection exits from the buccal aspect of the bone below the attachment of the buccinator muscle the superficial facial compartment is involved. The root apices are usually above the origin of this muscle, so that infections from these teeth commonly localize within the oral vestibule.

The attachment of the mylohyoid muscle roughly parallels the direction of the attachment of the buccinator muscle (Fig. 4.25). Lingual perforation of infections from the premolars and first molar are almost always above the attachment of the muscle into the sublingual space. Infections arising from the second molar have an equal potential to track either bucally or lingually and may pass either above or below the buccinator or mylohyoid muscles. Perforation below the mylohyoid muscle produces an infection of the submandibular space. An abscess in the superficial facial compartment is readily differentiated from involvement of the submandibular space. The former swelling begins at the lower border of the mandible and extends upwards to the level of the zygomatic arch, while the latter begins at the lower border of the mandible and extends downwards to the level of the hyoid bone.

Lower third molar. Infection arising in relation to the lower third molar may spread:

1. Through the mucous membrane covering the lingual side of the alveolar process into the back of the mouth cavity.

2. Below the mylohyoid into the back of the submandibular space.

3. Forwards along the lingual nerve beneath the mucous membrane to the sublingual space.

4. Backwards along the lingual nerve into the superficial pterygoid space between the medial pterygoid and the mandible.

5. Backwards along the styloglossus muscle to the parapharyngeal space (deep pterygoid space).

6. Upwards in the anterior pillar of the fauces into the soft palate.

7. Outwards into the back of the oral vestibule or the facial space, or outwards and backwards between the masseter and mandible.

Upper incisors. Infection arising in relation to the upper incisors may spread:

1. Upwards to involve the mucous membrane lining the floor of the nasal cavity.

2. Outwards to the vestibule of the mouth or the root of the upper lip.

3. Backwards into the palate. This type of spread is most common in relation to the lateral incisors. Infection arising from these teeth may form an abscess which discharges at the junction of the hard and soft palate.

The apices of the upper incisors usually lie closer to the labial aspect of the alveolar process, so that infections from these teeth exist most commonly in the oral vestibule. Spread of infection is influenced by the orbicularis oris muscle and the dense connective tissue at the base of the nose. Because of this, tissues around the nose can be very painful and tender. Infections which spread into the palate tend to be limited by the dense connective tissue of the mucoperiosteum.

Upper canines. Infections arising in relation to the upper canines may spread:

1. Outwards into the vestibule of the mouth, or, if at a higher level, into the front of the superficial facial space above the attachment of the buccinator. From here infection may involve the lower eyelid or the orbital cavity along the infra-orbital canal.

2. Upwards and inwards to involve the nasolacrimal duct and nasal cavity.

3. Backwards to involve the antrum, if large, or the palate.

Infections originating from the upper canines point most commonly on the labial aspect on the alveolar bone. The levator anguli oris muscle is the influencing factor for the subsequent path taken by the infection. If the infection perforates below the muscle attachment, the resulting swelling will be intra-oral. If it is located above the muscle attachment, the infection will spread into the canine region of the superficial facial compartment. The superficial muscles of the face may restrict further progress of the infection but, usually, the gaps between the muscles of the upper lip allow the infection and the resulting swelling to extend upwards to the vicinity of the medial corner of the eye.

Upper premolars and molars. Infection arising in relation to the upper cheek teeth may involve (Fig. 9.13):

1. The maxillary antrum.
2. The oral vestibule.
3. The hard palate.
4. The superficial facial region above the attachment of the buccinator.
5. The pterygopalatine fossa from which infection may spread into the orbital cavity or the cranial cavity, along the ophthalmic veins, or into the pterygoid space.
6. The paratonsillar region and soft palate (along the tendon of the tensor palati muscle).

Periapical infections from the upper molars usually perforate the outer side of the alveolar process, because the bone on the buccal aspect of the teeth is thinner than that on the palatal side. If a palatal abscess occurs it is well localized by the dense nature of the palatal mucoperiosteum. As is the case in the lower jaw, infections which exit through the buccal bone are influenced by the attachment of the buccinator muscle. If the perforation is below the attachment of the muscle, the swelling will be found in the oral vestibule. Infections which track above the attachment of the muscle form abscesses in the superficial facial space. Because of the relatively loose nature of the tissues of the cheek an abscess in this region can be widespread, extending from the zygomatic arch to the lower border of the mandible.

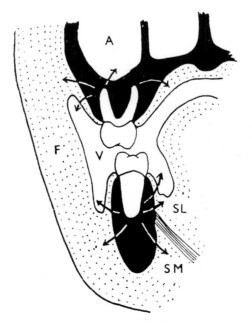

Fig. 9.13 Spread of infection from upper and lower cheek teeth. A, maxillary antrum; F, superficial facial space; SL, sublingual space; SM, submandibular space; V, oral vestibule. (By courtesy of *British Dental Journal*.)

FRACTURES OF THE JAWS

With the significant number of cases resulting from road accidents, the student should have some knowledge of the anatomical features involved in fractures of the facial skeleton.

A simple fracture is one in which there is no communication between the site of fracture and the skin or any body cavity. A compound fracture is one in which there is communication between the site of fracture and the skin surface or some body cavity. Compound fractures involving the skin or mucous membrane of the mouth are liable to infection from these regions. Fractures as well as being simple or compound may be single or multiple according to the number of fragments.

Fractures of the jaws, especially those involving the mandible and maxilla, are very often compound owing to the close relationship of the skin to the facial surfaces of these bones; the adherent nature of the mucoperiosteum of the gums and hard palate; and because of the size and position of the nasal cavities and the air sinuses, especially the maxillary antrum.

Fractures of the mandible are usually single or bilateral but may be multiple. Fractures of the maxilla are more often multiple. The direction and extent of the displacement of the fragments depends on the force and direction of the blow and on the action of the muscles on the fragments. Muscle action is especially important in the consideration of mandibular fractures.

In oblique fractures of the mandibular body, particularly in the region of the angle of the jaw, the combined elevating forces of the masseter, medial pterygoid and temporalis muscles displaces the proximal segments upwards. At the same time the suprahyoid muscles which depress the mandible pull the distal segment downward. When the neck of the condyle is fractured the lateral pterygoid muscle tips the condylar head medially and anteriorly into the infratemporal fascial space.

Teeth may be involved in the following manner:

1. They may be fractured.
2. They may be dislodged.
3. Developing teeth may be damaged in their crypts.

The inferior dental vessels and nerves in the lower jaw, the superior dental vessels and nerves, the infra-orbital vessels and nerves, and the palatine vessels and nerves in the upper jaw, are all liable to be damaged in fractures involving the facial bones.

Mandibular fractures (Fig. 9.14)
The most common sites of fracture in the mandible are:

At the neck of the condyle. Such fractures may be unilateral but are often bilateral, especially when they are the result of a blow on the chin. The line of fracture is usually below the line of attachment of the joint capsule.

Across the coronoid process.

Through the ramus. These are not very common owing to the protection afforded by the temporal and masseter muscle.

Fractures of the condylar neck, the coronoid process, or of the ramus are usually simple and do not as a rule involve the skin or the mouth cavity.

Between the body and the ramus behind the last molar tooth. This is a common site of fracture.

Through the body of the mandible. The most common sites are between the second and third molars, and in relation to the canine. Vertical fractures of the symphysial region are less frequent.

Common examples of *multiple fractures* of the mandible are:

1. Fracture of the condylar neck on one side and the ramus-body junction region on the other.

2. Fracture of the body of the mandible on both sides; the site of fracture, however, usually differs on the two sides.

In the mandible, displacement of the fragments usually depends upon the action of the muscles of mastication as well as the site, direction and force of the blow. Here only a brief summary of the effect of the muscles will be given.

Fracture of the condylar region
The attachments of the capsule and collateral ligaments usually hold the head of the condyle in close relation to the glenoid fossa or articular eminence. The lower end of the condylar fragment is usually drawn forwards and inwards by the lower fibres of the lateral pterygoid muscle. It is more rarely displaced laterally, medially or posteriorly. The remainder of the mandible, especially in bilateral cases, may be displaced backwards by the direction of the blow. The line of fracture is usually outside the line of attachment of the capsule, but it may be in part intracapsular.

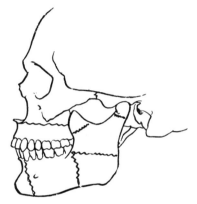

Fig. 9.14 Diagram illustrating the common sites for fracture of the facial bones.

Fractures of the body of the mandible
The amount of displacement resulting from muscle action depends on the direction of the fracture line. The maximum degree of displacement is found when the plane of the fracture runs from below upwards and forwards and from without inwards and forwards (see Fig. 9.14). In these cases the posterior fragment is drawn inwards and upwards by the medial pterygoid and temporal muscles. The anterior fragment, especially in bilateral fractures, is drawn downwards and backwards by the digastric and geniohyoid muscles. If, however, the line of fracture is in the opposite direction, i.e. from below backwards and upwards, and from without inwards and backwards, the overlapping of the fragments in the vertical and horizontal plane tends to prevent displacement by muscle action.

Displacement of the posterior fragment will depend to some extent on the presence or absence of teeth. The amount of displacement is usually greater in edentulous patients.

Fracture of the body of the mandible anterior to the last molar teeth usually results in some tearing of the mylohyoid muscle. Comminuted fractures (fractures in which there are a number of fragments at the site of fracture) of the symphysial region may result in loss of attachment of the digastric, geniohyoid and genioglossus muscles.

Fractures of the upper facial skeleton (Fig. 9.14)
The more common types are:
  Fracture of the zygomatic arch.

Fracture of the zygomatic bone. In this type the lateral wall of the orbital cavity and the outer wall of the maxillary antrum are often involved.

Bilateral transverse fracture of the maxilla. In this fracture, which often results from trauma applied to the side or front of the face, the alveolar process with the teeth is separated from the supporting facial skeleton. This compound fracture usually involves the oral and nasal cavities and the maxillary air sinuses. The displaced fragment may be mobile or impacted among the bones of the upper facial skeleton.

Complex fractures of the upper facial skeleton. These usually show two or more fragments involving the maxilla and other bones and parts of the orbital and nasal cavities. The mouth cavity may or may not be involved. The overlying skin is usually torn.

Severe fractures of the upper facial skeleton are often accompanied by fractures of the cranial base. These may involve the anterior, middle or posterior cranial fossa. If the anterior fossa is involved cerebrospinal fluid may enter the nasal cavities via the cribriform plate; if the middle fossa is involved fluid may enter the nasopharynx.

ANATOMICAL CONSIDERATIONS OF CRANIOFACIAL PAIN

Many patients come to the dental surgeon with complaints regarding pain in the face. Facial pain has many causes and even when there is evidence of dental disease the teeth may not be the cause of the pain. On the other hand dental disease may be responsible for pain which appears to have its origin in other tissues and sites. An understanding of the anatomical basis of craniofacial pain is necessary in the diagnosis of many dental conditions.

Sensory distribution

The *great auricular nerve* (C.2.3.), a branch of the cervical plexus (Fig. 8.33) supplies the skin over the angle of the mandible and the lower part of the auricle. The skin of the neck and of the scalp behind the ears is supplied by other branches of the cervical plexus and by branches of the posterior rami of the upper cervical nerves. Branches of the cervical plexus also supply the dura mater of the posterior cranial fossa (via the meningeal branch of the hypoglossal nerve).

The *trigeminal nerve* (Fig. 9.15).

1. The mandibular division supplies the skin of the lower and posterior part of the face, the lower lip, the cheek, the front of the ear, the external auditory meatus, the tympanic membrane (outer surface), the lower teeth, their periodontal ligaments, and the gums, the mucous membrane of the vestibule of the mouth, the floor of the mouth, the anterior two-thirds of the tongue, the mandibular joint, and the dura mater of the middle cranial fossa. As well as supplying skin and mucous membrane and teeth, it gives sensory (proprioceptive) fibres to the muscles of mastication.

2. The maxillary division supplies the skin of the middle part of the face, the upper lip, the nose, the lower eyelid, part of the cheek and the roof of the oral vestibule, the upper teeth, their periodontal ligaments and gums, the hard and soft palate, the greater part of the nasal cavity, the maxillary antrum, the roof of the nasopharynx and the dura mater of the middle cranial fossa.

3. The ophthalmic division supplies the skin of the forehead and scalp to a line between the ears, the upper eyelids and the skin at the base of the nose, the eyeball and orbital contents, the frontal, ethmoidal, and sphenoidal air sinuses, the roof of the anterior part of the nasal cavity, and the dura mater of the anterior cranial fossa.

The *glossopharyngeal nerve* supplies the mucous membrane of the posterior (pharyngeal) third of the tongue, the great part of the nasopharynx and oral pharynx, the tonsils, the auditory tube, middle ear, tympanic antrum and tympanic membrane (inner surface).

The *vagus nerve* supplies through its auricular branch (nerve of Arnold) part of the external auditory meatus and tympanic membrane and through its laryngeal branches the mucous membrane of the laryngeal part of the pharynx and the larynx, the upper part of the oesophagus and trachea. A meningeal branch of the vagus supplies the dura mater of the posterior cranial fossa.

It should be remembered that sensory nerves commence not only in end organs situated in the skin and mucous membranes but in muscles, articular structures (capsules, ligaments, intra-articular disks), the periosteum of bones and in the deep fasciae.

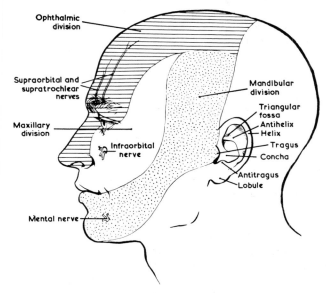

Fig. 9.15 The distribution of the trigeminal nerve on the face and the parts of the external ear.

Areas with a dual nerve supply

There are a number of regions where there is some overlapping in the distribution of the sensory nerves supplying the head and neck. These include;

1. On the face; where there is overlapping in the distribution of all three divisions of the trigeminal nerve and of the upper cervical spinal nerves.

2. The cheeks and vestibule of the mouth; where there is an overlap between branches of the mandibular and maxillary divisions of the trigeminal nerve.

3. At the angle of the lower jaw and in the floor of the mouth cavity; where there is overlapping in the sensory distribution of the cervical spinal nerves (great auricular and anterior cutaneous of the neck) and the mandibular division of the trigeminal nerve.

4. The auricle, external auditory meatus and tympanic membrane derive sensory fibres from the mandibular division of the trigeminal nerve, the auricular branch of the vagus nerve, branches of the cervical plexus (great auricular and small occipital) and of the glossopharyngeal nerve.

5. The oropharyngeal isthmus and dorsum of the tongue; where there is a limited amount of overlap in the distribution of the trigeminal nerve (mandibular and maxillary divisions) and the glossopharyngeal nerve.

6. The nasal cavity; where there is some overlapping in the distributions of the maxillary and ophthalmic divisions of the fifth nerve with one another and with the area of distribution of the olfactory nerve.

7. The roof of the nasopharynx; where the maxillary division of the trigeminal nerve and the glossopharyngeal nerve overlap in the distribution.

8. Between the oropharynx and laryngeal pharynx, the epiglottis and the base of the tongue, where there is some overlapping in the distribution of the glossopharyngeal and vagus nerves.

9. On the scalp, which is supplied by branches of the ophthalmic and mandibular divisions of the trigeminal nerve (in front of the ear) and branches of the upper cervical nerves (behind the ear).

10. The dura mater of the three cranial fossae receives sensory nerve fibres from all three divisions of the trigeminal nerve, the vagus nerve and the upper cervical spinal nerves.

As well as the sensory nerves belonging to the somatic nervous system, sensory fibres of the autonomic system arise in the walls of the blood vessels (arteries, veins and cranial sinuses) in the alimentary canal, the respiratory system and in other viscera.

THE CAUSES OF CRANIOFACIAL PAIN

Pain may be produced by stimulation of nerve endings or pressure exerted on nerve trunks (neuralgia), by inflammation of nerves (neuritis), or may follow on damage or tearing of nerves (causalgia and post-traumatic nerve injuries). Many patients suffer from functional disorders of the nervous system (neurosis) in which there is no evidence of injury or inflammation or disease of any tissue.

Pain resulting from the stimulation of nerve endings
Teeth. Pain may arise as the result of the stimulation of sensory nerve endings in the dental pulp by acute or chronic inflammation. This is often the result of dental caries. Pain may also arise from the pressure exerted on the nerves of the pulp by small calcified nodules—pulp stones. The nerve endings in the pulp may be stimulated by the effects of heat or cold being transmitted by large metal fillings. The pain endings in the gums and periodontal ligament are stimulated by gingivitis and by abscesses developing around the apices of teeth. The pain is usually of a dull throbbing nature as opposed to the sharp acute pain of pulpitis. The pain from a dental abscess is usually reduced when it breaks into the mouth cavity, the maxillary antrum, or one of the circumoral fascial spaces.

Pain may arise from buried unerupted teeth (e.g. upper canines), or impacted third molars. The pain may be produced by pressure on a nerve trunk, but is more usually the result of an infective process associated with the unerupted tooth.

Pathological conditions of the mouth cavity. These include inflammation of the oral mucous membrane (stomatitis), ulcers, cancer of the tongue, floor of the mouth or oropharyngeal region.

Tumours of the jaws. These may be of dental origin such as dental cysts, odontomes, odontoblastomas; or epitheliomas (epithelial cancers) arising from the oral mucous membrane and invading the jaws; or they may arise from the mucous membrane of the maxillary antrum or nasal cavity. They also include cysts of non-dental origin, and sarcomas (mesodermal cancers) arising from bone or cartilage (osteosarcomas and chondrosarcomas). Tumours and cysts may cause pain by pressure on nerve trunks or by the inflammation and ulceration which is often associated with them.

Sinusitis. Infection of the frontal sinus is felt as forehead pain, involvement of the ethmoidal air cells is felt as pain between the eyes, of the sphenoid sinus as basal headache. Infection of the maxillary sinus may be appreciated as pain in the maxillary region but is sometimes felt as toothache involving one or more of the teeth of the upper jaw. This is because the same nerves (superior dental) supply the mucous membrane of the sinus and the upper teeth (referred pain).

Headache. Inside the cranial cavity nerve endings sensitive to pain are found in the great venous sinuses, in the dura mater, especially in the basal parts related to the three cranial fossae, in the adventitia of the dural and cerebral arteries and the coverings of the cranial nerves within the cranial cavity. The pia mater, arachnoid, and the dura mater (other than at the base and adjacent to the venous sinuses) are insensitive to pain as is brain substance and the lining of the cerebral ventricles. Headache of intracranial origin may result from:

1. Traction on the walls of the venous sinuses, the middle meningeal arteries, and the branches of the circle of Willis at the base of the brain.

2. Distension of intracranial arteries (i.e. blood pressure changes).

3. Inflammation involving the dura mater, the venous sinuses, and arteries, or the sheaths of the sensory nerves.

4. Pressure and traction effects produced by intracranial tumours.

5. Contraction of cranial foramina with pressure on emerging nerves, resulting from various bone diseases such as acromegaly, osteitis deformans, and osetopetrosis.

Earache. Pain in the ear may arise in the inner, middle or external compartments, and may be the result of infective processes (especially in the middle ear, tympanic antrum or mastoid air cells) or tumours (more common in the inner ear), or bone diseases such as osteosclerosis. As the ear region is supplied by a large number of separate nerves (fifth, ninth and tenth cranial, and branches of the upper cervical nerves), pain arising in the ear may appear to come from other regions supplied by these nerves such as the head, face, pharynx, larynx and the back of the neck. Conversely, pain which is caused by disease in other regions (e.g. lower teeth and the mandibular joint) may appear to be located in the ear region (referred pain).

Orbital cavities. Diseases of the eyeball such as corneal ulceration or glaucoma (increased intraocular pressure) or strain produced in the ocular muscles in visual disturbances produced pain which may radiate over the whole area of distribution of the ophthalmic division of the fifth nerve. Mention has already been made to the interorbital pain associated with infection of the ethmoidal air sinuses.

Muscle pain. Small inflammatory nodules sometimes develop in muscle tissue (myositis). They are relatively frequent in the masseter and temporal muscles. More chronic inflammation may produce fibrous tissue and the pain of 'fibrositis'.

Mandibular joint pain. This may result from inflammatory conditions within the joint cavity (arthritis). A frequent cause of pain in the mandibular joints and ear region follows on the extraction of teeth and over-closure of the jaw so that the condyle presses against the posterior part of the articular disk which contains nerve endings. There is also a stretching of the capsule and ligaments of the joint. The condition can sometimes be treated by inserting dentures which restore the normal space between the jaws (opening the bite). Patients in a state of psychological strain often indulge in habitual tooth clenching which may produce pain in the joints.

A great deal of confusion exists about the cause of temporomandibular joint pain. Pain originating from muscles and associated structures is properly called myofascial pain. Many investigators have shown that what was previously referred to as temporomandibular joint syndrome, implying a direct involvement of the joint in the initiation of pain, is due in fact to a primary involvement of the muscles of mastication. Thus, the term *myofascial pain dysfunction syndrome* has become widely used to indicate this origin and the lack of, or secondary involvement of the temporomandibular joint. In any event, the differential diagnosis of pain in this region is extremely difficult.

Pain resulting from disease of peripheral nerves

Trigeminal neuralgia (tic douloureux) may involve one or more of any of the divisions of the trigeminal nerve, although it is usually confined to the maxillary or mandibular divisions. The pain is acute, of sudden onset and of short duration, at least in the early stages. It is strictly unilateral and occurs in sharp paroxysms which cause the patient to wince and withdraw from any stimulus which appears to cause onset of the pain. It may be brought on by quite trivial stimulation, such as a draught on the face, light touch on the face, shaving, brushing the teeth, chewing or even swallowing. A small area of skin, which is quite often localized around the region of the lips, becomes identified by the sufferer as the trigger zone, which when stimulated will result in a paroxysm of pain. The natural history of trigeminal neuralgia is for a short series of paroxysms to decrease and allow the patient a period of freedom from the pain. Inevitably, the pain returns and the periods of relief become shorter and shorter, until the patient suffers from persisting and continuous paroxysms.

The cause of trigeminal neuralgia is uncertain. It may be produced by spasm of the small blood vessels supplying the trigeminal, or Gasserian ganglion, localized inflammatory conditions in the region of the ganglion, pressure on the ganglion by the closely related internal carotid artery, or by disturbance of secondary connections of the pathway in the brain stem.

Until quite recently the treatment of this form of neuralgia was surgical injection of the trigeminal ganglion by alcohol or phenol, or section of the sensory root of the trigeminal nerve between the ganglion and the brain stem. This procedure is called rhizotomy of the sensory root. The development of new anticonvulsant drugs, such as carbamazepine, has demonstrated that the condition can be relieved in many patients by medical means.

Glossopharyngeal neuralgia. A somewhat similar condition may involve the area of distribution of the glossopharyngeal nerve, the base of the tongue, the side of the throat, and the ear. The pain is sharp, lancinating and momentary. Glossopharyngeal neuralgia does not respond so well to drugs and surgical treatment consists of removing a section of the glossopharyngeal nerve in the neck, or intracranial division of the ninth nerve and the upper few rootlets of the vagus nerve.

Postherpetic neuralgia. This is the result of a virus infection of the nerve cells in the trigeminal ganglion, clinically known as herpes zoster. There is a skin eruption, commonly called shingles, over the area supplied by the nerve. The extent depends on how many of the divisions of the nerve are involved—usually the ophthalmic division. Pain is felt along the course of the appropriate branches of the trigeminal nerve and may precede the onset of the rash, reaching its most severe form when the typical skin vesicles have formed. Local surgical measures are not effective and the pain may be so severe and intractable as to lead to suicide. One of the most successful forms of treatment is the application of vibration to the affected area.

Causalgia. This term is applied to the burning type of pain which sometimes follows peripheral nerve injury. It may occur in the face after extraction of teeth or following fractures of the facial bones involving nerves such as the inferior dental and infra-orbital.

THE ANATOMY OF ORAL CANCER

At the present time about 5 per cent of cancer deaths in men and 1 per cent in women result from cancer of the mouth and pharynx. In the oral cavity the order of frequency is:

1. The tongue.
2. The upper and lower jaws.
3. The oral mucous membrane.
4. The lips; especially the lower lip.

It is important that the dental surgeon should be able to identify cancerous lesions of the oral region early when radiotherapy or surgery can save the patient. A sound rule is that 'any ulcer which fails to heal within a month should be considered to be malignant until proved innocent'.

Cancers may originate from the epithelium of the skin or mucous membranes (epitheliomas) of the oral cavity or lips or from mesodermal tissues (sarcomas). The former are more common in adults, the latter in children. Occasionally oral cancers are derived from cancer cells which have reached the mouth region from other parts of the body via the blood stream. In these latter cases the most common primary sites are the prostate gland, the ovaries, the breasts, the thyroid gland or the lungs. These 'metastatic' cancers usually occur in the upper and lower jaw (including the condylar cartilage) from which they may spread to adjacent tissues.

Epitheliomas (carcinomas) tend to spread along the lymphatics as well as by direct invasion of adjacent tissues, and hence in carcinoma of the lips the submental and submandibular nodes may be involved early, while in cancer of the tongue the same nodes, and also the juglo-omohyoid node, may become invaded by cancer cells. This makes the nodes hard and more readily palpable.

Cancer of the tongue may invade the floor of the mouth or the mandible; in both cases there will be a progressive fixation and limitation of movement of the tongue as well as involvement of lymph nodes, ulceration and pain. It is very important to diagnose the condition before it has progressed to this extent. Cancer may also originate in the floor of the mouth, the inner surface of the cheek, in the retromolar region, the palate, or in relation to the tonsils.

Sarcomas most commonly commence within the mandible or maxilla (osteosarcomas and chondrosarcomas). In both cases they may produce loosening of the teeth which may be incorrectly diagnosed as 'pyorrhea'. In the lower jaw a sarcoma may spread readily through the inner cancellous tissue without at first involving the overlying cortical bone. The growth may ultimately spread from the bone through the mental or mandibular foramina. In the upper jaw a developing sarcoma may fill up the maxillary sinus before invading the oral, nasal or orbital cavities. Pain of dental or sinus origin may be the earliest symptom, due to involvement of the dental nerves.

SURGICAL ANATOMY OF THE PAROTID GLAND

A knowledge of the anatomy of the seventh cranial nerve in relationship to the parotid gland is necessary for surgical procedures involving the gland tissue; including biopsy, removal of gland tissue for a variety of parotid tumours, and for making incisions to drain abscesses of this region.

The facial nerve is the most important structure in surgical procedures involving the gland because injury to it results in loss of facial expression. This includes inability to close the eyelids, which may lead to damage to the cornea. The nerve may be injured in a variety of ways during surgery; it may be cut, tied, stretched, crushed or dehydrated. Because of these pitfalls, surgeons exercise extreme care in dissection of the gland, establish a blood free field during the operation and take steps to avoid dehydration. The surgeon is less concerned about the arteries and veins which pass through the gland or in relation to it, because their loss does not produce harmful effects. Branches of the external carotid artery and posterior facial vein traverse the gland.

The facial nerve passes downward, forward and outward from its exit at the stylomastoid foramen and enters the posterior part of the parotid gland (page 249). When the parotid gland is retracted forwards a short length of the nerve can be seen. Shortly after entering the gland, after about 1 cm, it bifurcates into upper and lower divisions. The branches which arise from these divisions are shown in Figure 4.16 and they emerge from the gland on the medial aspect of the superficial part near its anterior border, where they lie on the masseter muscle. The parotid gland is sometimes considered as a bilobed structure, but this is of little importance as the path of the facial nerve seldom follows a plane of easy separation through the gland tissue.

The parotid gland is firmly enclosed by its fascial sheath, derived from an upward extension of the investing layer of deep cervical fascia (page 137). The fascia is sufficiently dense and unyielding to cause considerable discomfort or pain if there is moderate swelling of the gland. Swelling can result from obstruction to the parotid duct, abscess formation in the gland and inflammatory swelling of the gland which is a frequent occurrence during mumps.

The parotid duct lies about 1 cm below and parallel to the zygomatic arch and turns medially around the anterior margin of the masseter muscle to penetrate the buccinator muscle and the oral mucosa opposite the second maxillary molar. The duct may be severed as the result of trauma and it is important that the cut ends be sutured accurately to avoid the formation of a salivary cyst or the secretion of saliva onto the face through a salivary fistula.

Appendix

The illustrations which follow are intended for use by the student for revision purposes. These illustrations appear elsewhere in the text complete with labels and the following list gives the figure numbers of the labelled illustrations which correspond to those contained in this appendix.

Fig. A.1	corresponds to	Fig. 4.5
A.2	,,	4.27
A.3	,,	5.29
A.4	,,	4.24
A.5	,,	4.26
A.6	,,	4.30
A.7	,,	4.32
A.8	,,	4.33
A.9	,,	4.34
A.10	,,	8.10
A.11	,,	8.12
A.12	,,	8.14

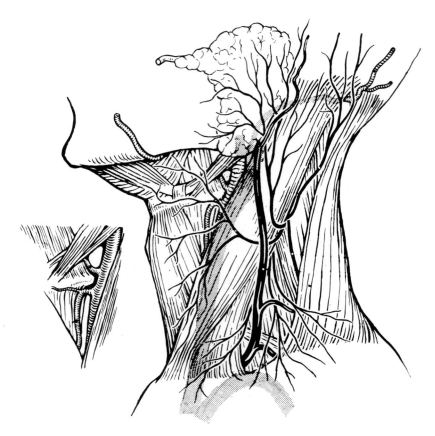

Fig. A.1

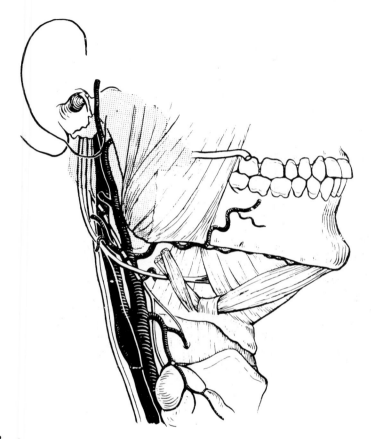

Fig. A.2

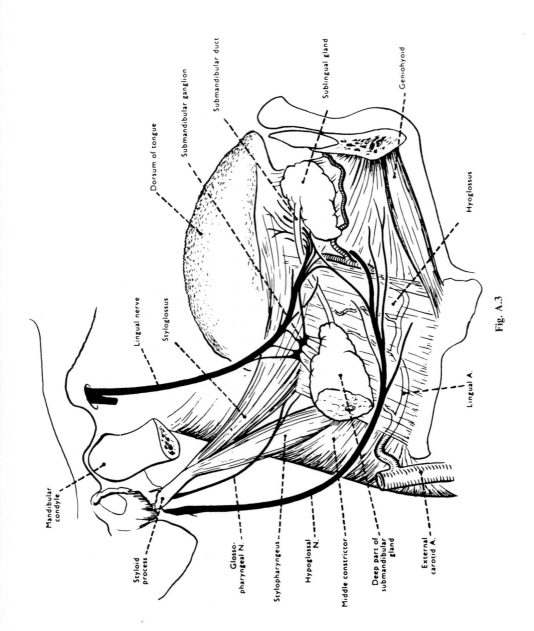

Fig. A.3

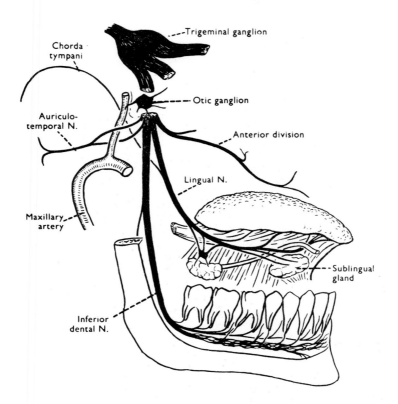

Fig. A.4

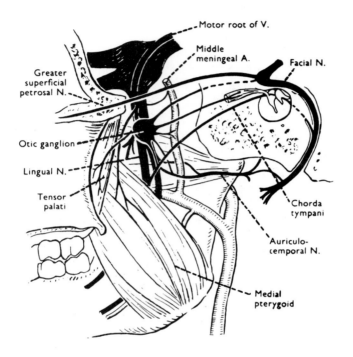

Fig. A.5

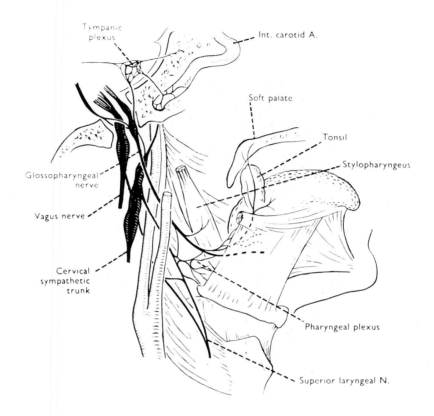

Fig. A.6

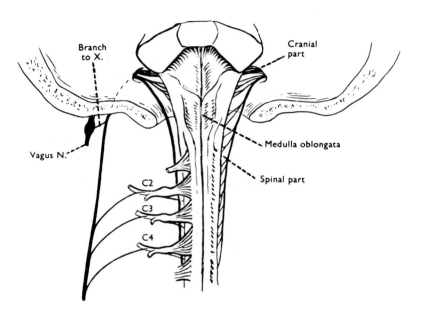

Fig. A.7

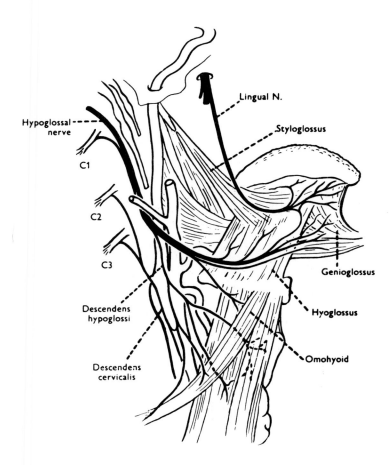

Fig. A.8

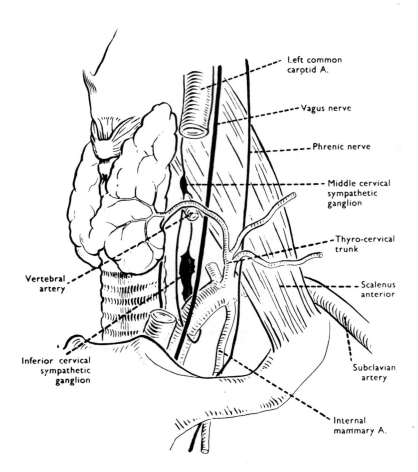

Left common carotid A.

Vagus nerve

Phrenic nerve

Middle cervical sympathetic ganglion

Thyro-cervical trunk

Scalenus anterior

Subclavian artery

Internal mammary A.

Vertebral artery

Inferior cervical sympathetic ganglion

Fig. A.9

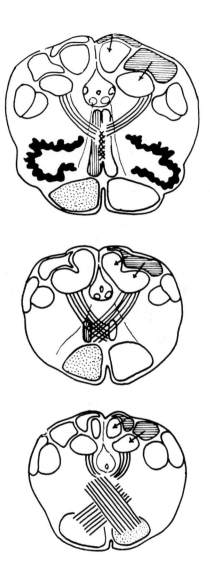

Fig. A.10

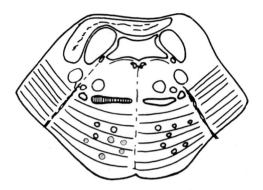

Fig. A.11

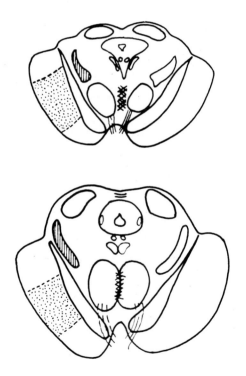

Fig. A.12

Selected Reading List

Bone and cartilage

Adams, P. H. *et al.* (1967) Effects of hyperthyroidism on bone and mineral metabolism in man. *Quarterly Journal of Medicine*, **36**, 1.

Amprino, R. (1963) On the growth of cortical bone and mechanism of osteon formation. *Acta anatomica*, **52**, 177.

Anderson, C. E. (1962) The structure and function of cartilage. *Journal of Bone and Joint Surgery*, **44A**, 777.

Ascenzi, A. & Bonucci, E. (1964) The ultimate tensile strength of single osteons. *Acta anatomica*, **58**, 160.

Bernard, G. W. & Pease, D. C. (1969) An electron microscopic study of initial intramembranous osteogenesis. *American Journal of Anatomy*, **125**, 271.

Boyde, A. & Hobdell, M. H. (1969) Scanning electron microscopy of lamellar bone. *Zeitschrift für Zellforschung und mikroskopische Anatomie*, **93**, 213.

Brash, J. C. (1934) Some problems in the growth and developmental mechanics of bone. *Edinburgh Medical Journal*, **41**, 305.

Brookes, M. (1971) *The Blood Supply of Bone*. London: Butterworth.

Cooper, R. *et al.* (1966) Morphology of the osteon. *Journal of Bone and Joint Surgery*, **48A**, 1239.

Copp, D. H. (1969) Endocrine control of calcium homeostasis. *Journal of Endocrinology*, **43**, 137.

Davies, T. G. H. & Fearnhead, R. W. (1960) The neurohistology of mammalian bone. *Archives of Oral Biology*, **2**, 263.

Enlow, D. H. (1962) A study of postnatal growth and remodelling of bone. *American Journal of Anatomy*, **110**, 79.

Enlow, D. H. (1962) Functions of the Haversian system. *American Journal of Anatomy*, **110**, 269.

Epker, B. N. & Frost, H. M. (1966) Biochemical control of bone growth and development: histological and tetracycline study. *Journal of Dental Research*, **45**, 364

Foster, G. V. *et al.* (1966) Effect of thyrocalcitonin on bone. *Lancet*, **ii**, 1428.

Frazier, P. D., Zipkin, I. & Mills, L. F. (1967) X-ray diffraction study of human bone. *Archives of Oral Biology*, **12**, 73.

Fullmer, H. M. (1966) Histochemical studies of mineralized tissues. *Annales d'histochimie*, **11**, 369.

Fullmer, H. M. & Lazarus, G. S. (1969) Collagenase in bone of man. *Journal of Histochemistry and Cytochemistry*, **17**, 793.

Glenister, T. W. (1976) An embryological view of cartilage. *Journal of Anatomy*, **122**, 323.

Goldhaber, P. (1966) Remodelling of bone in tissue culture. *Journal of Dental Research*, **45**, 490.

Hoskins, W. E. & Asling, C. W. (1977) Influence of growth hormone and thyroxine on endochondral osteogenesis in the mandibular condyle and proximal tibial epiphysis. *Journal of Dental Research*, **56**, 5, 509.

Hoyte, D. A. N. (1968) Alizarin red in the study of the apposition and resorption of bone. *American Journal of Physical Anthropology*, **29**, 157.

Lacroix, L. (1951) *The Organization of Bones*. London: Churchill.

Lang, S. B. (1969) Elastic coefficients of animal bone. *Science*, **165**, 287.

MacConaill, M. A. (1951) The mechanical structure of articulating cartilage. *Journal of Bone and Joint Surgery*, **33B**, 251.

Mjor, I. A. (1969) Bone lamellae. *Acta anatomica*, **73**, 127.

Murray, P. D. F. (1936) *Bones. A Study of the Development and Structure of the Vertebrate Skeleton*. Cambridge University Press.

Pritchard, J. J. (1961) Ossification. *Scottish Medical Journal*, **6**, 177.

Ralph, J. P. & Caputo, A. A. (1974) Analysis of stress patterns in the human mandible. *Journal of Dental Research*, **54**, 814.

Reid, L. (1976) Visceral cartilage. *Journal of Anatomy*, **122**, 349.

Richardson, A. (1965) Pattern of alveolar bone resorption following extraction of anterior teeth. *Dental Practitioner and Dental Record*, **16**, 77.

Sarnat, B. G. (1968) Growth of bones as revealed by implant markers in animals. *American Journal of Physical Anthropology*, **29**, 255.

Scott, J. H. (1957) The mechanical basis of bone formation. *Journal of Bone and Joint Surgery*, **39B**, 134.

Sherman, M. S. (1963) The nerves of bone. *Journal of Bone and Joint Surgery*, **45A**, 522.

Sognnaes, R. F. (ed.) (1960) Calcification in biological systems. *American Association for the Advancement of Science, Publication no. 64*.

Tammisalo, E. H. & Tammisalo, R. (1969) Correlation between density of the alveolar trabecular pattern and the actual mineral content of human mandibles. *Acta odontologica Scandinavica*, **27**, 417.

Toto, P. D. & Magon, J. J. (1966) Histogenesis of osteoclasts. *Journal of Dental Research*, **45**, 225.

Weinmann, J. P. & Sicher, H. (1955) *Bone and Bones: Fundamentals of Bone Biology*, 2nd ed. London: Kimpton.

Yeomans, J. D. & Urist, M. R. (1967) Bone induction by decalcified dentine implanted into oral, osseous and muscle tissues. *Archives of Oral Biology*, **12**, 999.

Muscles and muscle action

Abed-el-Malek, S. (1955) The part played by the tongue in mastication and deglutition. *Journal of Anatomy*, **89**, 250.

Achari, N. K. & Thexton, A. J. (1974) A masseteric reflex elicited from the oral mucosa in man. *Archives of Oral Biology*, **19**, 299.

Anderson, D. J. (1955) The physiology of mastication. *Dental Practitioner and Dental Record*, **5**, 389.

Anderson, D. J. (1956) Measurement of stress in mastication. *Journal of Dental Research*, **35**, 664.

Andran, G. M. & Kemp. F. H. (1951) The mechanism of swallowing. *Proceedings of the Royal Society of Medicine*, **44**, 1037.

Andran, G. M. & Kemp, F. H. (1960) Biting and mastication. *Dental Practitioner and Dental Record*, **11**, 23.

Atkinson, H. F. & Shepherd, R. W. (1973) Masticatory movement in the absence of teeth in man. *Archives of Oral Biology*, **18**, 855.

Basmajian, J. U. & Dutta, C. R. (1961) Electromyography of the pharyngeal constrictors and levator palati in man. *Anatomical Record*, **139**, 561.

Bosma, J. F. (1957) Deglutition: pharyngeal stage. *Physiological Reviews*, **37**, 275.

Brill, N., Lammie, G. A., Osborne, J. & Perry, H. T. (1959) Mandibular positions and mandibular movements. *British Dental Journal*, **106**, 391.

Brodie, A. G. (1950) Anatomy and physiology of head and neck musculature. *American Journal of Orthodontics*, **34**, 831.

Dewel, B. F. (1941) Dentofacial musculature. *American Journal of Orthodontics*, **27**, 469.

Gill, H. I. (1971) Neuromuscular spindles in human lateral pterygoid muscles. *Journal of Anatomy*, **109**, 157.

Goss, C. M. (1944) The attachment of skeletal muscle fibers. *American Journal of Anatomy*, **74**, 259.

Haines, R. W. (1932) The laws of muscle and tendon growth. *Journal of Anatomy*, **66**, 578.

Hannam, A. G., Matthews, B. & Yemm, R. (1970) Receptors involved in the response of the masseter muscle to tooth contact in man. *Archives of Oral Biology*, **15**, 17.

Harpman, J. A. & Woollard, H. H. (1938) The tendon of the lateral pterygoid muscle. *Journal of Anatomy*, **73**, 112.

Hiiemae, Karen M. (1967) Masticatory function in the mammals. *Journal of Dental Research*, **46**, 883.

Hoyte, D. A. & Enlow, D. H. (1966) Wolff's law and the problem of muscle attachment on resorptive surfaces of bone. *American Journal of Physical Anthropology*, **24**, 205.

Huxley, H. E. (1969) The mechanism of muscular contraction. *Science*, **164**, 1356.

Huxley, H. & Hanson, J. (1960) The molecular basis of contraction in cross-striated muscles. In *Structure and Function of Muscle*, ed. J. Bourne, Vol. 1, p. 183. New York: Academic Press.

Jerge, Charles R. (1964) The neurologic mechanism underlying cyclic jaw movements. *Journal of Prosthetic Dentistry*, **14**, 667.

Kawamura, Yojira. (1964) Recent concepts of the physiology of mastication. *Advances in Oral Biology*, **1**, 77.

Kawamura, Y. (1967) Neurophysiologic background of occlusion. *Periodontics*, **5**, 175.

Last, R. J. (1954) The muscles of the mandible. *Proceedings of the Royal Society of Medicine*, **47**, 571.

Le Gros Clark, W. E. (1958) *Tissues of the Human Body*, 4th ed. London: Oxford University Press.

Malpas, P. (1962) Anomalies of the mylohyoid muscle. *Journal of Anatomy*, **61**, 64.

MacDougall, J. D. B. (1955) The attachments of the masseter muscle. *British Dental Journal*, **98**, 526.

MacDougall, J. D. B. & Andrews, B. L. (1953) An electromyographic study of the temporalis and masseter muscle. *Journal of Anatomy*, **87**, 37.

McKenzie, J. (1955) The morphology of the sternomastoid and trapezius muscles. *Journal of Anatomy*, **89**, 527.

Merkeley, H. J. (1954) The labial and buccal accessory muscles of mastication. *Journal of Prosthetic Dentistry*, **4**, 327.

Merkeley, H. J. (1955) The lingual accessory muscles of mastication. *Journal of Prosthetic Dentistry*, **5**, 101.

Molnar, S. (1968) Mechanical simulation of human chewing motions. *Journal of Dental Research*, **47**, 559.

Mommaerts, W. F. H. M. (1969) Energetics of muscular contraction. *Physiological Reviews*, **49**, 427.

Moyers, R. E. (1950) An electromyographic analysis of certain muscles involved in temporomandibular movement. *American Journal of Orthodontics*, **36**, 487.

Nairn, R. I. (1975) The circumoral musculature: structure and function. *British Dental Journal*, **138**, 49.

Picton, D. C. A. (1962) Distortion of the jaws during biting. *Archives of Oral Biology*, **7**, 573.

Roche, A. F. (1963) Functional anatomy of the muscles of mastication. *Journal of Prosthetic Dentistry*, **13**, 548.

Rohan, R. F. & Turner, L. (1956) The levator palati muscle. *Journal of Anatomy*, **90**, 153.

Rushmer, R. F. & Hendron, J. A. (1951) The act of deglutition. *Journal of Applied Physiology*, **3**, 622.

Scott, J. H. (1954) The growth and function of the muscles of mastication in relation to the development of the facial skeleton and of the dentition. *American Journal of Orthodontics*, **40**, 429.

Simons, D. G. (1975) Muscle pain syndromes—part 1. *American Journal of Physical Medicine*, **51**, 289.

Symons, N. B. B. (1954) The attachment of the muscles of mastication. *British Dental Journal*, **96**, 76.

Walsh, J. P. (1951) Neurophysiological aspects of mastication. *Dental Journal of Australia*, **23**, 49.

Wardill, W. E. M. & Whillis, J. (1936) Movement of the soft palate. *Surgery, Gynecology and Obstetrics*, **62**, 836.

Weddell, G., Harpman, J. A., Lambley, D. G. & Young, L. (1940) The innervation of the musculature of the tongue. *Journal of Anatomy*, **74**, 255.

Whillis, J. (1946) Movement of the tongue in swallowing. *Journal of Anatomy (London)*, **80**, 115.

Yu, S.-K. J., Schmitt, A. & Sessle, B. J. (1973) Inhibitory effects on jaw muscle activity of innocuous and noxious stimulation of facial and intraoral sites in man. *Archives of Oral Biology*, **18**, 861.

Joints—mandibular joint

Barnett, C. H., Davies, D. V. & MacConaill, M. A. (1961) *Synovial Joints. Their Structure and Mechanics*. London: Longmans.

Baume, L. J. (1963) Ontogenesis of the temporomandibular joint: development of the condyles. *Journal of Dental Research*, **41**, 1327.

Blackwood, H. J. J. (1966) Cellular remodelling in articular tissue. *Journal of Dental Research*, **45**, 480.

Blackwood, H. J. J. (1969) Pathology of the temporomandibular joint. *Journal of the American Dental Association*, **79**, 118.

Blackwood, H. J. J. (1965) Vascularization of the condylar cartilage of the human mandible. *Journal of Anatomy*, **99**, 551.

Boyer, C. C., Williams, T. W. & Stevens, F. H. (1964) Blood supply of the temporomandibular joint. *Journal of Dental Research*, **43**, 224.

Butler, J. H., Folke, L. E. & Bandt, C. L. (1975) A descriptive survey of signs and symptoms associated with the myofascial pain-dysfunction syndrome. *Journal of the American Dental Association*, **90**, 635.

Choukas, N. C. & Sicher, H. (1960) The structure of the temporomandibular joint. *Oral Surgery, Oral Medicine and Oral Pathology*, **13**, 1263.

Dixon, A. D. (1962) Structure and functional significance of the intra-articular disc of the human temporomandibular joint. *Oral Surgery, Oral Medicine and Oral Pathology*, **15**, 48.

Durkin, J. F., Irving, J. T. & Heeley, J. D. (1969) A comparison of the circulatory and calcification patterns in the mandibular condyle with those found in the tibial epiphyseal and articular cartilages. *Archives of Oral Biology*, **14**, 1365.

Fish, S. F. (1963) The rest position of the mandible. *Dental Progress*, **3**, 211.

Furstman, L. (1965) Effect of loss of occlusion upon the mandibular joint. *American Journal of Orthodontics*, **51**, 245.

Gardner, E. (1950) Physiology of mandible joints. *Physiological Reviews*, **30**, 127.

Greenfield, B. E. & Wyke, B. (1966) Reflex innervation of the temporomandibular joint. *Nature*, **211**, 940.

Hankey, G. T. (1954) Temporomandibular arthrosis. *British Dental Journal*, **97**, 249.

Herring, S. W. (1972) Sutures—a tool in functional cranial analysis. *Acta anatomica*, **83**, 222.

Hoyte, D. A. N. (1973) Basicranial elongation: 2. Is there differential growth within a synchondrosis? *Anatomical Record*, **175**, 347.

Moffett, B. C. (1957) The prenatal development of the human temporomandibular joint. *Contributions to Embryology. Publication no*. 243. **36**, 21.

Moffett, B. C., Johnson, L. C., McCabe, J. B. & Askew, H. C. (1964) Articular remodelling in the adult human temporomandibular joint. *American Journal of Anatomy*, **115**, 119.

Moffett, B. (1966) Morphogenesis of the temporomandibular joint. *American Journal of Orthodontics*, **52**, 401.

Murphy, T. R. (1965) The movement of translation at the temporomandibular joint as it occurs in mastication. *British Dental Journal*, **118**, 163.

Parsons, M. T. and Boucher, L. J. (1966) The bilaminar zone of the meniscus. *Journal of Dental Research*, **45**, 59.

Rees, L. A. (1954) The structure and function of the mandibular joint. *British Dental Journal*, **96**, 126.

Sarnat, B. G. (1961) *The Temporomandibular Joint*. Springfield, Ill.: Thomas.

Sarnat, B. G. (1969) Developmental facial abnormalities and the temporomandibular joint. *Journal of the American Dental Association*, **79**, 108.

Schmid, F. (1969) On the nerve distribution of the temporomandibular joint capsule. *Oral Surgery, Oral Medicine and Oral Pathology*, **28**, 63.

Scott, J. H. (1955) A contribution to the study of mandibular joint function. *British Dental Journal*, **98**, 345.

Sharpe, C. J.; Gee, E. J. & Griffin, C. J. (1965) Osteogenic potential of the human condyle. *Australian Dental Journal*, **10**, 287.

Shore, N. A. (1962) Head pain of temporomandibular joint origin. *New York State Journal of Medicine*, **62**, 3580.

Stanier, D. I. (1977) Anatomical Note, the function of muscles around a simple joint. *Journal of Anatomy*, **123**, 3, 827.

Symons, N. B. B. (1952) The development of the human mandibular joint. *Journal of Anatomy*, **86**, 326.

Thilander, B. (1964) Innervation of the temporomandibular joint. *Acta odontologica Scandinavica*, **22**, 151.

Thomson, H. (1959) Mandibular joint pain. *British Dental Journal*, **107**, 243.

Toller, P. A. (1961) The synovial apparatus and temporomandibular joint function. *British Dental Journal*, **111**, 355.

Yuodelis, R. A. (1966) The morphogenesis of the human temporomandibular joint and its associated structures. *Journal of Dental Research*, **45**, 182.

Yuodelis, R. A. (1966) Ossification of the human temporomandibular joint. *Journal of Dental Research*, **45**, 192.

Skull growth
Babula, W. J. Jr., Smiley, G. R. & Dixon, A. D. (1970) The role of the cartilaginous nasal septum in midfacial growth. *American Journal of Orthodontics*, **58**, 250.

Backlund, E. (1963) Facial growth, and the significance of oral habits, mouthbreathing and soft tissues for malocclusion. *Acta odontologica Scandinavica*, **21**, (suppl. 36), 7.

Baume, L. J. (1961) The nasal septum: An endochondral growth centre. *Journal of Dental Research*, **40**, 625.

Bergersen, E. O. (1966) The directions of facial growth from infancy to adulthood. *Angle Orthodontist*, **36**, 18.

Bjork, A. (1955) Cranial base development. *American Journal of Orthodontics*, **41**, 198.

Bjork, A. (1955) Facial growth in man, studied with the aid of metallic implants. *Acta odontologica Scandinavica*, **13**, 9.

Bjork, A. (1966) Sutural growth of the upper face studied by the implant method. *Acta odontologica Scandinavica*, **24**, 109.

Brodie, A. G. (1953) Late growth changes in the human face. *Angle Orthodontist*, **23**, 146.

Burdi, A. R. (1965) Sagittal growth of the nasomaxillary complex. *Journal of Dental Research*, **44**, 112.

Burdi, A. R. (1969) Cephalometric growth analyses of the human upper face region during the last two trimesters of gestation. *American Journal of Anatomy*, **125**, 113.

Cobb, W. M. (1940) Craniofacial union in man. *American Journal of Physical Anthropology*, **26**, 87.

Dixon, A. D. (1961) Studies of the growth of the upper facial skeleton using radioactive calcium. *Journal of Dental Research*, **40**, 204.

Dixon, A. D. & Hoyte, D. A. N. (1963) A comparison of autoradiographic and alizarin techniques in the study of bone growth. *Anatomical Record*, **145**, 101.

Enlow, D. H. & Bang, S. (1965) Growth and remodelling of the human maxilla. *American Journal of Orthodontics*, **51**, 446.

Enlow, D. H. (1966) Morphogenetic analysis of facial growth. *American Journal of Orthodontics*, **52**, 283.

Enlow, D. H. & Harris, D. B. (1964) Postnatal growth of the human mandible. *American Journal of Orthodontics*, **50**, 25.

Enlow, D. H. (1975) *Handbook of Facial Growth*. Philadelphia: W. B. Saunders Co.

Ford, E. H. R. (1956) The growth of the foetal skull. *Journal of Anatomy*, **90**, 63.

Health, M. R. (1966) Cephalometric study of the middle third of the adult human face related to age and loss of dental tissue. *Archives of Oral Biology*, **11**, 677.

Hoyte, D. A. N. (1966) Experimental investigations of skull morphology and growth. *International Review of General and Experimental Zoology*, **2**.

Hoyte, D. A. N. (1971) Mechanisms of growth in the cranial vault and base. *Journal of Dental Research*, **50**, 1447.

Hoyte, D. A. N. (1975) A critical analysis of the growth in length of the cranial base. *Birth Defects: Original Article Series*, **XI**, 7, 255.

Israel, H. (1969) Pubertal influence upon the growth and sexual differentiation of the human mandible. *Archives of Oral Biology*, **14**, 583.

Knott, V. B. (1971) Change in cranial base measures of human males and females from age 6 years to early adulthood. *Growth*, **35**, 145.

Koski, K. (1968) Cranial growth centers: facts or fallacies. *American Journal of Orthodontics*, **54**, 566.

Koski, K. (1971) Some characteristics of cranio-facial growth cartilages. In Moyers, E. E., and Krogman, W. M. (eds.), *Cranio-facial Growth in Man*, Oxford: Pergamon Press, p. 125.

Latham, R. A. (1969) The pathogenesis of the skeletal deformity associated with unilateral cleft lip and palate. *Cleft Palate Journal*, **6**, 404.

Latham, R. A. (1971) The development, structure and growth pattern of the human mid-palatal suture. *Journal of Anatomy*, **108**, 31.

Latham, R. A., Burston, W. R. (1966) The postnatal pattern of growth at the sutures of the human skull. *Dental Practitioner and Dental Record*, **17**, 61.

Latham, R. A. (1972) The sella point and postnatal growth of the human cranial base. *American Journal of Orthodontics*, **61**, 156.

Meredith, H. V. (1960) Changes in form of the head and face during childhood. *Growth*, **24**, 215.

Moore, W. J. (1977) Associations in the hominoid facial skeleton. *Journal of Anatomy*, **123**, 111.

Moss, M. L. & Greenberg, S. N. (1955) Postnatal growth of the human skull base. *Angle Orthodontist*, **25**, 77.

Moss, M. L., Noback, C. R. & Robertson, G. G. (1956) Growth of certain foetal cranial bones. *American Journal of Anatomy*, **98**, 191.

Moss, M. L. & Simon, M. R. (1968) Growth of the human mandibular angular process: a functional cranial analysis. *American Journal of Physical Anthropology*, **28**, 127.

Moss, M. L. & Young, R. W. (1960) A functional approach to craniology. *American Journal of Physical Anthropology*, **18**, N.S., 281.

Moss, M. L. (1968) The role of the functional matrix in mandibular growth. *Angle Orthodontist*, **39**, 95.

Moss, M. L. and Bromberg, B. (1968) The passive role of the nasal septal cartilage in mid-facial growth. *Plastic and Reconstructive Surgery*, **41**, 536.

Murphy, T. (1957) Changes in mandibular form during postnatal growth. *Australian Dental Journal*, **2**, 267.

Powell, T. V. & Brodie, A. G. (1963) Closure of the spheno-occipital synchondrosis. *Anatomical Record*, **147**, 15.

Sarnat, B. G. (1963) Postnatal growth of the upper face: some experimental considerations. *Angle Orthodontist*, **33**, 139.

Sarnat, B. G. (1971) Surgical experimentation and gross postnatal growth of the face and jaws. *Journal of Dental Research*, **50**, 1462.

Sarnat, B. G. (1976) The postnatal maxillary-nasal-orbital complex: some considerations in experimental surgery. Factors Affecting the Growth of the Midface, McNamara, J. A. Jr., (ed.) Ann Arbor: Center for Human Growth and Development.

Savara, B. W. & Tracy, W. E. (1967) Norms of size and annual increments for five anatomical measures of the mandible in boys from three to sixteen years of age. *Archives of Oral Biology*, **12**, 469.

Scott, J. H. (1955) Craniofacial regions. *Dental Practitioner and Dental Record*, **5**, 208.

Scott, J. H. (1958) The cranial base. *American Journal of Physical Anthropology*, **16**, N.S., 319.

Scott, J. H. (1962) The growth of the craniofacial skeleton. *Irish Journal of Medical Science*, 6th ser., 276.

Scott, James. (1969) The doctrine of functional matrices. *American Journal of Orthodontics*, **56**, 38.

Siegel, M. I. (1974) The role of the cartilaginous nasal septum in midfacial growth. *American Journal of Physical Anthropology*, **41**, 503.

Silverman, F. N. (1965) Growth of the face in developmental defects. *Journal of Dental Research*, **44**, 209.

Subtehny, J. D. (1957) A cephalometric study of the growth of the soft palate. *Plastic and Reconstructive Surgery*, **19**, 49.

Sullivan, P. G. (1972) A method for the study of jaw growth using a computer-based three-dimensional recording technique. *Journal of Anatomy*, **112**, 457.

Symons, N. B. B. (1951) Studies on the growth and form of the mandible. *Dental Record*, **71**, 41.

Takagi, Y. (1964) Human postnatal growth of vomer in relation to base of cranium. *Annals of Otology, Rhinology, and Laryngology*, **73**, 238.

Tracy, W. E., Savara, B. S. & Brant, J. W. A. (1965) Relation of height, width and depth of the mandible. *Angle Orthodontist*, **35**, 269.

Warwick, R. (1950) The relation of the direction of the mental foramen to the growth of the human mandible. *Journal of Anatomy*, **84**, 116.

Yen, P. K. U. & Shaw, J. H. (1963) Studies of the skull sutures of the rhesus monkey by comparison of the topographic sampling technique, autoradiography and vital staining. *Archives of Oral Biology*, **8**, 349.

Developmental anatomy

Bagnall, K. M., Harris, P. F. & Jones, P. R. M. (1977) A radiographic study of the human fetal spine. 1. The development of the secondary cervical curvature. *Journal of Anatomy*, **123**, 3, 777.

Barry, A. (1951) The aortic arch derivatives in the human adult. *Anatomical Record*, **111**, 221.

Baume, L. J. (1961) Principles of cephalofacial development revealed by experimental biology. *American Journal of Orthodontics*, **47**, 881.

Baume, L. J. (1968) Patterns of cephalofacial growth and development. *International Dental Journal*, **18**, 489.

Bosma, J. F. (1963) Oral and pharyngeal development and function. *Journal of Dental Research*, **42**, 375.

Bowman, A. J. & Latham, R. A. (1967) Midpalatal mucosa: persistence of fetal pattern. *Journal of Dental Research*, **46**, 295.

Boyd, J. D. (1933) The classification of the upper lip in animals. *Journal of Anatomy*, **67**, 409.

Brachet, J. (1961) The living cell. *Scientific American*, **205**, 50.

Burdi, A. R. & Faist, K. (1967) Morphogenesis of the palate in normal human embryos with special emphasis on the mechanisms involved. *American Journal of Anatomy*, **120**, 149.

Burke, G. W. *et al.* (1966) Some aspects of the origin and fate of midpalatal cysts in human foetuses. *Journal of Dental Research*, **45**, 159.

Crouse, G. S. & Cucinotta, A. J. (1965) Progressive neuronal differentiation in the submandibular ganglia of a series of human foetuses, *Journal of Comparative Neurology*, **125**, 259.

Curtis, E. J., Fraser, F. C. & Warburton, D. (1961) Congenital cleft lip and palate. *American Journal of Diseases of Children*, **102**, 853.

Davis, C. L. (1927) Development of the human heart from its first appearance to the stage found in embryos of 20 paired somites. *Contributions to Embryology*, **19**, 245.

Davis, W. B. (1935) Congenital deformities of the face. *Surgery, Gynecology and Obstetrics*, **61**, 201.

De Beer, G. R. (1937) *The Development of the Vertebrate Skull*. Oxford: Clarendon Press.

Dixon, A. D. (1953) The early development of the maxilla. *Dental Practitioner and Dental Record*, **3**, 331.

Dixon, A. D. (1958) The development of the jaws. *Dental Practitioner and Dental Record*, **9**, 10.

Fawcett, E. (1905) Ossification of the lower jaw in man. *Journal of the American Medical Association*, **45**, 696.

Fawcett, E. (1910) Development of the human sphenoid. *Journal of Anatomy*, **44**, 156.

Fawcett, E. (1911) The development of the human maxilla, vomer and paraseptal cartilages. *Journal of Anatomy*, **45**, 378.

Ford, E. H. R. (1956) The growth of the foetal skull. *Journal of Anatomy*, **90**, 63.

Friede, J. (1975) A histological and enzyme-histochemical study of growth sites of the premaxilla in human foetuses and neonates. *Archives of Oral Biology*, **20**, 809.

Gamble, H. J. (1966) Further electron microscope studies of human foetal peripheral nerves. *Journal of Anatomy*, **100**, 3, 487.

Gardner, E. (1956) Osteogenesis in the human embryo and foetus. In *The Biochemistry and Physiology of Bone*, ed. Bourne, G. H. New York: Academic Press.

Garn, S. M. *et al.* (1970) Prenatal dental development as a reference standard for embryologic status. *Journal of Dental Research*, **49**, 894.

Gasser, R. F. (1967) The development of the facial muscles in man. *American Journal of Anatomy*, **120**, 357.

Gaunt, W. A. (1967) Quantitative aspects of the developing tooth germ. *Journal of Dental Research*, (suppl.), **46**, 851.

Gilbert, P. W. (1952) Origin and development of head cavities in the human embryo. *Journal of Morphology*, **90**, 149.

Glasstone, S. (1967) Morphodifferentiation of teeth in embryonic mandibular segments in tissue culture. *Journal of Dental Research*, **46**, 611.

Haines, R. W. (1947) The development of joints. *Journal of Anatomy*, **81**, 33.

Hirschorn, K. & Cooper, M. L. (1961) Chromosomal aberrations in human disease. *American Journal of Medicine*, **31**, 442.

Hollman, K. (1969) Theoretical study of the inheritance of cleft lips, jaws and palates. *Plastic and Reconstructive Surgery*, **44**, 167.

Humphrey, T. (1966) Development of trigeminal nerve fibres to oral mucosa, compared with development to cutaneous surfaces. *Journal of Comparative Neurology*, **126**, 317.

Humphrey, T. (1969) The relation between human foetal mouth opening reflexes and closure of the palate. *American Journal of Anatomy*, **125**, 317.

Humphrey, T. (1968) The development of mouth opening and related reflexes involving the oral area of human fetuses. *Alabama Journal of Medical Science*, **5**, 126.

Jacobs, M. J. (1970) The development of the human motor trigeminal complex and accessory facial nucleus and their topographic relations with the facial and abducens nuclei. *Journal of Comparative Neurology*, **138**, 161.

Johnson, P. A., Atkinson, P. J. & Moore, W. J. (1976) The development and structure of the chimpanzee mandible. *Journal of Anatomy*, **122**, 467.

Jordan, R. E., Kraus, B. S. & Neptune, C. M. (1966) Dental abnormalities associated with cleft lip and/or palate. *Cleft Palate Journal*, **3**, 22.

Kernahan, D. A. & Stark, R. B. (1958) A new classification for cleft lip and cleft palate. *Plastic and Reconstructive Surgery*, **22**, 435.

King, T. S. (1954) The anatomy of hare-lip in man. *Journal of Anatomy*, **88**, 1.

Kraus, B. S. (1960) Prenatal growth and morphology of the human palate. *Journal of Dental Research*, **39**, 1177.

Kraus, B. S. & Decker, J. D. (1960) The prenatal inter-relationships of the maxilla and premaxilla in the facial development of man. *Acta anatomica*, **40**, 278.

Latham, R. A. & Deaton, T. G. (1976) The structural basis of the philtrum and the contour of the vermillion border: a study of the musculature of the upper lip. *Journal of Anatomy*, **121**, 151.

Lee, C. K. (1973) Ultrastructure of the dorsal lingual epithelium in human embryos and foetuses. *Archives of Oral Biology*, **18**, 265.

Monie, I. W. & Cacciatone, A. (1962) The development of the philtrum. *Plastic and Reconstructive Surgery*, **30**, 313.

Padget, D. H. (1948) The development of the cranial arteries in the human embryo. *Contributions to Embryology*, **32**, 205.

Pearson, A. A. (1977) The early innervation of the developing deciduous teeth. *Journal of Anatomy*, **123**, 3, 563.

Poswillo, D. (1966) Observations of foetal posture and causal mechanisms of congenital deformity of palate, mandible, and limbs. *Journal of Dental Research*, **45**, 584.

Poswillo, D. & Roy, L. J. (1965) Pathogenesis of cleft palate: animal study. *British Journal of Surgery*, **52**, 902.

Poswillo, D. (1975) Casual mechanisms of craniofacial deformity. *British Medical Bulletin*, **31**, 101.

Pritchard, J. J., Scott, J. H. & Girgis, F. G. (1956) The structure and development of cranial and facial sutures. *Journal of Anatomy*, **90**, 73.

Pruzansky, S. (1961) *Congenital Anomalies of the Face and Associated Structures*. Springfield, Ill.: Thomas.

Shepherd, W. M. & McCarthy, M. D. (1955) Observation on the appearance and ossification of the premaxilla and maxilla in the human embryo. *Anatomical Record*, **121**, 13.

Smiley, G. R. (1975) A histological study of the formation and development of the soft palate in mice and man. *Archives of Oral Biology*, **20**, 297.

Smith, D. W. (1970) Recognizable Patterns of Human Malformation. Philadelphia: W. B. Saunders Co.

Sperber, G. H. (1976) Craniofacial Embryology. Bristol: John Wright & Sons Ltd.

Stark, R. B. (1954) The pathogenesis of hare-lip and cleft palate. *Plastic and Reconstructive Surgery*, **13**, 20.

Tondury, G. (1958) Concerning the development of the human face. *International Dental Journal*, **8**, 496.

Tonge, C. H. (1957) Basic aspects of mouth development. *Proceedings of the Royal Society of Medicine*, **50**, 185.

Tonge, C. H. (1966) Advances in dental embryology. *International Dental Journal*, **16**, 328.

Warbrick, J. G. (1960) The early development of the nasal cavity and upper lip in the human embryo. *Journal of Anatomy*, **94**, 351.

Warbrick, J. G. (1963) Aspects of facial and nasal development. *Scientific Basis of Medicine. Annual Reviews*, **99**.

Warkany, J. & Kalter, H. (1961) Congenital malformations. *New England Journal of Medicine*, **265**, 993.

Wood, N. D. *et al.* (1969) Osteogenesis of the human upper jaw: proof of the non-existence of a separate premaxillary centre. *Archives of Oral Biology*, **14**, 1331.

Wood, P. J. & Kraus, B. S. (1962) Prenatal development of the human palate. Some histological observations. *Archives of Oral Biology*, **7**, 137.

Yokoh, Y. (1967) Development of the palate in man. *Acta anatomica*, **68**, 1.

Yokoh, Y. (1969) The early development of the nervous system in man. *Acta anatomica*, **71**, 492.

Nervous system

Alexander, R. W. & Fitzgerald, M. J. T. (1967) An example of transmedian neuromuscular innervation. *Journal of Dental Research*, **46**, 261.

Anson, B. J., Harper, D. G. & Warpeha, R. L. (1963) Surgical anatomy of the facial canal and facial nerve. *Annals of Otology, Rhinology and Laryngology*, **72**, 713.

Axford, M. (1928) Some observations on the cervical sympathetic in man. *Journal of Anatomy*, **62**, 301.

Beaver, David L. *et al.* (1965) Electron microscopy of the trigeminal ganglion. II. Autopsy study of human ganglia. *Archives of Pathology*, **79**, 557.

Blatt, I. M. & Bunto, W. G. (1960) The structure of nerve elements in the major salivary glands of the human. *Annals of Otology, Rhinology and Laryngology*, **69**, 375.

Blunt, M. J. (1954) The blood supply of the facial nerve. *Journal of Anatomy*, **88**, 520.

Bohn, A. (1961) The course of the premaxillary nerves and blood vessels. *Acta odontologica Scandinavica*, **19**, 179.

Bosma, J. A. (ed.) (1970) Oral sensation and perception; second symposium. Springfield, Ill.: C. C. Thomas.

Brown, J. W. (1958) The development of subnucleus caudalis of the nucleus of the spinal tract of V. *Journal of Comparative Neurology*, **110**, 105.

Burr, H. S. & Robinson, G. B. (1925) An anatomical study of the gasserian ganglion with particular reference to the nature and extent of Meckel's cave. *Anatomical Record*, **29**, 269.

Carmichael, E. A. & Woollard, H. H. (1933) Some observations on the fifth and seventh cranial nerves. *Brain*, **56**, 109.

Casey, K. L. (1973) Pain: A current view of neural mechanisms. *American Scientist*, **61**, 194.

Christensen, G. (1969) A biometrical approach to the gasserian ganglion. *Medical Journal of Australia*, **1**, 13.

Corkin, K. B. (1940) Observations on the peripheral distribution of fibres arising in the mesencephalic nucleus of the fifth cranial nerve. *Journal of Comparative Neurology*, **73**, 153.

Cushing, H. (1904) The sensory distribution of the fifth cranial nerve. *Bulletin of the Johns Hopkins Hospital*, **15**, 213.

Dandy, W. E. (1934) Concerning the cause of trigeminal neuralgia. *American Journal of Surgery*, **24**, 447.

Dastur, D. K. (1961) The relationship between terminal lingual innervation and gustation. *Brain*, **84**, 499.

Davenport, J. C. (1969) Pressure-pain thresholds in the oral cavity. *Archives of Oral Biology*, **14**, 1267.

Dillworth, T. F. M. (1921) The nerves of the human larynx. *Journal of Anatomy*, **56**, 46.

Dixon, A. D. (1961) The innervation of hair follicles in the mammalian lip. *Anatomical Record*, **140**, 147.

Dixon, A. D. (1961) Sensory nerve terminations in the oral mucosa. *Archives of Oral Biology*, **5**, 105.

Dixon, A. D. (1962) The position, incidence and origin of sensory nerve terminations in oral mucous membrane. *Archives of Oral Biology*, **7**, 39.

Dixon, A. D. (1963) The ultrastructure of nerve fibres in the trigeminal ganglion. *Journal of Ultrastructure Research*, **8**, 107.

Dixon, J. S. (1966) The fine structure of parasympathetic nerve cells in the otic ganglia of the rabbit. *Anatomical Record*, **156**, 239.

Dubner, R. & Kawamura, Y. (1971) *Oral-Facial Sensory and Motor Mechanisms*. New York: Appleton-Century-Crofts.

Eccles, J. (1965) The synapse. *Scientific American*, **212**, 56.

Fitzgerald, M. J. T. & Scott, J. H. (1958) Observations on the anatomy of the superior dental nerves. *British Dental Journal*, **104**, 205.

Foley, J. O. & Du Bois, F. S. (1943) An experimental study of the facial nerve. *Journal of Comparative Neurology*, **79**, 79.

Gairns, F. W. (1956) The sensory nerve endings of the human palate. *Quarterly Journal of Experimental Physiology*, **40**, 40.

Gasser, R. F. (1970) The early development of the parotid gland around the facial nerve and its branches in man. *Anatomical Record*, **1672**, 63.

Gomez, H. (1961) The innervation of lingual salivary glands. *Anatomical Record*, **139**, 69.

Gregg, J. M. & Dixon, A. D. (1973) Somatotopic organization of the trigeminal ganglion in the rat. *Archives of Oral Biology*, **18**, 487.

Henderson, W. R. (1965) The anatomy of the gasserian ganglion and the distribution of pain in relation to injections and operations for trigeminal neuralgia. *Annals of the Royal College of Surgeons of England*, **37**, 346.

Horstadius, S. (1950) *The Neural Crest*. London: Oxford University Press.

Jefferson, G. (1950) Localization of function in the cerebral cortex. *British Medical Bulletin*, **6**, 333.

Jerge, Charles R. (1963) The function of the nucleus supratrigeminalis. *Journal of Neurophysiology*, **26**, 393.

Jerge, Charles R. (1963) Organization and function of the trigeminal mesencephalic nucleus. *Journal of Neurophysiology*, **26**, 279.

Jones, F. W. (1939) The anterior superior alveolar nerve and vessels. *Journal of Anatomy*, **73**, 583.

Kerr, F. W. L. (1963) The divisional organization of afferent fibres of the trigeminal nerve. *Brain*, **85**, 721.

Kerr, F. W. L. (1964) Somatotopic organization of trigeminal ganglion neurones. *Archives of Neurology*, **ii**, 593.

Kingsbury, B. F. (1922) The fundamental plan of the vertebrate brain. *Journal of Comparative Neurology*, **34**, 461.

Koch, S. L. (1916) The structure of the third, fourth, fifth, sixth, ninth, eleventh and twelfth cranial nerves. *Journal of Comparative Neurology*, **26**, 54.

Kuntz, A. & Richens, C. A. (1946) Components and distribution of the nerves of the parotid and submandibular glands. *Journal of Comparative Neurology*, **85**, 21.

Laskin, D. M. (1969) Etiology of the pain-dysfunction syndrome. *Journal of the American Dental Association*, **79**, 147.

Leake, C. D. (1968) Historical aspects of the autonomic nervous system. *Anesthesiology*, **29**, 623.

Lim, R. K. S. (1967) Pain mechanisms. *Anesthesiology*, **28**, 106.

Marlow, C. D., Winkelmann, R. K. & Gibilisco, J. A. (1965a) General sensory innervation of the human tongue. *Anatomical Record*, **152**, 503.

Marlow, C. D., Winkelmann, R. K. & Gibilisco, J. A. (1965b) Special sensory innervation of the human tongue. *Journal of Dental Research*, **44**, 1381.

Martin, W. D., Jr. & Smith, R. D. (1966) Weight changes of the submandibular glands in the rat resulting from denervation. *Anatomical Record*, **156**, 325.

Martin, M. R., Caddy, K. W. T. & Biscoe, T. J. (1977) Numbers and diameters of motoneurons and myelinated axons in the facial nucleus and nerve of the albino rat. *Journal of Anatomy*, **123**, 3, 579.

Matthews, B. (1976) The mechanisms of pain from dentine and pulp. *British Dental Journal*, **140**, 57.

McDaniel, W. L. (1956) Variations in nerve distribution of the maxillary teeth. *Journal of Dental Research*, **35**, 916.

McKenzie, J. (1948) The parotid gland in relation to the facial nerve. *Journal of Anatomy*, **82**, 183.

Mitchell, G. A. G. (1953) *Anatomy of the Autonomic Nervous System*. Edinburgh: Livingstone.

Mitchell, G. A. G. (1956) *Cardiovascular Innervation*. Edinburgh: Livingstone.

Miyoshi, S., Nishijima, S. & Imanishi, I. (1966) Electron microscopy of myelinated and unmyelinated nerve fibres in human dental pulp. *Archives of Oral Biology*, **11**, 845.

Mongkolluqsana, D. & Edwards, L. F. (1957) The extra-osseous innervation of the gingivae. *Journal of Dental Research*, **36**, 516.

Moses, H. L. (1967) Comparative fine structure of the trigeminal ganglia, including human autopsy studies. *Journal of Neurosurgery*, **26**, 112.

Moses, H. L. *et al.* (1965) Electron microscopy of the trigeminal ganglion. 1. Comparative ultra-structure. *Archives of Pathology*, **79**, 541.

Mumford, J. M. (1965) Pain perception threshold and adaptation of normal human teeth. *Archives of Oral Biology*, **10**, 957.

Mumford, J. M. & Newton A. V. (1974) Trigeminal convergence from human teeth: influence of con-tralateral stimulation and stimulus frequency on the pain perception threshold. *Archives of Oral Biology*, **19**, 145.

Olsen, N. H., Teuscher, G. W. & Verke, K. L. (1955) A study of the nerve supply to the upper anterior teeth. *Journal of Dental Research*, **34**, 413.

Pearson, A. A. (1977) The early innervation of the developing deciduous teeth. *Journal of Anatomy*, **123**, 3, 563.

Penfield, W. (1956) Some observations on the functional organization of the human brain. *Smithsonian Institute Report for 1955*, p. 433.

Rieske, E. (1969) Effects of the nerve growth factor (NFG) on the differentiation of trigeminal ganglia in cell cultures. *Zeitschrift für Zellforschung und mikroskopische Anatomie*, **95**, 546.

Rhoton, A. L., Jr., O'Leary, J. L. & Ferguson, J. P. (1966) The trigeminal, facial, vagal, and glosso-pharyngeal nerves in the monkey. Afferent connections. *Archives of Neurology*, **14**, 530.

Romasch, J. & Malpass, J. (1958) The human motor-trigeminal nucleus. *Anatomical Record*, **130**, 9.

Strassburg, M. (1967) Morphologic reaction of the trigeminal ganglion after experimental surgery on the maxillodental region. *Journal of Oral Surgery*, **26**, 107.

Stones, H. H. (1956) Facial pain: review of aetiological factors. *Proceedings of the Royal Society of Medicine*, **49**, 39.

Symington, J. (1911) The relations of the main divisions of the trifacial nerve. *Journal of Anatomy*, **45**, 183.

Ten Cate, A. R. & Shelton, L. (1966) Cholinesterase activity in human teeth. *Archives of Oral Biology*, **11**, 423.

Tolman, D. E., Winkelmann, R. K. & Gibilisco, J. A. (1965) Nerve endings in gingival tissue. *Journal of Dental Research*, **44**, 657.

Weiss, Paul A. (1969) 'Panta Rhei'—And so flow our nerves. *American Scientist*, **57**, 287.

Wrete, M. (1959) The anatomy of the sympathetic trunks in man. *Journal of Anatomy*, **93**, 448.

Wyke, B. D. (1958) The surgical physiology of facial pain. *British Dental Journal*, **104**, 153.

Descriptive and clinical anatomy

Abbie, A. A. (1947) Headform and human evolution. *Journal of Anatomy*, **81**, 233.

Aitchison, J. (1963) Sex differences in teeth, jaws and skulls. *Dental Practitioner and Dental Record*, **14**, 52.

Alexander, R. W. & Fitzgerald, M. J. T. (1967) An example of transmedian neuromuscular innervation. *Journal of Dental Research*, **46**, 295.

Al Yassin, T. M. & Toner, P. G. (1977) Fine structure of squamous epithelium and submucosal glands of human oesophagus. *Journal of Anatomy*, **123**, 3, 705.

Beaver, D. L. (1973) Electron microscopy of the gasserian ganglion in trigeminal neuralgia. *Journal of Neurosurgery*, **35**, 444.

Bernick, S. (1967) Age changes in the blood supply to human teeth. *Journal of Dental Research*, **46**, 694.

Binnie, W. H., Stacey, A. J., Davis, R. & Cawson, R. A. (1975) Applications of xeroradiography in dentistry. *Journal of Dentistry*, **3**, 3, 99.

Birn, H. (1966) The vascular supply of the periodontal membrane. *Journal of Periodontal Research*, **1**, 51.

Botelho, S. Y. (1964) Tears and the macrimal gland. *Scientific American*, **211**, 78.

Boyer, C. C. & Neptune, C. M. (1962) Patterns of blood supply to teeth and adjacent tissues. *Journal of Dental Research*, **41**, 158.

Brachet, J. L. (1969) General introduction to cytoplasm. *Proceedings of the Royal Society. Series B*, **173**, 3.

Breathnach, A. A. (1965) *Frazer's Anatomy of the Human Skeleton*, 6th ed. London: Churchill.

Brown, I. A. (1972) Scanning electron microscopy of human dermal fibrous tissue. *Journal of Anatomy*, **113**, 2, 159.

Bunyan, John (1966) The history of the microscope. *British Dental Journal*, **121**, 55.

Burton, R. C. (1969) The problem of facial pain. *Journal of the American Dental Association*, **79**, 93.

Campbell, E. H. (1933) The cavernous sinus. Anatomical and clinical considerations. *Annals of Otology, Rhinology and Laryngology*, **42**, 51.

Carney, L. R. (1968) Considerations on the cause and treatment of trigeminal neuralgia. *Neurology*, **17**, 1143.

Castelli, W. A. (1963) Vascular architecture of the human adult mandible. *Journal of Dental Research*, **42**, 786.

Costen, J. B. (1932) Anatomic phases involved in the surgery of the naso-antral wall and the floor of the mouth. *Annals of Otology, Rhinology and Laryngology*, **41**, 820.

Craddock, F. W. (1953) Retromolar region of the mandible. *Journal of the American Dental Association*, **47**, 453.

Davenport, J. C. (1969) Pressure–pain thresholds in the oral cavity in man. *Archives of Oral Biology*, **14**, 1267.

Demester, W. T. & Enlow, D. H. (1959) Patterns of vascular channels in the cortex of the human mandible. *Anatomical Record*, **135**, 189.

Dixon, A. D. & Torr, J. B. D. (1956) Sex chromatin in oral smears. *British Medical Journal*, **ii**, 799.

Ducker, T. B., Kempe, L. G. & Hayes, G. J. (1969) The metabolic background for peripheral nerve surgery. *Journal of Neurosurgery*, **30**, 270.

Duinkerke, A. S. H., Van De Poel, A. C. M. & Doesburg, W. H. (1973) Variations in measurements of human periapical structures in radiographs. *Archives of Oral Biology*, **18**, 745.

Eloff, F. C. (1952) On the relations of the human vomer to the anterior paraseptal cartilages. *Journal of Anatomy*, **86**, 16.

Fickling, B. W. (1957) Oral surgery involving the maxillary sinus. *British Dental Journal*, **103**, 199.

Foster, J. B. (1969) Facial pain. *British Medical Journal*, **iv**, 667.

Garn, S. M., Lewis, A. B. & Blizzard, R. M. (1965) Endocrine factors in dental development. *Journal of Dental Research*, **44**, 243.

Garrett, J. R. (1962) Some observations on human submandibular salivary glands. *Proceedings of the Royal Society of Medicine*, **55**, 488.

Garrett, J. R. (1967) The innervation of normal human submandibular and parotid salivary glands. *Archives of Oral Biology*, **12**, 1417.

Gaughran, G. R. L. (1957) Fasciae of the masticator space. *Anatomical Record*, **129**, 383.

Granite, E. L. (1976) Anatomic considerations in infections of the face and neck : review of the literature. *Journal of Oral Surgery*, **34**, 34.

Haines, R. W. & Barnett, S. G. (1959) The structure of the mouth in the mandibular molar region. *Journal of Prosthetic Dentistry*, **9**, 962.

Haugen, F. P. (1968) The autonomic nervous system and pain. *Anesthesiology*, **29**, 785.

Hedegard, B. & Lundberg, M. (1965) Temporomandibular joint (changes) in patients with immediate upper dentures. *Acta odontologica Scandinavica*, **23**, 163.

Henderson, W. R. (1965) The anatomy of the gasserian ganglion and the distribution of pain in relation to injections and operations for trigeminal neuralgia. *Annals of the Royal College of Surgeons of England*, **37**, 346.

Hollinshead, W. H. (1952) Anatomy of the endocrine glands. *Surgical Clinics of North America*, **32**, 1115.

Jamieson, J. K. & Dobson, J. F. (1920) The lymphatics of the tongue. *British Journal of Surgery*, **8**, 80.

Jefferson, G. & Schorstein, J. (1955) Injuries of the trigeminal nerve, its ganglion and its divisions. *British Journal of Surgery*, **42**, 36.

Kanagasuntheram, R., Wong, W. C. & Chan, H. L. (1969) Some observations on the innervation of the human nasopharynx. *Journal of Anatomy*, **104**, 361.

Kerr, F. W. L. & Miller, R. H. (1966) The pathology of trigeminal neuralgia. *Archives of Neurology*, **15**, 308.

Klein-Szanto, A. J. P. & Schroeder, H. E. (1977) Architecture and density of the connective tissue papillae of the human oral mucosa. *Journal of Anatomy*, **123**, 93.

Kostrubala, J. G. (1945) Potential anatomical spaces in the face. *American Journal of Surgery*, **68 N.S.**, 28.

Kozma, M., Poberal, L. V., Gelbert, A. & Foldi, M. (1962) The lymphatics of the tongue. *Acta anatomica*, **49**, 252.

Krippaehne, W. W. & Hunt, T. K. (1964) Surgical anatomy of the parotid gland. *Surgical Clinics of North America*, **44**, 1145.

Larson, D. L. *et al.* (1967) Lymphatics of the upper and lower lip. A clinical and experimental study. *American Journal of Surgery*, **114**, 525.

Last, R. J. (1964) Clinical anatomy in practice. *British Journal of Oral Surgery*, **1**, 80; **2**, 177.

Lindsay, B. (1969) Trigeminal neuralgia: a new approach. *Medical Journal of Australia*, **1**, 8.

Listgarten, M. A. (1966) Electron microscopic study of the gingivo–dental junction of man. *American Journal of Anatomy*, **119**, 147.

Melzack, R. & Wall, P. D. (1965) Pain mechanisms: a new theory. *Science*, **150**, 971.

Manson, J. D. & Lucas, R. B. (1962) A microradiographic study of age changes in the human mandible. *Archives of Oral Biology*, **7**, 761.

Marshall, D. (1964) Spheno-ethmoid complex. *Oral Surgery, Oral Medicine and Oral Pathology*, **17**, Suppl. 3, 1.

Mayerson, H. S. (1963) The lymphatic system. *Scientific American*, **208**, 80.

McAuley, J. E. (1966) The early development of local anesthesia. *British Dental Journal*, **139**.

Miles, A. E. W. (1958) The assessment of age from the dentition. *Proceedings of the Royal Society of Medicine*, **51**, 1057.

Moher, W. P. & Swindle, P. F. (1962) Submucosal blood vessels of the palate. *Dental Progress*, **2**, 167.

Montagna, W. (1965) The skin. *Scientific American*, **212**, 56.

Mumford, J. M. (1965) Pain perception threshold and adaptation of normal human teeth. *Archives of Oral Biology*, **10**, 957.

Negus, V. E. (1954) The functions of the paranasal sinuses. *Acta oto-laryngologica*, **44**, 408.

New, G. B. (1947) Congenital cysts of the tongue, the floor of the mouth, the pharynx and the larynx. *Acta oto-laryngologica*, **45**, 145.

Norman, J. E. de B. (1969) Facial pain—a dental viewpoint. *Practitioner*, **203**, 650.

Perskin, S. & Laskin, D. M. (1965) Contribution of autogenous condylar grafts to mandibular growth. *Oral Surgery, Oral Medicine and Oral Pathology*, **20**, 517.

Rapp, R., Avery, J. K. & Strachan, D. S. (1967) The distribution of nerves in human primary teeth. *Anatomical Record*, **159**, 89.

Rowe, N. L. & Killey, H. C. (1968) *Fractures of the facial skeleton*. 2 ed. Baltimore: Williams and Wilkins Co.

Schroeder, H. E. & Theilade, J. (1966) Electron microscopy of normal human gingival epithelium. *Journal of Periodontal Research*, **1**, 95.

Schulte, W. C. (1967) Effect of an osteogenic extract on the healing of extraction wounds. *Journal of Dental Research*, **46**, 656.

Schumacher, G-H. ed. (1972) Morphology of the maxillo-mandibular apparatus. Symposium of the IX International Congress of Anatomists. Leipzig: Veb Georg Thieme.

Scott, J. H. (1952) The spread of dental infection. Anatomical considerations. *British Dental Journal*, **93**, 236.

Scott, J. H. (1954) The position of the opening of the parotid duct. *Dental Record*, **74**, 344.

Scott, J. H. (1954) Heat regulating function of the nasal mucous membrane. *Journal of Laryngology and Otology*, **68**, 308.

Scott, J. H. & Symons, N. B. B. (1964) *Introduction to Dental Anatomy*, 4th ed. Edinburgh: Livingstone.

Semenov, E. K. (1969) Variability of the form and size of the human hard palate. *Stomatologiya* (*Moskva*), **48**, 57.

Seward, G. R. (1954) The detailed radiographic anatomy of the standard true occlusal view of the maxilla. *British Dental Journal*, **97**, 177.

Seward, G. R. (1962) The radiographic anatomy of the human lateral nasal wall in occlusal radiographs. *Archives of Oral Biology*, **7**, 235.

Shapiro, H. H., Sleeper, E. L. & Guralnick, W. G. (1950) Spread of infection of dental origin. Anatomical and surgical considerations. *Oral Surgery, Oral Medicine and Oral Pathology*, **3**, 1407.

Shiller, W. R. & Wiswell, O. B. (1954) Lingual foramina of the mandible. *Anatomical Record*, **119**, 387.

Siegel, M. I. (1976) Mechanisms of early maxillary growth—implications for surgery. *Journal of Oral Surgery*, **34**, 106.

Smith, C. J. & Cimasoni, G. (1967) Cholinesterases in buccal and gingival mucosa of monkey and man. Histochemical demonstration and effect of local anesthetics. *Archives of Oral Biology*, **12**, 349.

Smith, R. D. & Marcarian, H. Q. (1967) The neuromuscular spindles of the lateral pterygoid muscle. *Anatomischer Anzeiger*, **120**, 47.

Stanfield, J. P. (1960) The blood supply of the human pituitary gland. *Journal of Anatomy*, **94**, 257.

Steier, A. et al. (1967) Effect of vitamin D_2 and fluoride on experimental bone fracture healing in rats. *Journal of Dental Research*, **46**, 675.

Stern, I. B. (1965) Electron microscopic observations of oral epithelium. *Periodontics*, **3**, 224.

Tandler, B. (1962) Ultrastructure of the human submaxillary gland. *American Journal of Anatomy*, **111**, 287.

van Mens, P. R., Pinkse-Veen, M. J. & James, J. (1975) Histological differences in the epithelium of denture-bearing and non-denture-bearing human palatal mucosa. *Archives of Oral Biology*, **20**, 23.

Wever, E. G., Lawrence, M. & Smith, K. R. (1948) The middle ear in sound conduction. *Acta oto-laryngologica*, **48**, 19.

Williams, T. H. & Dixon, A. D. (1963) The intrinsic innervation of the soft palate. *Journal of Anatomy*, **97**, 259.

Winter, G. B. (1962) Abscess formation in connection with deciduous molar teeth. *Archives of Oral Biology*, **7**, 373.

Wood, N. K. et al. (1970) Prenatal observations on the incisive fissure and the frontal process in man. *Journal of Dental Research*, **49**, 1125.

Wood-Jones, F. (1940) The nature of the soft palate. *Journal of Anatomy*, **74**, 147.

Wynder, E. L. & Bross, I. J. (1957) Aetiological factors in mouth cancer. *British Medical Journal*, **100**, 1137.

Zolotko, V. S. (1966) The topography of vascular fields of the hard palate. *Stomatologiya* (*Moskva*), **45**, 64.

Index

1º FDS Edinburgh Oct 1981